Gilded Metals

Gilded Metals

History, Technology and Conservation

Edited by
Terry Drayman-Weisser

Archetype Publications
in association with

The American Institute for Conservation of Historic and Artistic Works

First published 2000 by
Archetype Publications Ltd
6, Fitzroy Square
London W1P 6DX

ISBN 1 873132 76 X

International Academic Projects, sponsors of this volume, gratefully acknowledge support from the Samuel H. Kress Foundation.

Typeset by Kate Williams, Abergavenny.
Printed and bound by Biddles Ltd., Guildford & King's Lynn.

Contents

Contributors

Carol Aiken
Aiken & Ramer
Baltimore, MD USA

Kilian Anheuser
School of History and Archaeology
University of Wales
Cardiff UK

Silvia A. Centeno
Sherman Fairchild Center for Objects Conservation
Metropolitan Museum of Art
New York, NY USA

W. T. Chase
Freer Gallery of Art/Arthur M. Sackler Gallery
Smithsonian Institution
Washington, D.C. USA

Brian Considine
Decorative Arts Conservation
Cleveland Museum of Art
Cleveland, OH USA

Pete Dandridge
Sherman Fairchild Center for Objects Conservation
Metropolitan Museum of Art
New York, NY USA

Jean Dommermuth
Private Conservator
New York, NY USA

Jack Flotte
Conservation Department
Cleveland Museum of Art
Cleveland, OH USA

Susanne Gänsicke
Conservation and Collections Management
Museum of Fine Arts
Boston, MA USA

Patricia Griffin
Conservation Department
Cleveland Museum of Art
Cleveland, OH USA

Michael Jamet
Private Conservator
Paris, France

Paul Jett
Department of Conservation and Scientific Research
Freer Gallery of Art/Arthur M. Sackler Gallery
Smithsonian Institution
Washington, D.C. USA

Michael Kramer
The Gilders' Studio, Inc.
Olney, MD USA

Andrew Lins
Conservation Department
Philadelphia Museum of Art
Philadelphia, PA USA

Deborah Long
Gerard R. Ford Conservation Center
Nebraska State Historical Society
Omaha, NE USA

Christopher Maines
Conservation Division
National Gallery of Art
Washington, D.C. USA

Sally Malenka
Conservation Department
Philadelphia Museum of Art
Philadelphia, PA USA

Linda Merk-Gould
Conservation Technical Associates LLC,
Westport, CT USA

Ryu Murakami
National Cultural Properties Research Institute
Nara, Japan

Richard Newman
Conservation and Collections Management
Museum of Fine Arts
Boston, MA USA

Peter Northover
Materials Science-based Archaeology Group
Department of Materials
University of Oxford
Oxford UK

Andrew Oddy
Department of Conservation
The British Museum
London UK

Alice Boccia Paterakis
Agora Excavations
American School of Classical Studies
Athens, Greece

Deborah Schorsch
Sherman Fairchild Center for Objects Conservation
Metropolitan Museum of Art
New York, NY USA

David A. Scott
Museum Research Laboratory
The Getty Center
Los Angeles, CA USA

Lyndsie Selwyn
Department of Canadian Heritage
Canadian Conservation Institute
Ottawa, ONT Canada

Donna K. Strahan
Conservation Department
Asian Art Museum
San Francisco, CA USA

Jonathan Thornton
Art Conservation Department
Buffalo State College
Buffalo, NY USA

Preface

In June 1995 approximately 100 individuals gathered in St. Paul, Minnesota, brought together by a shared interest in gilded metal surfaces. The event was a three-day symposium sponsored by the Objects Specialty Group of the American Institute for Conservation of Historic and Artistic Works. Speakers and participants travelled from all parts of the USA and from abroad to share insights on the history, chemistry, technology, deterioration and care of gilded metal objects. This book has as its core the papers presented at that symposium. It has been supplemented with the contributions of others currently carrying out significant studies on gilded metal.

Gilded metal objects, from jewellery to monumental sculpture, are found throughout the world from ancient through modern times. When asked, 'Why did almost all societies gild metal surfaces?' the most common response assumes a desire to falsify or give the impression of solid gold, a rare and valued commodity. Although this may indeed be the intention of most cultures, or individuals within a culture, other motives should not be ruled out. Gilding was also applied because the colour was thought to be attractive, or to have spiritual meaning. In some instances only parts of a surface were gilded to create a design or a colour contrast. In addition, gilding could give permanence to a less permanent material by creating a corrosion-resistant surface.

Whatever the impetus for gilding metal surfaces, the techniques employed tell us something about a society and its possible ties to other cultures. For example, complex, yet identical gilding techniques found amongst disparate cultures each with their own characteristic artifact forms, may indicate trade and technological links. On the other hand, a society producing artifacts that appear to share an outward appearance and form of another culture, while employing different gilding techniques, may indicate familiarity and a shared aesthetic, without a technological connection.

The purpose of this book is to bring together in one place many aspects of gilded metal surfaces. Some chapters review previously published material, while others bring to light new observations and conclusions about gilding practices throughout history. Documented here is gilding technology from many parts of the world, including Europe and the United Kingdom, Greece, Russia, the Americas, China, Japan, Thailand, Tibet and the Middle East. Also included is a review of imitation gilding. Case studies and recommendations for the care and treatment of gilded metal objects, from intricate Fabergé creations to modern day sculpture and architectural elements, form a significant aspect of this publication.

By its very nature any study of materials from ancient to modern day contexts can be only a work in progress. Therefore, it is not the intention here to present an exhaustive survey on the subject of gilded metals. With ever-increasing sophistication in archaeological and scientific examination techniques, the body of knowledge will continue to grow. This book is intended to serve as a foundation for placing new information in context for many years to come.

Support and assistance from a number of sources have made this publication possible. I am grateful for the encouragement and financial support of the Objects Specialty Group of the American Institute for Conservation of Historic and Artistic Works (AIC) and for the assistance provided by the staff of the AIC National Office. I also would like to acknowledge the commitment and hard work of the planning committee members, who contributed to the success of the symposium on which the core of this book is based. In addition to the editor, members of the committee included Arthur Beale, Paul Jett, David Scott and Thomas Taylor, Jr.

In order to meet today's professional standards, a book such as this must undergo peer review. I am indebted to the following reviewers who participated in this process by giving their time and sharing their expertise without reward: Arthur Beale, Ann Boulton, Andreas Burmester, Bruce Christman, Meg Craft, Ellen Howe, Paul Jett, Simeon Krumbein, Julie Lauffenburger, Henry Lie, Andrew Lins, Linda Schiefler Marks, Albert Marshall, Rika Smith McNally, Linda Merk-Gould, Pieter Meyers, Deborah Schorsch, David Scott, Carol Snow, Donna Strahan, Shelley Sturman, Jonathan Thornton and Jane Williams.

Of course this book would not exist at all if not for the authors. I am especially grateful that these individuals were willing to share their knowledge and endure the long and arduous process required to make this book a reality. Special appreciation goes to Donna Strahan and Julie Lauffenburger, my colleagues at the Walters Art Gallery, who often acted as my sounding board for issues relating to this publication. And finally, there are all those little things that must be done on a daily basis to keep a project on track. I would like to thank my secretary, Paul Smith, who assisted me so patiently and with good humour.

Terry Drayman-Weisser
Editor

1 A History of Gilding with Particular Reference to Statuary

Andrew Oddy

Abstract

Gilding is defined and its origins explored. The techniques of gilding discussed are: foil gilding; leaf gilding; diffusion bonding; cold mercury gilding; fire gilding; electrochemical gilding; gilding with gold powder; electroless gilding and the decomposition of organic gold compounds; and electrolytic gilding. This is followed by a discussion of the relationship between alloy composition and gilding technology for Roman bronze statues, in which it is shown that fire-gilded statues were usually cast in fairly pure copper whereas leaf-gilded statues (which are generally earlier than the fire-gilded ones) were cast in a leaded bronze.

Introduction

For the purposes of the following discussion, gilding is defined as the *overlaying* of the surface of a baser metal with a layer of gold. It is immaterial whether the gold is held in place by mechanical, physical or chemical forces. However, the techniques of gilding must be differentiated from those of inlaying with gold. Inlaid objects have the gold applied in grooves or recesses cut into the surface so that the gold decoration is flush with that of the surrounding area of baser metal. Inlaid work will not be discussed in this paper, which is confined to a discussion of the history of gilding in Europe, the Middle East and the Far East.

On the basis of this definition, gilding of other metals can be traced back to at least 3000BC in the Middle East (Oddy 1981, 1991a, 1993) and has been continuously used since then until the present day. In Europe and the Far East, however, gilding began much later. Needless to say, the techniques of attaching the gold to the baser metal have undergone numerous changes and developments. The evidence for the study of these changes is of two types – descriptions in the contemporary literature and the results of scientific examination of securely dated gilded objects.

The earliest useful references in the classical literature to the process of gilding occur in the first century BC, but earlier authors frequently mentioned gold, golden or gilded objects, often ambiguously. References to gilding in Homer have been discussed by Gray (1954), but the texts give no indication of which gilding techniques were employed, and a similar extensive discussion would be useful for references in the later classical literature to 'gold' cult statues and the like.

If the study of historical texts from a technical point of view is still in its infancy, especially for those texts written in oriental languages, the scientific study of the objects themselves is hardly more advanced. Lechtman (1971, 1973) and Lins and Oddy (1975) were the first people to specialise in the study of gilded surfaces using scientific techniques, although numerous other analyses have been published as part of the examination of individual antiquities. Sometimes this data has described objects obtained from the 'art' market, and it cannot be too highly stressed that the compilation of a databank must employ only well dated and securely provenanced pieces.

The investigations of a large number of gilded objects have shown that several different methods have been used to attach a layer of gold to the surface of other metals and alloys since *c.*3000BC. Three main methods were used up to the Middle Ages: the attachment of gold leaf; the attachment of gold foil; and the application of a solution of gold dissolved in mercury (gold amalgam). There are variations on each main method. One technique not described is depletion gilding, which, until recently,[1] was not known to have been used in the Old World, although it is well known from published examinations of pre-Columbian antiquities from the Americas.

Foil gilding

Foil gilding involves wrapping gold foil[2] round an object and holding it in place by bending the gold foil over the edges of the object, or by riveting the foil in place, or by cutting grooves into the surface of the base metal and inserting the edges of the gold foil into the grooves, which are then hammered closed. A variation of the latter technique involved laying a piece of gold foil over an area of the surface and then hammering around the edge of the gold with a punch so that the gold was forced into the surface of the base metal.

The mechanical attachment of gold foil can be illustrated by a copper alloy boss, now much corroded, from the Sumarian levels (third millennium BC) at Ur (Oddy 1985: Plate 1; Oddy 1993: Figs 15.2, 15.3). A sheet of gold foil covers the front of the boss and the edge of the foil has been bent over the rim of the boss, where creases resulting from the bending over process are clearly visible. The same method was sometimes used to completely cover an object with gold, in which case the overlapping edges of the gold foil would be burnished together. Several examples of this type of gilding can be found among the objects in the Oxus treasure (Oddy 1981: Fig. 1; Oddy 1991a: Figs 5a, b), which dates to the middle centuries of the first millennium BC (Dalton 1964).

The riveting of gold foil to another metal does not appear to have been a common process, although the British Museum has several fragments of an Etruscan silver repoussé panel, showing two Amazons on horseback with a fallen comrade which has been partly gilded by attaching gold foil with rivets (Strong 1966: 72; Haynes 1959; Oddy 1988: Plates 1.3, 1.4).

Another unusual foil-gilding technique was the insertion of the edges of gold foil into grooves cut into the surface of the base metal. Although this technique was described by Pliny, it has not been identified on many objects.

> The emperor Nero was so delighted by this statue of the young Alexander that he ordered it to be gilt; but this addition to its money

value so diminished its artistic attraction that afterwards the gold was removed, and in that condition the statue was considered yet more valuable, even though still retaining scars from the work done on it and incisions in which the gold had been fastened.

(Rackham 1968: 175)

The statue in question has not survived, but fragments of a life-size equestrian bronze statue, which illustrate this technique, were recently found in Athens (Shear 1973: 165–8). They have been identified by Houser (1979, 1987, in press) as pieces from a statue of Demetrios Poliorketes and dated to the very end of the fourth century BC. The surviving fragments consist of a sword, some pieces of drapery and a leg. All are in good condition with a thin green patina and are cast in leaded bronze (3–4 per cent tin and 23–35 per cent lead), except for the sword (which contains only about 1.7 per cent of lead and 7.4 per cent of tin in the copper) (Tite and Craddock 1977). The surface is, however, 'scarred' by lines of 'gold', which are all that remains of a former layer of gilding. This was attached by cutting grooves in the bronze, inserting the edges of pieces of gold foil, and then pressing the foil to the surface of the statue until another groove was reached where the free edge of the foil could be trapped. Subsequently, the gold foil has been torn off the statue leaving the edges of the sheets protruding from the surface. The gilding of this statue has now been fully investigated by Paterakis (this volume, Chapter 5).

There can be little doubt that the technique of gilding exhibited on the pieces of sculpture from Athens is identical to that mentioned by Pliny as having been used on the statue of Alexander. However, this method of gilding statues does not appear to have been very common, as only one other example is known on a piece of sculpture, which is much smaller than life size (Thompson 1940). The reasons for the apparent unpopularity are probably twofold. First, gold foil is wasteful of gold when used to cover a surface, as the same decorative effect can usually be achieved by the use of gold leaf. Second, the thickness of gold foil obscures the detail of a sculpture, a problem that is not encountered when using gold leaf.

Leaf gilding

Once craftsmen had learned, probably early in the second millennium BC (Forbes 1971), how to refine the natural gold that was found in certain streams and rivers, they soon found that the purified gold could be hammered out into extremely thin sheets (now known as gold leaf). Pliny comments on gold leaf as follows:

> ... an ounce of gold can be beaten out into 750 or more leaves – 4 fingers square. The thickest kind of gold leaf is called Palestrina leaf, still bearing the name taken from the most genuinely gilded statue of Fortune in that place. The foil next in thickness is styled Quaestorian leaf. (Rackham 1968: 49)

In Egypt, the beating of gold seems to have been a specialised craft which was depicted on tomb paintings as early as the VIth Dynasty (around 2250BC) (Duell *et al.* 1938: Plates 30, 33), but it is impossible to know whether the product was 'foil' or 'leaf' (James 1962). However, by the New Kingdom, more information is available and the British Museum contains a funerary papyrus of Neferronpet (fourteenth century BC) who is described as 'chief of the makers of thin gold'. This papyrus is actually decorated with gold leaf (James 1965; Oddy 1981: Fig. 2).

The Egyptians used gold foil, and later gold leaf, from towards the end of the third millennium BC onwards for the decoration of furniture and coffins. Naturally, most of our evidence comes from tomb finds and may, therefore, be biased, but the use of gold leaf on metal appears to be unusual. However, the British Museum has a number of gilded bronze statuettes from Egypt on which the gold leaf appears to have been laid on top of a ground. This is the technique still in use today when gilding wood or stone, where the surface is coated with 'gesso', which is polished to give an absolutely smooth surface. The gold leaf is laid on top of this with an adhesive. A few Egyptian gilded bronzes have been studied scientifically, and the gesso has been shown to be mainly of two types: calcite or gypsum (Oddy *et al.* 1988). One object was found, however, to have a dolomite ground (Oddy 1981: Fig. 3), and some copper–bronze leaves from a funerary wreath excavated at Kameiros, on the Island of Rhodes, has a ground consisting of clay minerals (Oddy *et al.* 1979b).

The gesso would have been mixed with an adhesive, but neither this nor the adhesive used to fix the gold leaf has ever been identified by analysis, although Pliny mentions white of an egg and other glues for gilding on non-metallic objects:

> On marble and other materials incapable of being raised to a white heat, gold is laid with white of egg: on wood it is laid with glue according to a formula; it is called *leucophorum* ... .
>
> (Rackham 1968: 51)

Pliny subsequently describes *leucophorum* (Rackham 1968: 287) as a mixture of yellow ochre with a little red ochre and some marl (probably a naturally occurring mixture of clay and chalk). This, however, sounds much more like a ground for preparing the surface of the wood than an adhesive, and it is probable that the gold leaf was actually fixed in place with white of egg. Although Pliny does not mention the gilding of bronze with gold leaf, it is probably safe to assume that the principles of the method he describes for use on stone and wood was that frequently used by the Egyptians for gilding bronze.

By the time of Pliny, however, gold leaf was being applied directly to the surface of bronze objects without an intervening layer of gesso. Several examples of monumental, gilt-bronze sculpture are known on which the small squares of gold leaf that were used are still clearly visible, because the double thickness of gold leaf where the squares overlap has resisted the effect of time, resulting in a criss-cross pattern of gold on the surface. This is visible on the statuary group from Cartoceta of *c.*AD27 (now in the museum at Ancona) (Oddy 1981: Fig. 4), on the Apollo of Lillebonne of the first or second century AD (now in the Louvre) (Tailliez 1982: Fig. 1) and on the arm from a Roman statue found near Rheims (now in the British Museum) (Oddy 1985: Fig. 4; Oddy *et al.* 1990: Fig. 4).

Diffusion bonding

Whenever gold leaf is stuck in position with an adhesive, it is not very resistant to wear and tear. Hence objects, such as statues, that stood in the open air or were in daily use in temples, were unlikely to retain the gilding in a pristine condition for very long. The result was that from at least as early as *c.*1200BC, goldsmiths found that gold leaf could be bonded to silver by burnishing it directly on to the silver surface and then heating gently for a very short time in order to cause some interdiffusion of the gold and silver to take place. This technique has been labelled diffusion bonding (Oddy *et al.* 1981b), but although common on silver in the Hellenistic and early Roman periods, it does not appear to have been much used on bronze, unless the leaf-gilded Roman statues mentioned above are, in reality, diffusion bonded.

Unfortunately, the technique needs scientific investigation to determine how easy it is to stick gold leaf to bronze by this method and whether the composition of the bronze determines the efficiency of the process, as has been suggested (Vittori 1976).

Cold mercury gilding

In the first century BC, Vitruvius mentioned that mercury is used for the gilding of both silver and brass (Granger 1970: 116) but gave no details of the process, whereas Pliny, in describing the gilding of copper, says that it must be thoroughly cleaned and then it will 'take the gold leaf laid on with quicksilver' [ie, mercury] (Rackham 1968: 51). These sources are now interpreted as describing a variant of leaf gilding in which the mercury is used as an adhesive (Vittori 1978, 1979); in another passage Pliny says that a coat of quicksilver is applied underneath the gold leaf:

> when also things made of copper are gilded, a coat of quicksilver is applied underneath the gold leaf and keeps it in its place with the greatest tenacity: but if the gold leaf is put on in one layer or is very thin it reveals the quicksilver by its pale colour. (Rackham 1968: 77)

Scientifically, this method of gilding is difficult to distinguish from true fire gilding (see below) as, in both cases, a sensitive method of analysis will detect some mercury in the layer of gilding. It is very likely, however, that cold mercury gilding did not survive for long before craftsmen learned, or developed, the method of fire gilding in which the mercury is driven off by deliberate heating, rather than being allowed to dissipate slowly by evaporation.

Fire gilding

By the fourth century BC at the latest, a new method was in use for the gilding of copper in China (Wang 1984) in which gold was dissolved in hot mercury and the resulting mixture rubbed on to the cleaned metal surface. The object was then heated when almost all the mercury was evaporated off, leaving behind a very firmly bonded layer of gold. This process is still in use in Nepal (Oddy *et al.* 1981a).

Analysis of numerous gilded metal objects of the Roman period has shown that the presence of a trace of mercury in the gilding is common from the third century AD onwards, but rare earlier (Oddy 1988), which is surprising in view of the cold mercury gilding method apparently described by Pliny. It can only be assumed that the cold technique was not common, and that once the 'hot' method was developed its use became widespread. There is, however, another possible explanation of why the technique of fire gilding became common; it may be the result of supplies of mercury becoming more readily available. Mercury ore had been known for a long time, and was used as a pigment, but the large-scale extraction of metallic mercury depended on the invention of an efficient system for condensing the mercury vapour produced in the extraction process. Perhaps such a condenser was introduced in the Roman world in the second or third century AD.[3]

A variation on the usual method of fire gilding is to apply a layer of mercury to the metal surface to be gilded and then to lay pieces of gold leaf on top. The gold leaf dissolves in the mercury and, after heating, the result is indistinguishable from the application of a mercury/gold mixture to a clean surface. This technique is still in use in Japan (Oddy 1991b).

It is now clear that fire gilding was common in China in the late Chou and Han periods (Lins & Oddy 1975; Wu Kunyi 1981), and that the technique was widely known in the Roman world by the third century AD, but where and when was this method of gilding invented?

Analyses of miscellaneous objects have been reported from time to time that claim to have identified fire gilding as early as the Bronze Age in Europe, but these must be treated with caution until a firmly stratified gilded object of undoubted Bronze-Age type has been excavated and shown to contain mercury in the gold by analysis.

However, evidence is now beginning to appear that indicates a knowledge of fire gilding in the Celtic world, apparently by the first century BC at the latest! In all, six objects have been analysed that contain mercury in their gilding.

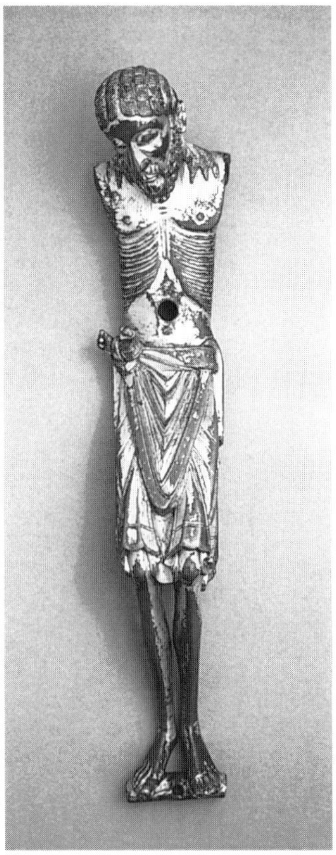

Figure 1.1 Romanesque crucifix (Angevin, first quarter of the twelfth century) (Copyright Trustees of the British Museum). (See Plate 1)

Some years ago, Zwicker (1973) published the analysis of a gilded stater of the Vendelici which he showed to contain mercury in the gold. Since then, two gilt Celtic staters have been identified in the British Museum, which also contain mercury in the gold (Oddy and Cowell 1993: Nos 24, 25). These coins were, however, struck in southern Britain at the end of the first century BC or in the early years of the first century AD by the Dobunni and the Corieltauvi respectively. The other Celtic objects that have been shown to contain mercury in the gold are a brooch of the Spanish Iron Age (Tait 1976: No. 139), which has been attributed to the second or first century BC, and fragments of two gilt bronze torcs of about 50BC from Snettisham, Norfolk (Northover 1995) The gilding on the Spanish brooch is patchy, presumably because of the uneven application of the amalgam to the surface, but does look more like an unskilled attempt at fire gilding than the result of a cold mercury gilding with gold leaf.

Fire gilding became the most widely used technique for gilding by the end of the Roman period and it continued in use throughout Dark Age and medieval Europe (Oddy *et al.* 1986). It was the gilding method *par excellence* for metal objects over more than 1500 years and examples of its use are extremely common in museums, cathedral treasuries and private collections all over Europe (Figures 1.1 and 1.2). However, craftsmen were well aware of the health risks from using mercury, and from the seventeenth century at the latest they were on the look-out for safer methods of plating gold on to other metals. Several techniques were published in encyclopaedias and books of receipts, but it was the invention of electroplating in the middle of the nineteenth century that rendered all other methods effectively obsolete for most purposes.

Electrochemical gilding

This is one of three methods of depositing gold from a solution, the other two being electroless gilding and decomposition of organic gold compounds (Wise 1964: 169).

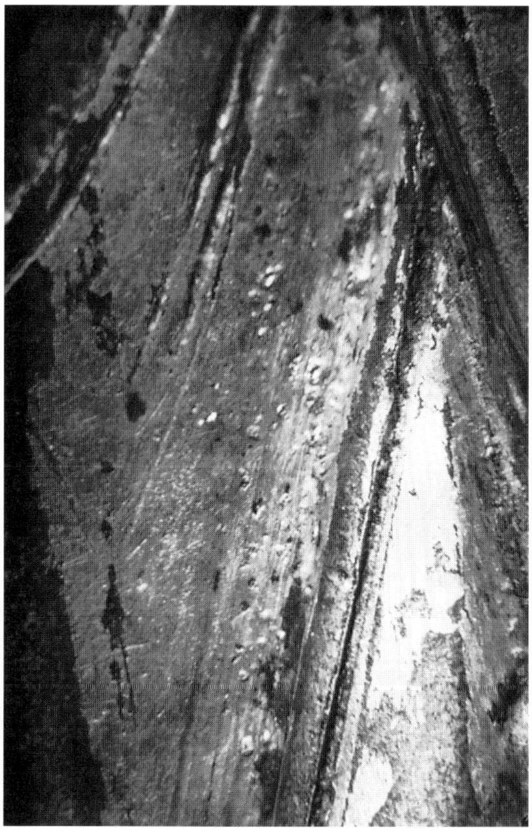

Figure 1.2 Detail of the surface of Figure 1.1. The crucifix is cast in 'impure' copper (containing 4.5% tin + zinc + lead) and has been fire gilded (Copyright Trustees of the British Museum). (See Plate 2.)

Electrochemical gilding, also known as immersion or replacement gilding, involves immersing the copper or silver in a solution containing a soluble salt of gold. An electrolytic reaction takes place at the surface of the base metal, which undergoes simultaneous anodic dissolution and deposition of gold. This technique will only deposit a very thin layer of gold because as soon as the base metal is completely covered the process stops (Simon 1993).

The gold layer produced by this technique is not very durable and because of the difficulty of getting gold into solution, it seems unlikely that it was used in antiquity, although it has been postulated, on the basis of the thinness of the gilding, that this technique was used in pre-Columbian Peru (Lechtman 1979; Lechtman *et al.* 1982). However, it is possible to achieve equally thin gilded layers by the technique of diffusion bonding.

Electrochemical plating is easier to carry out with silver than with gold, and is susceptible to mass production, so that it can be postulated that this is the technique by which the 'silver washed' antoniniani of the third century AD were produced.[4] This hypothesis remains to be proved, and there is no hint of electrochemical silvering being described in the late Roman Leiden and Stockholm papyri (Halleux 1981) or the early medieval *Compositiones variae* (Burnam 1920) and *Mappae Clavicula* (Smith and Hawthorne 1974), although both of these medieval sources do describe the plating of copper onto iron by deposition from a solution of copper ions.[5] This is a necessary first step in the gilding of iron with either gold leaf or by fire gilding and is, in fact, a process of electrochemical plating (Burnam 1920: 100, 128; Smith and Hawthorne 1974: Chs 146G, 146H, 219A, 245, 291, 292). Carradice and La Niece have postulated deposition of silver from solution as the technique of silvering of a Libyan coin of the third century BC (1988: 51), but did not specify which of the three 'dipping' mechanisms was thought to be involved.

Gilding with gold powder

One of the last of the 'medieval style' of craftsman's manual for silversmiths and goldsmiths was published in 1903 (Wilson 1903) and described only two methods of gilding: fire gilding and the use of gold powder. It is interesting that, even at this late date, the author did not describe electrolytic gilding!

The gold powder method involves soaking linen rags in a solution of gold chloride, allowing them to dry, and then burning them. The ashes, which contain finely divided gold, are then rubbed on to thoroughly clean silver with a piece of damp leather, the gold gradually forming a layer on the surface. The ashes are rinsed away and the gilding is finally burnished.

The method was earlier described by MacKenzie, where a piece of cork moistened with a solution of common salt was recommended for rubbing the ashes on the surface of silver (MacKenzie 1823: 28). The same technique had earlier been described by Rees (1819) who quotes a communication to the Royal Society by Sir Robert Southwell in 1698 as his source (Southwell 1698). Southwell implies that the technique had originated in Germany, perhaps not long before he was writing.

Other recipes for the manufacture and use of gold powder for gilding metals are of considerable antiquity. For example, the late Roman Leiden papyrus recommends mixing two parts of finely powdered lead and one of gold with a gum. This mixture is spread over the surface of the base metal and heated. The process is repeated several times until the lead has been 'consumed' by the fire and the object is completely gilded (Halleux 1981: 94, no. 37). The early medieval *Mappae Clavicula* similarly recommends coating the surface of silver or copper with filings of a 1:1 alloy of gold and lead made into a paste with water and natron (naturally occurring impure sodium carbonate) (Smith and Hawthorne 1974: 40). Strong heating of the object in a furnace is said to dissipate the lead and leave a gilded surface on the baser metal. Presumably the function of the lead is to act as a medium that holds the gold while some interdiffusion takes place with the silver or copper. It is then lost from the surface by oxidation (and perhaps by being partially absorbed into the substrate in the case of silver).

Electroless gilding and the decomposition of organic gold compounds

These are relatively modern techniques that give rise to weakly bonded layers of gold. Modern descriptions of the technology are given by Rapson and Groenewald (1978: 181ff.), Okinaka (1974) and Simon (1993).

Searching through nineteenth century (and earlier) encyclopaedias and books of receipts produces a number of methods of gilding which might belong to one of these headings. Examples are Grecian gilding (Anon. 1708; MacKenzie 1823: 28), Talbot's process (Cooley 1845: 465) and variations of a method for gilding iron and steel in which the gold is deposited from an organic solvent (Aitkin & Aitkin 1807: 477; Spon 1883: 307). However, it seems unlikely that any of these techniques would produce a durable layer of gilding as the gold is deposited on the surface in a finely divided form.

Electrolytic gilding

This technique only became possible after the invention of an electrolytic cell capable of generating a reasonably constant current. The process was developed during the first half of the nineteenth century and became commercially viable in the 1840s (Hunt 1973; Raub 1993) and was already being described (as voltaic gilding) in Cooley's encyclopaedia of 1845 (Cooley 1845: 465).

It has, however, been postulated that electrolytic gilding and silvering were known to the Parthians 2000 years ago, based on the excavation of some battery-like vessels near Baghdad in 1936 (König 1940: 166–9; Jansen *et al.* 1987). Since then, numerous opinions have been expressed, both for and against the Parthians having the ability to carry out electrolytic plating of silver and gold (Al-Haik 1964; Winton 1962; Pàszthory 1989; Eggert 1995), but until the existence of electroplating is proved by the metallurgical examination of gilded and silvered objects, the invention of the electrolytic cell at such an early date must remain questionable, particularly as there is no evidence for the knowledge of the electrolytic cell in subsequent centuries.

The relationship between alloy composition and gilding technology for Roman 'bronze' statuary

In the famous treatise on metal technology written under the pseudonym of Theophilus early in the twelfth century there is an excellent description of fire gilding (Hawthorne and Smith 1979: 110–15). In this work, Theophilus mentions in several places the importance of removing lead from copper alloys which are destined to be fire gilded: 'if brass is to be gilded, it should be completely pure and purged of lead' (1979: 139); 'It [ie, coarse brass] cannot be gilded, since the copper has not been completely purged of lead before the alloying' (1979: 144); 'silver and unalloyed copper can be gilded more easily than brass' (1979: 145). Theophilus also gives the reason why lead should not be present and why the zinc in brass should be evenly distributed.

> The amalgamation of brass must be done more scrupulously and carefully and it must be gilded more thickly and washed more often and dried for a longer time. When it begins to take on a yellow colour (during the heating process), if you see white spots emerging on it so that it refuses to dry evenly, this is the fault of the calamine, because it was not evenly alloyed, or of lead, because the copper was not purged and refined free of it. (Hawthorne and Smith 1979: 145–6)

These passages in Theophilus, together with a scientific examination of a statuette of *Herakles* in the British Museum (Brailsford 1958: 54) were the key to a new understanding of the technology of fire gilding in antiquity. When the *Herakles* figure, recently identified by Coulston and Phillips (Coulston and Phillips 1988: 77–8) as a statuette of the emperor Commodus, was analysed by Craddock as part of a study of bronze composition (Craddock 1975), he noted the unusual composition of the alloy (almost pure copper) and consulted with the author, who was independently engaged in a study of gilding (Lins and Oddy 1975). The fact that the statuette is fire gilded made sense of the unusual composition when reference is made to Theophilus.[6] The question must be asked, however, whether a text written in Germany in the early twelfth century AD has any relevance to a bronze cast in the Roman Empire.

For the Romanesque period in Europe, which is contemporary with the writings of Theophilus, Oddy *et al* (1986) have shown that the copper content of ungilded cast domestic and ecclesiastical metalwork ranged from 70 per cent to 91 per cent, while that of fire-gilded cast copper/ bronze objects ranges from 81 per cent to 99 per cent. Both lead and zinc contents were found to be lower in the gilded ones. For the early medieval period (before 1000AD), very few analyses of comparable gilded and ungilded objects are available, but what little evidence is available fails to show any significant difference in lead contents between the two groups.

In the Roman period, however, the difference is even more marked than for the twelfth or thirteen centuries, especially when comparisons are restricted to objects of a similar type. Gilded Roman bronze statues are a good example, and analyses of gilded examples fall into two groups according to whether the gilding was carried out with gold leaf or by fire gilding (Oddy 1982b; Oddy *et al.* 1990). If the analyses are plotted on a ternary diagram, the statues that are fire gilded are concentrated towards the apex of 100 per cent copper (open circles on Fig. 1.3), whereas those that are leaf gilded are much more widely spread (closed circles on Figure 1.3).

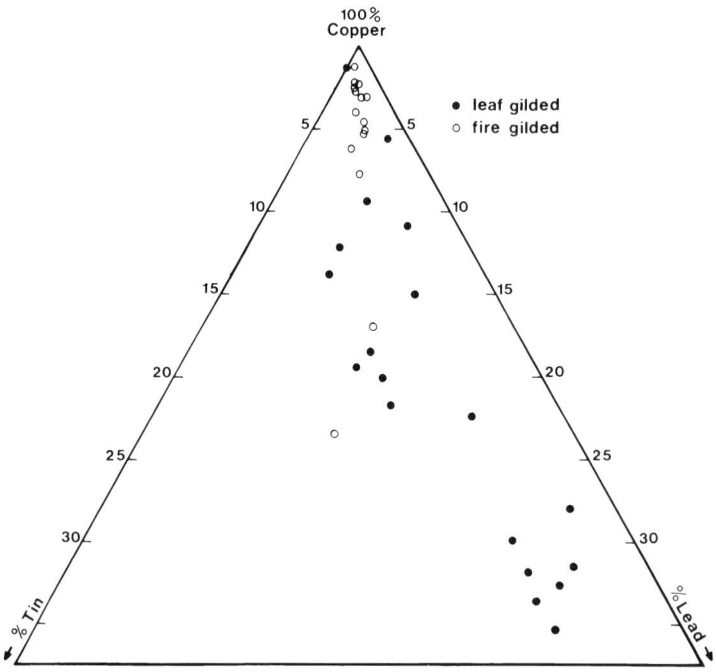

Figure 1.3 Ternary diagram to represent the composition of gilded Roman statuary.

The same is true if the composition of gilded and ungilded Chinese belt hooks of the late Chou and Han periods are plotted on a similar diagram (open circles gilded, closed circles ungilded on Fig. 1.4). These analyses were carried out by Chase (1980) and, although he did not analyse for mercury in the gilding, his analyses are entirely consistent with the type of low-lead copper alloy that is required as a base for fire gilding. Of about 150 examples analysed, 29 of the 40 with gilding contained more than 95 per cent copper and 26 of these contained less than 1 per cent of lead. Some of the ten gilded examples containing less than 95 per cent copper and between 10 per cent and 25 per cent lead may not be authentic Chou or Han pieces. Figure 1.4 also shows that a significant number of the

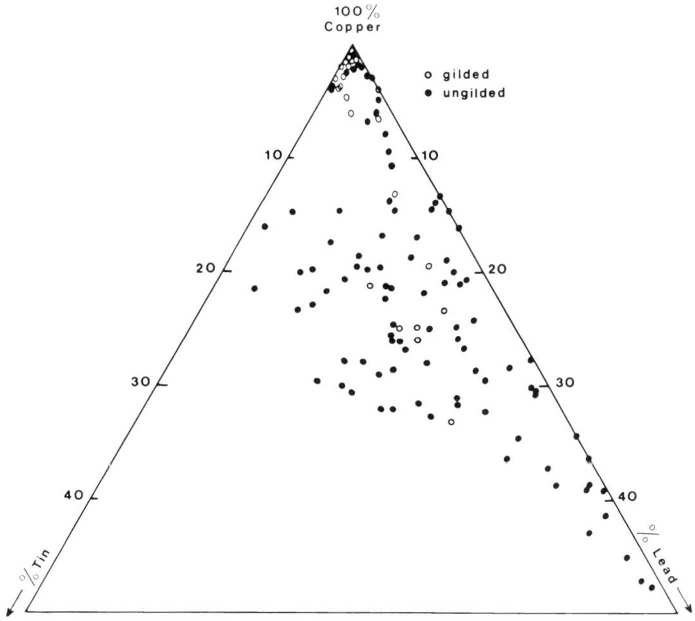

Figure 1.4 Ternary diagram to represent the composition of gilded and ungilded Chinese belt hooks of the late Chou and Han periods (after Chase 1980).

ungilded belt hooks were made of fairly pure copper, but this is not important. The important fact is that *few* of the *gilded* ones contain significant amounts of lead. It is thus evident that the importance of copper-alloy composition for fire gilding was known from the earliest emergence of the technique.[7]

In the past 25 years, knowledge of the composition of Roman statuary bronze has greatly increased, culminating in the recent publication of several thousand analyses of Greek, Etruscan, and Roman metalwork (Craddock 1985). In a review published in 1970, however, Caley listed only 17 analyses of statuary (Caley 1970), at least two of which were from gilded objects. However, both had high tin and lead contents and must either be presumed to be leaf gilded or are now known to be leaf gilded (Caley 1970: Tables VI.1, VII.5), so he did not come across the unusual composition for the 'bronze' substrate that is associated with fire gilding.

However, in a series of papers published in the period 1966 to 1973, Picon *et al.* did note that some gilded objects had unusually low levels of tin and lead (Picon *et al.* 1966, 1967). They attributed this to the need for the copper alloy to remain malleable so that sheets of gold could be used, as described above, by having their edges hammered into grooves in the bronze. These authors appear not to have extended their analysis programme to include the gilding layer, so they did not notice the presence of mercury in the gilding on low-tin–low-lead bronzes.

The recognition of the relationship between gilding technology and composition has important implications for authenticity and for the dating of certain objects. To return to the figure of Commodus, for instance; it has recently been suggested on stylistic and iconographic grounds that the statuette may be either Etruscan (Richardson 1989), Renaissance (Ortiz 1989), or nineteenth century (Nikolaus-Havé 1986). On technical grounds, an Etruscan date can be discounted as the method of gilding and composition of the alloy cannot be paralleled in the Mediterranean area at this period.[8] The technology is entirely consistent with a Roman date, but not enough scientific work has been published on Renaissance and nineteenth-century bronzes to allow a comparison to be made for these periods. A nineteenth-century date seems unlikely because of the date at which the figure came to light, but the fact that the figure is a solid cast and is in remarkably good condition may have some bearing on whether it is Renaissance or Roman. This needs further consideration.

Technology is similarly the clue to the dating of the four horses of San Marco. Nowadays, no one seriously suggests that they are Greek in origin (Hunt 1940–5: 47), but again the use of fire gilding and the composition of the alloy (impure copper) rule out a date before the second century AD, and, in view of the difficulty of casting large amounts of almost pure copper, a later date may be preferable (Oddy *et al.* 1979a, 1982b).

Because of this difficulty – caused by the higher temperature needed to melt copper than to melt bronze, and by the higher viscosity of molten copper, and by its tendency to oxidise rapidly – it may seem strange that fire-gilded copper statues were actually produced at all. The answer lies in their increased durability in the open. Fire gilding creates a continuous and strongly bonded layer of gold on the surface, which can be expected to protect the statue for many years from corrosion in the open air.

Nevertheless, regilding must be expected in the course of routine maintenance, and a number of metallographic examinations have shown that it did take place in antiquity. A good example is the head of Minerva (Toynbee 1963: No. 25) in the Roman Baths Museum at Bath, which has been shown to have at least six layers of gilding (Oddy *et al.* 1990: Fig. 10b). Analysis of a flake of the gilding by emission spectroscopy showed the presence of mercury in the gold, but when the individual layers were analysed on the scanning electron microscope with an X-ray analyser, the level of mercury was too low to be detected. However, there is a very clear physical difference between the inner two layers of gold, which are porous, and the outer layers, which are not. The technique of fire gilding gives rise to porosity in the gold, and it can thus be postulated that the Minerva figure was originally fire gilded, probably on two separate occasions, and subsequently regilded a number of times with gold leaf.

A similar result has been observed in the examination of a small sample from the tail of one of the horses of San Marco (Fig. 1.5) (Leoni 1979). At

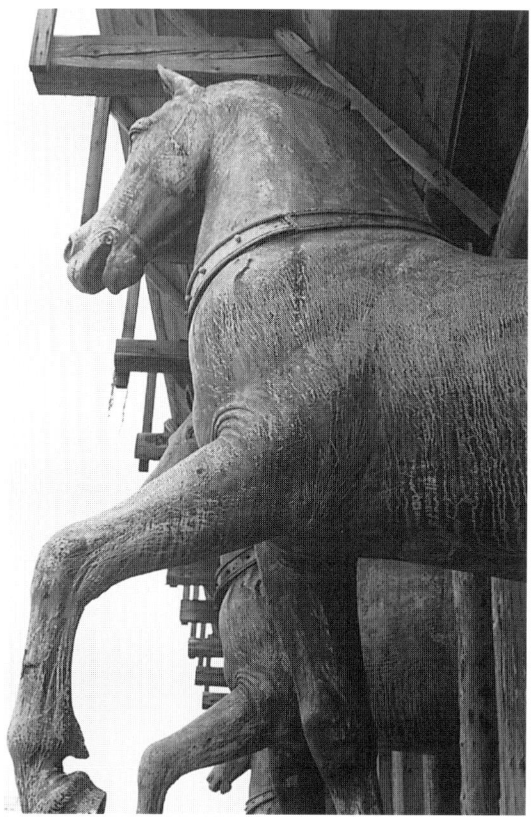

Figure 1.5 The gilded bronze horses on the facade of the Basilica di San Marco in Venice, photographed before their removal in the late 1970s (Copyright Trustees of the British Museum). (See Plate 3.)

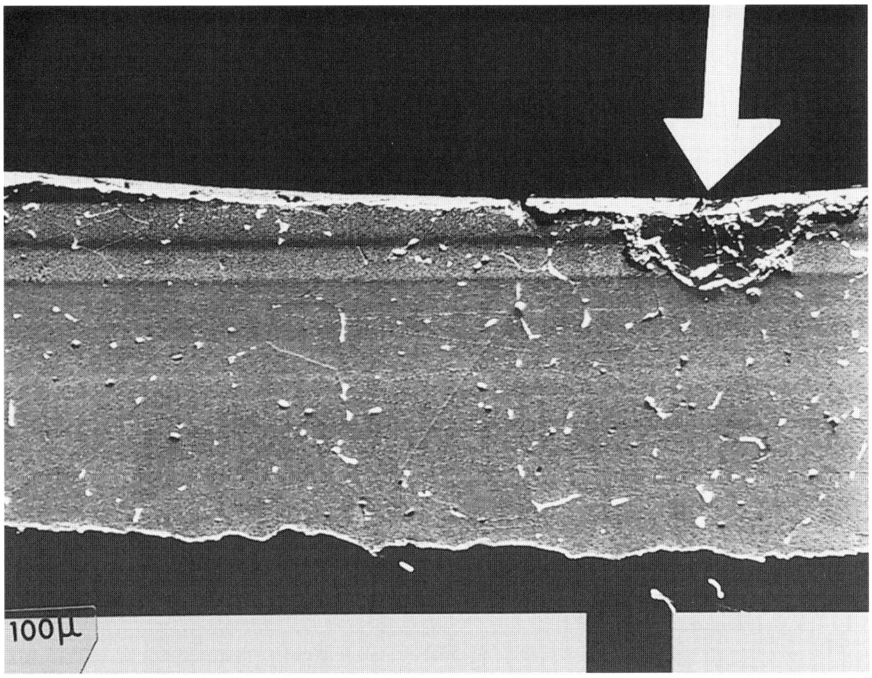

Figure 1.6 Metallographic section through the surface of a minute sample removed from the tail of one of the horses. The arrow indicates the gilded surface. A small amount of lead can be seen (as white globules) at the grain boundaries.

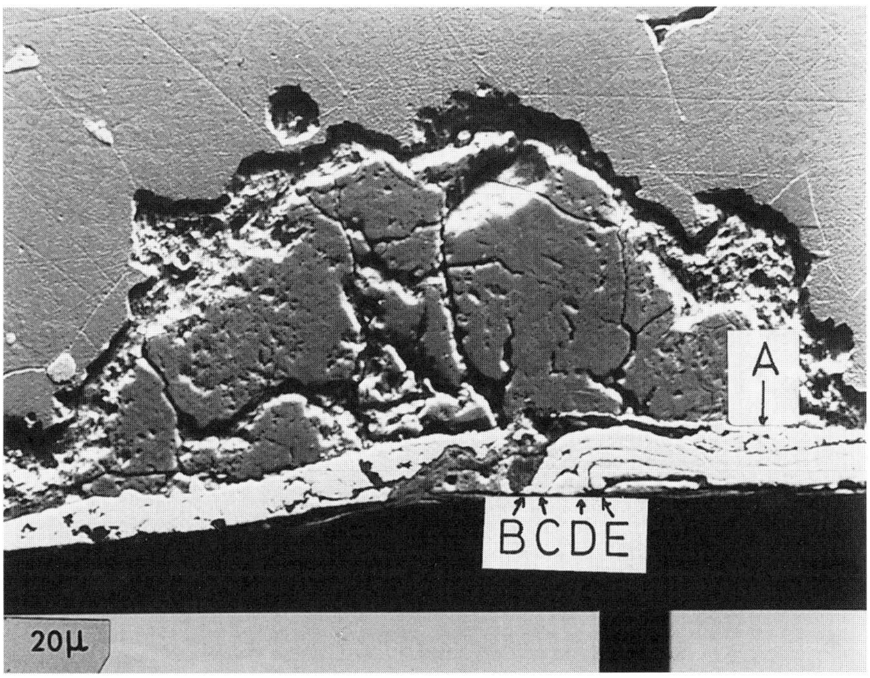

Figure 1.7 Detail of Figure 1.6 to show the five layers of gilding (A–E). The innermost layer can be seen to be porous and was shown to contain mercury by analysis.

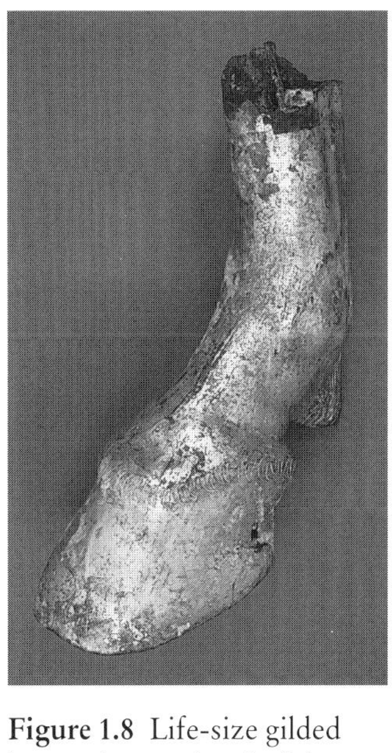

Figure 1.8 Life-size gilded bronze horse's hoof of the Roman period (Copyright Trustees of the British Museum). (See Plate 4.)

least five layers of gilding have been observed (Figs 1.5, 1.6 and 1.7), and the gold nearest the copper alloy of the horse is more porous than the outer layers (Fig. 1.7a). This inner layer also contains mercury. Thus again it can be suggested that the horses were originally fire gilded, but that they were subsequently regilded, probably by attaching gold leaf to the surface with an adhesive.

Regilding has also been observed on leaf-gilded statues. One of the well known sculpture group from Cartoceto has two layers of gold (Anon. 1987: Colour plate 18b), and a horse's hoof in the British Museum (Fig. 1.8) has at least four layers in one area (Figs 1.9 and 1.10).

From a practical point of view, regilding of a statue *in situ* can only be carried out with gold leaf and an adhesive, and not by the fire-gilding technique. This is because fire gilding will only work on a scrupulously clean metal surface, free from dirt and corrosion products. In addition,

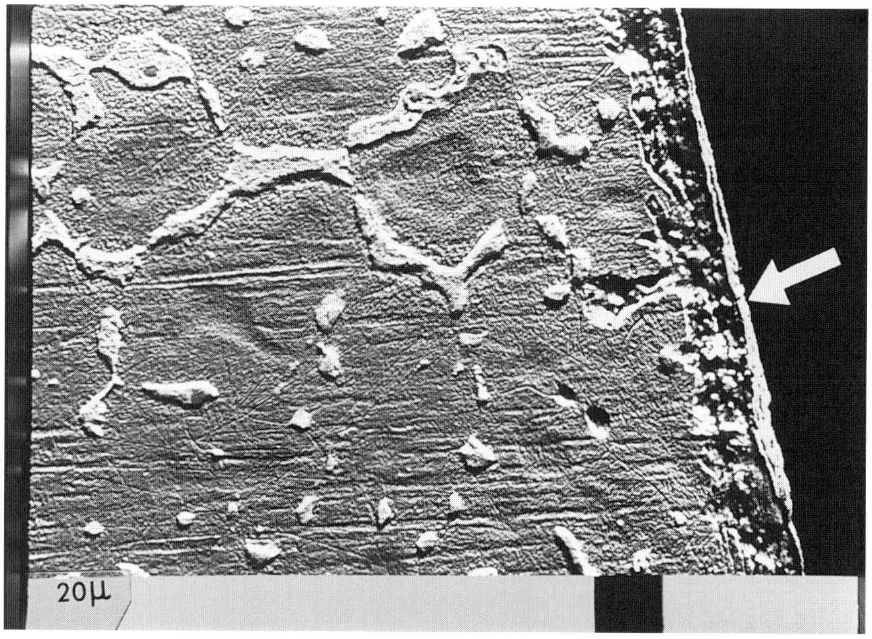

Figure 1.9 Metallographic section through the surface of a minute sample removed from the broken edge of the leg. The arrow indicates the gilded surface. A considerable amount of lead can be seen at the grain boundaries.

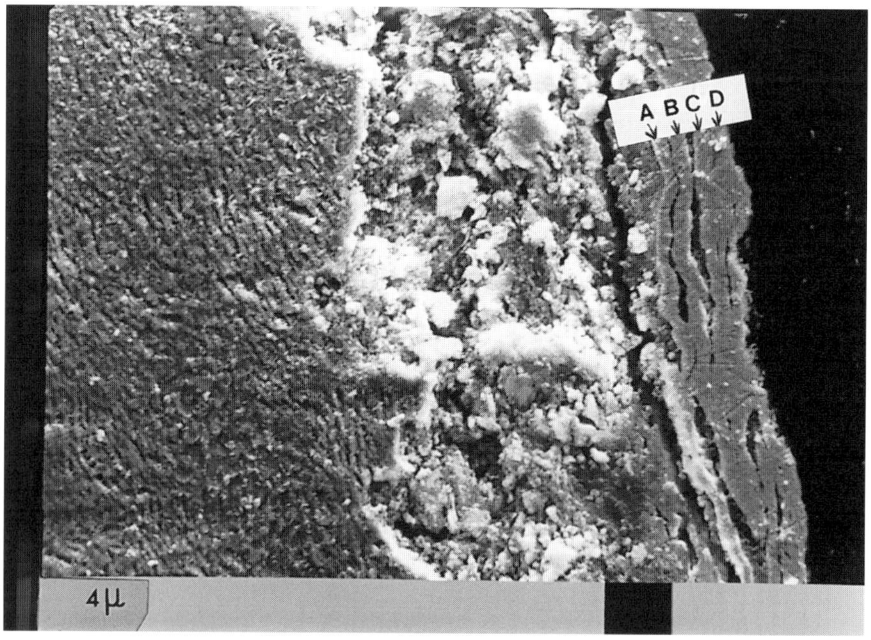

Figure 1.10 Detail of Figure 1.9 to show that the gilding on the surface consists of at least four layers (A to D).

controlled heating of the statue to drive off the mercury would be difficult. Leaf gilding, on the other hand, can be applied with an adhesive to any relatively smooth surface so the presence of corrosion products is not a problem, provided that any loose material is first removed by gentle abrasion.

Conclusions

It is still far too early to write a definitive history of the development of gilding technology as many more analyses of gilded objects, combined with a technological examination of the gilded surface, are required. However, it is possible to see that gilding with gold foil, held in place by mechanical forces, was in use in Mesopotamia by at least 3000BC, and that its use continued, with modifications to the method of holding the foil in position, into the medieval period. From soon after 2000BC, the use of gold leaf was introduced and this was held in place either physically (by using an adhesive) or chemically (by interdiffusion with the substrate).

Although the gilding of figure sculpture can be traced back to the beginning of the first millennium BC in Egypt, it seems to have been rare in the classical world before the Roman period. Even then it was not common. The gilding of sculpture in the Roman period was carried out by two different methods, and the composition of the copper alloy used to cast the sculpture varied according to the method of gilding to be used.

In the Far East, little is known about gilding before the late Chou period when fire gilding was introduced. It spread to Persia by the Sasanian period and to the Roman world at about the same time (second century AD), but the technique may have already been known to the Celts in Europe and Iberia.

Fire gilding was finally superseded by the introduction of electroplating in the 1840s.

Acknowledgements

Numerous colleagues in the British Museum have assisted me over the years by carrying out analyses and sharing with me the benefits of their experience. I am particularly grateful to Tom Chase, formerly of the Smithsonian Institution, for allowing me to illustrate his unpublished analyses of Chinese belt hooks.

My attendance at the Gilded Metals Symposium, where this paper was first presented, was made possible by a generous grant from the Gabo Trust to cover the cost of a transatlantic flight and by support from the conference organisers.

Notes

1. However, an example of depletion gilding has recently been identified from the mid-third millennium at Ur (La Niece 1995).
2. The word 'foil' is used to describe a thin sheet of gold that can be handled with ease and is strong enough to support its own weight. Further hammering will reduce the thickness and convert the foil into gold leaf, which is handled with difficulty using special techniques.
3. This is speculation as the known sites of mercury production in the ancient world (in Spain and Anatolia) have not been scientifically excavated and the processes of production have not been elucidated.
4. The 'fiction' that these coins were tin-plated seems to be due to Von Bibra (1869) and was repeated by Gowland (1900). That these coins are actually *silver-plated* was scientifically demonstrated in an analytical footnote by Le Gentilhomme (1962) and by Hedges and Robins, (1963).
5. Few silver washed antoniniani have been scientifically examined. However, La Niece and Craddock (1990) and La Niece (1993) have shown that some were certainly subjected to depletion silvering. Cope (1972) suggested an electro-chemical deposition from molten silver chloride, but without any proof that the method was used. The coins are so common and the plating so poorly bonded to the surface that electrochemical deposition must be a possibility.
6. The authenticity of this *Herakles* has recently been questioned and is discussed briefly below. Its composition and the technology of the gilding are, however, consistent with a Roman date, and, whether or not it is genuine, it was the 'catalyst' that led to our understanding of composition *vs.* gilding technology for Roman statuary
7. Wang (1984) hypothesises that the origin of fire gilding should be in the

late spring and autumn period (722–481 BC). In fact, however, the earliest fire-gilded objects to be scientifically examined are from the Warring States period (481–221 BC).

8. Although the three gilded Etruscan figures listed by Richardson (1983: 261/15, 320/29, 325/1) have not been scientifically examined, the scanty traces of gilding indicate the likely use of gold leaf with an adhesive.

References

Aitkin, A. and Aitkin, C. R. 1807. *A Dictionary of Chemistry and Mineralogy*, vol. 1. London: J. and A. Arch.

Al-Haik, A. 1964. 'The Rabbou'a galvanic cell', *Sumer* **20**: 103–4.

Anon. 1708. *Der Kunstliche Gold- und Silber-Arbeiter*. Nuremberg.

Anon. 1987. *Bronzi Dorati de Cartoceto: Un Restauro*. Florence: Cantini.

Brailsford, J. W. 1958. *Guide to the Antiquities of Roman Britain*, 2nd edn. London: British Museum.

Burnam, J. M. 1920. *A Classical Technology*. Boston: The Gorham Press.

Caley, E. R. 1970. 'Chemical composition of Greek and Roman statuary bronzes'. In *Art and Technology: A Symposium on Classical Bronzes,* S. F. Doeringer, D. G. Mitten and A. Steinberg (eds), 37–49. Cambridge, Mass.: MIT Press.

Carradice, I. and La Niece, S. 1988. 'The Libyan War and Coinage'. *Numismatic Chronicle* **148**: 33–52.

Chase, W. T. 1980. Letter to W. A. Oddy.

Cooley, A. J. 1845. *A Cyclopaedia of Practical Receipts,* 2nd edn. London: John Churchill.

Cope, L. H. 1972. 'Surface-silvered ancient coins'. In *Methods of Chemical and Metallurgical Investigation of Ancient Coinage,* E. T. Hall and D. M. Metcalf (eds), special publication no. 8, 261–78. London: Royal Numismatic Society.

Coulston, J. C. and Phillips, E. J. 1988. 'Hadrian's Wall West of the North Tyne and Carlisle', *Corpus of Sculpture of the Roman World: Great Britain*, vol. 1, fasc. 6. London: Oxford University Press for the British Academy.

Craddock, P. T. 1975. 'The composition of coper alloys used in the classical world'. Ph.D. dissertation. London: Institute of Archaeology, London University.

Craddock, P. T. 1985. 'Three thousand years of copper alloys'. In *Application of Science in Examination of Works of Art,* P. A. England and L. van Zelst (eds), 59–67 and microfiche. Boston, Mass.: Museum of Fine Arts.

Dalton, O. M. 1964. *The treasure of the Oxus*, 3rd edn. London: British Museum.

Duell, P. 1938. *The Mastoba of Mereruka*. Chicago, Ill.: University of Chicago, Oriental Institute, Publication No 31.

Eggert, G. 1995. 'On the origin of a gilding method of the Baghdad silversmiths'. *Gold Bulletin* **28**(1): 12–16.

Forbes, R. J. 1971. *Studies in Ancient Technology*, vol. 8, 2nd edn. Leiden: E. J. Brill.

Gowland, W. 1900. 'Remains of a Roman silver refinery at Silchester'. *Archaeologia* **57**(1): 113–24.

Granger, F. 1970 *Vitruvius on Architecture*, vol. 2. London: Loeb Classical Library.

Gray, D. H. F. 1954. 'Metal-working in Homer'. *Journal of Hellenic Studies* **74**: 1–15.

Halleux, R. 1981. *Les Alchimistes Grecs: Tome I: Papyrus de Leyde, Papyrus de Stockholm, Fragments de Recettes*. Paris: Société d'Édition 'Les Belles Lettres'.

Houser, C. 1979. 'An honorary statue in the Agora of Athens'. In *Greece and Italy in the Classical World, Acta of the 11th International Congress of Classical Archaeology* J. N. Coldstream and M. A. R. Colledge (eds), 222. London: The National Organizing Committee, XI International Congress of Classical Archaeology.

Houser, C. 1987. *Greek Monumental Bronze Sculpture of the Fifth and Fourth Centuries* BC, 255–81. New York and London: Garland Publishing.

Houser, C. (in press). *The Agora Horseman,* supplement to *Hesperia.*

Hawthorne, J. G. and Smith, C. S. 1979. *Theophilus: On Divers Arts*, 2nd edn. New York: Dover.

Haynes, S. 1959. 'Drei Silberreliefs im Britischen Museum'. *Jahrbuch des Deutschen Archäologischen Instituts* **73** (1958): 9–17.

Hedges, E. S. and Robins, D. 1963. 'Examination of some silver-coated Roman coins'. *Numismatic Chronicle* (7th series) **3**: 237–40.
Hunt, D. W. S. 1940–45. 'An archaeological survey of the classical antiquities on the Island of Chios . . . March and July 1938'. *Annual of the British School at Athens* **41**: 29–52.
Hunt, L. B. 1973. 'The early history of gold plating'. *Gold Bulletin* **6**(1): 16–27.
James, T. G. H. 1962. 'Craftsmanship in ancient Egypt'. *Whiley Monitor*, January: 9–11.
James, T. G. H. 1965. 'Addendum to S. Alexander, "Notes on the use of gold-leaf in Egyptian papyri"'. *Journal of Egyptian Archaeology* **51**: 48–52.
Jansen, W. *et al.*, Fickenfrerichs, H., Peper, R. and Flintjer, B. 1987. 'The battery of the Parthians and the gilding of the Baghdad goldsmiths' (in German). *Chemie für Labor und Betrieb* **38**(10): 528–30, 533.
König, W. 1940 *Neun Jahre Irak*. Brünn: Rudolf M. Rohrer Verlag.
La Niece, S. and Craddock, P. T. 1990. 'Silver plating on copper, bronze and brass'. *Antiquaries Journal* **LXX**(1): 102–14.
La Niece, S. 1993. 'Silvering'. In *Metal Plating and Patination*, S. La Niece and P. T. Craddock (eds), 201–10. Oxford: Butterworth-Heinemann.
La Niece, S. 1995. 'Depletion gilding from third millennium BC Ur'. *Iraq* **LVII**: 41–7.
Lechtman, H. N. 1971. 'Ancient methods of gilding silver: examples from the Old and New Worlds'. In *Science and Archaeology*, R. H. Brill (ed.), 2–30. Cambridge, Mass.: MIT Press.
Lechtman, H. N. 1973. 'The gilding of metals in pre-Columbian Peru'. In *Application of Science in Examination of Works of Art*, W. J. Young (ed.), 38–52. Boston; Mass.: Museum of Fine Arts.
Lechtman, H. 1979. 'A pre-Columbian technique for electrochemical replacement plating of gold and silver on copper objects'. *Journal of Metals* **31**(12): 154–60.
Lechtman, H., Erlij, A. and Barry, E. J. 1982. 'New perspectives on Moche metallurgy: techniques of gilding copper at Loma Negra, northern Peru'. *American Antiquity* **47**(1): 2–30.
Le Gentilhomme, P. 1962. 'Variations du titre de l'Antoninianus au IIIe siècle'. *Revue Numismatique* (6th series) **4**: 141–66.
Leoni, M. 1979. 'Metallographic investigations of the horses of San Marco'. In *The Horses of San Marco* (trans. J. Wilton-Ely and V. Wilton-Ely), 191–9. London: Royal Academy.
Lins, P. A. and Oddy, W. A. 1975. 'The origins of mercury gilding'. *Journal of Archaeological Science* **2**: 365–73.
MacKenzie, C. 1823. *Five Thousand Receipts in all the Useful and Domestic Arts*. London: Sir Richard Phillips and Co.
Nikolaus-Havé, F. 1986. 'Der Herkules von Birdoswald – Unikum oder Fälschung?' *Archäologischer Anzeiger* Heft 3: 571–81.
Northover, P. 1995. Personal communication.
Oddy, A. 1981. 'Gilding through the ages'. *Gold Bulletin* **14**(2): 75–9.
Oddy, W. A. 1982a. 'Gold in antiquity: Aspects of gilding and of assaying'. *Journal of the Royal Society of Arts* **130**: 730–43.
Oddy, W. A. 1982b. 'Scientific dating of the San Marco Horses'. *MASCA Journal* **2**(2): 45–7.
Oddy, W. A. 1985. 'Vergoldungen auf prähistorischen und klassischen Bronzen'. In *Archäologische Bronzen: Antike Kunst, Moderne Technik*, H. Born (ed.), 64–71. Berlin: Staatliche Museen Preussischer Kulturbesitz.
Oddy, W. A. 1988. 'The gilding of Roman silver plate'. In *Argenterie Romaine et Byzantine*, F. Baratte (ed.), 9–21. Paris: De Boccard.
Oddy, W. A. 1991a. 'Gilding: an outline of the technological history of the plating of gold on to silver or copper in the Old World'. *Endeavour* (new series) **15**(1): 29–33.
Oddy, A. 1991b. 'Gilding of metals in Japan'. *Conservation News* **45**: 37–8.
Oddy, A. 1993. 'Gilding of metals in the Old World'. In *Metal Plating and Patination*, S. La Niece and P. Craddock (eds), 171–81. Oxford: Butterworth-Heinemann.
Oddy, W. A. and Cowell, M. R. 1993 'The technology of gilded coin forgeries'. In *Metallurgy in Numismatics*: vol. 3, M. M. Archibald and M. R. Cowell (eds), 199–221. London: Royal Numismatic Society, special publication no. 24.

Oddy, W. A., Borrelli Vlad, L. and Meeks, N. D. 1979a. 'The gilding of bronze statues in the Greek and Roman world'. In *The Horses of San Marco, Venice* (trans. J. Wilton-Ely and V. Wilton-Ely), 182–6. London: Thames and Hudson.
Oddy, W. A., Padley, T. G. and Meeks, N. D. 1979b. 'Some unusual techniques of gilding in antiquity'. *Archaeo-Physika* **10**: 230–42.
Oddy, W. A., Bimson, M. and La Niece, S. 1981a. 'Gilding Himalayan images: history, tradition and modern techniques'. In *Aspects of Tibetan Metallurgy*, W. A. Oddy and W. Zwalf (eds), 87–101. London: British Museum, Occasional paper no. 15.
Oddy, W. A., La Niece, S., Curtis, J. E. and Meeks, N. D. 1981b. 'Diffusion-bonding as a method of gilding in antiquity'. *MASCA Journal* **1**: 239–41.
Oddy, W. A., La Niece, S. and Stratford, N. 1986. *Romanesque Metalwork: Copper Alloys and their Decoration*. London: British Museum.
Oddy, A. Pearce, P. and Green, L. 1988. 'An unusual gilding technique on some Egyptian bronzes'. In *Conservation of Ancient Egyptian Materials*, S. C. Watkins and C. E . Brown (eds), 35–9. London: United Kingdom Institute for Conservation, Archaeology Section.
Oddy, W. A., Cowell, M. R., Craddock, P. T. and Hook, D. R. 1990. 'The gilding of bronze sculpture in the classical world'. In *Small Bronze Sculpture from the Ancient World*, M. True and J. Podany (eds), 103–24. Malibu: Getty Museum.
Okinaka, Y. 1974. 'Electroless solutions'. In *Gold Plating Technology*, F. H. Reid and W. Goldie (eds), 82–102. Ayr: Electrochemical Publications Ltd.
Ortiz, G. 1989. Personal communication.
Pàszthory, E. 1989. 'Electricity generation or magic? The analysis of an unusual group of finds from Mesopotamia'. *MASCA Research Papers* **6**: 31–8.
Paterakis, A. B. 1999. This volume, Chapter 5.
Picon, M., Boucher, S. and Condamin, J. 1966. 'Recherches Techniques sur des Bronzes de Gaule Romaine'. *Gallia* **24**: 189–215.
Picon, M., Boucher, S. and Condamin, J. 1967. 'Recherches Techniques sur des Bronzes de Gaule Romaine'. *Gallia* **25**: 153–68.
Rackham, H. 1968. *Pliny: Natural History*, vol. 9. London: Loeb Classical Library.
Rapson, W. S. and Groenewald, T. 1978. *Gold Usage*. London: Academic Press.
Raub, C. 1993. 'The history of electroplating'. In *Metal Plating and Patination* S. La Niece and P. Craddock (eds), 284–90. Oxford: Butterworth-Heinemann.
Rees, A. 1819. 'Gilding of metals', In *The Cyclopaedia or Universal Dictionary of Arts, Sciences and Literature*, vol. 16. London: Longman, Hurst, Rees, Orme & Brown.
Richardson, E. 1983. *Etruscan Votive Bronzes.* Mainz: P. von Zabern.
Richardson, E. 1989. Personal communication.
Shear, T. L. 1973. 'The Athenian Agora: Excavations of 1971'. *Hesperia* **42**: 121–79.
Simon, F. 1993. 'Deposition of gold without external current source'. *Gold Bulletin* **26**(1): 14–23.
Smith, C. S. and Hawthorne, J. G. 1974. '*Mappae Clavicula*: A little key to the world of medieval techniques'. *Transactions of the American Philosophical Society* (new series) **64**(4): 1–128.
Southwell, R. 1698 'A way of guilding gold upon silver'. *Philosophical Transactions of the Royal Society* **20**: 296.
Spon, E. 1883. *Workshop Receipts for the use of Manufacturers, Mechanics and Scientific Amateurs*. London: E. and F. N. Spon.
Strong, D. E. 1966. *Greek and Roman Gold and Silver Plate*. London: Methuen.
Tailliez, B. 1982. 'Trente ans d'histoire de l'Apollon de Lillebonne'. *Revue du Louvre*, no. 2: 81–8.
Tait, H. (ed.) 1976. *Jewellery through 7000 Years*. London: British Museum.
Thompson, H. A. 1940. 'A golden Nike from the Athenian Agora'. *Harvard Studies in Classical Archaeology*, supplementary volume 1: 183–210.
Tite, M. S. and Craddock, P. T. 1977. 'Report on the composition of eight fragments of bronze sculpture from the Athenian Agora'. Unpublished report, 9 November 1977. British Museum Research Laboratory, London.
Toynbee, J. M. C. 1963. *Art in Roman Britain*, 2nd edn. London: Phaidon.
Vittori, O. 1976. Personal communication.
Vittori, O. 1978. 'Interpreting Pliny's gilding: archaeological implications'. *Rivista di Archeologia* **2**: 71–81.

Vittori, O. 1979. 'Pliny the elder on gilding'. *Gold Bulletin*, **12**(1): 35–9.
von Bibra, E. F. 1869. *Die Bronzen und Kupferlegirungen der alten und ältesten Völker*, 57–68. Erlangen.
Wang, H. 1984. 'Survey of gilding' [in Chinese]. *Gugong Bowuyuan Yuankan* [*Journal of the Palace Museum*, *Beijing*] (1984/2): 50–58, 84, plates 6–7.
Wilson, H. 1903. *Silverwork and Jewellery*. London: John Hogg [revised edition 1951, London: Sir Isaac Pitman & Sons].
Winton, W. 1962. 'Baghdad batteries BC'. *Sumer* **18**: 87–8.
Wise, E. M. (ed.) 1964. *Gold: Recovery, Properties and Applications*. Princeton, New Jersey: Van Nostrand.
Wu Kunyi 1981. 'Chemical distribution of mercury in ancient Chinese gilding' [in Chinese]. *Zhongguo Keji Shiliao* (*Material about the History of Chinese Science and Technology*), **1**: 90–94, 157–9.
Zwicker, U. 1973. 'Untersuchungen an goldplattierten keltischen und griechischen Münzen'. *Jahrbuch für Numismatik und Geldgeschichte* **23**: 115–17.

2 Corrosion Chemistry of Gilded Silver and Copper

Lyndsie Selwyn

Abstract

This paper summarises information about the corrosion chemistry of gold, gold alloys, and gilded silver and copper. It is intended as a review to provide conservators of gold or gilded objects with relevant information about corrosion and corrosion prevention. Through the ages, the attraction of gold has always been its excellent chemical stability; in the natural environment, gold does not, in general, corrode. The first section covers the physical and chemical properties of pure gold, including the special conditions under which gold does corrode. The next section deals with porosity in a gilding layer and the conditions responsible for the corrosion of the underlying (substrate) metal. Typical corrosion problems of two common substrate metals, copper and silver, are addressed. The third section describes the corrosion of ternary alloys of gold, silver, and copper. Such a ternary alloy may be present in the gilding layer, which is rarely pure gold, or in the metal under the gilding. The last section briefly deals with corrosion prevention.

Introduction

Gold, one of eight noble (or precious) metals, has long been known for its remarkable chemical stability in the natural environment. The other seven precious metals are ruthenium, rhodium, palladium, silver, osmium, iridium, and platinum. In ancient times, gold was valued because it kept its original color and could be easily hammered into thin sheets. It was recovered from placer deposits and mines, and made into ornaments for personal adornment or, later, into coins for trading. Naturally occurring gold (Scott 1983a) is usually an alloy with silver (5 weight per cent (wt%) to 45 wt%) and copper (0.1 wt% to 5 wt%), and special names exist for certain gold alloys. Gold–silver alloys containing over about 40 wt% silver are called electrum, or white gold, and special high copper content gold alloys containing 10 wt% to 90 wt% copper are called tumbaga alloys (Scott 1983a).

Gold was often used to make entire ornaments. For example, the Sumerians made pendants for necklaces by hammering gold into the shape of leaves (Ogden 1992). This, however, required large quantities of gold. To allow for the greater use of limited gold supplies, the application of a gold layer to the surface of another material, a process known as gilding, came into favour. The earliest forms of metal gilding date to about 3000BC and involve mechanically attaching relatively thick sheets of gold foil to other metals (Oddy 1993). Techniques for applying thinner layers of gold emerged as gold refining techniques developed. Throughout the ages, copper and silver objects have been gilded and examples can be found in collections around the world. These include Roman gilded silver (earrings, rhytons, statuettes, cups, bowls, medallions, brooches) (Oddy 1988), Central and South American gilded silver and copper (disks, figures, mummy masks) (Lechtman 1971; Bray 1993) and gilded bronze statues (Oddy *et al.* 1990; Jett 1993).

In modern times, people have exploited the special properties of gold, particularly its corrosion resistance, high density, good reflectance, and high electrical and thermal conductivity (Robertson 1990). Jewellers take advantage of the remarkable malleability and ductility of gold that allow it to be drawn to fine wire or hammered into thin gold leaf (Drost and Haußelt 1992). Even after cold-working, gold still remains very soft (Raykhtsaum and Agarwal 1993). Miners capitalise on gold's density – 19.32 $g\,cm^{-3}$ at 20°C (Robertson 1990) – being greater than most rocks to recover placer gold from the bottom of stream beds. Dentists use gold for fillings and crowns (Kempf and Haußelt 1992). The electronics industry uses gold in electronic devices, particularly in printed circuit boards and connectors, where long term reliability is required (Jacobson and Humpston 1992). Gold's ability to reflect radiation above 500 nm is exploited in architecture and engineering with the use of thin gold films on glass windows, space suits, and fire fighters' masks (Grimwade 1992). These thin films are capable of transmitting visible light while still reflecting higher wavelength radiation, especially infrared radiation. Gold compounds are even used in medicine to treat ailments such as rheumatoid arthritis (Puddephatt 1978).

This paper summarises the corrosion chemistry of gilded copper and silver objects. The first section focuses on pure gold and covers the rare situations when pure gold does corrode. The next section moves to the next level of complication and addresses the corrosion of copper or silver beneath a layer of pure gold. The third section covers the complicated corrosion problems associated with gold alloy objects or metals gilded with a gold alloy. The final section deals briefly with corrosion prevention. This paper is intended as a review to provide conservators of gold or gilded objects with information relevant to corrosion and corrosion prevention.

Corrosion of pure gold

The remarkable corrosion resistance of gold is based on its high values of standard reduction potential in the electrochemical series relative to other metals. Table 2.1 contains some standard reduction potentials (Puddephatt 1978; Weast 1984–85) for reactions of gold (Au), silver (Ag) and copper (Cu), as well as other reactions to be discussed later in this section. The more positive the reduction potential for a metal, the greater is its resistance to corrosion. Equations (i) and (ii) in Table 2.1 show high reduction potentials of 1.69 and 1.50 volts for gold in equilibrium with the Au^{1+} (aurous) or the Au^{3+} (auric) ions. (The chemical symbol for gold, Au, and the names of the common oxidation states aurous (Au^{1+}) and auric (Au^{3+}) are derived from *aurum*, the Latin name for gold.) Silver has a lower potential of 0.80 volts and copper even lower at 0.52 volts (ie, for the cuprous ion Cu^{+}). Because gold is above other metals in the electrochemical series, it is

Table 2.1 Standard reduction potentials, E°, for some reduction reactions relevant to gold (Weast 1984–85). The standard conditions include 25°C, 1 atmosphere pressure, and concentrations of 1 mole per litre for dissolved species.

No.	Reaction	E° (volts)
(i)	$Au^+ + 1e^- \leftrightarrows Au$	1.69
(ii)	$Au^{3+} + 3e^- \leftrightarrows Au$	1.50
(iii)	$Cl_2 + 2e^- \leftrightarrows 2Cl^-$	1.36
(iv)	$[AuCl_4]^- + 3e^- \leftrightarrows Au + 4Cl^-$	1.00
(v)	$Ag^+ + 1e^- \leftrightarrows Ag$	0.80
(vi)	$Cu^+ + 1e^- \leftrightarrows Cu$	0.52
(vii)	$O_2 + 2H_2O + 4e^- \leftrightarrows 4OH^-$	0.40
(viii)	$[Au(CS(NH_2)_2)_2]^+ + 1e^- \leftrightarrows Au + 2CS(NH_2)_2$	0.38
(ix)	$[Au(CN)_2]^- + 1e^- \leftrightarrows Au + 2CN^-$	−0.61

called ‘the most noble of the noble metals’ (Cohn and Stern 1980) or ‘the king of the metals’ (Schmidbaur 1992).

Figure 2.1 shows a potential–pH (or Pourbaix) diagram for the gold–water system at 25°C (Pourbaix 1974). Water is stable within the boundaries set by lines (a) and (b), which are the hydrogen-equilibrium and oxygen-equilibrium lines, respectively. At or above line (b), water can be oxidised to oxygen (O_2) gas according to $2H_2O \rightarrow O_2 + 4H^+ + 4e^-$. At or below line (a), the H^+ component of water can be reduced to hydrogen (H_2) gas according to $2H^+ + 2e^- \rightarrow H_2$. In Figure 2.1, the most important feature is that in water, which is stable between lines (a) and (b), gold remains in its elemental (native) state.

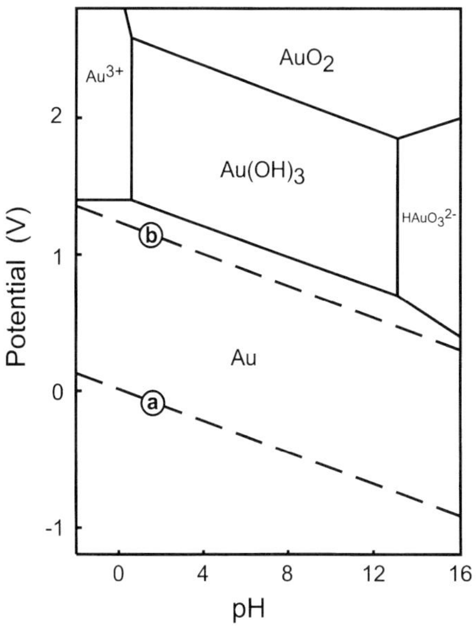

Figure 2.1 A potential–pH (or Pourbaix) diagram for the gold–water system at 25°C and 1 atmosphere pressure. All gold species are at concentrations of 10^{-4} M. The pH dependence of the potential for line (b) is $E = 1.228 - 0.0591$ pH at an oxygen partial pressure of 1 atmosphere, and for line (a) is $E = 0.000 - 0.0591$ pH at a hydrogen partial pressure of 1 atmosphere. (Adapted from Pourbaix 1974.)

Another way of demonstrating the stability of gold in aqueous solutions is through the use of standard reduction potentials. These can be used to predict the direction in which a reaction should proceed. If gold were to corrode in water containing dissolved oxygen, the net reaction would be

$$4Au + 3O_2 + 6H_2O \rightarrow 4Au^{3+} + 12OH^- \quad [1]$$

This reaction will only proceed in the forward direction (from left to right) if $\Delta E = E°(\text{cathode}) - E°(\text{anode})$ is positive. Here, E° is a standard reduction potential, E°(cathode) is the cathodic half reaction of oxygen reduction given by equation (vii) in Table 2.1, and E°(anode) is the anodic half reaction of gold oxidation given by equation (ii) in Table 2.1. For equation (1), $\Delta E = 0.40 - 1.50 = -1.10$ volts. This reaction will not proceed in the direction as written because ΔE is negative and so there is no driving force to proceed. Therefore, gold will not corrode in water containing dissolved oxygen.

The result of gold's high reduction potential is its ability to remain unaffected by many environments, including most natural environments. It is not tarnished by sulphur, or sulphur-containing gases, and is not attacked by most inorganic salts in solutions (Wise 1964). Gold is not oxidised by dissolved oxygen in water or in strong acids or strong alkalis. It is not attacked by concentrated hydrochloric acid up to the boiling point or to boiling nitric acid at concentrations up to 50 per cent (Smith 1987).

Although gold will not corrode in a solution containing just an oxidising agent (such as dissolved oxygen) or in a solution containing just a complexing agent (such as chloride ions), it will corrode in a solution containing both (Puddephatt 1978). A complexing agent attaches itself to a central metal atom forming a complex ion. Complexing agents, such as chloride ions (Cl^-) or cyanide ions (CN^-), lower the reduction potential of gold dissolution below that of water (Cohn and Stern 1980). Standard reduction potentials can again be used to demonstrate this. For the dissolution of gold in cyanide solutions, the net reaction is

$$4Au + O_2 + 2H_2O + 8CN^- \rightarrow 4OH^- + 4[Au(CN)_2]^- \qquad [2]$$

Using the two half reactions given by equations (vii) and (ix) in Table 2.1, $\Delta E = 0.40 - (-0.61) = +1.01$ volts. This reaction proceeds in the direction as written because ΔE is positive.

Gold (I) complexes

Gold recovery processes in the mining industry take advantage of the formation of stable gold (I) complexes by using them as intermediates. Alkaline sodium cyanide (NaCN) solutions are used to recover gold, mainly for economic reasons despite their high toxicity (Tran 1992). In this process, gold is extracted from finely crushed ore using aerated solutions (oxygen is essential). The net reaction is (Gasparrini 1983):

$$4Au + 8NaCN + O_2 + 2H_2O \rightarrow 4Na[Au(CN)_2] + 4NaOH. \qquad [3]$$

Gold is recovered by the addition of either zinc (Zn) dust:

$$2Na[Au(CN)_2] + Zn \rightarrow Na_2[Zn(CN)_4] + 2Au \qquad [4]$$

or activated charcoal, which adsorbs gold (Rapson 1992).

Acidic solutions of thiourea ($CS(NH_2)_2$) are also used to recover gold via the formation of a gold (I) complex. If a suitable oxidant, such as hydrogen peroxide, sodium peroxide or ferric ions, is added to an acidic thiourea solution, gold will corrode (Hiskey 1981). When the oxidant is ferric sulphate ($Fe_2(SO_4)_3$), the dissolution reaction is (Groenewald 1976; Schulze 1984):

$$2Au + 4CS(NH_2)_2 + Fe_2(SO_4)_3 \rightarrow [Au(CS(NH_2)_2)_2]_2SO_4 + 2FeSO_4 \qquad [5]$$

Although thiourea is less toxic than cyanide, it is more expensive and less stable in the leaching process; therefore, cyanide solutions remain the preferred extractant for gold (Torma 1992).

Gold (III) complexes

Gold purification processes are based on the formation of stable gold (III) complexes. These are most often formed when gold reacts with halide ions in an oxidising media. The best example of a gold (III) complex is the $[AuCl_4]^-$ complex (the tetrachloroaurate ion). This complex is formed when gold is dissolved in aqua regia which is a mixture of three volumes of concentrated hydrochloric acid (HCl) and one volume of concentrated nitric acid (HNO_3). It is said that alchemists named aqua regia, translated 'royal water' or the 'kingly water', after its ability to dissolve gold, the very king of metals (Parkes 1967; Cohn and Stern 1980).

When aqua regia is prepared by mixing the two acids, the nitric acid oxidises some of the hydrochloric acid to chlorine (Cl_2) gas, according to (Parkes 1967):

$$HNO_3 + 3HCl \rightarrow NOCl + Cl_2 + 2H_2O \qquad [6]$$

Nitrosyl chloride (NOCl) is also formed. Aqua regia thus contains the necessary oxidizing power (Cl_2) and complexing agents (Cl^-) to dissolve gold. The net reaction for gold dissolution in aqua regia is based on:

$$2Au + 3Cl_2 + 2Cl^- \rightarrow 2[AuCl_4]^- \qquad [7]$$

Using the half reactions (iii) and (iv) in Table 2.1, $\Delta E = 1.36 - (1.00) = +0.36$ volts. This reaction proceeds in the forward direction because ΔE is positive. The action of aqua regia is, however, probably more complex than this (Parkes 1967).

Solutions containing the $[AuCl_4]^-$ complex can be used to prepare other gold chemicals. Fulminating gold is one example. When ammonia or ammonium salts are added to a solution containing the $[AuCl_4]^-$ complex, explosive powders known as fulminating gold are formed (Johnson and Davis 1973). The precise composition and structure of these compounds is unknown, but they are considered to contain Au–N bonds (Johnson and Davis 1973). They are extremely dangerous compounds and, on moderate heating or on impact (even when wet), detonate to give gold, nitrogen, ammonia and water (Puddephatt 1978). Because of the possibility of forming explosive gold compounds, caution is recommended when using mixtures of aqua regia and ammonium salts to dissolve gold–silver–copper alloys (Scott 1983b).

Another example of the use of the $[AuCl_4]^-$ complex is for the formation of colloidal (or finely divided) elemental gold. When the $[AuCl_4]^-$ complex is mixed with a suitable reducing agent, elemental gold is formed and the resulting colour may be brown, red, blue or violet, depending on the average particle size and shape (Puddephatt 1978). Historically important is the purple-violet precipitate of colloidal gold called 'purple of Cassius', which was discovered in the mid 1600s and used to produce ruby-coloured glass (Wise 1964). (Although modern texts attribute the discovery of 'purple of Cassius' to Andreas Cassius Junior in 1685, its preparation was described by Johann Rudolph Glauber in 1659; Rapson and Groenewald 1978.) 'Purple of Cassius' is formed when stannous chloride ($SnCl_2$) containing some stannic chloride ($SnCl_4$) is added to a dilute solution of gold chloride and then heated; stannic chloride is essential because pure stannous chloride produces only a brown precipitate of gold (Rapson and Groenewald 1978). The purple precipitate is thought to be a mixed colloid of hydrated stannic oxide (SnO_2) and gold (Puddephatt 1978).

Relevance to the conservation of gold

The formation of colloidal gold from a solution containing the $[AuCl_4]^-$ complex can be used by conservators in a spot test for gold. Laver devised a simple test for gold (Laver 1978), and Townsend contributed additional information (Townsend 1988). First, a drop of sodium chloride is placed in an inconspicuous area on the object. Next, the metal is forced to corrode by connecting one electrode of a battery to the object, and filter paper attached to the other electrode is placed in the sodium chloride. The metal corrodes and any gold on the surface of the object dissolves and forms the $[AuCl_4]^-$ complex. Finally, the filter paper is removed and a drop of stannous chloride is added. If the gold complex is present, a dark precipitate of colloidal gold forms on the filter paper. Other metals, such as copper or silver, do not interfere with gold identification (Townsend 1988). This spot test must be used with caution because it can leave a small mark on the gold where the test was carried out.

Corrosion of metal beneath pure gold

This section discusses several gilding techniques and their associated corrosion problems. It also deals with the corrosion reactions of two common substrate metals, silver and copper, under an inert layer of pure gold. The information covered here is applied in the next section, which deals with the more complicated subject of gold alloyed with silver and/or copper.

Gilding of a metal is the application of a gold layer to the surface of a less noble metal (Oddy 1993). The stability of a gilded object depends on the thickness, adherence and porosity of the gold, all of which depend on the technique used to apply it. Any technique that produces a continuous, well secured layer of pure gold will protect the underlying metal from corrosion in a natural environment. Tarnishing or more severe corrosion happens when the layer is imperfect.

Adhesion of gold in different gilding techniques

Foil gilding involves attaching gold foil to a metal surface. Gold foil (Lechtman 1971) is usually thicker than a micron (1 micron = 1 μm = 10^{-6} m). Because the foil is mechanically attached to a metal by crimping, riveting, or inserting in grooves (Oddy *et al.* 1990), it is not well bonded to the metal surface. The schematic diagram in Figure 2.2a shows a gold layer poorly secured to a metal surface. Wherever there is an opening through the gold such as at a joint or crimp, water can penetrate and cause corrosion of the underlying metal. The growth of corrosion products can then force off the gold foil (Fig. 2.3a).

In leaf gilding, gold leaf is applied directly to a clean surface and burnished (Oddy 1991), or is glued to a prepared surface. Gold leaf is thin, less than about 1 μm (Lechtman 1971) and is so fragile that it cannot

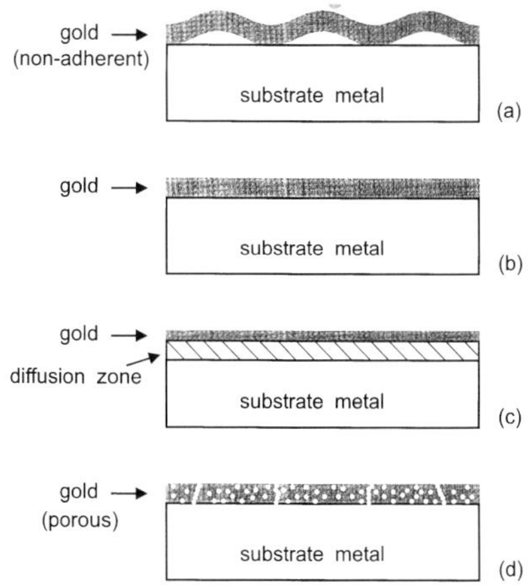

Figure 2.2 Schematic diagrams showing four variations in gilding layers that are (a) poorly secured, e.g., mechanically attached foil, (b) weakly secured, e.g., burnished or glued gold leaf, (c) strongly secured, e.g., diffusion bonded by heating burnished gold leaf, and (d) strongly secured but porous, e.g., fire-gilded or depletion gilded.

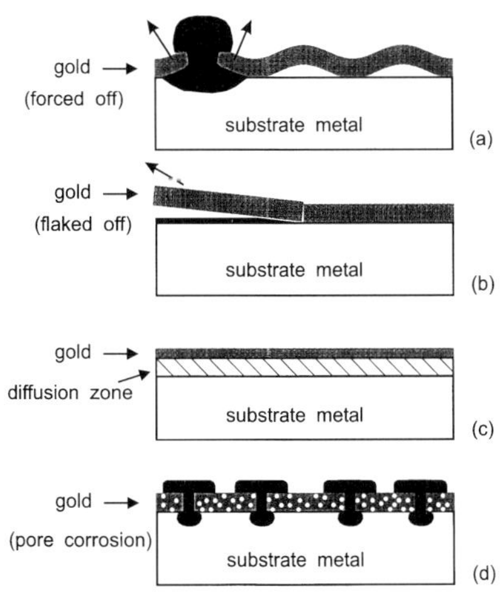

Figure 2.3 Schematic diagrams showing corrosion possible with the four types of gilding in Fig. 2.2. (a) Gold forced off by corrosion beneath it, (b) flaking of gilding after deterioration of glue or corrosion beneath gilding, (c) no corrosion as long as gold is continuous and well secured, and (d) corrosion through pores in gold.

support its own weight (Oddy 1981). Oddy *et al.* discuss the unusual but deliberate application of an intermediate layer of fill (ie, a ground called gesso today) to a bronze surface prior to adhering gold leaf (Oddy *et al.* 1988). Usually, glued gold leaf does not remain adherent because the adhesive deteriorates with time. Figure 2.2b illustrates schematically a thin layer of gold leaf glued or burnished onto a metal surface. If the bond formed by gluing or burnishing fails, then the gold layer can flake off (Fig. 2.3b).

After gold leaf is burnished on a copper or silver surface, a stronger bond can be created by gently heating the metals to produce a diffusion bond (Oddy *et al.* 1981). Heating causes interdiffusion of gold with silver or copper (Fig. 2.2c). Interdiffusion requires good contact between the gold and the base metal, and its extent depends on the temperature and time of heating (Lechtman 1971). (In modern electrical contacts, nickel is placed between the copper and gold to prevent the copper from diffusing to the surface where corrosion can lower the contact's electrical conductivity; Antler 1970.) A continuous, well adhered gold layer remains corrosion free.

Mercury gilding (fire gilding) depends on the formation of a gold amalgam – an alloy of gold and mercury – as an intermediate. To gild a surface using this method, a gold amalgam is first spread on a cleaned metal surface (usually silver or copper) and then heated to remove the mercury (Lins and Oddy 1975). The remaining gold forms a well bonded but porous layer (Oddy 1993) (Fig. 2.2d). Because the surface is heated, a certain amount of diffusion bonding takes place. After cooling, the surface is burnished. Depletion gilding, another technique for surface gilding, starts with a gold alloy, such as a gold–copper tumbaga alloy from South America (Lechtman 1971; Scott 1983a). The gold alloy surface is finished by chemically stripping the alloying element and then burnishing it to produce a gilded surface. Like fire gilding, depletion gilding produces a well bonded but porous gold layer. The base metal can corrode in areas where pores form a channel through the gold layer (Fig. 2.3d). The final burnishing step in mercury or depletion gilding helps to block these channels.

Galvanic corrosion

Corrosion begins when imperfections in a gold layer provide openings through which atmospheric corrodants can attack the underlying metal. The result can be unsightly as corrosion products creep over the gold. Pores are created in the gold layer during mercury or depletion gilding, or are present in very thin gold (Lins 1991). Abrasion can also reveal the underlying metal, as can cracks formed by flexing or differential thermal expansion (Guttenplan 1987). Today, the number of imperfections in a gold layer is determined by exposure of the gold-plated metal to corrosive atmospheres (Krumbein and Holden 1987). Such tests demonstrate that porosity, defined as the number of pores per square centimeter (Garte 1974) decreases with increasing thickness of gold. The increase in porosity with decreasing gold thickness is shown in Figure 2.4, using averaged results for electrodeposited gold (Russell 1976).

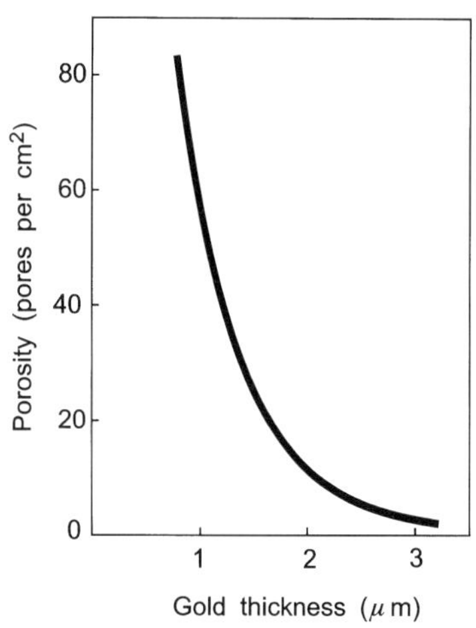

Figure 2.4 The increase in porosity in gold with decreasing thickness using averaged results for electrodeposited gold. Adapted from (Russell 1976).

When an imperfection exists through a gold layer, corrosion of the less noble substrate metal can occur. This is shown schematically in Figure 2.5. An electrical connection exists as long as the gold and substrate metal are in direct electrical contact and not separated by a non-conducting layer, such as an adhesive. Enough moisture must be present to provide an ionic pathway between the two metals. The substrate metal is the anode and corrodes (represented here simply as $M \rightarrow M^+ + 1e^-$). The electrons produced by the anodic reaction readily move to the gold surface, where they can participate in a cathodic reaction. Under most natural conditions, gold does not participate in the cathodic reaction. Instead, gold acts as an inert electrode and supports the cathodic reaction, which is usually oxygen reduction (equation (vii) in Table 2.1) although other reactions,

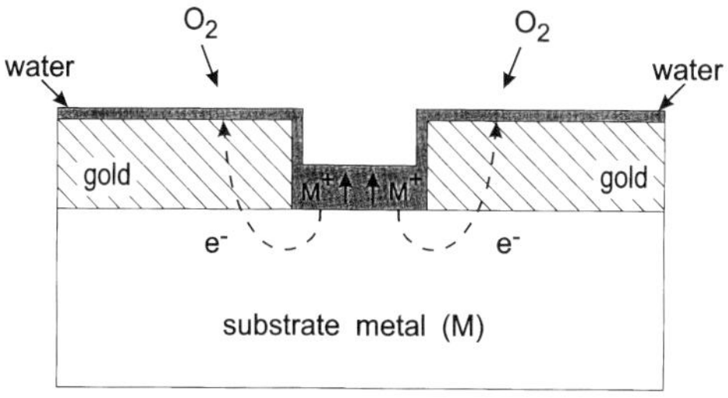

Figure 2.5 Schematic diagram of galvanic corrosion of substrate metal (M) through pore in gold. Water acts as the electrolyte. Corrosion of the substrate metal (the anode) is represented by $M \rightarrow M^+ + 1e^-$. Gold acts as an inert electrode supporting the cathodic reaction of oxygen reduction (equation vii in Table 1). Adapted from (Krumbein and Holden 1987).

such as reduction of dissolved acidic pollutants, are possible. As long as the gold surface remains electrically conducting, (ie, not covered with an insulating deposit), corrosion continues.

The corrosion process just described is known as galvanic corrosion (Baboian and Pohlman 1987). It occurs when one metal is electrically connected to a different metal in the same electrolyte. A galvanic series tabulates the relative reactivities (potentials) of metals under conditions governed by a specified electrolyte, such as sea water (Hack 1987). Of the three metals of interest here and in most electrolytes, gold is most noble, silver falls in the middle and copper is most active. A metal with a more negative (anodic or active) potential tends to suffer accelerated corrosion when electrically connected to a metal of more positive (cathodic or noble) potential. Galvanic corrosion of the metal beneath a gilded surface can be severe because the gold provides a large surface for cathodic reactions, such as oxygen reduction. The anodic reaction is concentrated where the more active metal is exposed through pores in the gilding layer. Pits often form as corrosion proceeds.

For silver or copper covered with pure gold, corrosion is associated with the underlying metal and not the gold, and so corrosion on gilded silver or copper is likely to be the same as that on silver or copper in similar environments. Therefore, we consider next the corrosion reactions common to silver and copper in indoor, outdoor and burial environments, because these reactions are equally relevant to gilded silver and copper. (Because the corrosion reactions of copper alloys such as bronze and brass are predominantly those of copper, the following discussion is also relevant to gilded copper alloys.) Table 2.2 contains the chemical formulas, and chemical and mineral names for the common corrosion products on silver and copper in indoor, outdoor and burial environments.

Corrosion in an indoor environment

Many silver and copper objects, such as silver trophies or brass candlesticks, are displayed or stored in an indoor museum environment. Their surfaces are intended to be bright and shiny. In an indoor environment, silver and copper objects are susceptible to tarnishing by sulphur-containing gases. The main tarnish-producing sulphur compounds are those containing sulphur in a reduced state (Franey *et al.* 1985) such as hydrogen sulphide (H_2S) or carbonyl sulphide (COS), and elemental sulphur (S). Silver reacts to form silver sulphide (Thompson 1973; Schmalzried 1981; Graedel 1992) and copper reacts to form copper sulphide (Graedel *et al.* 1985; Rickett and Payer 1993). The exact crystal structure of the copper sulphide has not been identified, but it is probably chalcocite. The tarnishing rate of silver and copper depends strongly on the relative humidity (Graedel *et al.* 1985), with the rate being more rapid under humid conditions than in dry air. Gaseous pollutants such as

Table 2.2 Common corrosion products identified on copper and silver objects from indoor, outdoor, and burial environments.

Chemical name	Mineral name	Chemical formula	Reference
Indoor environment			
Silver sulphide*	acanthite	Ag_2S	Graedel 1992
Copper sulphide	chalcocite	Cu_2S	Graedel *et al.* 1985; Rickett and Payer 1993
Outdoor environment			
Silver sulphide*	acanthite	Ag_2S	Gal-Or 1992
Silver chloride**	chlorargyrite	$AgCl$	Gal-Or 1992
Copper oxide	cuprite	Cu_2O	Graedel *et al.* 1987; Selwyn *et al.* 1996
Copper sulphate hydroxide	brochantite	$Cu_4(SO_4)(OH)_6$	Graedel *et al.* 1987; Selwyn *et al.* 1996
Copper sulphate hydroxide	antlerite	$Cu_3(SO_4)(OH)_4$	Graedel *et al.* 1987; Selwyn *et al.* 1996
Copper chloride hydroxide	paratacamite†	$Cu_2(Cl)(OH)_3$	Graedel *et al.* 1987; Selwyn *et al.* 1996
Copper chloride hydroxide	atacamite	$Cu_2(Cl)(OH)_3$	Graedel *et al.* 1987; Selwyn *et al.* 1996
Burial environment			
Silver sulphide*	acanthite	Ag_2S	Šrámek *et al.* 1987
Silver chloride**	chlorargyrite	$AgCl$	Šrámek *et al.* 1987
Copper sulphide	chalcocite	Cu_2S	Duncan and Ganiaris 1987
Copper sulphide	geerite	$Cu_{1.6}S$	Duncan and Ganiaris 1987
Copper sulphide	covellite	CuS	Duncan and Ganiaris 1987
Copper oxide	cuprite	Cu_2O	Gettens 1970
Copper chloride	nantokite	$CuCl$	Gettens 1970
Copper carbonate hydroxide	azurite	$Cu_3(CO_3)_2(OH)_2$	Gettens 1970
Copper carbonate hydroxide	malachite	$Cu_2(CO_3)(OH)_2$	Gettens 1970
Copper chloride hydroxide	atacamite	$Cu_2(Cl)(OH)_3$	Gettens 1970
Copper chloride hydroxide	paratacamite†	$Cu_2(Cl)(OH)_3$	Gettens 1970

* Silver sulphide exists in nature as the minerals acanthite (stable below 175°C, α-Ag_2S,) and argentite (stable above 175°C, β-Ag_2S) (Thompson 1973; Schmalzried 1981).

** Silver chloride is also called cerargyrite or horn silver; the latter name comes from the horn-like lustre of the mineral.

† Recent work (Jambor *et al.* 1996) has shown that the phase identified as paratacamite is probably clinoatacamite.

nitrogen dioxide (NO_2), chlorine (Cl_2), and ozone (O_3) also increase the tarnishing rate because they catalyse the tarnishing reactions (Graedel *et al.* 1984; Fiaud and Guinement 1986; Muller 1991).

As Graedel points out, silver does not always form a uniform tarnish film when it corrodes (Graedel 1992). Instead, when conditions are right, needle-like crystals of black silver sulphide can grow and such crystals have been reported on museum objects (Weisser 1990; Sease 1994; Sease *et al.* 1997). They have also been produced under controlled laboratory conditions using high levels of tarnishing gases. In the late 1950s, Fischmeister and Drott observed black whiskers growing perpendicular to a silver surface exposed to 100 per cent relative humidity and 2 per cent H_2S (Fischmeister and Drott 1959). Drott identified these whiskers as single crystals of Ag_2S (Drott 1960a) and studied their formation at relative humidities below 100 per cent (Drott 1960b,c). The sulphide crystals are thought to originate at sites where water condenses by capillary action, and growth is stimulated by high concentrations of reduced-sulphur gases. The formation of these needles may be related to the high mobility of silver atoms in pure silver and in silver sulphide crystals (Wagner 1953).

On a gilded silver or copper surface, dark silver or copper sulphides form in pores or other imperfections in the gold layer. These sulphides have the ability to creep over an adjacent gold surface (Egan and Mendizza 1960; Tierney 1981). Egan and Mendizza studied the tarnishing of silver plated with gold, rhodium, or palladium and they observed silver sulphide spreading from the silver surface over adjacent gold surfaces, but not over rhodium or palladium (Egan and Mendizza 1960). They also observed that the silver sulphide adhered more strongly to gold than it did to silver. This stronger adherence may be related by a chemical interaction between the silver sulphide and the gold to form mixed silver–gold sulphides such as AuAgS or $AuAg_3S_2$ (Dick *et al.* 1993). Furthermore, this stronger adherence means that tarnish on gold may be more difficult to remove than tarnish on silver. Copper sulphide films can creep over a variety of metal (and even non-metal) surfaces (Frant 1960), and so the process of copper sulphide creep probably does not involve any chemical interaction between the copper sulphide and those surfaces. The copper sulphide spreads over gold at a slower rate than does silver sulphide (Antler 1974).

Silver is inert to other indoor atmospheric contaminants, but copper is not. Green or blue corrosion typical of copper (II) salts is sometimes observed on copper or plated copper. Copper corrosion can be caused by volatile organic acids, such as acetic acid released from wood (Evans 1951) or from food, or by ammonia. When the relative humidity is high enough to form a thin film of water on the metal surface, dissolved acids or ammonia react to form soluble copper salt such as copper acetate or a soluble ammonia complex $[Cu(NH_3)_4]^{2+}$ (Massey 1973). As these soluble species diffuse through the water film, they can be converted to insoluble compounds, such as basic copper sulphates, thereby releasing the original ions to cause further damage.

Corrosion in an outdoor environment

In an outdoor environment, bronze statues are fairly common, but silver objects are not. Despite the extreme rarity of outdoor silver, corrosion studies have been made and the reported corrosion products are mainly silver sulphide, but also some silver chloride (Gal-Or 1992). Copper is more reactive than silver in an outdoor environment. As it corrodes, copper usually forms layers of different corrosion products. Next to the metal surface there is an inner layer composed of copper (I) compounds covered by an outer layer of copper (II) compounds. Copper (I) compounds often separate the copper and copper (II) compounds because many copper (II) compounds are not stable in the presence of copper metal (Opila 1987). During outdoor exposure, copper usually forms an inner layer of cuprite and an outer layer of basic copper sulphates and basic copper chlorides, although many other copper corrosion products are possible (Graedel *et al.* 1987; Selwyn *et al.* 1996).

For a gilded copper object in an outdoor environment, it is not long before the abrasive action of wind-blown material damages the gilding layer and copper corrosion starts. As the copper corrodes, the blue-green copper corrosion products form and spread over the gilded surface. Antlerite has been identified on the gilded bronze doors of the Baptistery in Florence (Fiorentino *et al.* 1982), brochantite on the gilded bronze equestrian statues near Lincoln Memorial in Washington, DC (Ogburn *et al.* 1977), and cuprite, brochantite, antlerite, and atacamite or paratacamite (probably clinoatacamite; Jambor *et al.* 1996) on the gilded horses of St Mark's Basilica in Venice (Alunno-Rossetti and Marabelli 1976). In these examples, the bronzes were originally fire gilded. In fact, during the Roman period, gilded bronze statuary was either cast in leaded bronze and leaf gilded with an adhesive, or cast (with difficulty) in almost pure copper, typically greater than 95 per cent copper, and fire gilded. Fire gilding could only be successfully carried out on fairly pure copper and not on leaded bronzes because of the problems created by the greater solubility of lead, tin and zinc in mercury relative to copper in mercury (Oddy *et al.* 1990).

Corrosion during burial

Gilded metal artifacts recovered from archaeological sites are often made of gilded silver or copper. When a gilding layer is present, the silver or copper corrosion rate can be accelerated through galvanic corrosion, especially if the gilding layer contains many imperfections and the environment is wet. During burial, silver reacts either with chloride ions to form silver chloride or with hydrogen sulphide to form silver sulphide (Šrámek *et al.* 1978). Chloride ions are usually dissolved in groundwater and hydrogen sulphide is produced by sulphate-reducing bacteria or by decaying organic matter. Depending on the length of burial and the burial conditions, a silver object may be superficially or completed converted to AgCl. Silver chloride is a stable compound that does not cause further corrosion to any remaining silver in an object (Plenderleith and Werner 1971).

As copper corrodes during burial, the inner corrosion layer that forms is often red cuprite and gray-green nantokite, and the outer corrosion layer is usually blue-green basic copper carbonates or basic copper chlorides (Gettens 1970). If there is a nearby source of hydrogen sulphide, then the corrosion layers can also contain copper sulphides (Duncan and Ganiaris 1987). The potentially harmful corrosion product, nantokite, precipitates within the inner corrosion layer because, during burial, chloride ions concentrate at anodic sites. Because nantokite is unstable in the presence of water and oxygen, the outer surface of CuCl reacts with water and oxygen to form basic copper chlorides. Nantokite plays a key role in bronze disease – the active corrosion of copper and copper alloys. When nantokite is present in an inner corrosion layer and it becomes exposed to water and oxygen through cracks in the outer layer, the bronze disease reaction of hydrolysis and oxidation occurs via (Scott 1990):

$$4CuCl + O_2 + 4H_2O \rightarrow 2Cu_2Cl(OH)_3 + 2HCl \qquad [8]$$

The reaction produces hydrochloric acid (HCl), which promotes further copper corrosion.

Reports of the positive identification of cuprous chloride on archaeological gilded copper are rare. Scott notes that nantokite is absent on depletion-gilded tumbaga alloys (Scott 1983a). One example, however, is given by Oddy and Meeks in their study of a gilded Roman bronze spoon (Oddy and Meeks 1982). In a cross-section through the spoon, they identified copper chloride, presumably cuprous chloride, above the gold layer as well as below it in the cuprite layer.

Relevance to the conservation of gilded silver and copper

Pure gold does not corrode in most environments and so no conservation is required. This is, however, not the case for gilded silver or copper. Conservation treatments for gilded silver or copper objects are often based on the conservation treatments available for silver or copper objects (Plenderleith and Werner 1971; Scott 1983a). For a gilded object, the degree of intervention depends largely on how thick the gilding is, how well it is adhered to the metal surface, and by which method it was produced. If the gilding is thick foil, mechanically attached by crimping, then it should be strong enough to withstand a chemical treatment needed to dissolve outer layers of copper or silver corrosion products. If the gilding was glued to a metal surface and is flaking off the surface, then it is unlikely that the object can withstand any chemical treatment. Thus, the decision about an appropriate conservation treatment needs to be made for each object by the conservators and curators responsible for the artifacts.

Corrosion of metals gilded with gold alloys

Using current technologies, it is possible to produce and apply pure gold to substrate metals. In the past, this was not possible because gold was naturally alloyed with other elements, especially silver and copper. Even today, gold leaf can be purchased in a range of colours created by alloying gold with other elements. Because the gilding alloy is often not pure gold but gold alloyed with silver and copper, the gilding alloy is susceptible to corrosion, either from tarnishing when exposed to an indoor atmosphere or from severe corrosion if buried. This section begins with a description of the concentrations and colours in ternary gold–silver–copper alloys. Then, the phases found in binary mixtures of gold with silver, gold with copper, and silver with copper are introduced to illustrate the complicated phase mixtures and microstructures possible in a ternary mixture of gold, silver and copper. Finally, the tarnish and corrosion properties of the ternary mixture, including the influence of microstructure on corrosion, are described.

Gold alloy concentrations and colours

There are several ways to describe the concentrations of elements in a gold alloy. They can be given as weight per cent (wt%) or as atomic per cent (at%). These percentages can be quite different because gold has a relatively high atomic weight (197.0 g mol^{-1}) compared to silver (107.9 g mol^{-1}) or copper (63.5 g mol^{-1}). For example, an alloy containing 50 at% of gold and 50 at% copper has equal numbers of gold and copper atoms, but has gold making up 76 per cent of the weight of the alloy and copper making up only 24 wt%. Formulas for the conversion between wt% and at% are available (Baker 1992).

Gold alloys can also be described by the carat (ct) system or in terms of fineness. In the carat system, the weight fraction of gold is expressed in parts per 24 or carats. The word carat arises from an Arabic word, *Kîrât*, for the seed of the locust bean tree. Because the seeds have the characteristics of similar size and weight, they were used by African and Arab traders for weighing gold. Formerly, one gold coin weighed 24 carats, but now it is taken to indicate how many twenty-fourth parts of the metal are gold (Grimwade 1985). Fineness is the weight fraction of gold expressed in parts per thousand. Thus, pure gold is 24 carat and has a fineness of 1000. A 14 carat gold alloy contains 14/24 or 58.3 wt% gold, and has a fineness of 583. Gold alloys higher than 18 ct become progressively softer with increasing gold content and so are susceptible to scratching and wear.

The colour of gold alloys varies with the content of the alloying elements, as outlined on the gold–silver–copper concentration diagram (the large triangle) in Figure 2.6 (Wise 1964). In such a ternary diagram, the corners represent the pure metals. Each axis along the sides of the triangle defines the concentrations of binary mixtures of just two elements. For example, the axis along the bottom of the triangle represents binary mixtures of silver and copper. Points along this axis are shown as wt% of copper; the silver wt% is just one hundred minus the copper wt%.

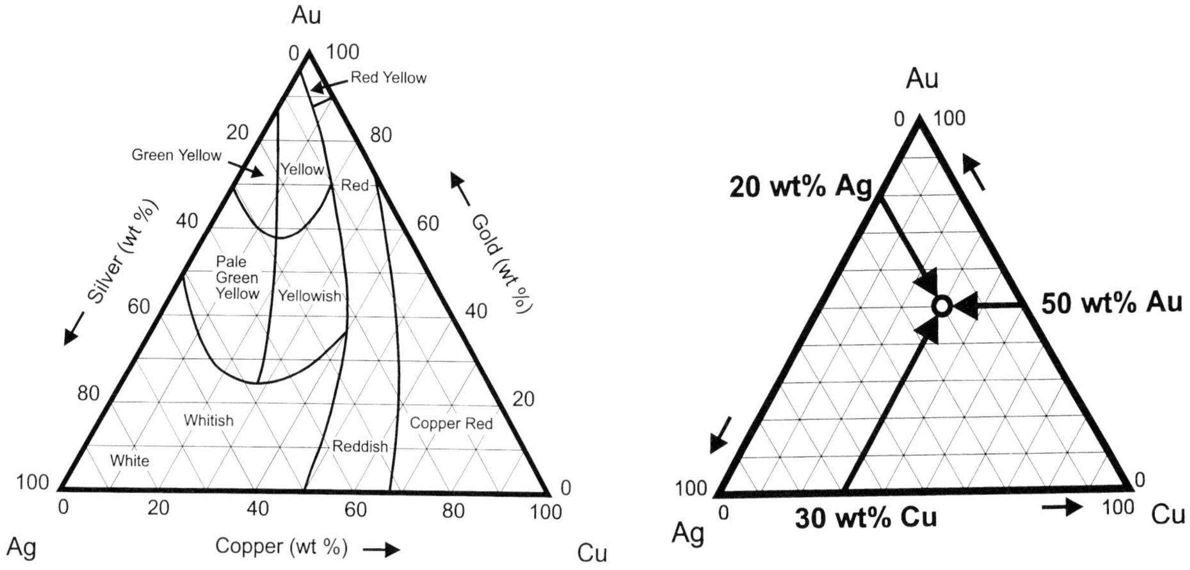

Figure 2.6 The composition ranges for various colours in gold–silver–copper alloys. The weight per cent concentrations for each element are defined along the sides. In the upper right corner, the point in the concentration triangle identifies an alloy with 20 wt% silver, 30 wt% copper, and 50 wt% gold. Adapted from (Wise 1964).

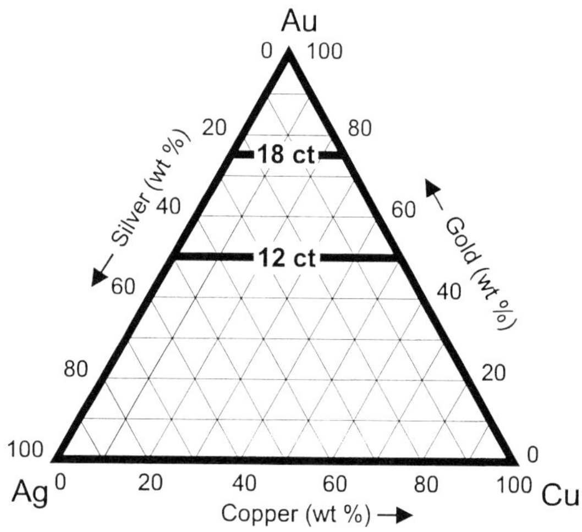

Each point within the triangle corresponds to a different alloy composition. In the upper right-hand corner of Figure 2.6, the point in the concentration triangle identifies an alloy with 50 wt% gold, 20 wt% silver and 30 wt% copper. The concentration ranges of 18 ct and 12 ct gold are highlighted in the triangle in the lower right-hand corner of Figure 2.6.

In gold–silver–copper alloys, the colour depends on the ratio of the three elements and ranges from the white of silver to the red of copper and the yellow of gold. For example, the addition of silver to gold progressively changes the colour through shades of yellow and green-yellow to white. Similarly, the color of 18 carat gold (75 wt% gold) varies from red (when gold is mixed with mainly copper) to yellow (when gold is mixed with copper and silver) to green-yellow (when gold is mixed with mainly silver).

Phases in gold–silver alloys

Figure 2.7 contains the gold–silver phase diagram, which shows that these two metals are completely miscible with one another. When gold and silver atoms are mixed together and their melt is cooled slowly, the atoms randomly fill the crystal structure, forming a single-phase solid solution. The crystal structure is face-centered cubic (fcc), in which the basic building block is a cube with gold or silver atoms randomly occupying positions

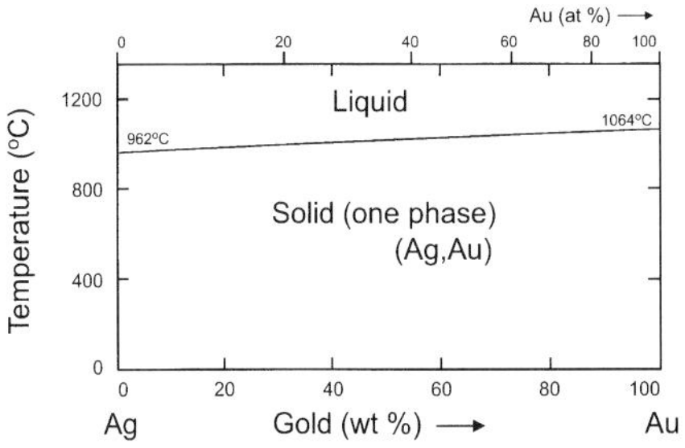

Figure 2.7 Equilibrium phase diagram for silver and gold. The scale is in weight per cent along the bottom and atomic per cent along the top. Solids exist below the line, melt at the temperature specified by the line, and form liquids above the line. The temperature range over which these alloys melt is so small (about 10°C maximum) that it cannot be seen on this scale. Adapted from Baker (1992).

at each corner and at each face. This random arrangement of gold and silver is considered 'disordered'.

A binary phase diagram gives the phases of a mixture of two metals at equilibrium. These diagrams apply when a melt is cooled very slowly to ensure enough time for the mixture to reach equilibrium, and, when the two metals are miscible, for a uniform composition to be reached. If the melt is cooled rapidly, the resulting solid may no longer have the uniform composition given in the phase diagram, but may contain compositional variations. For example, a compositional gradient may form between the inside and outside of the cooling alloy, or dendrites may form as the higher melting point component solidifies first (Scott 1991).

Phases in gold-copper alloys

The gold-copper phase diagram (Fig. 2.8) is similar to the one for gold–silver because gold and copper are also completely miscible. Gold–copper alloys form single-phase solid solutions with fcc structures across the full concentration range. The gold–copper phase diagram also contains features, representing regions of ordered structure, not found in the gold–silver phase diagram. In alloys containing 5–60 wt% copper, ordered structures can form either by slow cooling or by heating (ageing) between 150°C and 400°C. Rapid cooling (quenching) from above 410°C prevents their formation.

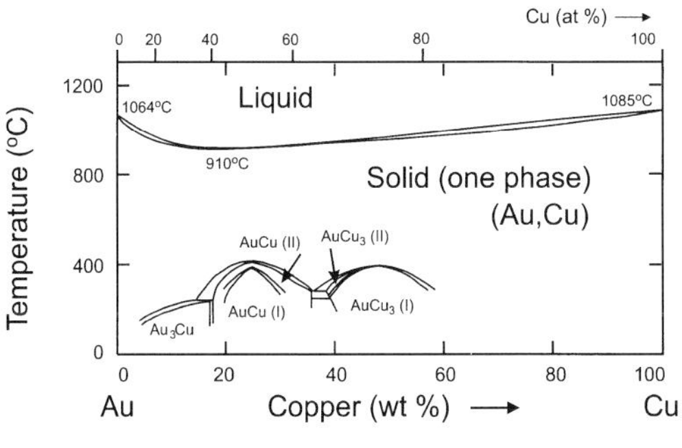

Figure 2.8 Equilibrium phase diagram for gold and copper. The scale is in weight per cent along the bottom and atomic per cent along the top. The temperature range over which these alloys melt is given by the upper ('liquidus') and lower ('solidus') lines. These lines represent the temperature at which any alloy becomes fully liquid on heating (upper line) or completely solid on cooling (lower line). Adapted from Baker (1992).

Ordered structures form as gold and copper atoms move from their random positions in 'disordered' structures into specific positions to form 'ordered' structures (Wells 1984; Grimwade 1985). For example, above 410°C in an alloy containing 25 wt% copper (50 at% each of gold and copper), there is a random arrangement of atoms in the unit cell (Fig. 2.9a). Below 380°C, the atoms rearrange and form an ordered AuCu(I) structure with alternating layers of gold and copper atoms (Fig. 2.9b). Between 410°C and 380°C, a more complicated ordered AuCu(II) structure forms. There are three other known ordered structures. In $AuCu_3$(I), gold occupies the corners of the cubes and copper occupies the

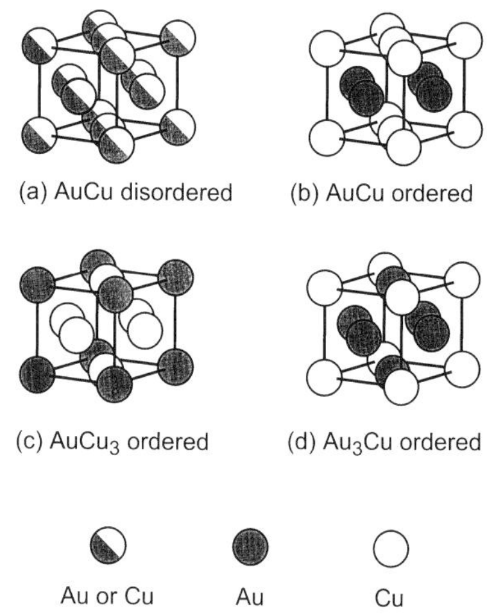

Figure 2.9 Some disordered and ordered unit cells with face-centered cubic structures in gold-copper alloys (a) disordered AuCu, (b) ordered AuCu (I), (c) ordered $AuCu_3$ (I), and (d) ordered Au_3Cu. Adapted from Wells (1984).

centers of the cube faces (Fig. 2.9c). The reverse is true for Au_3Cu because copper is in the cube corners and gold is in the cube faces (Fig. 2.9d). Finally, $AuCu_3$(II) is orthorhombic and it exists only over a narrow temperature range (Baker 1992).

Ordering may be accompanied by volume changes which can give rise to buckling or cracking and an increase in hardness. There are no serious volume changes as ordered $AuCu_3$(I) or Au_3Cu phases form because there is little distortion of the lattice; there is also no significant change in the hardness (Rapson and Groenewald 1978). On the other hand, as ordered AuCu forms, the lattice dimensions do change enough to cause severe internal strains and there is a significant increase in hardness. In fact, 18 ct gold–copper alloys are difficult to roll and apt to crack because they contain roughly equal numbers of gold and copper atoms and so are susceptible to the strains produced if ordered AuCu forms (Wise 1964). Over long times, ordered compounds are expected to form in gold–copper alloys, but their presence in ancient gold alloys has not been reported (Scott 1983a; Ogden 1985).

Phases in silver–copper alloys

In contrast to gold–silver and gold–copper mixtures, silver–copper alloys are usually two-phase mixtures because silver and copper have only limited solubility in one another. Figure 2.10 contains the silver–copper binary phase diagram. Many of these alloys melt over a wide temperature range, as shown by the large regions where the solid (either a_1 or a_2) and liquid (L) can coexist. The exception is the eutectic silver–copper alloy containing 72 wt% silver, which melts sharply at 779°C.

When small amounts of copper are added to silver, the copper atoms replace some of the silver atoms in the fcc lattice. The lattice remains a

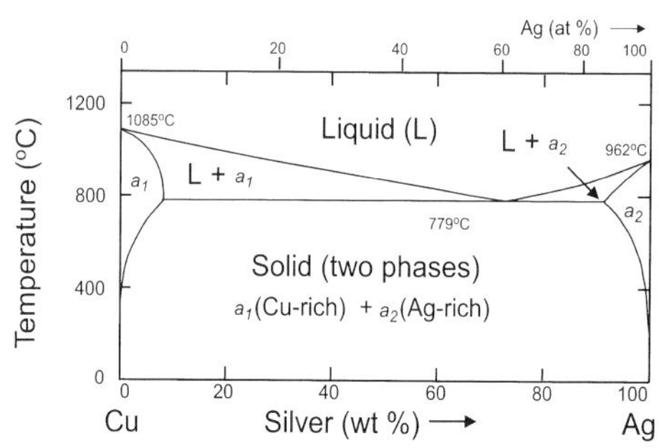

Figure 2.10 Equilibrium phase diagram for copper and silver. The scale is in weight per cent along the bottom and atomic per cent along the top. Adapted from Baker (1992).

silver-rich single phase (a_2). Similarly, as small amounts of silver are added to copper, silver replaces copper to form a copper-rich single phase (a_1). Alloys to the left of 9 wt% copper or to the right of 92 wt% copper are two phase under equilibrium (slow cooling) conditions. This is shown as the 'solid' region in Figure 2.10 with the alloy containing a mixture of the two phases, a_1 and a_2. If, however, an alloy is cooled rapidly, it may not have time for the two phases to separate and so will be single phase; this single phase can be heat-treated to allow the two phases to separate (precipitate).

Phase separation in silver–copper alloys results in a harder alloy because two-phase solids are substantially harder and less malleable than single-phase alloys. Because phase separation in an alloy that has initially been cooled rapidly is promoted by ageing an alloy, either by heating gently or naturally ageing for a long time, the process is called age hardening. Because hardening is caused by precipitation of two phases, it is also known more specifically as age hardening by precipitation. Sterling silver (92.5 wt% Ag, 7.5 wt% Cu) offers an example. It can be prepared as a soft single-phase alloy by heating to 750°C for 30 minutes followed by quenching to lock in the a_2 phase. This alloy can then be age hardened by heating it at 300°C for one hour so that a fine dispersion of a_1 phase particles form in the a_2 structure (Grimwade 1985).

Phases in gold–silver–copper alloys

The binary phase diagrams of gold–silver, gold–copper, and silver–copper were discussed above to identify the key microstructural features in each pair under equilibrium conditions. Gold and silver form single-phase solid solutions at all concentrations, as do gold and copper. In addition, certain gold–copper alloys can form ordered structures at low temperatures and, if formed, can sometimes result in a harder structure (age hardening by ordering). Finally, silver and copper have limited solubility in one another and usually exist as two phases; one silver-rich and the other copper-rich. This phase separation leads to a harder alloy (age hardening by precipitation). The influence of the binary phase properties, especially the limited solubility of silver and copper, shows up when the three metals are combined. Both hardening mechanisms, by ordering and by precipitation, can be present concurrently in gold–silver–copper alloys.

Figure 2.11 contains a gold–silver–copper ternary phase diagram (Rapson and Groenewald 1978). In the upper region of the triangle, a single-phase solid solution (Au, Ag, Cu) exists, in which gold, silver, and copper atoms are randomly mixed. As the gold content decreases from the

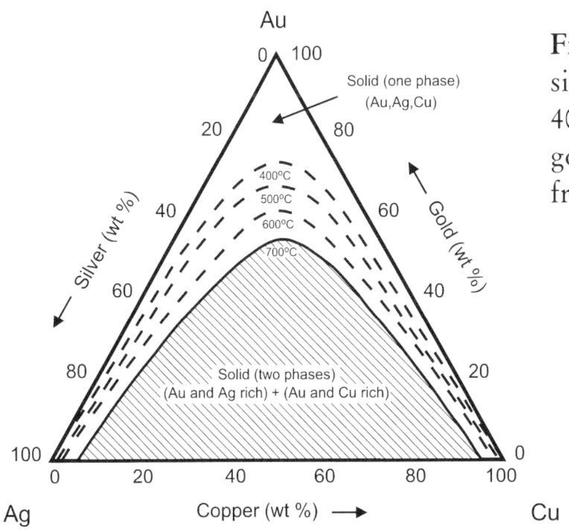

Figure 2.11 Boundaries between single-phase and two-phase regions at 400°C, 500°C, 600°C, and 700°C in gold–silver–copper alloys. Adapted from Rapson and Groenewald (1978).

top of the triangle, the single phase separates into two phases, one rich in gold and silver (Au and Ag rich) and the other rich in gold and copper (Au and Cu rich). The concentrations where phase separation occurs changes with temperature because the solubility of silver and copper in one another decreases with decreasing temperature. The shaded area in Figure 2.11 covers the alloys with two phases at 700°C. The dashed lines at 600°C, 500°C, and 400°C separate the single-phase region from the two-phase region at those temperatures. Not shown in Figure 2.11 is the existence of ordered gold–copper structures below 400°C.

Alloys from different regions in the ternary phase diagram can be grouped into three types according to microstructure (McDonald and Sistare 1978). One type falls within the shaded area in Figure 2.11, where the solid phase naturally separates. These alloys are difficult to quench, harden noticeably during air cooling, and harden further upon annealing, making them the hardest and the most difficult ones to work. Another type falls within the dashed lines in Figure 2.11, and heat treatment determines whether the microstructure is single-phase or two-phase. These alloys are moderately soft in the annealed condition, but are age-hardened by precipitation following a suitable heat treatment. For example, rapid cooling of a 14 ct (58 wt%) gold alloy from 650°C yields a workable alloy (Rapson and Groenewald 1978) and subsequent heating to 300°C hardens the alloy because the two phases separate. Formation of ordered AuCu in the (Au and Cu rich) phase can also contribute to age hardening. The last type are single-phase solid solutions found along the binary edges of gold–copper or gold–silver. These alloys are soft in the annealed condition and cannot be age hardened by precipitation although certain alloys, such as 18 ct (75 wt%) gold alloys with high copper, can be age hardened by ordering.

Tarnishing of gold–silver–copper alloys

In an earlier section, the discolouration or tarnish on pure gold was attributed to tarnish forming from the underlying metal through pores in the gold. If the gilding layer is not pure gold, then it is also susceptible to tarnishing. This susceptibility depends primarily on the gold content (Randin *et al.* 1992). Tuccillo and Nielson, in their accelerated tarnish testing of gold alloys, found that even a 22 ct (92 wt%) gold alloy can tarnish (Tuccillo and Nielsen 1971). In general, the lower the gold content, the more easily that alloy tarnishes.

For alloys with the same gold content, the alloys containing more silver than copper are more susceptible to tarnishing. Randin *et al.* studied tarnishing and tarnish thickness in two sets of gold alloys, each set with varying silver and copper contents (Randin *et al.* 1992). Within each set, tarnishing increased with increasing silver content. Even binary gold–silver alloys are more prone to tarnishing than gold–copper alloys (Kuhn 1981). This effect is unexpected because pure copper is more sensitive to tarnishing than is pure silver (Graedel *et al.* 1985; Randin *et al.* 1992). The gold in gold–silver and gold–silver–copper alloys seems to influence the tarnishing (sulphidation) of silver over copper and this effect may be related to the formation of mixed gold–silver sulphides in the tarnish film. Courty *et al.* identified the mixed sulphides AuAgS and $AuAg_3S_2$ in their study of gold–silver and gold–silver–copper alloys (Courty *et al.* 1991). Also, Dick *et al.* detected AuAgS, Ag_3AuS_2, and Ag_2S in their tarnishing study of gold–silver alloys (Dick *et al.* 1993), and Frantz and Schorsch identified AuAgS and possibly $AuAg_3S_2$ on red-coloured Egyptian gold objects (Frantz and Schorsch 1990). On copper-rich gold alloys, the tarnish consists of copper sulphides (Randin *et al.* 1992).

The microstructure of a gold alloy also plays a role in its resistance to tarnishing. German and colleagues studied the effects of microstructure on the tarnishing of gold alloys (German 1981; Fioravanti and German 1988). They determined that two-phase alloys are more susceptible to tarnishing than single-phase alloys. Single-phase alloys have a uniform microstructure because each crystal or grain within the solid has the same composition and the same relative potential. On the other hand, two-phase alloys have a non-uniform microstructure; they can be grains of the two phases in contact with one another, or dendrites or grains of one phase surrounded by the second phase. Because these two phases contain different concentrations of metals, they have different potentials and this results in galvanic corrosion on a microscopic scale. The more active phase in the two-phase microstructure tarnishes at an accelerated rate. German observed tarnish along the dendritic casting pattern in his tarnishing tests with gold alloys (German 1981). In addition to the microstructures predicted by the equilibrium phase diagrams, there are also microstructures created by rapid cooling. Many ancient gold objects were cooled rapidly, then worked and further annealed. Such treatment can result in non-uniform microstructures, compositional gradients, and microscopic galvanic cells, and, as in cases of two-phase alloys, can produce regions, often those enriched with the more active metal, that are more susceptible to tarnishing.

Corrosion of gold–silver–copper alloys

Single-phase gold alloys have been divided into four regions, each with different susceptibilities to corrosion and attack by acids as illustrated in Figure 2.12 (Rapson and Groenewald 1978; Drost and Haußelt 1992). The separation between regions depends on the atom per cent of gold in the alloys. Region 1 contains gold alloys with greater than 50 at% gold. (An 18 ct gold–copper alloy or a 15.5 ct gold–silver alloy contains 50 at% gold.) These alloys are corrosion resistant and are normally attacked only by aqua regia. Region 2 contains alloys with between 50 at% and 37 at% gold, and these alloys are lightly attacked by strong acids, such as nitric acid. Region 3 contains alloys with between 37 at% and 25 at% gold, and these are attacked by strong acids, which leach copper or silver from the surface. Finally, region 4 contains alloys with less than 25 at% gold, which are the most susceptible to corrosion. These alloys are easily decomposed by acids because the alloying elements are selectively dissolved.

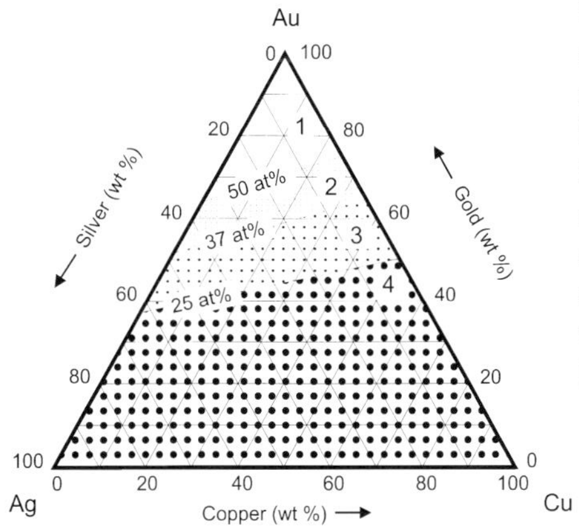

Figure 2.12 Four regions in gold-silver-copper alloys with different susceptibilities to corrosion and attack by acids. The four regions are as follows: (1) attacked only by aqua regia, (2) lightly attacked by strong acids (e.g. nitric acid), (3) attacked by strong acids, and (4) susceptible to tarnishing. Adapted from Drost and Haußelt (1992).

Over long periods of time in a burial environment, gold alloys can suffer selective dissolution of their less noble elements, leaving a gold-rich residue (Forty 1981). For example, Scott discusses extensive corrosion during burial of gold–copper tumbaga alloys beneath a gilding layer (Scott 1983a). He identified the black powder as finely divided gold mixed with cuprite (Cu_2O). This selective dissolution is also the basis for finishing the tumbaga alloys because the alloying elements are chemically removed from the outer surfaces, leaving a thin gold layer.

Single-phase gold alloys, especially those below about 14 ct (58 wt%) gold, may be susceptible to extensive cracking caused by a local corrosion process known as stress corrosion (Dugmore and DesForges 1979; Cohn *et al.* 1980). Certain combinations of local stress and a corrosive environment cause gold alloys to be attacked along grain boundaries (Rapson and Groenewald 1978). On a microscopic scale, stress corrosion occurs as local stress progressively opens up grain boundaries to attack, forming cracks. The local stress can be external or internal, and internal stresses are often created during shaping or burnishing of an object. As stress corrosion occurs, the less noble metals in the gold alloy are selectively dissolved (Forty 1981). The nature of the corrosive environment is not critical with alloys containing below 40 at% gold, as long as the environment dissolves copper and silver (Wise 1964). Stress-corrosion cracking occurs primarily with single-phase alloys, but not in alloys with a two-phase microstructure. A new crack growing in the more active phase in a two-phase alloy is blocked off as soon as it encounters the more noble phase (Rapson and Groenewald 1978).

Any corrosion process, whether it be selective leaching of alloying constituents or stress-corrosion cracking, is accelerated by the galvanic effect created when different metals are in contact with one another. Galvanic corrosion can occur on a macroscopic scale. For example, in depletion-gilded tumbaga alloys, there is the galvanic effect created by the contact between the gold layer and the underlying gold-copper alloy which accelerated the corrosion of the gold–copper alloy (Scott 1983a). Galvanic corrosion can also occur on a microscopic scale, particularly when alloys have a non-uniform or a two-phase microstructure (Cohn and Stern 1980; Fioravanti and German 1988). The more active of these phases will undergo accelerated corrosion.

Relevance to the conservation of gold-silver-copper alloys

Although gilding is often thought of as pure gold, it is more likely to be a ternary gold–silver–copper alloy. The gold may have been used as is, from a placer deposit, or crudely refined to a fairly pure alloy. Whenever the gilding is a gold alloy and not pure gold, the alloy is susceptible to tarnishing. Caution is therefore recommended when deciding whether or not to remove the tarnish with an abrasive polish. Pure gold and high gold content alloys are soft and are easily damaged by abrasive polishing, and gilding layers are usually extremely thin and easily removed.

Gilded objects that have been buried for long periods of time in moist soil may have changed superficially because of surface depletion of alloying silver and copper, or may have altered extensively because of stress-corrosion cracking. Selective leaching of copper and silver from a gold alloy leaves a purer gold surface and slightly matt finish (Ogden 1985). Corrosion removal or even abrasive polishing of this surface enrichment will remove evidence of selective leaching. Heating for repair work can also alter the surface concentrations because of silver or copper diffusion from the bulk of the object into the surface gold, re-forming a gold alloy and altering the surface colour. Heating might also cause ordering in certain gold–copper alloys, which could then produce strains and buckling in the gilding. Extensive stress-corrosion cracking may not be apparent but, if present, will have left the object in an extremely fragile state. Gilded objects that have been buried should be handled with extreme care and any repair work, such as soldering, should be carried out with caution.

Corrosion prevention

Corrosion prevention is based on keeping harmful materials away from the metal surface. Water is an essential ingredient for corrosion and so it is not surprising to encounter the general recommendation to keep metals dry. Other factors influencing corrosion include particulate matter, especially those containing chloride ions or sulphate ions, and pollutant gases, particularly hydrogen sulphide. The following briefly summarizes some simple procedures for preventing or at least minimizing corrosion problems on gilded silver and copper objects in storage or on display. More detailed information can be found elsewhere on the general care of metals (Drayman-Weisser 1992; Canadian Conservation Institute 1995).

For ideal storage of gilded silver and copper, the relative humidity should be as low as possible. However, this is rarely practical. Instead, it is reasonable to store stable gilded metals with the rest of a collection as long as the relative humidity does not exceed about 55 per cent. If the relative humidity exceeds 55 per cent, then consideration should be given to moving the gilded artifacts to proper humidity-controlled storage facilities. Any objects displaying signs of active corrosion, such as bronze disease, need to be stored in a separate area with a relative humidity below 35 per cent and treated if possible to remove the source of the active corrosion (Scott 1990).

During storage, objects should be covered with acid-free tissue paper to avoid contamination by dust and other particulate matter. Within the storage area, materials to be avoided include wooden cabinets, particle board, and plywood because they can give off volatile organic acids (acetic or formic), which are especially harmful to copper. Ammonia-based cleaning products should also be avoided because they can also harm copper. Objects should not be allowed to touch other metal objects, metal supports or conducting non-metals, such as black plastics containing carbon black. This avoids the harmful effects of galvanic corrosion. Finally, sulphur-containing materials should be avoided because they cause tarnishing. Sources of sulphur-containing material, particularly hydrogen sulphide, are polluted industrial environments and well water. Other sources include rubber bands, latex gloves, rubber floor mats, wool, certain latex paints, some adhesives, cardboard and certain foods like mayonnaise and eggs (Brimblecombe *et al.* 1992).

The atmosphere within a display cabinet can be controlled to prevent tarnishing (Selwyn 1990). Silica gel can be used to keep the relative humidity as low as possible, because this minimizes the tarnishing rate. Activated charcoal can be used to remove harmful pollutants and tarnishing gases. Regular replacement or regeneration of the activated charcoal is necessary to keep it from becoming fully loaded with contaminants. Activated charcoal saturated with contaminants can do more harm than good because it becomes a source of tarnishing gases.

Gilded objects with polished surfaces that are displayed in open areas, such as in period rooms, are exposed to tarnishing gases and so may require frequent cleaning and possibly polishing. The application of a barrier coating, such as a wax or lacquer, offers an alternative to frequent cleaning. However, coating a gilded object should only be considered if the gilding is well adhered and can withstand future removal procedures. Acceptable waxes for coating silver and copper objects are hard microcrystalline waxes and carnauba wax (Barclay and Hett 1988). Although wax coatings are said to be easily removed with mineral spirits, this may not be the case with porous or ornate surfaces. Also, residual wax in crevices can make it impossible to achieve good adhesion of any other subsequent coatings. As a result, one is usually committed to a wax coating once an object has been waxed. Clear lacquers can also be effective in preventing tarnish if applied properly. A few clear lacquers recommended for silver are the acrylics Incralac (Moncrieff 1966) and Acryloid B-72 (De Witte 1973–74), and the nitrocellulose lacquers Frigilene (Child 1987) and Agateen (Heller 1983). Incralac is also widely used for coating outdoor bronze statues. Clear lacquers can yellow and may shrink and crack as they age, so they need to be removed and reapplied on a regular basis.

Summary

Gold's most important chemical property is its resistance to environmental effects. Its appealing yellow colour and the ease with which it can be hammered into thin sheets lead to its widespread use for decoration and personal adornment, and the accumulation of wealth. This paper reviews three aspects of gold. First, the chemical properties of pure gold are described, including the conditions when gold does corrode to form gold (I) or gold (III) complexes in solution. Second, the corrosion problems associated with a gilding layer of pure gold on the surface of silver and copper are summarised. A layer of pure gold is resistant to corrosion and will protect the underlying metal as long as the layer is continuous. Imperfections in pure gold allow underlying silver or copper to corrode. Typical corrosion products are summarised for gilded silver and copper in indoor, outdoor and burial environments. Third, the tarnishing and corrosion of ternary gold–silver–copper alloys are discussed in terms of the influence of single-phase, two-phase and ordered-phase microstructures. The final section deals briefly with simple procedures for preventing corrosion.

Acknowledgements

The author wishes to thank Marie-Claude Corbeil for help with the chemistry of aqua regia, Stefan Michalski for help with gold porosity, and Vasilike Argyropoulos for comments on the manuscript. She would also like to thank Terry Drayman-Weisser, Andrew Lins, David Scott, and Tom Chase for help in looking for evidence of bronze disease on a gilded bronze.

References

Alunno-Rossetti, V. and Marabelli, M. 1976. 'Analyses of the patinas of a gilded horse of St Mark's Basilica in Venice: Corrosion mechanisms and conservation problems'. *Studies in Conservation* **21**: 161–70.

Antler, M. 1970. 'Gold-plated contacts: Effect of heating on reliability'. *Plating* **57**: 615–18.

Antler, M. 1974. 'Contact properties'. In *Gold Plating Technology*, F. H. Reid and W. Goldie (eds), 277–94. Ayr, Scotland: Electrochemical Publications.
Baboian, R. and Pohlman, S. L. 1987. 'Galvanic corrosion'. In *Metals Handbook*, volume 13: *Corrosion*, 9th edn, J. R. Davis (ed.), 83–7. Metals Park, Ohio: American Society for Metals.
Baker, H. (ed.) 1992. *ASM Handbook*, volume 3: *Alloy Phase Diagrams*. Metals Park, Ohio: American Society for Metals.
Barclay, B. and Hett, C. 1988. *The Cleaning, Polishing and Protective Waxing of Brass and Copper* (CCI Notes 9/3). Ottawa, Canada: Canadian Conservation Institute.
Bray, W. 1993. 'Techniques of gilding and surface-enrichment in pre-Hispanic American metallurgy'. In *Metal Plating and Patination: Cultural, Technical and Historical Developments*, S. La Niece and P. Craddock (eds), 182–92. Oxford, UK: Butterworth-Heinemann.
Brimblecombe, P., Shooter, D. and Kaur, A. 1992. 'Wool and reduced sulphur gases in museum air'. *Studies in Conservation* **37**: 53–60.
Canadian Conservation Institute 1995. *Storage of Metals* (CCI Notes 9/2). Ottawa, Canada: Canadian Conservation Institute.
Child, B. 1987. 'Cleaning of silverware'. *Conservation News* **32**: 9.
Cohn, J. G. and Stern, E. W. 1980. 'Gold and gold compounds'. In *Kirk-Othmer Encyclopedia of Chemical Technology*, volume 11, 3rd edn, 972–95. New York: John Wiley & Sons.
Courty, C., Mathieu, H. J. and Landolt, D. 1991. 'Tarnishing of Au–Ag–Cu alloy studied by Auger electron spectroscopy and coulometry'. *Werkstoffe und Korrosion* **42**: 288–95.
De Witte, E. 1973–74. 'The protection of silverware with varnishes'. *Bulletin de l'Institut Royal du Patrimoine Artistique* **XIV**: 140–51.
Dick, L. F. P., Kaiser, H. and Kaesche, H. 1993. 'The tarnishing of Ag and Ag–Au alloys in aqueous sulfide solutions and in H_2S atmospheres'. In *Corrosion and Reliability of Electronic Materials and Devices, Toronto, Canada, 1992*, 406–14. Pennington, New York: The Electrochemical Society.
Drayman-Weisser, T. 1992. 'Metal objects'. In *Caring for Your Collections*, H. Whelchel (ed.), 108–21. New York: Harry N. Abrams.
Drost, E. and Haußelt, J. 1992. 'Uses of gold in jewellery'. *Interdisciplinary Science Reviews* **17**: 271–80.
Drott, J. 1960a. 'Growth of silver sulphide whiskers'. *Acta Metallurgica* **8**: 19–22.
Drott, J. 1960b. 'Reaction rate and growth forms in the reaction between silver and hydrogen sulfide: I. The influence of water vapor pressure'. *Arkiv för Kemi* **15**: 181–94.
Drott, J. 1960c. 'Reaction rate and growth forms in the reaction between silver and hydrogen sulfide: II. The influence of temperature and gas pressure'. *Arkiv för Kemi* **16**: 333–46.
Dugmore, J. M. M. and DesForges, C. D. 1979. 'Stress corrosion in gold alloys'. *Gold Bulletin* **12**: 140–44.
Duncan, S. J. and Ganiaris, H. 1987. 'Some sulphide corrosion products on copper alloys and lead alloys from London waterfront sites'. In *Recent Advances in the Conservation and Analysis of Artifacts*, J. Black (ed.), 77–93. London: University of London Summer Schools Press.
Egan, T. F. and Mendizza, A. 1960. 'Creeping silver sulfide'. *Journal of the Electrochemical Society* **107**: 353–4.
Evans, U. R. 1951. 'The corrosion situation: past, present and future'. *Chemistry and Industry* 706–11.
Fiaud, C. and Guinement, J. 1986. 'The effect of nitrogen dioxide and chlorine on the tarnishing of copper and silver in the presence of hydrogen sulphide'. *Proceedings of the Electrochemical Society* **86**: 280–304.
Fioravanti, K. J. and German, R. M. 1988. 'Corrosion and tarnishing characteristics of low gold content dental casting alloys'. *Gold Bulletin* **21**: 99–110.
Fiorentino, P., Marabelli, M., Matteini, M. and Moles, A. 1982. 'The condition of the "Door of Paradise" by L. Ghiberti: Tests and proposals for cleaning'. *Studies in Conservation* **27**: 145–53.

Fischmeister, H. and Drott, J. 1959. 'Reaction rate and growth forms of reaction product in the system Ag–H_2S'. *Acta Metallurgica* **7**: 777–81.

Forty, A. J. 1981. 'Micromorphological studies of the corrosion of gold alloys'. *Gold Bulletin* **14**: 25–35.

Franey, J. P., Kammlott, G. W. and Graedel, T. E. 1985. 'The corrosion of silver by atmospheric sulfurous gases'. *Corrosion Science* **25**: 133–43.

Frant, M. S. 1960. 'Copper sulfide creep on porous electroplate'. *Journal of the Electrochemical Society* **107**: 1009–11.

Frantz, J. H. and Schorsch, D. 1990. 'Egyptian red gold'. *Archeomaterials* **4**: 133–52.

Gal-Or, L. 1992. 'Tarnishing and corrosion of silver and gold alloys'. In *Santa Fe Symposium on Jewellery Manufacturing Technology, 1990*, 19–37. Boulder, Colo.: Met-Chem Research.

Garte, S. M. 1974. 'Porosity'. In *Gold Plating Technology*, F. H. Reid and W. Goldie (eds), 295–314. Ayr, Scotland: Electrochemical Publications.

Gasparrini, C. 1983. 'The mineralogy of gold and its significance in metal extraction'. *Canadian Mining and Metallurgical Bulletin* **76**: 144–53.

German, R. M. 1981. 'The role of microstructure in the tarnish of low-gold alloys'. *Metallography* **14**: 253–66.

Gettens, R. J. 1970. 'Patina: noble and vile'. In *Art and Technology: A Symposium on Classical Bronzes*, S. Doeringer, D. G. Mitten and A. Steinberg (eds), 57–72. Cambridge, Mass.: MIT Press.

Graedel, T. E., Franey, J. P. and Kammlott, G. W. 1984. 'Ozone- and photon-enhanced atmospheric sulfidation of copper'. *Science* **224**: 599–601.

Graedel, T. E., Franey, J. P., Gualtieri, G. J. *et al.* 1985. 'On the mechanism of silver and copper sulfidation by atmospheric H_2S and OCS'. *Corrosion Science* **25**: 1163–80.

Graedel, T. E., Nassau, K. and Franey, J. P. 1987. 'Copper patinas formed in the atmosphere: I. Introduction'. *Corrosion Science* **27**: 639–57.

Graedel, T. E. 1992. 'Corrosion mechanisms for silver exposed to the atmosphere'. *Journal of the Electrochemical Society* **139**: 1963–70.

Grimwade, M. 1985. 'Alloys of gold: the carat golds'. In *Introduction to Precious Metals*, 47–62. London: Butterworth & Co.

Grimwade, M. 1992. 'The metallurgy of gold'. *Interdisciplinary Science Reviews* **17**: 371–81.

Groenewald, T. 1976. 'The dissolution of gold in acidic solutions of thiourea'. *Hydrometallurgy* **1**: 277–90.

Guttenplan, J. D. 1987. 'Corrosion in the electronics industry'. In *Metals Handbook*, volume 13: *Corrosion*, 9th edn, J. R. Davis (ed.), 1107–12. Metals Park, Ohio: American Society for Metals.

Hack, H. P. 1987. 'Evaluation of galvanic corrosion'. In *Metals Handbook*, volume 13: *Corrosion*, 9th edn, J. R. Davis (ed.), 234–8. Metals Park, Ohio: American Society for Metals.

Heller, B. B. 1983. 'The coating of metal objects at Winterthur'. In *Preprints to the 11th Annual Meeting of the AIC, Baltimore, Maryland*, 57–64. Washington, D.C.: American Institute for Conservation of Historic and Artistic Works.

Hiskey, J. B. 1981. 'Thiourea as a lixiviant for gold and silver'. In *Proceedings of the 110th AIME Meeting*, 83–91. Chicago, Ill.: American Institute of Mining, Metallurgical and Petroleum Engineers.

Jacobson, D. M. and Humpston, G. 1992. 'Gold in the electronics industry'. *Interdisciplinary Science Reviews* **17**: 244–50.

Jambor, J. L., Dutrizac, J. E., Roberts, A. C. *et al.* 1996. 'Clinoatacamite, a new polymorph of $Cu_2(OH)_3Cl$, and its relationship to paratacamite and "anarakite"'. *The Canadian Mineralogist* **34**: 61–72.

Jett, P. 1993. 'A study of the gilding of Chinese Buddhist bronzes'. In *Metal Plating and Patination: Cultural, Technical and Historical Developments,* S. La Niece and P. Craddock (eds), 193–200. Oxford: Butterworth-Heinemann.

Johnson, B. F. G. and Davis, R. 1973. 'Gold'. In *The Chemistry of Copper, Silver and Gold*, 129–86. New York: Pergamon Press.

Kempf, B. and Haußelt, J. 1992. 'Gold, its alloys and their uses in dentistry'. *Interdisciplinary Science Reviews* **17**: 251–60.

Krumbein, S. J. and Holden, C. A., Jr. 1987. 'Porosity testing of metallic coatings'. In *Testing of Metallic and Inorganic Coatings*, ASTM STP 947, W. B. Harding and G. A. Di Bari (eds), 193–210. Philadelphia, Pa.: American Society for Testing and Materials.

Kuhn, A. T. 1981. 'Anodic dissolution, oxygen reduction and corrosion of gold alloys'. *Surface Technology* **13**: 17–31.

Laver, M. E. 1978. 'Spot tests in conservation: metals and alloys'. In *ICOM Committee for Conservation, 5th Triennial Meeting, Zagreb, Yugoslavia*, 78/23/8/1–78/23/8/11. Paris: ICOM.

Lechtman, H. N. 1971. 'Ancient methods of gilding silver: examples from the old and the new worlds'. In *Science and Archaeology*, R. H. Brill (ed.), 2–30. Cambridge, Mass.: MIT Press.

Lins, P. A. and Oddy, W. A. 1975. 'The origins of mercury gilding'. *Journal of Archaeological Science* **2**: 365–73.

Lins, A. 1991. 'Basic properties of gold leaf'. In *Gilded Wood: Conservation and History*, D. Bigelow (ed.), 17–23. Madison, Conn.: Sound View Press.

Massey, A. G. 1973. 'Copper'. In *The Chemistry of Copper, Silver and Gold*, 1–78. New York: Pergamon Press.

McDonald, A. S. and Sistare, G. H. 1978. 'The metallurgy of some carat gold jewellery alloys'. *Gold Bulletin* **11**: 66–73.

Moncrieff, A. 1966. 'Protecting silver from tarnishing'. *IIC News* **4**: 6–7.

Muller, C. O. 1991. 'Multiple contaminant gas effects on electronic equipment corrosion'. *Corrosion* **47**: 146–51.

Oddy, A. 1981. 'Gilding through the ages'. *Gold Bulletin* **14**: 75–9.

Oddy, A. 1993. 'Gilding of metals in the old world'. In *Metal Plating and Patination: Cultural, Technical and Historical Developments*, S. La Niece and P. Craddock (eds), 171–81. Oxford: Butterworth-Heinemann.

Oddy, A., Pearce, P. and Green, L. 1988. 'An unusual gilding technique on some Egyptian bronzes'. In *Conservation of Ancient Egyptian Materials*, S. C. Watkins and C. E. Brown (eds), 35–9. London: United Kingdom Institute for Conservation.

Oddy, W. A. 1988. 'The gilding of Roman silver plate'. In *Argenterie Romaine et Byzantine*, F. Baratte (ed.), 9–25. Paris: De Boccard.

Oddy, W. A. 1991. 'Gilding: an outline of the technological history of the plating of gold on to silver or copper in the old world'. *Endeavour* **15**: 29–33.

Oddy, W. A. and Meeks, N. D. 1982. 'Unusual phenomena in the corrosion of ancient bronzes'. In *Science Technology in the Service of Conservation*, 119–24. London: International Institute for Conservation of Historic and Artistic Works.

Oddy, W. A., Cowell, M. R., Craddock, P. T. and Hook, D. R. 1990. 'The gilding of bronze sculpture in the classical world'. In *Small Bronze Sculpture from the Ancient World*, 103–24. Malibu, Calif.: J. Paul Getty Museum.

Oddy, W. A., La Niece, S., Curtis, J. E. and Meeks, N. D. 1981. 'Diffusion-bonding as a method of gilding in antiquity'. *MASCA Journal* **1**: 239–41.

Ogburn, F., Passaglia, E., Burnett, H. C. *et al.* 1977. 'Restoration of large gilded statues using various electrochemical and metallurgical techniques'. In *Corrosion and Metal Artifacts*, NBS Special Publication 479, B. F. Brown (ed.), 167–78. Washington, D.C.: US National Bureau of Standards.

Ogden, J. 1985. 'Potentials and problems in the scientific study of ancient gold artifacts'. In *Application of Science in Examination of Works of Art*, P. A. England and L. van Zelst (eds), 72–5. Boston, Mass.: Research Laboratory Museum of Fine Arts.

Ogden, J. 1992 'Gold in antiquity'. *Interdisciplinary Science Reviews* **17**: 261–270

Opila, R. L. 1987. 'Copper patinas: an investigation by Auger electron spectroscopy'. *Corrosion Science* **27**: 685–94.

Parkes, G. D. (ed.) 1967. *Mellor's Modern Inorganic Chemistry*, 6th edn, 541. London: Longmans, Green, & Co.

Plenderleith, H. J. and Werner, A. E. A. 1971. *The Conservation of Antiquities and Works of Art*, 2nd edn. London: Oxford University Press.

Pourbaix, M. 1974. *Atlas of Electrochemical Equilibria in Aqueous Solutions*, 402. Houston, Tex.: National Association of Corrosion Engineers.

Puddephatt, R. J. 1978. *The Chemistry of Gold*. New York: Elsevier Scientific.
Randin, J. P., Ramoni, P. and Renaud, J. P. 1992. 'Tarnishing of AuAgCu alloys: Effect of the composition.' *Werkstoffe und Korrosion* **43**: 115–23.
Rapson, W. S. and Groenewald, T. 1978. *Gold Usage*. London: Academic Press.
Rapson, W. S. 1992. 'Mining, extraction and refining of gold'. *Interdisciplinary Science Reviews* **17**: 203–12.
Raykhtsaum, G. and Agarwal, D. P. 1993. 'Gold in jewelry'. *Precious Metals* **17**: 169–80.
Rickett, B. I. and Payer, J. H. 1993. 'X-ray photoelectron spectroscopic analysis of copper tarnish films formed by exposure to sulfur containing pollutant gases'. In *Corrosion and Reliability of Electronic Materials and Devices, Toronto, Canada, 1992*, 437–48. Pennington, New York: The Electrochemical Society.
Robertson, A. R. 1990. 'Precious metals and their uses'. In *ASM Handbook*, volume 2, *Properties and Selection: Nonferrous Alloys and Special-Purpose Materials*, 688–95. Metals Park, Ohio: American Society for Metals.
Russell, R. J. 1976. 'Inlay-clad gold alloys'. *Gold Bulletin* **9**: 2–6.
Schmalzried, H. 1981. 'Ag_2S – the physical chemistry of an inorganic material'. *Progress in Solid State Chemistry* **13**: 119–57.
Schmidbaur, H. 1992. 'Gold chemistry is different'. *Interdisciplinary Science Reviews* **17**: 213–20.
Schulze, R. G. 1984. 'New aspects in thiourea leaching of precious metals'. *Journal of Metals* **36**: 62–5.
Scott, D. A. 1983a. 'The deterioration of gold alloys and some aspects of their conservation'. *Studies in Conservation* **28**: 194–203.
Scott, D. A. 1983b. 'A note on the dissolution of ancient gold alloys'. *Archaeometry* **25**: 214–16.
Scott, D. A. 1990. 'Bronze disease: a review of some chemical problems and the role of relative humidity'. *Journal of the American Institute for Conservation* **29**: 193–206.
Scott, D. A. 1991. *Metallography and Microstructure of Ancient and Historic Metals*. Marina del Rey, Calif.: Getty Conservation Institute.
Sease, C. 1994. 'The cases of the black fuzzies'. In *Exhibitions and Conservation*, 125–30. Edinburgh: Scottish Society for Conservation and Restoration.
Sease, C., Selwyn, L. S., Zubiate, S. *et al.*. 1997 'Problems with coated silver: whisker formation and possible filiform corrosion'. *Studies in Conservation* **42**: 1–10.
Selwyn, L. S. 1990. 'Historical silver: storage, display, and tarnish removal'. *Journal of the International Institute for Conservation – Canadian Group* **15**: 12–22.
Selwyn, L. S., Binnie, N. E., Poitras, J. *et al.* 1996. 'Outdoor bronze statues: analysis of metal and surface samples'. *Studies in Conservation* **41**: 205–28.
Smith, G. D. 1987. 'Corrosion of the noble metals'. In *Metals Handbook*, volume 13, *Corrosion*, 9th edn, J. R. Davis (ed.), 793–807. Metals Park, Ohio: American Society for Metals.
Šrámek, J., Jakobsen, T. B. and Pelikán, J. B. 1978. 'Corrosion and conservation of a silver visceral vessel from the beginning of the seventeenth century'. *Studies in Conservation* **23**: 114–17.
Thompson, N. R. 1973. 'Silver'. In *The Chemistry of Copper, Silver and Gold*, 79–128. New York: Pergamon Press.
Tierney, V. 1981. 'The nature and rate of creepage of copper sulfide tarnish films over gold'. *Journal of the Electrochemical Society* **128**: 1321–26.
Torma, A. E. 1992. 'Bioenhanced gold extraction from refractory ores'. *Interdisciplinary Science Reviews* **17**: 366–9.
Townsend, J. H. 1988. 'The identification of metals: chemical spot tests'. In *Modern Metals in Museums*, R. E. Child and J. H. Townsend (eds), 15–22. London: Institute of Archaeology Publications.
Tran, T. 1992. 'The hydrometallurgy of gold processing'. *Interdisciplinary Science Reviews* **17**: 356–65.
Tuccillo, J. J. and Nielsen, J. P. 1971. 'Observations of onset of sulfide tarnish on gold-base alloys'. *Journal of Prosthetic Dentistry* **25**: 629–37.
Wagner, C. 1953. 'Investigations on silver sulfide'. *Journal of Chemical Physics* **21**: 1819–27.

Weast, R. C. (ed.) 1984–85. *CRC Handbook of Chemistry and Physics*, D159-D161. Boca Raton, Fla.: CRC Press.

Weisser, T. D. 1990. 'Some practical considerations and solutions for preservation in the museum environment'. In *Cultural Property and its Environment*, 81–96. Tokyo, Japan: Tokyo National Research Institute of Cultural Properties.

Wells, A. F. 1984. *Structural Inorganic Chemistry*, 5th edn, 1295–1300. Oxford: Clarendon Press.

Wise, E. M. 1964. *Gold: Recovery, Properties, and Applications*. Princeton, N.J.: Van Nostrand.

3 The Selective use of Gilding on Egyptian Polychromed Bronzes

Patricia S. Griffin

Abstract

The selective use of gilding on Egyptian figural bronzes is discussed using several illustrative and provocative examples in the collection of the Cleveland Museum of Art. Examples of leaf gilding applied to the metal substrate and leaf gilding applied to a gesso layer are discussed. These objects also demonstrate the polychromed approach to the finishing of the bronze surfaces, which include gilding, metal wire and cut sheet inlays and patination. These techniques of embellishment were all an important part of the original aesthetic.

Introduction

The Egyptian collection of the Cleveland Museum of Art (CMA) has been the focus of study in conjunction with an upcoming publication of the collection. (Berman *et al.* 1999). The collection ranges from Predynastic to Roman times and is relatively small but quite distinguished in terms of the quality of its objects. More than fifty bronze objects are included in the collection. As was typical for archaeological bronze finds at the beginning of this century, the majority were stripped of their overlying corrosion using chemical, electrochemical or electrolytic processes; many were subsequently repatinated chemically and/or using pigments and waxes.[1] Therefore, it is extremely fortunate that several of the most interesting bronzes in terms of their original construction and surface polychromy have been preserved with their burial patinas intact. The selective use of gilding was part of this original surface polychromy. This appears to have been used in combination with metallic inlays and surface patination for rich and varied decorative effects.

This chapter will focus on several relatively well preserved examples of Egyptian gilt bronzes in the museum's collection. These include examples of leaf gilding applied directly to the metal substrate and leaf gilding applied to a gesso layer. Since the gilding is an integral component of the original polychromed aesthetic of these objects, some attention will be given to their construction as well as to evidence for their original polychromy or patination. Three objects will be discussed in detail. The first, a bronze statuette of Amen-re, is distinctive for its curious construction, which includes a separate head that was cast in black copper, as well as for the extensive use of gilding to embellish dress accoutrements. The gilding utilises thin gold leaf, which was adhered to a thick gesso layer applied to a textured bronze substrate or to a textile interleaf layer. The second, a set of bronze figural mounts from a shawabty bundle, and the third a hawk-head mount, both exhibit gold leaf or foil applied directly to a metal substrate. In particular, the excellent state of preservation of the figural mounts enables some sense of their original polychromed surfaces to be appreciated. The hawk head is unusual in that the gilding was applied to added copper elements that had been inlaid or applied to the surface of the bronze. The gilt surfaces appear to be highly burnished. They are well preserved due to the excellent cohesion between the gilding and the copper. This suggests either the use of a more durable adhesive or that some diffusion of the gold into the copper may have taken place. Finally, the original polychromed and gilt surfaces on two stripped bronzes will be partially reconstructed based on surviving evidence and analysis.

Review of gilding in the Old World

The use of gold to adorn bronze and other metals in Egypt was generally employed in a non-metallurgical manner. The bronze surface was gilded using the same techniques that were employed on a variety of different substrates, including stone, wood and plaster (Hatchfield and Newman 1991). One method frequently used was leaf gilding, in which the leaf was adhered to the surface of the bronze using adhesive. This method was also in use in the classical world to gild metal work, although comparatively few examples have survived (Oddy 1993). A variation of leaf gilding on metal substrates appears to be uniquely Egyptian – the application of the leaf onto a gesso that had been applied to the surface of the metal.[2] Texturing of the bronze substrate or the application of a textile interleaf was sometimes employed in preparation for the gesso layer.

Leaf gilding requires the use of adhesive to affix the gold in place. Both hide glue and albumin have been suggested as possible adhesives for the gilding of classical bronzes; a historical text by Pliny indicates that a pigmented mixture of albumin (egg white) was used by the Romans (Oddy *et al.* 1990). Lucas suggested that glue (animal or fish) and gum were used for the application of gold leaf by the Egyptians, based on the water solubility of the samples that he tested (Lucas and Harris 1989). To date, published analyses of adhesives used in Egyptian gilding have focused on identifying the binders in the underlying gesso.[3] Albumin has been identified as the binder in one gesso sample, and both gum and glue have been found in gesso layers underlying gilt layers on a variety of substrates (Lucas and Harris 1989; Hatchfield and Newman 1991). Gum and glue are also believed to be the materials most commonly employed as binders for pigments and gessoes used in ancient Egyptian painting (Lucas and Harris 1989).

Another gilding method used in antiquity was the application of metal foil as cladding, which was inserted into carved channels and crimped. Although few examples survive, the practice was documented by Pliny in reference to a statue of Alexander gilded in this manner during Nero's reign (Oddy *et al.* 1990; Oddy 1993). One surviving Egyptian example of this technique is a silver cult figure in the Miho Museum (Shigaraki, Japan), which retains fragments of gold foil in and adjacent to the channels used to crimp it into place.[4] In addition, Lucas mentions several examples of 'thin sheet gold' being affixed to copper by hammering, all of which date to the Middle Kingdom or earlier (Lucas and Harris 1989). These may be variations of foil gilding, which includes folding, crimping or hammering the gold foil around the contours of an object, and punching the foil onto an object, as well as inserting the foil into carved channels. In all cases adhesion might be augmented by roughening the surface of the substrate. Examples of these variations have been noted in the classical world and were used to attach silver foil as well as gold. (Oddy *et al.* 1990; La Niece 1993; Oddy 1993).

To date only one example of diffusion gilding (Newman 1990) – which effects a metallurgical attachment using heat and burnishing – and no examples of mercury gilding have been identified on Egyptian metalwork, although the former has been documented on objects from the ancient Near East and classical world, and the latter was in use by Roman times (Oddy 1993).

Egyptian leaf gilding onto gesso

The gilding of Egyptian metals by applying gold leaf to gesso over linen was first noted by Lucas for a copper diadem and eight copper feathers, both from the Old Kingdom (Lucas and Harris 1989). The technique has been well documented in several recent studies.

This technique was also used on sculpture. A seated statuette of Amenre in the National Museum in Krakow, datable to the late New Kingdom, was the focus of a detailed paper that outlined the importance of a measured approach to the cleaning of archaeological bronzes (Jedrezejewdka 1976). This study is a model in the information that can be revealed by the careful examination and cleaning of an object by a skilled conservator. Treatment revealed the selective gilding of the statuette using gold leaf applied to a gypsum gesso. The gilding as reconstructed was originally applied to bracelets, anklets, collar, crown and 'the front of the apron'. A bone inlay remained in one eye and the beard straps were described as 'reddish', possibly the result of unidentified metal inlays. Published photographs suggest that the eyebrows and cosmetic lines were also originally inlaid.

A number of examples of gilded metals were noted by Hatchfield and Newman in their comprehensive study of Egyptian gilding conducted at the Museum of Fine Arts, Boston, including a late period bronze Ptah, a Saite uraeus made from tin, a silver girdle element and a copper lily pendant of Nefertari (both 19th Dynasty), and a Hellenistic copper ivy wreath (Hatchfield and Newman 1991). Of these the Ptah, girdle element and lily pendant have gessoes made from calcium carbonate; the ivy wreath exhibits a yellow clay-based bole. Plainly woven linen was applied to the girdle element and lily pendant prior to the application of the gesso and gold leaf.

Another example of the incorporation of a textile interleaf layer between the gesso and the bronze was noted on a fine statue of a female striding figure in the collection of the Musée du Louvre that was dated to Dynasty 22–26 based on a number of stylistic parallels to well known examples of statuary and sculpture (Delange *et al.* 1995, 1998). The gilding is fragmentary but appears to have been applied overall to the surface. The extensive corrosion prevented a conclusive determination as to whether the bronze surface had been textured in addition to the application of a textile interleaf. The gesso was made from calcium carbonate bound with a small amount of a protein-based medium. The eyes were inlaid using glass and a calcareous stone as well as the gilding. Three other examples of gilding applied onto gesso with a textile interleaf in the Louvre and the Egizio Museo, Turin, were also noted. All are sculptural elements.

The study of the gilding of metals in antiquity has been an ongoing project at the British Museum for many years under the direction of Andrew Oddy (Oddy 1981, 1993; Oddy *et al.* 1988, 1990) A study of Egyptian gilt bronzes documented thirteen examples from the New Kingdom with preserved areas of gold leaf applied to gesso. (Oddy *et al.* 1988). In three of these cases, the surface of the bronze exhibited a cast, engraved or cast and engraved texture consisting of rows of raised dots, raised lines and/or engraved lines. Three additional bronzes with no visible gilding were identified with cast or engraved textured surfaces, suggesting that they had originally been gilded. Analysis of the gessoes indicates that both calcium carbonate and calcium sulphate-based gessoes were commonly used. One gesso consisted of crushed dolomite (Oddy 1981). An example of the Hellenistic use of bole also was noted. Fifteen additional bronzes were identified that exhibited traces of gold leaf that had been applied directly to the metal.

It is logical to attribute the origins of the gilding of metals using gold leaf applied to a gesso layer to Egyptian woodworking practices (Oddy *et al.* 1988; Hatchfield and Newman 1991). Egyptian technology is marked by the tendency to employ similar techniques in different materials to achieve the same effect. The general construction of many statuettes with their separate arms and mortise and tenon joints was certainly influenced by the standard for statue construction as established in the art of woodworking. However, an equally strong characteristic was the desire to mimic. Ancient Egyptian technology was driven in part by a desire to create something precious in a different or more economic way. Stone vessels were mimicked in painted ceramic and wood. Faience technology probably was developed to mimic brightly coloured stones such as turquoise or lapis (Friedman 1998).

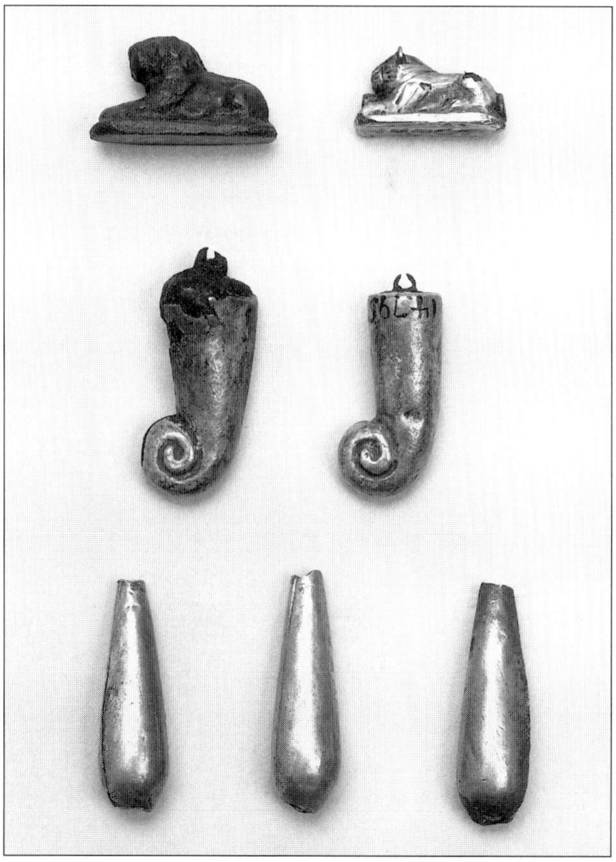

Figure 3.1 Examples of Middle Kingdom jewellery in the collection of the CMA made by applying gold, silver or electrum leaf or foil to a gesso or clay core. From top to bottom: Recumbent lion beads (1914.784 and 1914.787), Pair of sidelocks (1914.795.a and .b) and Teardrop shaped pendants (1914.797). (Photo by Janet Burke, © Cleveland Museum of Art.)

In turn, paint and varnish were used to mimic faience (Griffin1999). Gold leaf, when applied to a substantial ground layer, would give the appearance of thicker gold sheet, rather than functioning as a flat, purely decorative layer. The desire for a realistic representation was an important consideration in the fabrication of these polychromed bronzes. This point is aptly illustrated by the Krakow Amen-re and in a Seated Amen-re in the CMA, which will be described below. Both statuettes display inlaid eyes and the effective dressing of the deity by the gilding process. A similar approach was used for many of the finer wooden figures from the Middle and New Kingdoms. For instance, on the refined, almost life-sized Middle Kingdom sculpture of Nakht in the Miho Museum, the use of multiple layers of linen textile below the paint layer is limited to the kilt. These and the thickly applied paint effectively create the appearance of a worn garment rather than a decorative layer.[5]

There was a well developed precedent for the use of gold leaf or foil on gesso to mimic the look of solid gold in Egyptian jewellery manufacture. Some ancient hollow silver and gold jewellery was filled with clay or gesso as a means of minimising explosions from soldering; however, in other cases the filling clearly functioned as a substrate for gilding or silvering (Ogden 1982). Several pieces of Middle Kingdom jewellery in the CMA illustrate how well metal leaf or foil over a gesso or clay (est.) core created an object with the appearance of solid gold or silver (Fig. 3.1).[6] This practice is especially meaningful when considering the rarity of Egyptian silver, which they considered to be the most precious of metals (Lucas and Harris 1989).

The seated statuette of Amen-re

The statuette (1914.574) represents the deity in a seated position (Figs 3.2 and 3.3), dressed in a short kilt and adorned with a broad collar, arm bands, wrist cuffs, ankle cuffs and a flat-topped crown which was originally surmounted by a sun-disc, double plumes or a combination of the two (Berman *et al.* 1999).[7] The statuette originally had a separate beard, which is now missing. All dress accoutrements were gilded. The figure is currently attached with a modern adhesive to a wooden base, which has fragmentary remains of polychromy. The base appears to be ancient but not original to the sculpture. The reverse of the figure has carbonised wood splinters adhered to it in several areas, suggesting that the original base for the figure was wood. The statuette is unprovenanced but is probably datable to Dynasty 25–26 based on the apparent stylistic and technical similarities to another dated example that preserves both the beard and a sun-disc headdress surmounting the crown (Berman *et al.* 1999).

The current condition of the statuette makes interpretation of the surface difficult. The gilding is fragmentary. Most of the figure, except for the face, is heavily corroded with adhered dirt and isolated deposits of powdery white material, which appears to be gesso that has transferred from the gilded areas. The original parameters of the gilding can be reconstructed using evidence from areas of the textured bronze substrate, the preserved areas with underlying textile interleaf and textile pseudomorphs, as well as the remaining gilding. The appearance of the corrosion may also provide some clues (Figs 3.4 and 3.5). The corrosion is complex and varied in the gilded areas. Highly crystalline, granular forms of cuprite (copper oxide) and atacamite (basic copper chloride) are common, as well as light green and blue-green corrosion products consisting of sampleite (a complex copper phosphate also containing sodium, calcium and chlorine),

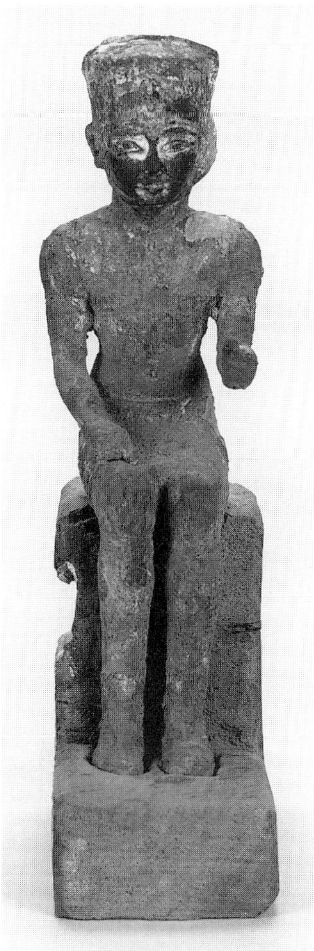

Figure 3.2 Front view of the Seated Amen-re (1914.574). The compact dark surface of the face is readily evident. Also note the thickening on the proper right side of the arm and the ankle in the locations of the arm band and ankle cuff. The striated texture of the kilt is visible below the waist. (Photo by Janet Burke, © Cleveland Museum of Art.)

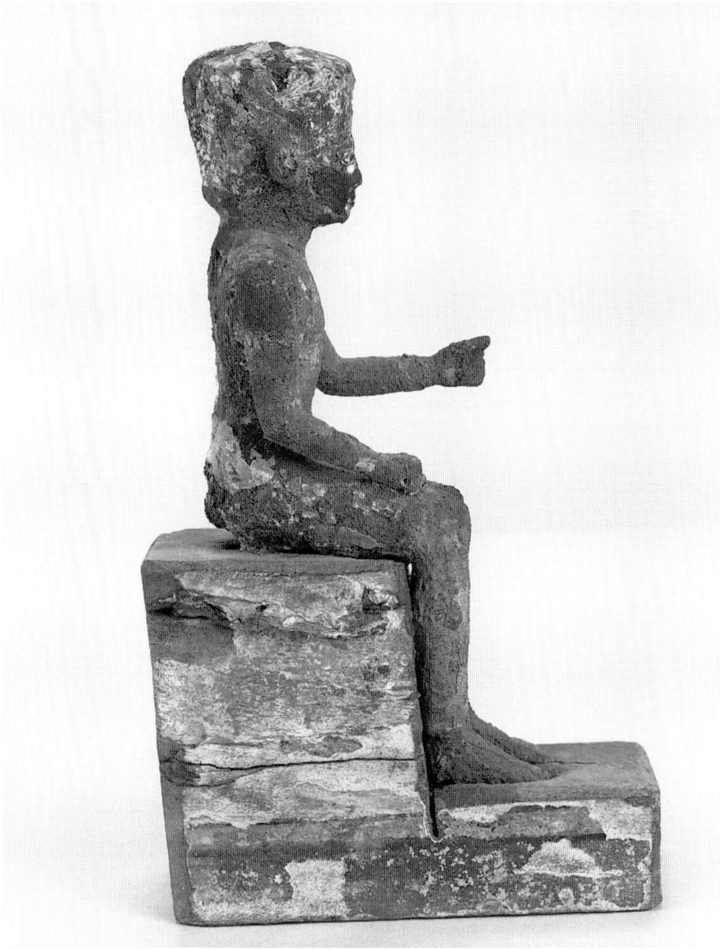

Figure 3.3 Side view of the Seated Amen-re (1914.574). Many areas of preserved gilding are visible on the crown, collar and kilt, at the wrists and on the proper right ankle. Signs of mechanical cleaning are visible on the crown. (Photo by Janet Burke, © Cleveland Museum of Art.)

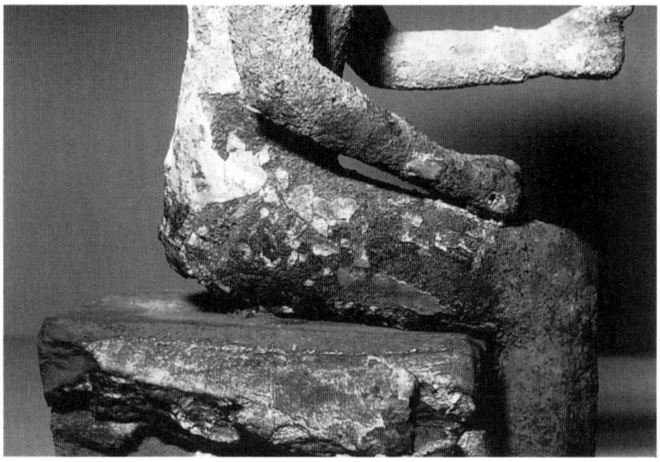

Figure 3.4 Detail of corrosion associated with fragmentary gilding in the area of the kilt. Note the granular green crystals of atacamite. Paratacamite and cuprite were also identified from this area. (Photo by Janet Burke, © Cleveland Museum of Art.) (See Plate 5.)

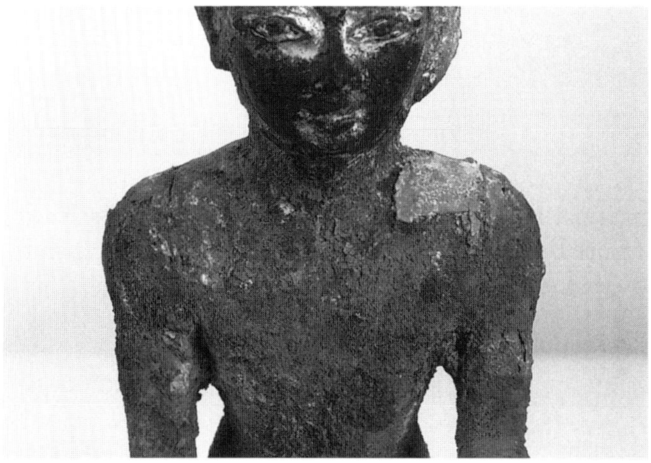

Figure 3.5 Detail of corrosion associated with fragmentary gilding in the area of the collar. Note the pale powdery green and blue-green corrosion. Sampleite and chalconatronite were identified in this area in addition to malachite and cuprite. Also note the dark, compact lead-rich corrosion on the surface of the face. (Photo by Janet Burke, © Cleveland Museum of Art.) (See Plate 6)

paratacamite (dimorph of atacamite) and chalconatronite (double carbonate of sodium and copper).[8] Bright green malachite (basic copper carbonate) and compact bright red cuprite, more commonly found on archaeological bronzes, were also identified (Gettens 1970).

In contrast to the body, the face appears to be in a better state of preservation overall and exhibits a compact dark grey-brown corrosion that occurs as a discrete layer, overlying a compact red cuprite layer. Salmon-coloured bare metal is visible where the cuprite is breached. The dark grey corrosion consists primarily of lead-based corrosion products, including scrutinyite (lead dioxide), litharge (lead monoxide) and cerussite (lead carbonate) as well as cuprite, according to X-ray diffraction (XRD).[9] Finely divided white powder, different in appearance from the gesso, is visible around the eyes and on the proper left side of the face; this was identified as a combination of phosgenite (lead carbonate chloride) and cotunnite (lead chloride) using XRD.

Construction of the seated Amen-re

The statuette was constructed from multiple components that were cast in a leaded tin-bronze alloy[10] using the lost wax process. X-radiography indicates that the torso, including the chest and skirt, is hollow and was cast with a core; the legs were solid wax appendages that were attached to the torso prior to casting (Fig. 3.6). The arms were cast separately as solid appendages and were attached to the torso using mortise and tenon joins. The head was a separate component which was hollow cast with a core and fitted onto a tenon extending from the neck using a central mortise. The tenon for the headdress element was inserted into a mortise in the top of the head (Fig. 3.6). Hollows or channels were cut into the wax for the inlaid eyes, cosmetic lines, eyebrows and beard straps. The eyebrows, cosmetic lines and pupils were inlaid by hammering silver wire[11] into the cast channels. The eyes are obscured by corrosion but were inlaid probably using a compact white material such as limestone (Lucas and Harris 1989).[12] The beard strap inlays are missing. The beard was originally fitted into a mortise on the underside of the chin by an extended tenon, which can still be seen in the X-ray.

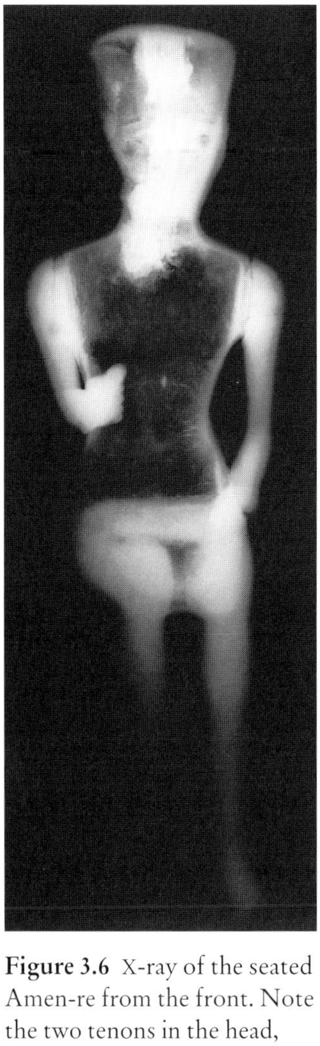

Figure 3.6 X-ray of the seated Amen-re from the front. Note the two tenons in the head, the mortise and tenon joints at the arms, the solid arms and legs, and the corrosion of metal adjacent to the cores in the torso and head. The X-ray was taken at XRI Testing, Cleveland using a 2 minute exposure at 300 kV, 10 mA and 48" stf. (Photo by Joan Neubecker, © Cleveland Museum of Art.)

Many Egyptian bronzes exhibit this composite approach to construction. It is a reflection of the Egyptian approach to sculpture on the whole as is seen by numerous examples in polychromed wood.[13] However, the construction of the Amen-re head is highly unusual, and must be related to a desire for the face to appear different. Casting it in a different alloy would provide a greater contrast with the inlaid silver elements. The preservation and surface characteristics of the face support this conclusion. Proton induced X-ray emission spectroscopy (PIXE) was performed on an area on the lower cheek. The analysis indicates that the alloy is a different composition from the rest of the body and contains small amounts of silver and gold, as well as arsenic, in sufficient quantity to characterize the alloy as Egyptian black copper.[14] However, analysis indicates that the dark compact patina on the surface is not the original patination but is the result of burial. Analysis of the face using PIXE demonstrated high percentages of lead on its surface. It contained more than 40 wt% lead in contrast to the chest, shoulder and back, which contained approximately 5 wt% lead or less. A sample of the corrosion and a small scraping of the bare metal were taken from the edge of the face for analysis using a scanning electron microscope with energy dispersive spectroscopy (SEM/EDS). This confirmed the X-ray diffraction analysis showing that lead was the primary component in the grey surface skin. The underlying red cuprite and bare metal are copper-rich with traces of lead and other components.[15]

Lead sweat has been reported to occur in tandem with tin sweat in bronzes. However, it forms discrete globules concentrated at the surface rather than a distinct layer (Meeks 1993). The lead-rich surface covering the face is clearly a separate layer, which occurs over the entire surface of the face when viewed with magnification, and is also imaged as an increased radiodensity in X-rays. The most likely cause of this lead-rich surface is that it was formed during burial.[16] It is also possible that a copper–lead alloy was applied to the surface after patination, perhaps to alter it or to correct unsatisfactory results. Young reported a copper–lead surface on a Late Period bronze situla and speculated that it was formed by dipping, followed by chasing and polishing (Young 1959). The surface layer and the underlying metal are completely mineralised on the Amen-re; they fractured and separated upon sampling, precluding the opportunity to examine a metallographic section or determine their structure and interface. In any case, it is clear based on the difference in the alloy composition alone, that the original intent was to alter the polychromy of the surface of the separate head.

Gilding the Amen-re

The selective gilding of the object was an important component of the polychromed effect. Both the crown and kilt were gilded; the contours of each were indicated in the casting. Gilding was used to articulate the collar, arm bands, bracelets and anklets; although these areas apparently were not delineated in the casting. The gilding was executed in thin gold leaf measuring less than a micron in thickness, which was applied to a substantial gesso layer measuring 1–2 mm in thickness. The gesso is ground calcium carbonate (identified by XRD and EDS), bound with an unidentified organic binder. EDS analysis indicates that the gold is extremely pure except for small amounts of copper.[17]

The surface of the kilt exhibits a striated texture in relief, which was probably introduced in the wax and cast into the metal (Fig. 3.7). The striations are oriented vertically except for the front tab where they are horizontal. This texture might be confused with stylized pleating in the kilt; however, it probably served to help adhere the gesso to the bronze. This texture is absent from the other gilded areas of the bronze. These were covered with textile as indicated by textile pseudomorphs of a plain woven fabric (Fig. 3.8). Fragments of linen are preserved in a very few

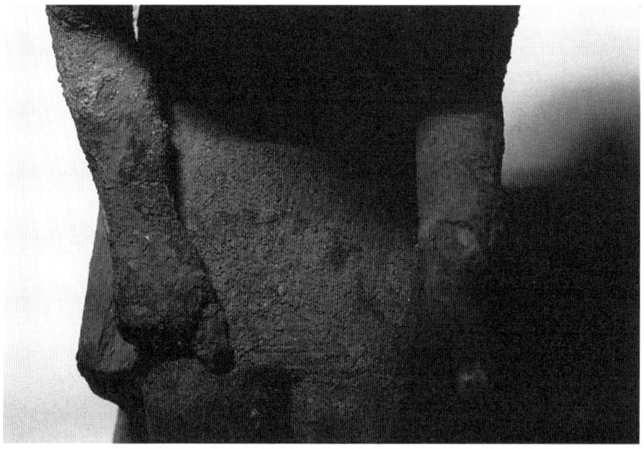

Figure 3.7 Detail of the striated texture preserved in the corroded bronze of the kilt. (Photo by Janet Burke, © Cleveland Museum of Art.) (See Plate 7.)

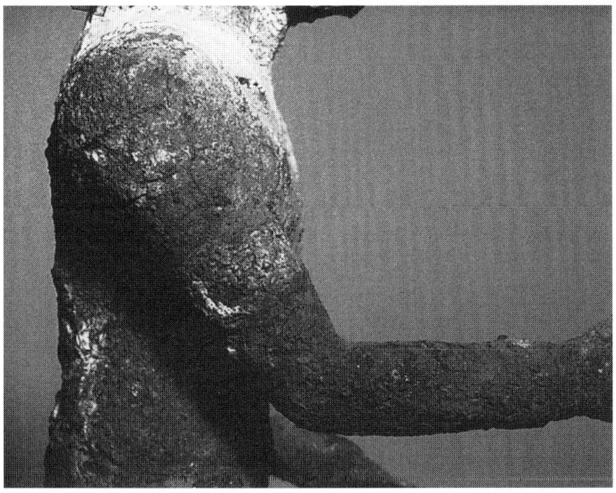

Figure 3.8 Detail of a plain woven fabric pseudomorphically preserved in the corrosion below the gesso and gold leaf on the proper left arm band. (Photo by Janet Burke, © Cleveland Museum of Art.) (See Plate 8.)

areas. These are visible only with magnification. It would appear that the linen was applied to the untextured surfaces of the bronze in preparation for gilding.

Polychromed bronze figural mounts

Two sets of bronze mounts, dating to the New Kingdom, are well preserved and exhibit surfaces largely unaltered by burial. The mounts consist of two hollow heads, two single arms, two single feet and a pair of crossed arms (Fig. 3.9). Their original function is unknown; however, documentary evidence suggests that they were probably mounted onto viscera bundles to make shawabty-like forms for funerary offerings (Berman *et al.* 1999). One set (1914.554, 558-560) comprising one head with two separate feet and arms, is in excellent condition and preserves what appears to be an original patination over a majority of the surfaces.[18] It illustrates a polychromed effect which must be close to its original appearance.

The head is hollow with a black wig; an inset depicting the face, neck and collar, appears brown (Fig. 3.10). The hands hold the implements of a scribe – the palette and feather of truth (Berman *et al.* 1999). They are similar in appearance to the face. The feet appear darker, probably due to corrosion.

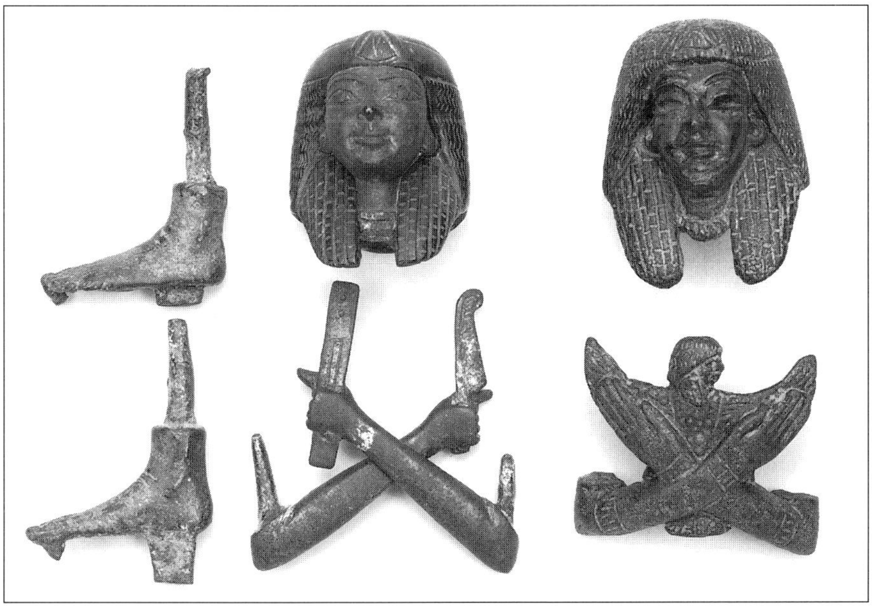

Figure 3.9 Two complete sets of figural mounts from 'shawabty bundles'. From left to right: feet (1914.560), head (1914.554), arm with feather (1914.558), arm with scribe's palette (1914.559), head (1914. 555) and crossed arms (1914.556). (Photo by Janet Burke, © Cleveland Museum of Art.)

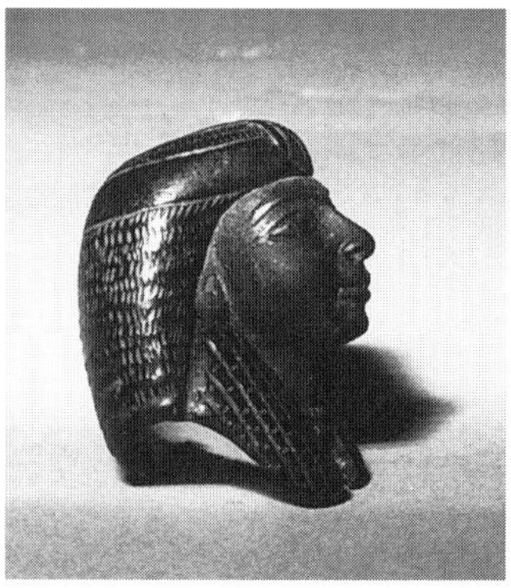

Figure 3.10 The lustrous dark surface of the wig contrasts to the more subdued tone and surface of the face. The high degree of polish and fineness of the incised details on the wig are both readily apparent. (Photo by Janet Burke, © Cleveland Museum of Art.)

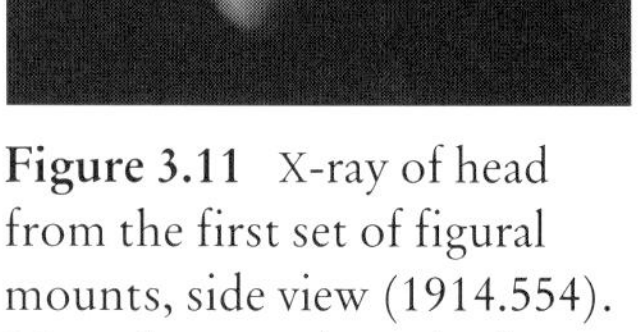

Figure 3.11 X-ray of head from the first set of figural mounts, side view (1914.554). Note the gap where the face and wig meet and the interior supports. The X-ray was taken at XRI Testing, Cleveland using a 2 minute exposure at 300 kV, 10 mA and 48" stf. (Photo by Joan Neubecker, © Cleveland Museum of Art.)

Although the gilded surfaces are fragmentary, the mounts are in excellent condition. The blackened surface of the wig is extremely well preserved with very few patches of compact green corrosion visible. The bronze face, hands and feet exhibit thin compact brown patinas which appear to be the result of copper oxides. These areas were previously coated with a pigmented wax to mimic the black patination of the wig.[19] A thin layer of surface dirt is visible beneath the wax. The interior of the head exhibits more dirt and corrosion, including green (malachite) and red (cuprite) corrosion products.

Facial features, and details of the wig, headband and collar were articulated by finely incised details in the wax, which transferred to the bronze when cast. All parts were cast solid. The surfaces appear to have been originally polished. In particular, the surface of the patinated wig retains a fine polish and exhibits a lustrous surface.[20] The fingers, toes and linear details of the feather and palette were finely incised in the wax prior to casting. The feet terminate in pins on the flat surface above the ankle and flat tenons below the feet. Casting flash (probably sprues or gates) appears below each big toe. The arms are flat on their reverses. Each terminates in a pin on the flat surface above their elbows. They probably would have been arranged on the mounting substrate with the arms crossed as is typical for mummiform funerary figures.

The two-part construction of the head is unusual, and illustrates that the original intention was to achieve a polychromed effect created by the juxtaposition of parts made from two different alloys. The color differences in the patinas of the two surfaces is striking (Fig. 3.11). The hair appears black and lustrous in contrast to the brown patina on the face. The purpose of the polychromy was clearly to render the figure in a more naturalistic manner. The face was inset into an opening in the wig. The fit was not tight and gaps between the two pieces are clearly visible. The attachment was partly mechanical as indicated by protruding braces attached to the face which fit inside the wig (Fig. 3.11). There are also traces of a yellow granular material on the interior of the head which is concentrated around the perimeter of the face. This material appears to be a mortar which was applied on the interior to augment the attachment.[21]

Several areas on the surface of the face and hair were analysed nondestructively by placing the object in an SEM. Analysis demonstrated that the alloys for the two components were different.[22] The wig exhibited appreciably less tin than the face and trace amounts of arsenic, gold and silver. Examination of the wig in backscatter electron imaging revealed globules of high atomic weight material scattered within the bronze matrix; these proved to be a gold-rich gold–silver alloy when analysed. An equiaxed grain structure was imaged in the SEM on the surface of the wig, indicating the bronze was heated to annealing temperatures during post-cast working or patination.

The alloy used in the fabrication of the face, arms and feet was determined to be a relatively high-tin bronze, with small amounts of lead and arsenic based on surface analysis using PIXE.[23] In fact several of the pieces exhibited a cast dendritic structure on their surfaces when examined using low magnification (6× or higher) similar to that published by Meeks for a bronze with a tin-enriched surface due to tin sweat.[24] The combination of patinated and tin-rich surfaces may in part account for the excellent preservation of the mounts.

The final aspect of the polychromy was the gilding, which was used in both a naturalistic manner to indicate precious jewellery adorning the figure as well as to accent the important implements of the scribe. Gold leaf made from an electrum alloy containing approximately 30 wt% silver, according to EDS, was selectively applied to the collar below the neck, in bands around the wrists and to adorn the feather and palette held in the hands. The gilding was executed by leaf gilding adhered with an unknown adhesive.

The second set of figural mounts (Fig. 3.9) consists of two components: a hollow head, which originally contained inlaid eyes, and a pair of crossed arms over a 'ba' bird (1914.555-556).[25] This set is more corroded and exhibits a compact dark green patina underlying light green powdery corrosion and dirt. Careful examination of the surface of the wig reveals patches of a lustrous black patina identical in appearance to the patinated surface of the wig on 1914.554. The head exhibits similarities in construction, surface finishing, patination and gilding to the previous set of mounts. Once again the construction relates to the desire to differentiate the wig from the areas corresponding to skin, much as the gilding was applied to make the collar appear like gold.

The construction of the head also includes a separate face and wig (Fig. 3.12). The face fits tightly within the opening of the wig and appears to have been adhered with a white granular mortar. Analysis of the surface of the wig, face and crossed arms using PIXE indicates that the wig was cast in black copper and the face and arms were cast in tin-bronze.[26] The collar with cast incised details was also gilded by applying gold leaf to the metal using an unknown adhesive, although only traces of this layer remain.

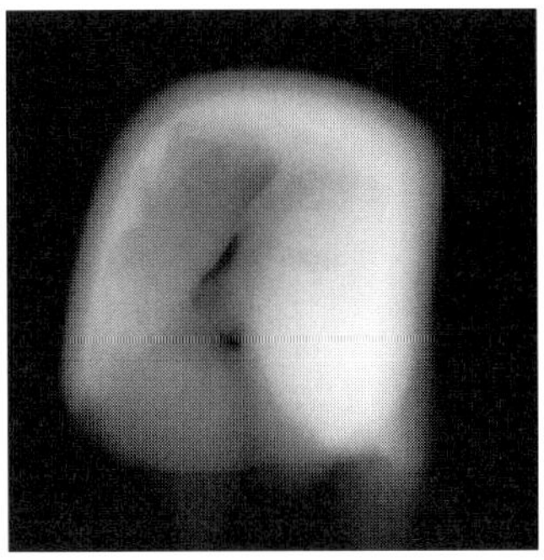

Figure 3.12 X-ray of head from the second set of figural mounts, side view (1914.555). Note the gap where the face joins the head. The X-ray was taken at XRI Testing, Cleveland using a 2 minute exposure at 300 kV, 10 mA and 48" stf. (Photo by Joan Neubecker, © Cleveland Museum of Art.)

Falcon head mount

The third mount is a hollow falcon head (1914.708) with decorative gilding applied to accent realistic features and the raised striations and lappets of the wig (Fig. 3.13).[27] This object is an unusual type with no known parallels; however, technical similarities to the figural mounts suggest that it served a related purpose (Berman *et al.* 1999). The features of the falcon were rendered with attention to anatomical detail. The head was originally surmounted by a headdress which is now missing.

The face and wig were cast separately and fitted tightly together (Figs 3.14 and 3.16). They appear to have been adhered using white granular mortar in a similar manner to that used for the second figural head, 1914.555. A tenon fragment remains in the head from the missing head-

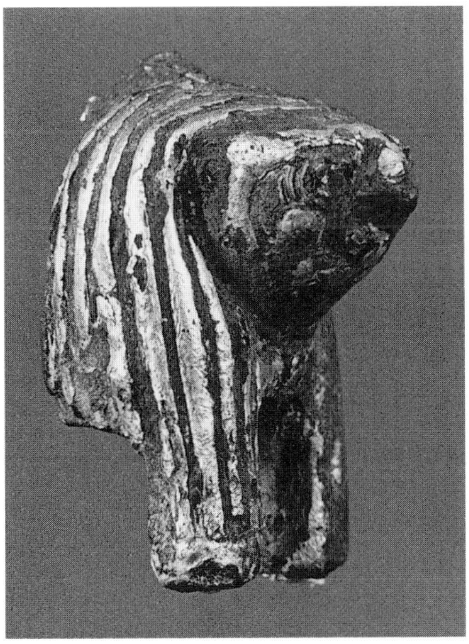

Figure 3.13 Three-quarter view of falcon head from the proper right side (1914.708). The gilding is very well preserved. Note the three electrum inlays to the proper right of the eye. (Photo courtesy of the © Cleveland Museum of Art (CMA neg. 51108).)

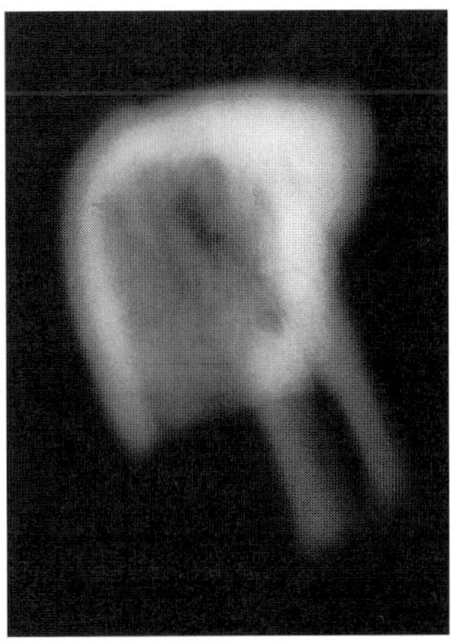

Figure 3.14 X-ray of falcon head, side view (1914.708). Note the faint radio-transparent seam where the face and the wig meet. The X-ray was taken at XRI Testing, Cleveland using a 2 minute exposure at 300 kV, 10 mA and 48" stf. (Photo by Joan Neubecker, © Cleveland Museum of Art.)

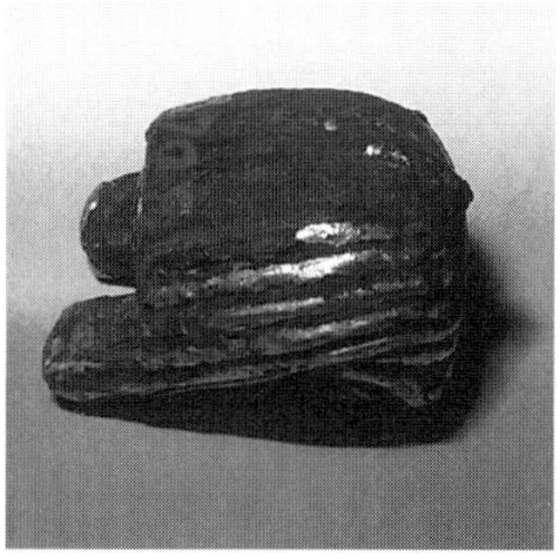

Figure 3.15 Detail showing the reverse of falcon head. The gilding is extant in most areas that the copper additions or inlays are preserved. Both the gilding and additions are absent on the reverse of the wig revealing the pairs of raised ribs on the surface of the wig used to hold the additions in place. Also note the powdery red corrosion on the surface of the wig and the compact red and black corrosion on the copper additions. (Photo by Janet Burke, © Cleveland Museum of Art.) (See Plate 9.)

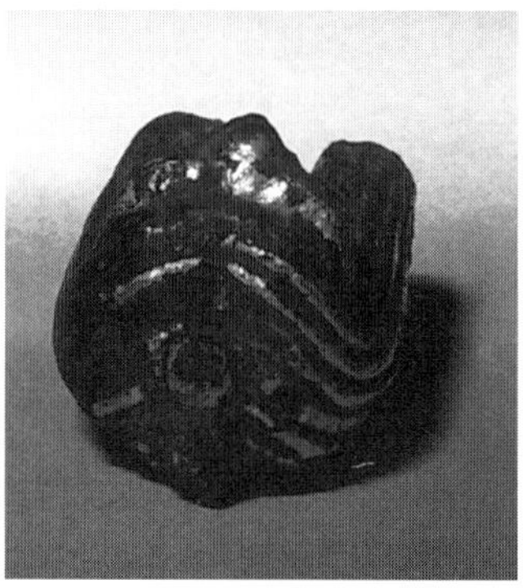

Figure 3.16 Detail showing the top of the falcon head. The seam between the face and wig is tight and difficult to see. The mortise which retains the tenon from the missing head-dress is readily visible. (Photo by Janet Burke, © Cleveland Museum of Art.) (See Plate 10.)

dress which was originally fitted into a square mortise on the top of the wig (Fig. 3.16). Markings on either side of each eye were delineated by three recessed channels. These were inlaid with electrum wire containing approximately 20 wt% silver according to EDS, which was hammered into place. Although it is corroded and details have been lost due to previous chemical and mechanical cleaning, a significant amount of the selectively applied gilding remains.

Some significant information is retained in the remaining corrosion. The wig and the separate face are both covered with a powdery orange red cuprite (est.) layer. The added striations and the facial inlays retain a compact red cuprite below a black surface; the thickness of the black layer varies and sometimes nearly the entire thickness appears black. Analysis of the black corrosion using XRD identified a complex mixture of corrosion products, including cuprite, tenorite (copper oxide), malachite and paratacamite. EDS analysis indicates that both the black and red layers consist primarily of copper with traces of gold and silver (less than one per cent). This area was subsequently gilded. The black layers may have formed as a by-product of the gilding process or as a result of natural corrosive processes.

The similarities in fabrication to the figural mounts suggests that the separate wig and face of the falcon were cast in different alloys to achieve a polychromed effect. Compositional analysis of the surfaces of the wig and face was not as conclusive as for the two figural mounts. Inconclusive analysis may be due to interference from the burial contaminants or impurities introduced in cleaning contained within the cuprite layer, as well as the close proximity of several different alloys in the face. However, analysis of the face and wig using EDS did indicate some compositional differences between the two components and suggests that the wig may have been made with black copper. The wig appears to contain less tin and slightly more gold and arsenic than the face. No silver was detected. It was noted that the black copper wigs on the two figural mounts also contained appreciably less tin than the bronze alloy used to cast the face and hands or feet. Research on Egyptian black copper by Giumlia-Mair and Quirke (1997) has also indicated that the black copper-alloys tend to be low in tin, generally in the range of less than three per cent by weight.

The gilding consists of metal foil, approximately 4 microns in thickness, which was made from a gold-rich alloy containing approximately 90 wt% gold with small amounts of silver and copper according to EDS. The gold foil was not adhered directly to the substrate metal from which the face and wig were cast. Instead it was applied to a metal interleaf consisting of almost pure copper. This in turn was applied to the surface of the bronze as cut sheet inlays on the face or as applied strips to form the lappets and striations of the wig. These copper additions were probably attached to the surface of the bronze by hammering. The cut sheet inlays were hammered into shallow recesses on the surface of the face. The applied strips were hammered onto narrow raised ribs cast into the surface of the bronze as is visible on the wig where the copper-rich strips have been removed (Fig. 3.15). The attachment of the lappet pieces is not clear; however, they probably were attached in a similar manner. The leaf appears to have been adhered after attachment of the copper as indicated by excess or overlapped gold; this is most visible on the reverse of the wig lappets.

Examination of the surface of the object with a binocular microscope indicates that the gold is surprisingly well adhered and exhibits a bright burnished surface. EDS analysis of a detached fragment indicates that the area below the gold is extremely carbon-rich, probably due to the organic binder used to attach the leaf. The fragment was taken by fracturing a small section with the point of the scalpel. Sampling distorted the gold which partially cleaved from the surface; however, instead of cleaving cleanly away from the copper substrate, a small amount of the copper remained adhered to the underside of the leaf in a number of areas. EDS analysis suggests that this zone of adhered copper contains small amounts of gold. The gold leaf with adhered copper proved difficult to maneuver into section. However, examination of the exposed surface of the copper that was previously in contact with the gold leaf also suggested that gold was present in the copper substrate. Diffuse opaque islands of gold-rich material were visible when the surface was examined in an SEM using backscattered electron imaging. Their diffuse quality indicated that they were below the surface of the copper rather than adhered bits of gold leaf which were also visible as distinct opaque particles. Evidence of diffusion of the gold was not visible on the rough fracture surface when this sample was examined in section. Diffusion gilding is identified by characteristic 'fingers' of gold penetrating into the substrate metal as visible in polished sections examined with backscatter electron imaging in the SEM. (Oddy *et al.* 1981). Although fingers were not visible on the rough fracture surface, the gold-rich islands that appear as diffuse opaque spots within the copper matrix are suggestive of this phenomenon.

Altogether, the accumulated evidence suggests that the gold foil was fixed in place with an organic binder, rich in carbon and also containing small amounts of sulphur. The surface was heavily burnished; some heating probably occurred at this stage resulting in diffusion of the gold into the copper substrate. Heating would also account for the presence of tenorite as a corrosion product in the black layer.[28] Examination of a polished metallographic section would be the best means of confirming the degree of diffusion and the depth of penetration of the gold into the copper.

The added copper elements – in particular the facial inlays – which consist of red cuprite with a blackened surface, are intriguing. Shearman has studied a bronze statuette of a falcon-headed deity in the British Museum. This object also exhibits a selectively blackened surface over a red cuprite layer (Shearman 1988). The areas also appeared to be inlays. The positioning of the blackened areas corresponds precisely to the facial inlays on the CMA hawk. Shearman determined by X-ray diffraction that the primary component of the black surface was cuprite. Surface analysis using X-ray fluorescence indicated trace amounts of arsenic and sulphur in some areas leading to speculation that the blackening was due to the application of an arsenic compound applied to darken the surface for a naturalistic effect. A survey of twenty additional bronze falcons in the collection found one other bronze with recessed areas that originally held inlays and one in which these areas were gilded. The similarity between the blackened cuprite described by Shearman and our falcon head and the reported analyses suggests that the British Museum falcon may also originally have been gilded. Another possibility for Shearman's black inlays was postulated by Craddock and Giumlia-Mair, who surmised that the inlays might consist of black copper (Craddock and Giumlia-Mair 1993a). Finally, it should be noted that black 'tarnish' will form on the surfaces of arsenical bronze.[29]

Black cuprite inlays on other bronzes in the CMA

Two additional statuettes exhibit surface characteristics and features of construction that indicate that their original appearance was intended to be polychromed. Unfortunately, both have been stripped and exhibit relatively compact surfaces consisting primarily of reddish cuprite (est.) corrosion. Both statuettes have been waxed. These objects are of particular interest because they retain selectively applied inlays of cut sheet metal, which appear to consist of a reddish cuprite with a black surface similar in appearance to those noted on the CMA falcon head mount and described for the falcon-headed deity in the British Museum.

The statuette of a mummiform Osiris (1914.571) is datable to the Late Period (Berman *et al.* 1999).[30] Although the surface is roughened and pitted from corrosion, some of the fine details of the casting are preserved. The statuette was hollow cast with a clay core, except for the beard, which was solid cast separately and attached by a long tang set into the underside of the chin, which penetrates almost to the nose. The head and hood of the uraeus was also a separate element, which is now missing. The facial features and the collar were articulated by finely incised details in the wax, which were transferred to bronze. PIXE analysis indicates that the statuette was cast from a tin-bronze that contained small amounts of silver (less than three per cent by weight), which suggests that the figure was cast in black copper.[31] The collar was originally gilded, as indicated by small traces of gold leaf that are visible on the lower proper right. Microscopic traces of gilding are also visible on the face. The eyes, chin straps and the body of the uraeus were originally inlaid. The crook and flail were decorated with alternating black and white square inlays which were applied to the cast surface (Fig. 3.17). The white squares appear to be a fine-grained limestone, probably travertine. The black squares appear to be pieces of cut copper-rich metal sheet.[32] These are more or less completely mineralised and consist of a reddish cuprite with a thin, spotty black surface. PIXE analysis indicates that the black surface is rich in silver. XRD analysis identified acanthite (silver sulphide) and silver in the black layer, as well as cuprite from the underlying substrate. It appears that copper inlays were silvered instead of gilded, with only a thin layer of silver and silver sulphide remaining due to a combination of corrosion and/or previous electrochemical treatment.

The second statuette, a standing Amen-re figure (1942.772), wears a short kilt, broad collar and flat-topped crown surmounted by a sun-disc

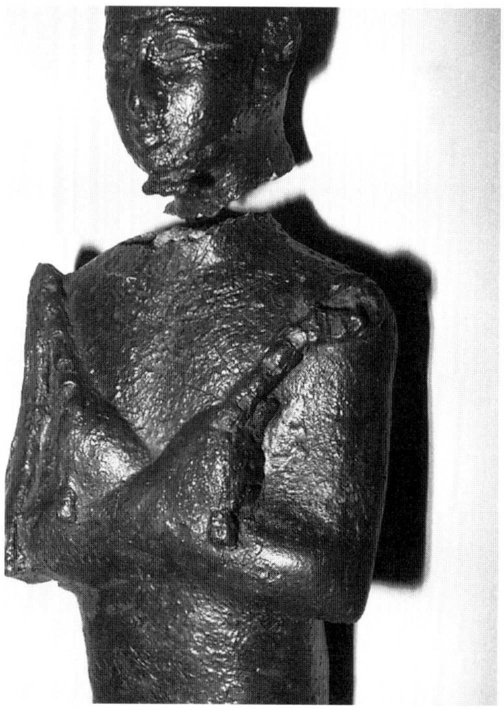

Figure 3.17 Detail of crook and flail on the statuette of Osiris (1914.571). White travertine inlays (est.) and the black inlays, which were probably originally silvered over a copper substrate, are visible on the proper left as a thin black surface over a compact red cuprite. Also note the uneven texture and tone of the stripped surface of the figure. A pigmented wax, coloured by carbon black according to XRD, was applied to unify the surface. (Photo by Janet Burke, © Cleveland Museum of Art.) (See Plate 11.)

and double plumes.[33] The statuette is datable to the Late Period. This statuette was apparently mechanically cleaned, probably in conjunction with chemical cleaning, to reveal a relatively compact patina consisting of red cuprite with some islands of preserved green corrosion characteristic of malachite.

The statuette was composed of numerous solid cast components, all attached using mortise and tenon joints; these are the headdress, the figure, the beard, two separate arms and the hollow base. All of these separate components were cast from a leaded-tin bronze.[34]

The figure was intended to be polychromed and the effect appears to have been created by a combination of gilding, inlaid wire and inlaid cut sheet. The wire inlays which decorate the face and beard are well preserved. The eyes were inlaid using silver (est.) wire. The eyebrows and beard straps were inlaid using gold (est.) wire. The braided beard was inlaid in electrum (est.) wire. The surface of the crown was clearly textured with a series of depressions created by short chisel strokes, probably in preparation for the application of a gilded gesso ground (Fig. 3.18). No evidence of gilding or gesso remains. The surface of the sun-disc also appears to be rough, although it is not clear whether it was originally textured. However, careful examination of its surface with magnification (18×) revealed microscopic traces of gold preserved at the edges in several places, indicating that it was also gilded. Copper-rich inlays are preserved in some recessed areas of the plumes. For the most part these have been converted to a reddish cuprite with a black surface (Fig. 3.19). Analysis of the inlays using both PIXE and EDS indicates that they were made using black copper.[35]

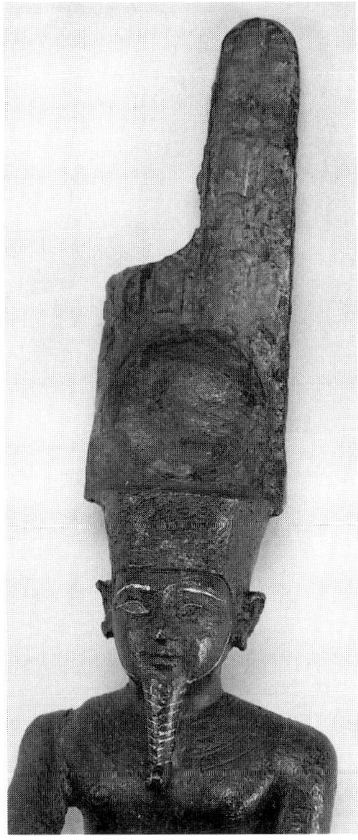

Figure 3.18 Detail of the crown, head and upper torso of a statuette of Amen-re (1942.772). Note the well preserved inlays and the textured surface of the crown, which was introduced prior to casting. This was probably originally gilt over a gesso layer. The inlays delineating feathers on the double plumes alternate with incised decoration. The collar was also finely incised. (Photo by Janet Burke, © Cleveland Museum of Art.)

Conclusion

Individually, the bronzes which have been discussed represent three different examples of the uses of gilding in Egypt: leaf gilding onto a gesso interleaf, leaf gilding onto the metal substrate, and foil gilding onto a metal interleaf. This last technique has not been previously noted, and appears to

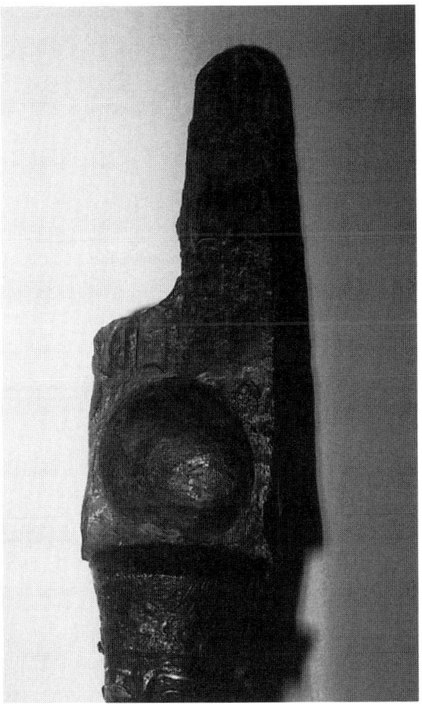

Figure 3.19 This detail of the double plumes surmounting the crown illustrates the use of the black copper inlays. The surface has been heavily waxed, which imparts a visible sheen to most areas. (Photo by Janet Burke, © Cleveland Museum of Art.) (See Plate 12.)

represent an example of diffusion gilding. It seems most likely that the use of the copper interleaf, applied as localised inlays and additions to the bronze substrate, was employed for technical reasons, presumably to produce a more durable product by enabling diffusion of the gold or silver into the copper to take place. The gilding has withstood the test of time and remains well adhered to the copper. It is also possible that the copper was used in a similar manner to the gesso, that is to create the appearance of a thicker gold sheet. The copper would undoubtedly have been more durable than the gesso and easier to apply in the localised manner in which it was used. A third possibility may be a change in aesthetic taste for the object, which resulted in the subsequent gilding of the added copper elements to change their appearance.[36]

As a group, the objects demonstrate that the selective use of gilding was an important component in the aesthetic of Egyptian polychromed bronzes. This effect was achieved by an intricate combination of materials and methods including wire and cut sheet inlays, inlaid elements in non-metallic materials, gilding, and the use of a combination of bronze and copper alloys, which may have been patinated for dramatic contrasts. In many cases the desired effect appears to have been to achieve a more naturalistic appearance as well as being decorative. For instance on the bronze figural mounts it is clear that the bronze was fabricated to create naturalistic coloration. The contrast of the black lustrous hair, with the brown skin and gold collar was intentional.

Reconstruction of the original appearance of the bronze or metal objects is often difficult because of burial corrosion and aggressive cleaning to remove burial layers. Evidence such as textured metal surfaces remaining on some bronzes that were originally gilt, or specific compositions related to original patination, allows the original surfaces to be reconstructed to some degree. Subtle differences in color and morphology of mineralised surfaces may also contain some clues. However, the interpretation of these surfaces is not always straightforward. For instance, three different examples of cuprite with a black surface in the CMA have developed from three different original surface treatments. In one case the inlays were gilt copper, in a second case they were silvered copper and in the third they were black copper.

Many of the bronzes discussed in this chapter use black copper as a part of the polychromed effect, which was either applied locally as an inlay or as separately cast components. This use is different from that more typically illustrated for black copper, in which the black copper itself functions as a substrate for the application of polychromed metal inlays or gilding.[37]

The CMA objects also demonstrate that in many cases a traditional bronze alloy was embellished with gilding and inlays of black copper and other metals. There may have been additional surface treatments to create contrast between the bronze and the other metals that were used as localised embellishment. In the case of the figural mounts, it is likely that bronze was intended to have a brown or red patina to approximate the colour of skin, which would provide a perfect complement to the naturalistic colors employed for the wig (black and lustrous) and the jewellery. Many researchers have argued persuasively for the colouration or patination of ancient bronzes based on the need for a contrast with bright metallic inlays (Cooney 1966, 1968; Born 1990, 1993). It may be that the polychromy of bronzes was more widespread and diverse than current research has revealed because of poor preservation of the original surfaces. A drapery fragment from a life-size Roman bronze sculpture is unique in both its preservation and the degree of patination that it received during manufacture. The highly polychromed surface includes areas of olive-brown and orange-yellow patination in addition to areas of black patination tinged with violet, which was used to offset bright metallic inlays in silver, brass and copper (Craddock and Giumlia-Mair 1993a,b). Current evidence suggests that a polychromed effect was also sought by goldsmiths. (Ogden 1993). In fact, this taste for a polychromed approach to metalworking is well documented in a surviving text from Roman Egypt, the Leyden Papyrus X. The papyrus contains over one hundred recipes, the majority of which were designed to achieve various polychromed effects in metal, either by alloying or by surface enrichment techniques (Caley 1926).

The question as to whether ancient bronzes were finished with a metallic surface or were polychromed and patinated has proven to be difficult to resolve in terms of classical metalwork;[38] however, it appears that a polychromed approach to bronze fabrication was employed in ancient Egypt. This taste is apparent by the careful study of surviving bronzes. Oddy has remarked on the scarcity of gilding on small-scale bronze sculpture in the classical world and contrasted it with the long-standing tradition in Egypt, which is evidenced by the numbers of surviving examples (Oddy *et al.* 1990). This dichotomy may be due in part to more favorable burial conditions in Egypt; however, it also may be a reflection of the Egyptian taste and skill for creating polychromed effects in metal.

Acknowledgements

This chapter was developed from work begun for a comprehensive catalogue project on the museum's Egyptian collection. I would like to thank my colleagues at the CMA, especially Lawrence M. Berman, Curator of Ancient Art and Bruce Christman, Chief Conservator, who both demonstrated a generosity of knowledge, spirit and their precious time during my examination of the collection. Some aspects of this discussion are based on their work which will be published in the forthcoming catalogue of the CMA's Egyptian collection. Thanks are also due to Judith DeVere for editing, and Joan Neubecker for developing all X-rays, X-ray diffraction film, and photographing the X-rays. Special thanks are due to Photo Technician Janet Burke for her excellent photography of the majority of the objects under discussion and printing of all negatives. We are privileged in Cleveland to be able to draw on excellent technical expertise outside the museum. Thanks are due to John Sears and Alan McIlwain, Center for Surface Analysis of Materials, Case Western Reserve University, for the SEM-EDS and PIXE work respectively. X-radiography was performed at XRI Testing. I would also like to thank several colleagues outside the CMA for their help. Thanks are due to Richard Newman, Museum of Fine Arts, Boston and Terry Drayman-Weisser, the Walters Art Gallery for sharing their unpublished research. Thanks are also due to Carol A. Grissom of the Smithsonian's Center for Materials Research and Education for providing key references.

Notes

1. For a discussion of earlier treatment protocols for Egyptian copper-alloy objects and the difficulties that this holds for interpretation, see Jaeschke and Jaeschke (1988), Norman (1988), Craddock and Giumlia-Mair (1993a) and Pearlstein (1995).
2. This technique was not used much outside Egypt. However, silver leaf applied to

the surface of a bronze helmet over a gesso consisting of calcium carbonate and an unidentified binder has been noted on an Illyrian helmet dating to the 6th century BC. Further study of related artifacts may yield more documented examples from this region (La Niece 1993).

3. Hatchfield and Newman examined the method of adhering gold leaf to parchment and concluded that the mucous inherent to the substrate functioned as the adhesive (Hatchfield and Newman 1991). Research by Newman on the binders in Egyptian gessoes has been ongoing and new data will be included in the forthcoming revision of Lucas's *Ancient Egyptian Materials and Industries* (personal communication).
4. Cult figure of a falcon-headed deity, cat. no. 5. (Miho Museum 1997).
5. Statue of Nakht, cat. no. 2. (Miho Museum 1997).
6. The beads contain calcium carbonate or clay (est.) cores adhered with unidentified binders. The clay cores appear to be bound with a dark shiny resinous material. The gesso was identified as calcium carbonate using microchemistry and polarising light microscopy. The binders in the gesso cores are not visible when examined using a binocular microscope; examination of the microstructure of the gesso in a scanning electron microscope may enable a determination as to whether the binder is an organic adhesive or recarbonated lime plaster. Although Lucas states that lime plaster was not used in Egypt before the Ptolemaic period, more recent research indicates that lime plaster may have been used as a binder for mortars as early as the Old Kingdom and as an adhesive by the 18th Dynasty. (Gourdin and Kingery 1975, Lucas and Harris 1989, Klemm and Klemm 1991 and Endlicher and Tillmann 1997). Extraction of the binder followed by organic analysis using a separating technique is the standard practice for identifying organic binders. This has proven problematic with some Egyptian materials because of the small amount of material typically introduced and the degradation over time (Hatchfield and Newman 1991). Binder analysis is not planned at this time. The construction of the beads is more fully discussed in Berman *et al.* (1999). See the following CMA numbers: 1914.784-.788 (recumbent lion beads), 1914.795.a,b (pair of sidelocks) and 1914.797 (teardrop-shaped pendants).
7. Statuette: Seated Amen-ra. Black copper and bronze, inlaid and gilt. H. 19, W. 5.5, D. 11.5 cm. Dynasty 25–26. Gift of the John Huntington Art and Polytechnic Trust, 1914.574.
8. Corrosion products were identified by powder X-ray diffraction patterns obtained using either a Debye-Scherrer or Gandolfi camera. Small samples were taken and mounted onto a glass filament using petroleum jelly. Patterns were obtained by running samples for 3 hours using 30 kV and 30 mA. *D*-spacing values were calculated and compared with known standards compiled by the JCPDS using a computerized search program (JPDPSwin). Sampleite has been identified as a corrosion product due to burial conditions on Egyptian bronzes from Memphis (Fabrizia and Scott 1987; Fabrizia *et al.* 1989) and on Pre-Columbian tumbaga by the author (unpublished), possibly as a result of previous cleaning. Chaconatronite was identified as a minor constituent. It has been identified by Gettens as a naturally occurring corrosion product on Egyptian bronzes; it also has been known to develop on bronzes that have been treated with sodium sesquicarbonate (Gettens and Frondel 1955; Horie and Vint 1982). Atacamite and paratacamite are dimorphic basic copper chlorides, which can form in burial due to the presence of chloride salts. (Gettens 1970).
9. The pattern obtained was poor and exhibited a strong background with many rings, which were difficult to resolve, creating problems in measuring. Several unmatched *d*-spacing values remained after subtraction of the reported compounds that could not be matched to a single compound. Therefore, it seems likely that the crust is a complicated mixture of products with only the strongest intensity lines remaining legible. Several different oxides and carbonates are possible matches. EDS analysis indicates that the crust contains primarily lead, carbon and oxygen with small amounts of copper and chlorine. An overview of the corrosion of lead is provided by Turgoose (1985). Scrutinyte was not noted among documented lead corrosion products; although another lead dioxide, plattnerite, has been identified previously.

10. Analysis was performed on the chest, shoulder and back. Percentages are not reported here due to contamination from the gesso, gilding, corrosion layers and burial soil, and relatively low copper values due to removal of the copper from the surface during burial; however, the data indicates that the alloy is a leaded tin bronze. All non-destructive proton induced X-ray emission spectroscopic analysis (PIXE) was performed by Alan McIlwain, Center for Surface Analysis of Materials, Case Western Reserve University. Data was obtained with broad beam bulk PIXE in a helium atmosphere, using a 3 meV proton beam at 1 nanoampere, and an acquisition time of up to 30 minutes. Data is summarised here, and in Berman *et al.* (1999). A complete report of the results will appear in a forthcoming technical supplement to the catalogue.
11. Analysis of a small amount of metal and corrosion scraped from the surface of the proper left cosmetic line using a scanning electron microscope with energy dispersive spectroscopy (SEM/EDS) indicates that the primary component is silver. All EDS analyses were performed using a Hitachi S-4500 field emission scanning electron microscope and a 5–30 kV beam by John Sears, Center for Surface Analysis of Materials, Case Western Reserve University. The beam energy used was dependent on the sample; however, most analyses were performed at 25–30 kV. In general, percentages for components were not derived due to surface irregularities of the samples, which were mounted as whole objects, surface scrapings or roughly broken surfaces on detached fragments. Any reported percentages should be considered standardless semi-quantitative data.
12. The method of attachment for the stone (est.) eye inlays is not visible; however, a fine-textured reddish mortar was used for this purpose on the statuette described by Delange *et al.*, which was identified as pigmented calcium carbonate. Similar mortars have been noted on other examples of inlaid eyes in the CMA, including a detached Middle Kingdom eye with a separate pupil (1915.505), and a wooden statuette from the New Kingdom, where it was used to attach eye, cosmetic line and brow inlays (1983.98). It is assumed based on their appearance and general composition that these mortars consist of clay and an unidentified binder. (Delange *et al.* 1995; Berman *et al.* 1999).
13. For a discussion of the construction of Egyptian bronzes, see for instance Jedrzejewska (1976), Shorsch (1988), Lucas and Harris (1989) and Bianchi (1990). The statuette of a seated Amen-re discussed by Jedrzejewska is of particular interest because of the similarities in construction to the CMA statuette, which includes the number and type of cast and added components as well as the technique and selective use of the gilding. The major differences of the Krakow Amen-re are that the head was cast in two parts and that some elements of the deity's dress were incised rather than gilt (the arm bands and cuirass).
14. Cooney was the first to identify several Egyptian bronzes exhibiting a lustrous black patinated surface with the term 'black bronze', a phrase that appeared in Egyptian texts as early as the 18th Dynasty. He also established that the practice was transferred to the classical world. (Cooney 1966, 1968). More recently, research by Craddock and Giumlia-Mair has determined that the mechanism for this process is essentially the same as that used to produce the Japanese alloy 'shakudo'. The addition of small amounts of gold and silver, as well as sometimes arsenic and iron, produces a lustrous black, blue or purple patinated surface when the bronze is chemically treated. A cuprite patina forms; however, its colour and appearance are altered due to optical interference from inclusions formed by the additional metals. The visual characteristics of the patina are affected by surface finishing of the alloy prior to patination as well as the specific metals added and their relative amounts. Recent scholarship has also determined that the term 'black copper' is more accurate and has pushed back the date of its use in Egypt to the Middle Kingdom (Craddock and Giumlia-Mair 1993a; Giumlia-Mair and Quirke 1997). The corpus of black copper artifacts continues to grow as more material is examined and analysed, including recent confirmation of its use on Mycenean daggers (Ogden 1993; Photos *et al.* 1994; Hill and Schorsch 1997). It should be noted that similar alloys are part of the Japanese and Chinese traditions of polychromed metalwork. The Japanese alloy 'shakudo' has received the most study (Notis 1988; Murakami 1993; Wayman and Craddock

1993). Previous studies have found small amounts of gold and/or silver in Egyptian bronzes without equating them with 'black copper' or a shakudo-type patination process (Jedrzejewska 1976; Fishman and Fleming 1980).

15. The ratio of copper to lead by weight as indicated by EDS was approximately 1:4 in the gray corrosion and 12:1 in the underlying layer.
16. The dealloying of copper alloys is complicated and dependant on burial conditions as well as trace elements within the alloy, which in different amounts might either inhibit or stimulate corrosion. This phenomena has been studied by Weisser as part of dissertation research at the Institute of Archaeology, University College of London. Her experimentation demonstrated that a lead-rich surface will form on copper-alloys containing lead that were exposed to certain corrosive solutions based on standard treatment protocols. This data also provides some insight into the behaviour of the alloys during corrosive burial conditions (personal communication). Some results of this work pertaining to tin-bronzes and brass have been published. See Weisser (1975) for an overview of the subject, which includes extensive references culled from corrosion science literature.
17. Two detached pieces of the gold leaf and gesso were examined in the SEM. Exact measurements of the leaf thickness were not obtained because the samples exhibited rough break edges and the gold was difficult to orient in section due to bending and distortion of the leaf. EDS analysis indicated that the gold was extremely gold-rich and contained less than 2 wt% copper. For a discussion of the range of compositions found in Egyptian gold leaf see Hatchfield and Newman (1991). An organic material was visible as a binder between particles when the sample was examined in a SEM using 5 kV and 5000× magnification. The material appeared to be a porous stringy film. Small amounts of sulphur and trace amounts of phosphorus were detected using EDS, as well as trace amounts of silica and magnesium, probably due to silica or clay impurities. The presence of sulphur suggests that the binder was an albumin product based on egg or casein (Kleiner 1995).
18. Head. Black copper and bronze, gilt. H. 3.3, W. 2.6, D. 2.9 cm. Arm with feather. Gilt bronze. L. 4.9 cm. Arm with scribe's palette. Gilt bronze. L. 4.9 cm. Feet. Bronze. L. 2.7 cm (each). Late Dynasty 18 or early Dynasty 19. Gift of the John Huntington Art and Polytechnic Trust, 1914.554, .588-.560.
19. A waxy coating containing black pigment which was applied to the skin-coloured surfaces, presumably to make them closer in tone to the wig. This is the opposite scenario from that reported by Craddock and Guimlia-Mair in the case of a Roman head with black hair and a skin-coloured cuprite layer on the face, which proved to be a result of selective cleaning (Craddock and Giumlia-Mair 1993b).
20. Research indicates that careful preparation of the metal surface, including polishing, is an important step in the patination process with shakudo-type alloys (Craddock and Giumlia-Mair 1993a; Giumlia-Mair and Quirke 1997).
21. The attachment of separate metal components using mortar or adhesive is not much reported in the literature. However, there is ample evidence that the Egyptians used atypical techniques in the construction of their bronzes, such as the gilding over gesso, the use of mortise and tenon joints and the use of mortars to adhere eye inlays, as previously discussed. A gilt silver cult statue (cat. no. 5) in the Miho Museum exhibits a mortar consisting of Egyptian blue, which was used to adhere stone inlays carved from lapis to decorate the striations of the wig (Miho Museum 1997). In addition to this mount, a granular white mortar was noted on the interior front surfaces of two other bronze mounts in the CMA (1914.555 and 1914.708, see following discussion). The heads on both of these are tightly fitted within the openings of the wig with no evidence of mechanical or metallurgical attachment. A yellow powdery material, described as a cement, consisting of calcium carbonate pigmented with iron oxide and possibly bound with an organic binder was also found on the interior of a bronze figural mount in the Metropolitan Museum of Art (Kopp 1936, unpublished report from curatorial files, Metropolitan Museum of Art; Berman *et al.* 1999). Although cast in bronze, the closest parallels for the construction of the shawabty mounts are the composite stone shawabtys also dating to the New Kingdom (Berman *et al.*

1999). These consist of red stone faces and hands on a white stone body; the various components would have been held by adhesive or mortar. Gesso mortars were also used to adhere elements on composite faience sculpture and relief in the New Kingdom (Friedman 1998).

22. The head was placed in the SEM and examined using a 25 kV beam. The major component of the black copper alloy was copper with small amounts of gold, lead and tin (less than 2–3 wt%). Less than 1 wt% silver and trace amounts of iron and arsenic were also noted.
23. According to PIXE, the compositions of the bronze alloys contained between 7–10 wt% tin, less than 2 wt% lead and less than 1 wt% gold, silver, arsenic and/or iron.
24. Meeks noted that the characteristic dendritic structure is often visible on the surface of ancient metals using low magnification (Meeks 1993).
25. Head. Black copper and bronze, inlaid and gilt. H. 3.6, W. 3.2, D. 2.7 cm. Crossed arms. Bronze, L. 3.5 cm. Late Dynasty 18 or early Dynasty 19. Gift of the John Huntington Art and Polytechnic Trust, 1914.555-.556.
26. The black copper wig contains approximately 2 wt% tin, less than 2 wt% gold and less than 1 wt% of silver, arsenic and lead. The bronze alloys used for the face and arms contain approximately 2–6 wt% tin with less than 1 wt% of lead, arsenic, iron and silver.
27. Falcon's head. Black copper and bronze, inlaid and gilt. H. 2.7, W. 1.8, D. 2.1 cm. Late Dynasty 18 or later. Gift of the John Huntington Art and Polytechnic Trust, 1914.708.
28. Tenorite is not typically found as a corrosion product on archaeological copper-alloys. Its presence implies that the bronze was heat treated.
29. This phenomenon was noted by Fink and Kopp on two bronzes treated by them, which were erroneously identified as antimony bronzes using microchemistry. They were correctly identified as arsenical bronzes with silvery arsenic-rich surfaces in a later publication by Smith (Fink and Kopp 1932–33; Smith 1973).
30. Statuette of Osiris. Bronze, hollow cast, inlaid and gilt. H. 28 (without tang), H. 30.6 (with tang), W. 6.8, D. 4.7 cm. Dynasty 26 or later. Gift of the John Huntington Art and Polytechnic Trust, 1914. 571.
31. Both the copper-alloy of the statuette and the copper-alloy used for inlays in the crook and flail exhibit high amounts of lead on the surface, which is most likely the result of previous electrochemical treatment. X-radiography of the thin-walled bronze exhibits no radio-opaque globules characteristic for high-lead bronzes due to the low solubility of the lead in copper. PIXE analysis of the CMA's collection of Egyptian bronzes is in progress; however, a number of apparently stripped bronzes are extremely lead-rich on the surface; up to 75 wt% lead has been noted thus far. Some results appear in Berman *et al.* (1999); a full report will appear in the forthcoming technical supplement.
32. Surface analysis of an inlay using PIXE detected less than 2 wt% tin and approximately 18 wt% silver in addition to high amounts of lead (approximately 24 wt%). The silver appears to be related to previous silvering of the surface. In contrast, more than 8 wt% tin was detected in the alloy used to cast the statuette, as well as approximately 38 wt% lead and the small amounts of silver previously noted. Copper was detected in amounts below 40 wt% in both areas, presumably due to corrosive processes during burial and/or treatment. See Weisser (1975) for an overview of the de-alloying of copper alloys.
33. Statuette of Amen-re. Bronze, solid cast, inlaid and gilt. H. 25.8, W. 5, D. 7.3 cm. Dynasty 26 or later. Bequest of John L. Severance, 1942.772.
34. PIXE analysis was performed on the reverse of the plumes, on the front of the plumes on a feather and the sun-disc, on the front leg, and on the back leg. The reported percentages for lead varied considerably from 6 to 30 wt%. The reported percentages for tin varied from 2 to 10 wt%.
35. PIXE analysis found less than 2 wt% gold and less than 1 wt% silver, arsenic and iron in the black inlay as opposed to other areas of the bronze which were analyzed; these generally had small amounts of arsenic and none to negligible amounts of gold or silver. EDS also found small amounts of gold in the inlays. XRD of a sample of the black material identified cuprite.

36. A well known example of a secondary decorative layer being applied over the original surfaces of a sculpture is the Head of Queen Tiye from the Ägyptisches Museum in Berlin. However, it appears that the alteration is unrelated to a subsequent restoration. A persuasive argument has been made which attributes the alteration to a change in social and religious climate (Arnold 1997). Another example from Dynasty 18 is the coffin of Henetwedjebu in the collection of the Washington University Gallery of Art (no. 2292), which is dated to the reign of Amenhotep III. (Kozloff *et al.* 1992). The striated wig is executed in alternating stripes of black bitumen and gilded gesso. However, blue pigment is visible below the black stripes indicating a change in aesthetic taste during fabrication. Losses in other areas of the coffin do not indicate the abandonment of a traditional polychromy for the bold use of glossy black and gold (Bruce Christman, personal communication).
37. Other examples of black copper that demonstrate its use as a substrate for inlays are presented by Craddock and Giumlia-Mair (1993a), Ogden (1993), Giumlia-Mair and Quirke (1997) and Hill and Schorsch (1997). Analysis of the bronzes in the CMA using PIXE is ongoing. However, based on the appearance of their well preserved dark lustrous patinas, two statuettes dated to Roman Egypt might prove to be cast in black copper with various metals used as inlays to contrast with the dark lustrous surface. These are a Harpocrates (1972.6) and a Beggar (1963.507). The Beggar was one of the corpus of black copper artifacts proposed by Cooney (Cooney 1968; Kozloff and Mitten 1990; Berman *et al.* 1999). Other bronzes in the CMA may also prove to be of this composition; however, this is not visually apparent due to previous stripping and or repatination of their surfaces.
38. See Born 1990, 1993, Craddock and Giumlia-Mair (1993b) and Ogden (1993) for an introduction to this complicated subject.

References

Arnold, D. 1997. *The Royal Women of Amarna: Images of Beauty from Ancient Egypt*. New York: Metropolitan Museum of Art.

Berman, L. M. with K. Bohac; Conservation Sections by P. S. Griffin with B. Christman 1999. *Catalogue of Egyptian Art*. Cleveland, Ohio: Cleveland Museum of Art.

Bianchi, R. S. 1990. 'Egyptian metal statuary of the Third Intermediate Period (circa 1070–656 BC), from its Egyptian antecedents to its Samian examples'. In *Small Bronze Sculpture from the Ancient World*, M. True and J. Podany (eds), 61–84. Malibu, Calif.: J. Paul Getty Museum.

Born, H. 1990. 'Patinated and painted bronzes: Exotic technique or ancient tradition?' In *Small Bronze Sculpture from the Ancient World,* M. True and J. Podany (eds), 179–96. Malibu, Calif.: J. Paul Getty Museum.

Born, H. 1993. 'Multi-coloured antique bronze statues'. In *Metal Plating and Patination: Cultural, Technical and Historical Developments*, S. La Niece and P. Craddock (eds), 19–29. Oxford: Butterworth-Heineman.

Caley, E. R. 1926. 'The Leyden papyrus X: an English translation with brief notes'. *Journal of Chemical Education* **3**: 1149–66.

Cooney, J. D. 1966. 'On the meaning of "black bronze"'. *Zeitschrift für Ägyptische Sprache und Altertumskunde* **93**: 43–7.

Cooney, J. D. 1968. 'Siren and Ba, birds of a feather'. *Bulletin of the Cleveland Museum of Art* **55**: 262–71.

Craddock, P. and Giumlia-Mair, A. 1993a. 'Hśmn-km, Corinthian Bronze, Shakudo: Black patinated bronzes in the ancient world'. In *Metal Plating and Patination: Cultural, Technical and Historical Developments*, S. La Niece and P. Craddock (eds), 101–27. Oxford: Butterworth-Heineman.

Craddock, P., and Giumlia-Mair, A. 1993b 'Beauty is skin deep: evidence for the original appearance of classical statuary'. In *Metal Plating and Patination: Cultural, Technical and Historical Developments*, S. La Niece and P. Craddock (eds), 30–38. Oxford: Butterworth-Heineman.

Delange, E., Di Mantova, A. and Taylor, J. H. 1998. 'Un Bronze Egyptien Méconnu'.

La Revue du Louvre et des Musées de France **5**: 67–75.
Delange, E., Di Mantova, A. and Meyohas, M-E. 1995. 'Une Extraordinaire statue de bronze Doré conservée au Louvre'. In *Conservation in Ancient Egyptian Collections,* C. Brown, F. Macalister and M. Wright (eds), 137–46. London: Archetype Publications.
Endlicher, G. and Tillmann, A. 1997. 'Lime plaster as an adhesive for hafting eighteenth-dynasty flint sickles from Tell el Dab'a, Eastern Nile Delta (Egypt)'. *Archaeometry* **39**: 333–42.
Fabrizia, M. and Scott, D. A. 1987. 'Unusual copper corrosion products and problems of identity'. In *Recent Advances in the Conservation and Analysis of Artifacts*, James Black (ed.), 131–3. London: Summer Schools Press.
Fabrizia, M., Ganiaris, H., Tarling, S. and Scott, D. A. 1989. 'Occurrence of sampleite, a complex copper phosphate, as a corrosion product on copper alloy objects from Memphis, Egypt'. *Studies in Conservation* **34**: 45–52.
Fink, C. G. and Kopp, A. H. 1932–33. 'Ancient Egyptian antimony plating on copper objects'. *Metropolitan Museum Studies* **4**: 163–7.
Fishman, B. and Fleming, S. J. 1980. 'A bronze figure of Tutankhamun'. *Archaeometry* **22**: 81–6.
Friedman, F. (ed.) 1998. *Gifts of the Nile.* Providence, R.I.: Rhode Island School of Design.
Gettens, R. J. 1970. 'Patina noble and vile'. In *Art and Technology*, S. Doeringer and D. S. Mitten (eds), 57–72. Cambridge, Mass.: MIT Press.
Gettens, R. J. and Frondel, C. 1955. 'Chalconatronite, an alteration product of some Egyptian bronzes'. *Studies in Conservation* **2**: 64–75.
Giumlia-Mair, A. and Quirke, S. 1997. 'Black copper in Bronze Age Egypt'. *Revue d'Egyptologie,* **48**: 95–108.
Gourdin, W.H. and Kingery, W. D. 1975. 'The beginnings of pyrotechnology: Neolithic and Egyptian lime plaster'. *Journal of Field Archaeology* **2**: 133–50.
Griffin, P. S. 1999. 'Observations on the use of surface coatings and varnishes in Egyptian art from the late 18th Dynasty'. In *The Materials, Technology and Art of Conservation – Studies in Honor of Lawrence J. Majewski*, R. A. Rushfield and M. W. Ballard (eds), 30–64. New York: Conservation Center of the Institute of Fine Arts.
Hatchfield, P. and Newman, R. 1991. 'Ancient Egyptian gilding methods'. In *Gilded Wood: Conservation and History*, D. Bigelow, E. Cornu, G. J. Landrey and C. van Horne (eds), 27–48. Madison, Conn.: Sound View Press.
Hill, M. and Schorsch, D. 1997. 'A bronze statuette of Thutmose III'. *Metropolitan Museum Journal* **32**: 5–18.
Horie, C.V. and Vint, J.A. 1982. 'Chalconatronite: A by-product of conservation?' *Studies in Conservation* **27**: 185–6.
Jaeschke, H. and Jaeschke, R. 1988. 'Early conservation techniques in the Petrie Museum'. In *Conservation of Ancient Egyptian Materials*, S. C. Watkins and C. E. Brown (eds), 17–24. London: UKIC Archaeology Section.
Jedrzejewska, H. 1976. 'A corroded Egyptian bronze: Cleaning and discoveries'. *Studies in Conservation* **21**: 101–14.
Kleiner, L. M. 1995. *Ancient Binding Media, Varnishes and Adhesives.* Rome: ICCROM.
Klemm, D. D. and Klemm, R. 1991. 'Mortar evolution in the old kingdom of Egypt'. In *Archaeometry '90*, E. Pernicka and G. A. Wagner (eds), 445–54. Basel: Birkhäuser Verlag Basel.
Kozloff, A. P., Bryan, B. M. and Berman, L. M. 1992. *Egypt's Dazzling Sun: Amenhotep III and His World.* Cleveland, Ohio: The Cleveland Museum of Art.
Kozloff, A. P. and Mitten, D. G. (eds.) 1990. *The Gods Delight: The Human Figure in Classical Bronzes*. Cleveland, Ohio: Cleveland Museum of Art.
La Niece, S. 1993. 'Silvering'. In *Metal Plating and Patination: Cultural, Technical and Historical Developments*, S. La Niece and P. Craddock (eds), 201–10. Oxford: Butterworth-Heineman.
Lucas, A and Harris, J. R. 1989. *Ancient Egyptian Materials and Industries*, 4th edn. [reprint 1962]. London: Histories and Mysteries of Man Ltd.
Meeks, N. 1993. 'Surface characterization of tinned bronze, high-tin bronze, tinned iron and arsenical bronze'. In *Metal Plating and Patination: Cultural, Technical and*

Historical Developments, S. La Niece and P. Craddock (eds), 247–75. Oxford: Butterworth-Heineman.
Miho Museum 1997. *Catalogue of the Miho Museum (The South Wing).* Shigaraki, Japan: The Miho Museum.
Murakami, R. 1993. 'Japanese traditional alloys'. In *Metal Plating and Patination: Cultural, Technical and Historical Developments*, S. La Niece and P. Craddock (eds), 85–94. Oxford: Butterworth-Heineman.
Newman, R. 1990. 'Technical examination of an ancient Egyptian royal pectoral'. *Journal of the Museum of Fine Arts, Boston* **2**: 19–29.
Norman, M. 1988. 'Early conservation techniques and the Ashmolean Museum'. In *Conservation of Ancient Egyptian Materials*, S. C. Watkins and C. E. Brown (eds), 7–16. London: UKIC Archaeology Section.
Notis, M. R. 1988. 'The Japanese alloy shakudo: Its history and its patination'. In *Metal Plating and Patination: Cultural, Technical and Historical Developments*, S. La Niece and P. Craddock (eds), 315–27. Oxford: Butterworth-Heineman.
Oddy, W. A. 1981 'Gilding through the ages: An outline history of the process in the old world'. *Gold Bulletin* **14**: 75–9.
Oddy, A. 1993. 'Gilding of metals in the old world'. In *Metal Plating and Patination: Cultural, Technical and Historical Developments*, S. La Niece and P. Craddock (eds), 171–81. Oxford: Butterworth-Heineman.
Oddy, W. A., La Niece, S., Curtis, J. E. and Meeks, N. D. 1981. 'Diffusion-bonding as a method of gilding in antiquity'. *MASCA Journal* **1**: 239–41.
Oddy, A., Pearce, P. and Green, L. 1988. An unusual gilding technique on some Egyptian bronzes'. In *Conservation of Ancient Egyptian Materials*, S. C. Watkins and C. C. Brown (eds), 35–40. London: UKIC Archaeology Section.
Oddy, W.A., Cowell, M. R., Craddock, P. T. and Hook, D. R. 1990. 'The gilding of bronze statues in the ancient world'. In *Small Bronze Sculpture from the Ancient World,* M. True and J. Podany (eds), 103–24. Malibu, Calif.: J. Paul Getty Museum.
Ogden, J. 1982. *Jewellery of the Ancient World.* London: Trefoil Books.
Ogden, J. 1993. 'Aesthetic and technical considerations regarding the colour and texture of ancient goldwork'. In *Metal Plating and Patination: Cultural, Technical and Historical Developments*, S. La Niece and P. Craddock (eds), 39–49. Oxford: Butterworth-Heineman.
Pearlstein, E. 1995. 'Conservation for the New Egyptian Galleries at the Brooklyn Museum'. In *Conservation in Ancient Egyptian Collections,* C. Brown, F. Macalister and M. Wright (eds), 93–102. London: Archetype Publications.
Photos, E., Jones, R. E. and Papadopolous, Th. 1994. 'The black inlay decoration on a Mycenean bronze dagger'. *Archaeometry* **36**: 267–75.
Shearman, F. 1988. 'An original decorated surface on an Egyptian bronze statuette'. In *Conservation of Ancient Egyptian Materials*, S. C. Watkins and C. C. Brown (eds), 29–34. London: UKIC Archaeology Section.
Shorsch, D. 1988. 'Technical examinations of ancient Egyptian theriomorphic hollow cast bronzes'. In *Conservation of Ancient Egyptian Materials*, S. C. Watkins and C. C. Brown (eds), 41–50. London: UKIC Archaeology Section.
Smith, C. S. 1973. 'An examination of the arsenic-rich coating on a bronze bull from Horoztepe'. In *Application of Science Examination of Works of Art*, W. J. Young (ed.), 96–102. Boston, Mass.: Museum of Fine Arts.
Turgoose, S. 1985. 'The corrosion of lead and tin: Before and after excavation'. In *Lead and Tin: Studies in Conservation and Technology*, occasional papers no. 3, 15–26. London: UKIC.
Wayman, M. and Craddock, P. 1993. 'Wu Tong, a neglected Chinese decorative technology'. In *Metal Plating and Patination: Cultural, Technical and Historical Developments*, S. La Niece and P. Craddock (eds), 128–34. Oxford: Butterworth-Heineman.
Weisser, T. D. 1975. 'The de-alloying of copper alloys'. In *Conservation in Archaeology and the Applied Arts*, D. Leigh, A. Moncrieff, W. A. Oddy and P. Pratt (eds), 207–14. London: IIC.
Young, E. 1959. 'Notes on a hitherto unknown technique in Egyptian bronze-working'. *Journal of Egyptian Archaeology* **XLV**: 104–6.

4 Gilded Silver from Ancient Nubia

Susanne Gänsicke and Richard Newman

Abstract

Gilded silver from ancient Nubia, Sudan, in the Collection of the Museum of Fine Arts, Boston (BMFA), was examined to identify gilding materials and techniques. The material dates primarily from the Twenty-Fifth Dynasty and Napatan Period (*c.* 750–300BC) to Meroitic times (300BC–AD350) with a few earlier pieces. The objects originate from the Harvard University–Museum of Fine Arts expeditions in the Sudan from the early part of this century and, thus, form an important body of provenanced ancient gilded silver. The first part of the article summarizes the use of gold, silver, and gilding in the Nile Valley, and ancient textual references to gilding. The second part discusses the technical examination of the objects and the results of analysis. Analytical techniques employed included reflected light microscopy, electron probe microanalysis, and metallographic studies on cross sections. Non-destructive energy-dispersive X-ray fluorescence was carried out on a few objects that could not be sampled. Diffusion bonding was apparent on most pieces with varying composition and thickness of gilding layers and silver substrates. Gilding of silver in the ancient Sudan and perhaps the entire Nile Valley appears to be much more widespread and elaborate than previously assumed.

Introduction

Nubia is the name given to all the Nile River borderlands extending between Aswan in Egypt and Khartoum in the Sudan. Lower Nubia, which stretches from Aswan to the Second Cataract, has been submerged since 1965 by the waters of Lake Nasser; and Upper Nubia, extending south from Wadi Halfa, has been similarly flooded in its extreme northern part (Fig. 4.1).

The ancient history of this region was virtually unknown until George A. Reisner, leading the Harvard University–Museum of Fine Arts Expedition, set out to excavate in this remote region between 1907 and 1932. As a result of these almost annual excavations the Museum of Fine Arts, Boston, houses one of the most significant collections of Nubian art and archaeological material in the world (Reisner 1923a,b; Dunham 1950, 1955, 1957, 1963; Kendall 1982).

Luxury goods such as ebony, ivory, ostrich feathers, leopard skins and gold formed the riches of Africa and were eagerly sought in trade from

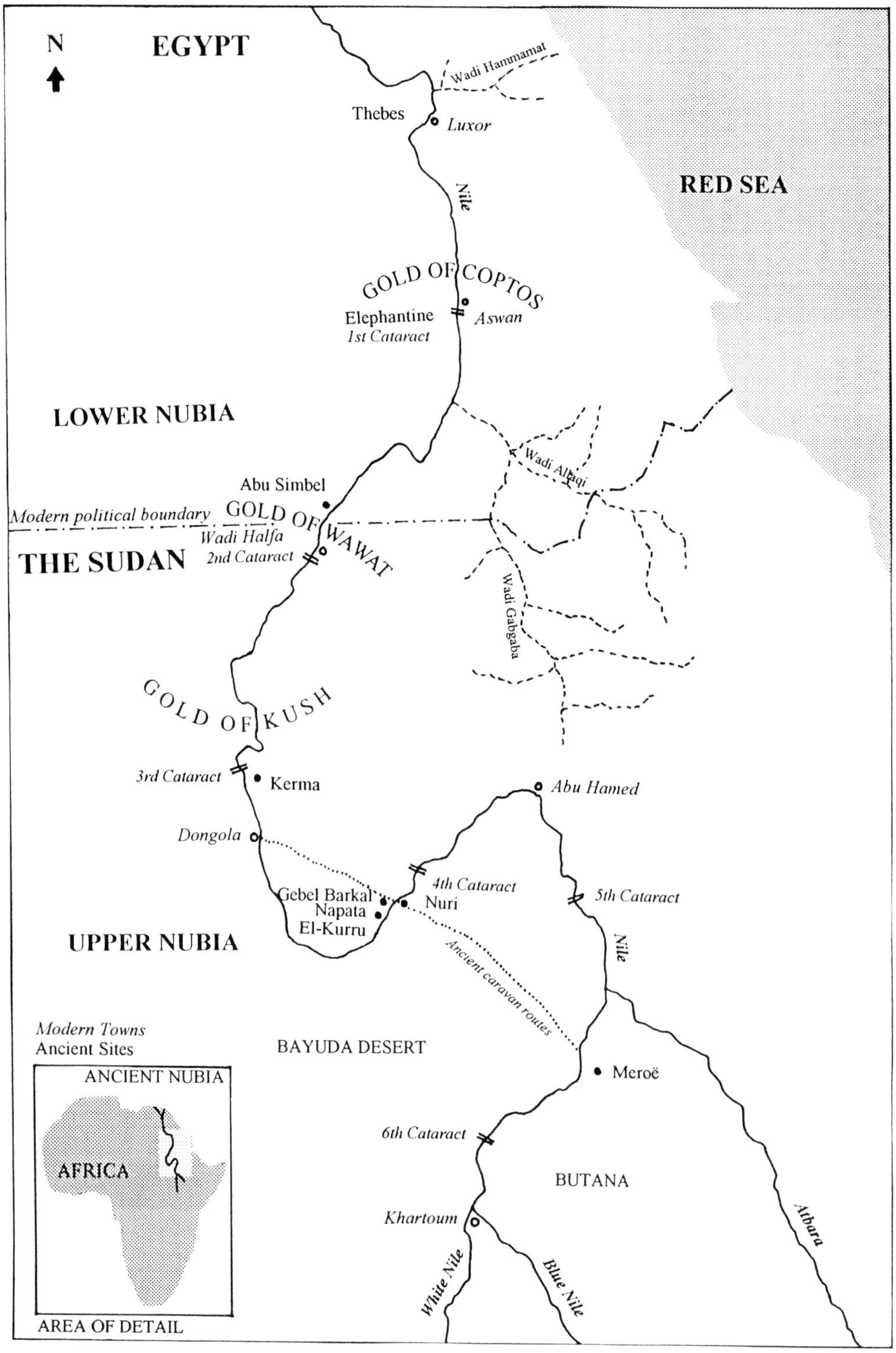

Figure 4.1 Map of Nubia: principal ancient sites and gold fields. (Graphics: Susanne Gänsicke.)

Nubia by the Egyptians. The first written reports of interaction with the southern peoples come from Egyptian expeditions of the Old Kingdom. When trade failed, Egypt pursued contact with Nubia by force, and the region was subjected to numerous Egyptian invasions over the millennia. One result of these campaigns was the Egyptian fortification of the Second Cataract during the Middle Kingdom (*c*.2000–1700BC). By controlling the important trade routes between the interior of Africa and the north, the peoples of Nubia and Egypt alternately gained great power and wealth.

Kerma in Upper Nubia near the Third Cataract was the capital of a highly developed state by the early second millennium BC. It was known as Kush by the ancient Egyptians, a name it also carries in the Old Testament. During the New Kingdom in the late sixteenth century BC this realm was conquered by the Egyptians, who established their frontier near the Fourth Cataract beside a small mountain today called Gebel Barkal, which is also the site of an important temple complex. In the eighth century BC a native Nubian dynasty rose to power here; the town was called Napata. Eventually this dynasty seized control over the entire Nile Valley, where its kings presided over Egypt as true pharaohs and were counted as Egypt's Twenty-Fifth Dynasty (716–661BC). Seventy years later the Nubian kings were forced by the Assyrian armies to withdraw once again to Nubia. In Napata and, later, at Meroë, the kings of Kush continued to rule over an empire in the Sudan until about the fourth century AD.

Solid gold and silver objects have been found in all the royal cemeteries of Kush (2500BC to AD400) at the sites of Kerma, El Kurru, Nuri, Gebel Barkal and Meroë.[1] Many objects from each cemetery are actually of gilded silver. The greatest numbers of gilded objects date to the Twenty-Fifth Dynasty, and the lavish use of gold leaf was remarked upon by Reisner, excavator of the royal cemeteries of this period. He thought that in some tombs perhaps all burial items (vessels, stone figurines and the like) had been wrapped in gold foil, much of which was recovered from the soil (Kendall 1982: 39). It is uncertain how representative these objects are of the original wealth of forms and materials because the Kushite cemeteries were plundered repeatedly in antiquity. Whether all of the precious metal objects found in the Nubian tombs were locally produced is not known. It is certainly possible that some were imported from Egypt or the Near East.

The study of these objects is important for several reasons. The few known gilded silver objects from Egypt and the Sudan have never been studied systematically. No such large number of gilded silver objects from the Nile Valley has been recognized previously, and gilding of silver in this area has not been thought a widely used technique. The frequent use of this technique in Nubia, the famous source of gold in pharaonic times, is unexpected. Is this a local development, or does it indeed indicate a rather widespread use of gilding in the entire Nile Valley? Does it reflect merely an economical use of gold on a baser material, or did it have some ritual significance?

In this chapter, we will examine the various gilding methods and materials exhibited by a group of these Nubian objects dating from the mid-second millennium BC to the first century AD, and we will compare them with a small group of gilded silver objects from Egypt. Although the corpus of objects is limited, we will examine different gilding methods, and the composition of gilding and silver alloys.

Gold in Egypt and Nubia

The extensive gold fields in Lower Nubia played an important role in the political and territorial struggles between Egypt and Kush. Nubia was one main source of Egypt's gold, the other being mines in Egypt's eastern desert (Vercoutter 1959). In many ancient Egyptian tomb paintings or temple reliefs Nubian offering bearers are depicted bringing quantities of gold in ring ingots or as dust in bags as well as other goods. Three major gold fields were identified by the Egyptians (Fig. 4.1): 'Gold of Coptos' in

Egypt, 'Gold of Wawat' in Lower Nubia and 'Gold of Kush' in Upper Nubia. The gold occurred naturally in two forms: in secondary deposits, found in alluvial sands in dry river beds or wadis; and as reef gold in gold-bearing veins in quartz rock. In these three areas more than one hundred ancient mining sites have been identified. This number is likely to increase as a consequence of recent increases in archaeological prospecting in the almost impassable eastern deserts of Egypt and Nubia. Grinding wheels, used to break the gold-bearing rock, and fragments of pottery jars used by the miners, give evidence of ancient workings as well as the quantities of rocks processed (Klemm and Klemm 1989, 1994; Sadr *et al.* 1994; Castiglioni and Castiglioni 1994; Castiglioni *et al.* 1995). It was recently estimated that in one mine in the eastern desert of Nubia, a total of 100,000–150,000 tons of stone was moved for the extraction of gold in antiquity (Gaskell 1985).

Ancient texts provide no significant information about working techniques or actual locations. A rare exception is the Turin Papyrus, dating to the twelfth century BC, which includes a map of a mine at Wadi Hammamat in the eastern Egyptian desert (Harrell and Brown 1992). More abundant are royal offering lists of the large quantities of precious metals presented by the pharaohs to the gods. During the time of Thutmosis III (c.1490–1436BC), for example, the yearly production of the Kush and Wawat mines is estimated to have been about 260 kg.

Native gold can be quite pure, but it also typically contains some silver, and a given mine can supply gold with variable silver content (Craddock *et al.* 1998). Such a natural alloy of gold and silver is harder than pure gold, withstands wear much better, and therefore could have been desirable. Consequently there might not have been an immediate need to purify the native silver-containing gold. Until today no firm date has been established for the introduction of gold refining in the ancient world, although it has been repeatedly suggested that it was the fifth century BC, when the production of accurate and constant alloys became necessary with the introduction of coinage (Forbes 1964: 167–77; Notton 1974; Craddock *et al.* 1998). Among the ancient writings on the subject of gold refining little is available before Greek times. One such text of Agatharchides from the second century BC describes the cementation process by which gold is purified with the use of salts in crucibles (Notton 1974: 52).

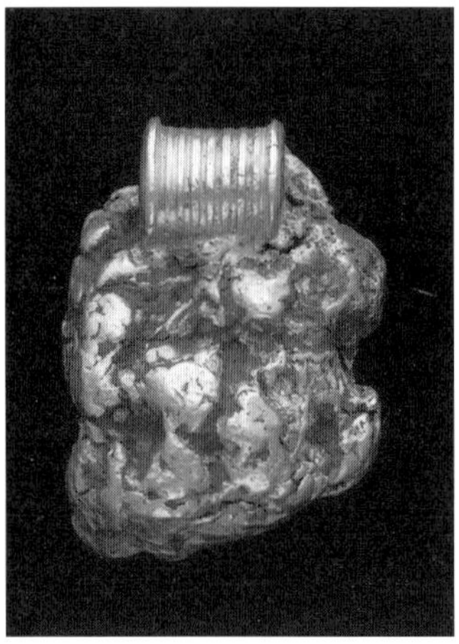

Figure 4.2 Nugget (seventh to sixth century BC), from Meroë, Tomb Beg. W. 859. Gold with attached loop, H: 3.2 cm, W: 2.3 cm, L: 2.2 cm. (Courtesy, Museum of Fine Arts, Boston. 23.311.)

The Nubian objects we have studied date before and after the period when refining is assumed to have been introduced, so it is possible that both refined and unrefined gold were used in the gilding of objects examined here. Virtually no analyses of native gold alloys from Egypt and Nubia are available; therefore the range of compositions of unrefined gold is not known. In the Nubian cemeteries of the seventh and sixth centuries BC natural gold nuggets have been found; these were adapted as jewellery by attached loops or drilled holes to allow their suspension (Fig. 4.2).[2] Since they very likely derived from within the Nubian gold mines, there is good reason to believe that they can shed some light on the compositions of Nubian alluvial gold. Our microprobe analysis of one nugget (BMFA 23.311) showed 91.2% gold and 8.8% silver, with no detectable copper.

Gilding in the ancient Nile Valley

In Egypt and Nubia gold was employed extensively from early times in the fabrication of jewellery and other decorative and religious items. It was also used frequently to cover a variety of substrates both for aesthetic and mythological reasons. The use of thin leaf or foil is probably almost as old as the use of gold itself.

Textual information

The ancient Egyptian language had a large vocabulary for gold and matters related to gold work (Forbes 1964: 179). Egyptian tomb scenes from the Old through the New Kingdoms frequently depict metal craftsmen, including scenes involving the beating and working of gold leaf (Scheel 1985, 1986, 1987). Unfortunately, such representations are not known from Nubia, but the work processes were probably similar. Gilding scenes are only known from two monuments: one in the 'Tomb of the Two Brothers' at Saqqara, dating to the Old Kingdom (about 2300BC) (Moussa and Altenmüller 1977: 135–7, pl. 62–4). The other one exists on the north wall of the main chamber in the tomb of Baqt III (Tomb 15 at Beni Hasan), an official of the eleventh Dynasty (2134–2061BC) (Newberry 1894: pl. 4). Both depict workmen heating gold in a crucible and applying gold leaf to funeral items such as a wooden sceptre and staff, and a knot-shaped girdle buckle. The accompanying hieroglyphic inscription in the tomb of Baqt III suggests that the belt buckle to be gilded was made from metal, possibly copper (Drenkhahn 1976: 34–5; Jüngst 1982). There also seems to have been a distinction between workmen who gilded organic materials and those who worked on metal (Scheel 1987: 202–3), but it remains uncertain how the gold was applied to the metal. The south wall in the tomb of the New Kingdom vizier Rekh-mi-Re depicts carpenters and metalsmiths at work (Garis Davies 1944: 52, pl. 55). One kneeling man appears to be occupied with the beating of gold; but his activity was previously described as plating gold on silver (Vernier 1925: 168).

Rare textual references to gilded silver can be found in the Egyptian royal correspondence known as the 'Tell el-Amarna Letters', where one item is described as 'silver, covered with gold' (Knudtzon 1907: 115). The funerary papyrus of Ramses III (Papyrus Harris I) also identifies one gift as 'silver overlaid with gold, a sacred eye for Thoth' among many hundreds of gifts of solid gold and silver (in translation by Breasted 1906: 186, 193). In the donation stela of Taharqa at Kawa both 'silver, – wrought in gold' and '2 silver, *šw*-vessel with gold rim' are mentioned (Eide *et al.* 1994: 169). Actual working directions on how to gild baser metals, mostly copper and copper alloys, can be found in much later texts. The Leyden Papyrus X, dating to the fourth century AD but also incorporating recipes from earlier centuries, mentions a gilding technique that involves the use of heat and lead to secure gold leaf. Pliny the Elder describes gilding techniques of copper and silver involving the use of mercury (Rackham 1961: 20, 32, 42; Hunt 1974).

General gilding techniques

All of the known techniques of covering base materials with gold foil described by Oddy can be seen in the Sudan and Egypt (Oddy 1993). Thicker gold foil was attached by riveting or cramping, as on small copper earrings or hair rings from Kerma (exp. no. 14-2-964), for example, (Reisner 1923b: 281, pl. 44.1). Furniture elements were lavishly encased with gold sheet, as indicated by a surviving golden case of a wooden bovine bed leg (Fig. 4.3); pierced holes along its edges suggest that nails were used to secure it (Reisner 1923b, 283, pl. 69. 2.). Even the tip of a sandstone pinnacle at Gebel Barkal, more than 90 metres above the ground, was covered with a monumental, inscribed gold sheet. The gold has long disappeared and today only copper alloy nails used to fasten it to the stone remain *in situ* (Kendall 1994). That also architectural elements of temples were once gilded is known today only from inscriptions (Eide *et al.* 1994: 172). Due to its fragility gold leaf (a material defined as being too thin to support its own weight) had to be secured to the underlying material by other means. In Egypt uneven surfaces were often smoothed with gesso and the leaf was attached with an adhesive (Nicholson 1979; Hatchfield and Newman 1991). The gilding of bronzes with the use of gesso has been observed on Egyptian objects (Oddy *et al.* 1988). Occasionally also Egyptian jewellery exhibits gilding with the use of gesso. Such fragile surfaces without metallic bonding were mostly prepared for burial purposes. Two such examples in Boston are a gilded copper diadem, inlaid with cornelian, from the Old Kingdom (Dunham 1946), and a gilded silver girdle fragment, also inlaid with various stones, of Queen Nefertari from the New Kingdom (Hatchfield and Newman 1991: 39; Markowitz *et al.* 1997).[3] Nubian bronze objects gilded in the same manner include inlay eyes from royal coffins in the royal tombs of the seventh century BC of Nuri (Dunham 1955: 79, 141, 162), uraeus-shaped attachments to headdresses (Dunham 1957: 60, Fig. 32), and at least on one known statue of an early Meroitic king of the third century BC from Kawa, now in Khartoum (Maystre 1986).[4]

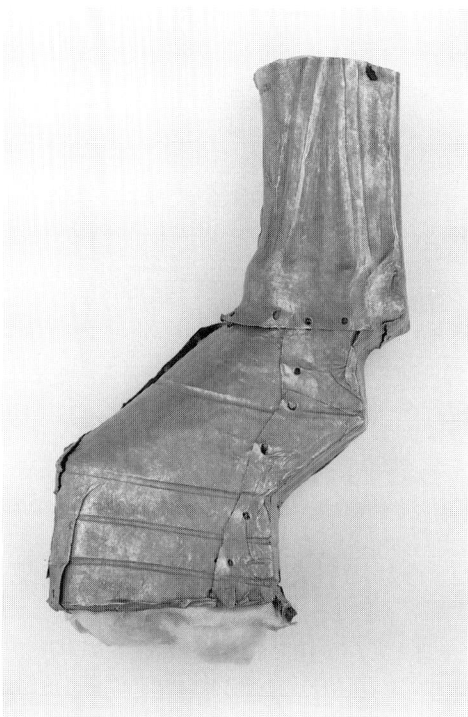

Figure 4.3 Casing from a wooden bed leg (1650–1550BC), from Kerma, Tumulus KX, Tomb K1035. Gold, H: 14.7 cm, W: 8.5 cm. (Courtesy, Museum of Fine Arts, Boston. 13.3987.)

Gilded silver in Egypt

Silver, originally called 'white gold' by the Egyptians in the Old Kingdom but later simply referred to as 'white', occurs much less frequently than gold in archaeological contexts in Egypt. The metal was more highly valued than gold until the Middle Kingdom (2061–1784BC), probably due to the lack of silver sources in Egypt and Nubia (Harris 1961: 42–4; Fuchs 1983). The precise geographical origin of the silver used in the Nile Valley is still unknown (Lucas 1928; Mishara and Meyers 1974; Stos-Fertner and Gale 1979; Gale and Stos-Gale 1981; Stos-Gale and Gale 1981). In the late New Kingdom silver seems to have been received as tribute and by trade, as revealed, for example, by numerous depictions in private tombs of Aegeans and Syrians carrying silver vessels (Garis Davies 1944: 17–30, pl. XVI–XXIII). Textual references indicate the extensive use of silver, although surprisingly few pieces have survived (Settgast 1986; Born 1986; Becker *et al.* 1994). Most prominent are the treasures from Tanis (Montet 1959). The Egyptians sometimes associated gold with the sun, the skin of the gods and eternity (Andrews 1991: 52; Vercoutter 1994: 403–10), while they identified silver with the gods' bones, the moon and purity (Hornung 1982: 37, line 5; Andrews (1991: 56).

Probably also reflecting greater availability of silver in later periods is the increase of gilding on electrum or silver substrates; but occasional examples of such gilding can be observed throughout history (Möller 1924: 32–5; Williams 1924; Vernier 1927). Few of these have been examined to date. Among the earliest gilded silver examples mentioned in the literature are tubular bracelets from Deir el-Bahri of the First Intermediate Period and the eleventh Dynasty (Andrews 1981: 60, pl. 29). Recent analysis, however, suggest that they may not have been gilded.[5] Several New Kingdom hoop earrings have been described as gilded (Williams 1924: 117–18, pl. XV a–b; Vernier 1927: pl. 29).[6] A winged vulture inlaid with glass and semi-precious stones is perhaps only one of a few elaborate pectorals with cloisons of gilded silver (Lacovara 1990; Newman 1990; Liliquist 1993).[7] A second jewel, previously described as gilded, was found on a mummified bull and carries the cartouche of Ramses II (Wilkinson 1971: 159, pl. LXI A). Recently it was published as having a gold-rich silver alloy (Jodicio 1995: 23).[8] Other examples of gilded silver are an inlaid funerary diadem of the New Kingdom (Wilkinson 1971: 113, pl. 38; Raven 1988), and an amulet of Queen Khama, 23rd Dynasty, from Tanis, on which two standing goddesses frame a large lapis inlay of a seated ram-headed god (Lucas 1924; Andrews 1991: 95), and a wooden hairpin with ram-headed ornament (Schäfer 1910: 68, pl. 11).[9] In a recent study of a Late Period silver mummy mask three thin layers of gold could be identified, but mercury was absent (Conrad 1993).[10]

Gold-foil embellishment is also known on larger silver statues. A monumental Horus was gilded by securing the gold foil in cut grooves (Roehrig 1996). A small solid-silver statue of Osiris of Dynasty Twenty-Five or Twenty-Six is entirely encased in gold (Eggebrecht 1990: 81).[11] The same form of mechanical gilding is visible on animal-shaped handles of a silver jug from Bubastis of the nineteenth Dynasty (Vernier 1927: 416, pl. VV).[12] Electrum coatings have been reported on large, solid silver pendants (Garside 1980: 37–9).[13] Other statues of Amun Re in London and of a king in Paris show partially gilded elements as skirts or collars, in which the gilding appears to have been attached with adhesive.[14] Finally, there is the large Ptolemaic temple treasure from Dendera in Cairo with extensive use of gilded silver for cult statues and sistra (Cauville 1987).

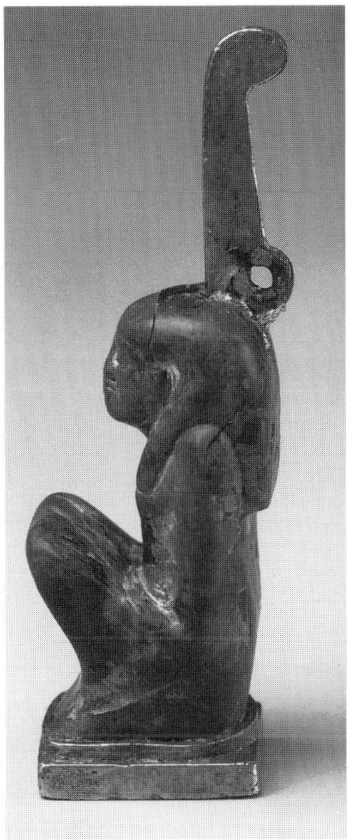

Figure 4.4 Maat figure amulet (*c.*716–702BC), from El Kurru, Tomb Ku. 55. Malachite, gilded silver, H: 6.0 cm, W: 2.0 cm. (Courtesy, Museum of Fine Arts, Boston. 21.323.) (See Plate 13.)

Gilded silver from Nubia at the Museum of Fine Arts

The earliest Nubian gilded silver we examined was a bracelet from Kerma, found in a grave dating to the second millennium BC (Reisner 1923b:

282–3, pl. 44.1), and embossed silver sheet fragments (Reisner 1923b: 285).[15]

Although the necropolis of the twenty-fifth Dynasty rulers of Egypt at El Kurru (850–650BC) was badly plundered in antiquity, a small number of very delicate jewels survived from the tomb of an unidentified queen of Piye (or Piankhy), the conqueror of Egypt (c.747–716BC), including a small malachite figure of the goddess Maat with metal attachments (Fig. 4.4) and a group of hollow-worked gilded ram-headed amulets (Fig. 4.5) (Dunham 1950: 93–4, Fig. 31f, pl. 66A, B, C) to which a remarkable parallel can be found in the pendant of Osorkon II, Twenty-Second Dynasty (Aldred 1971: 240, pl. 141).[16] These objects are believed to have been brought to Nubia as a hoard from Egypt (Markowitz 1993; Kendall 1998). The heavy bronze mirror (Fig. 4.6) of King Shabaqo (*c.*716–704BC) is supported by a gilded, solid silver handle decorated with four standing goddesses (Dunham 1950: 7, pl. 62). Another group of objects from El Kurru have their origin in the tombs of royal horses, among them gilded Hathor-headed trappings (Dunham 1950: 112, pl. 69B) and a plume holder (Fig. 4.7) in the form of the sun god Re (Dunham 1950: 115, Fig. 41b).[17]

The royal cemetery of Nuri lies just across the river from Gebel Barkal and Napata, once the Kushite capital from the ninth to the fourth century BC. From some kings' tombs of the seventh and sixth centuries originate approximately twenty cylindrical objects (Figs. 4.8 and 4.9) whose purpose remains unclear (Dunham 1955: pl. 93–113; Wildung 1996: 226–30).[18] One was of gold; all the others were gilded silver with thick gold foil

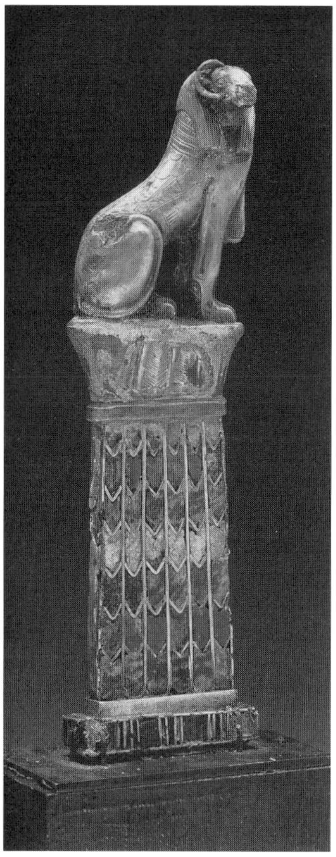

Figure 4.5 Ram-headed sphinx amulet on inlaid column (*c.*716–702BC), from El Kurru, Tomb Ku. 55. Gilded silver, inlays in lapis, red and turquoise glass, H: 10.3 cm, W: 3.0 cm, D: 1.5 cm. (Courtesy, Museum of Fine Arts, Boston. 24.972.) (See Plate 14.)

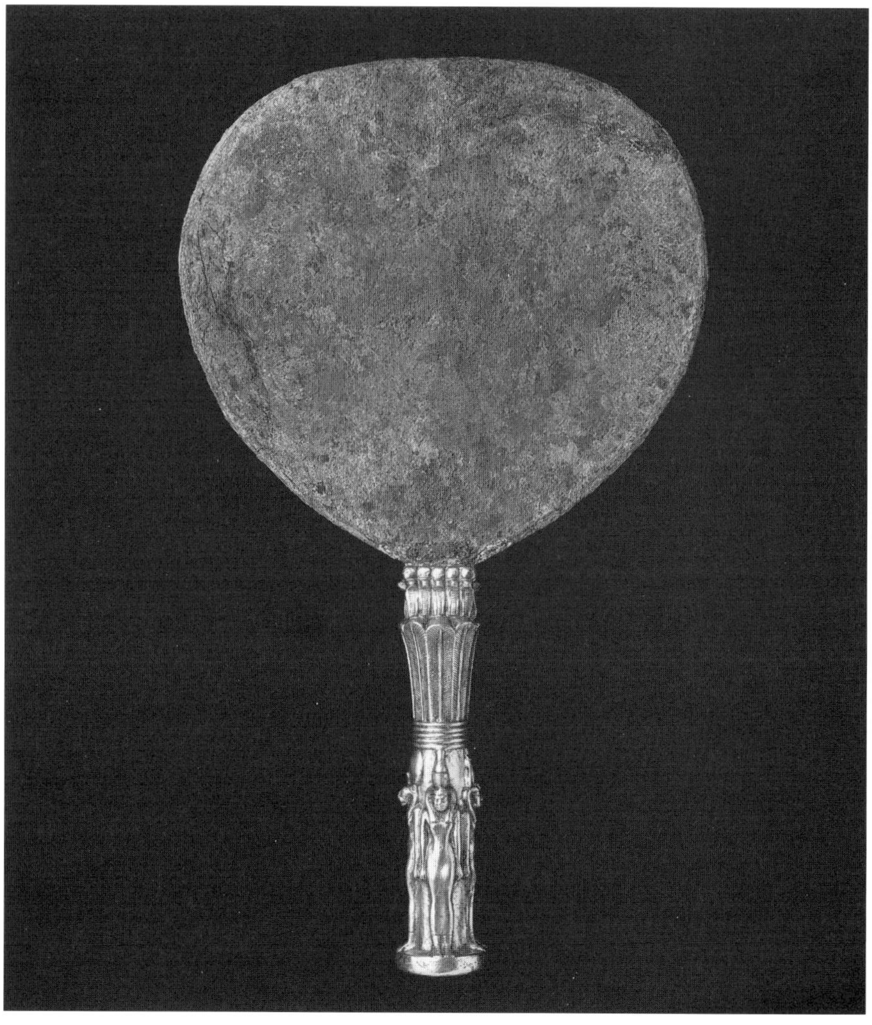

Figure 4.6 Mirror of King Shabako (*c.*716–702BC), from El Kurru, Tomb Ku. 15. Bronze, gilded silver, H: 33.0 cm, W: 20.0 cm, D: 5.0 cm. (Courtesy, Museum of Fine Arts, Boston. 21.318.) (See Plate 15.)

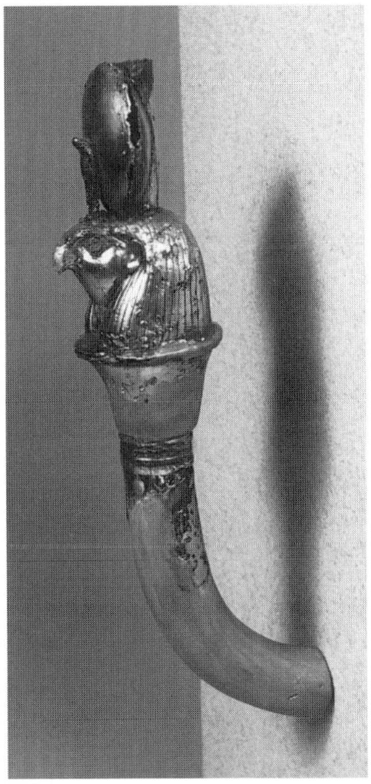

Figure 4.7 Plume holder from the burial of one of King Tanwetamani's horses (*c*.664–653BC), from El Kurru, Tomb Ku. 219. Gilded silver, H: 14.5 cm, W: 3.5 cm, D: 5.0 cm. Museum of Fine Arts, Boston. 23.877. (Photograph: Susanne Gänsicke.) (See Plate 16.)

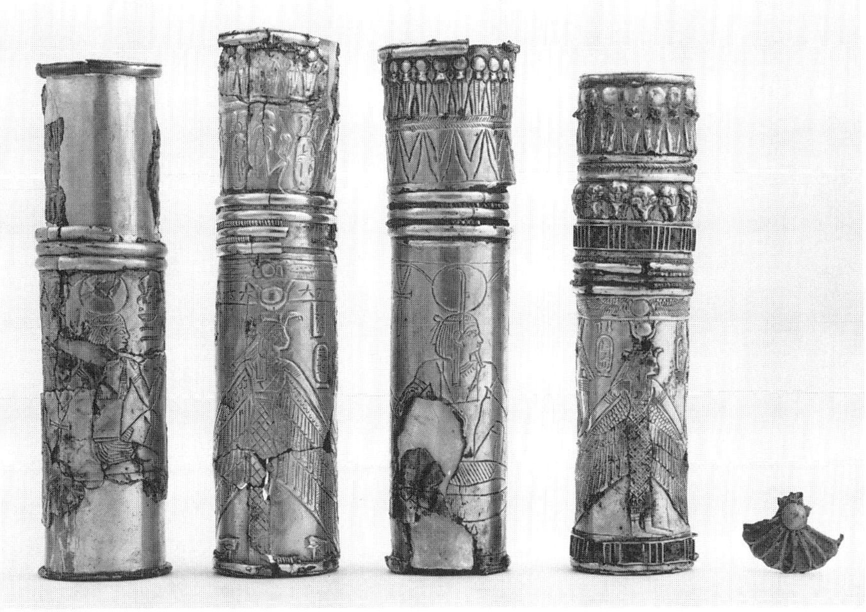

Figure 4.8 Four cylinder sheaths (*c*.643–513BC), from Nuri, Pyramids Nu. 3, Nu. 27, Nu. 26, Nu. 8. Gilded silver. (Courtesy, Museum of Fine Arts, Boston.) 21.11742, 21.11746, 21.11745, 20.275 (from left to right).

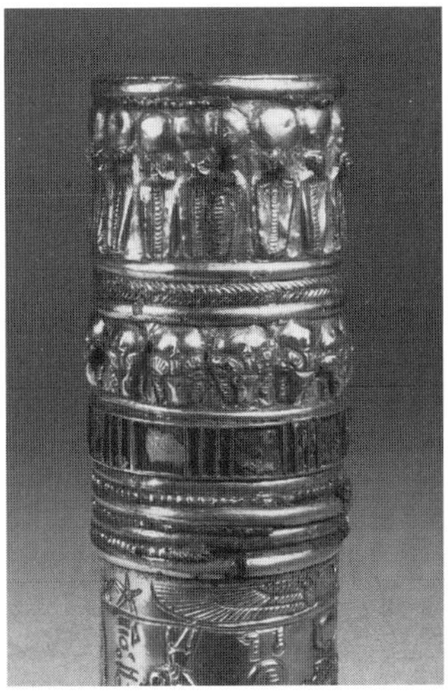

Figure 4.9 Detail of a cylinder from Nuri, Pyramid Nu. 10. Gilded silver, cloisons filled with paste. 20.275. (Photograph: Susanne Gänsicke.) (See Plate 17.)

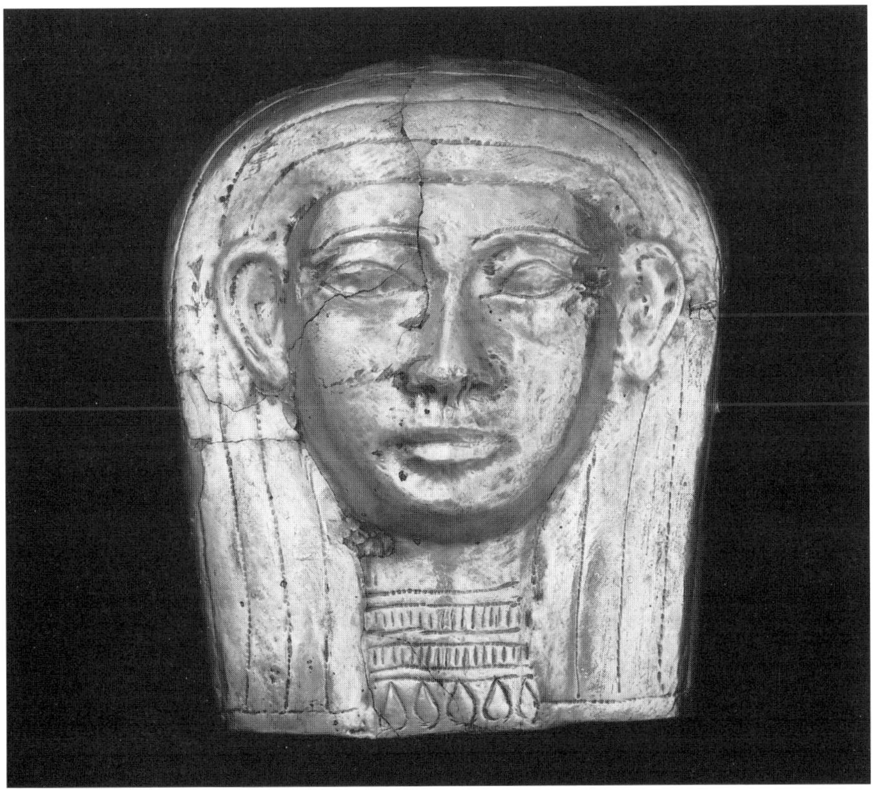

Figure 4.10 Mummy mask of Queen Malaqaye (*c*.664–653BC), Nuri, Pyramid Nu. 59. Gilded silver, H: 13.9 cm, W: 11.5 cm, D: 1.1 cm. (Courtesy, Museum of Fine Arts, Boston. 20.1059.) (See Plate 18.)

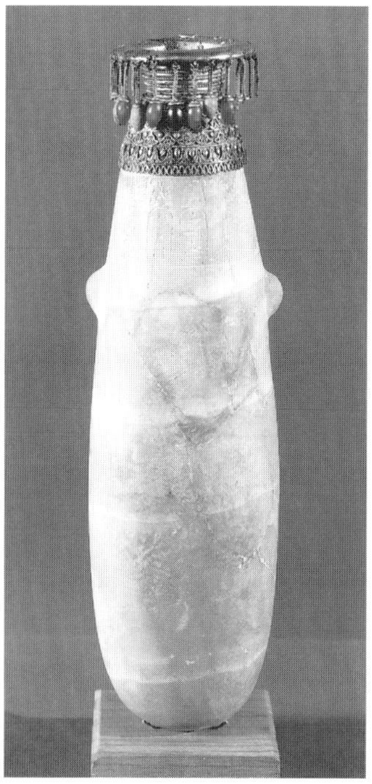

Figure 4.11 Alabastron of King Aspelta (*c*.600–580BC), Nuri, Pyramid Nu. 8. Gilded silver, calcite (Egyptian alabaster), cornelian, feldspar, magnetite, Egyptian blue, H: 25.2 cm, W: 7.2. (Courtesy, Museum of Fine Arts, Boston. 20.1070.)

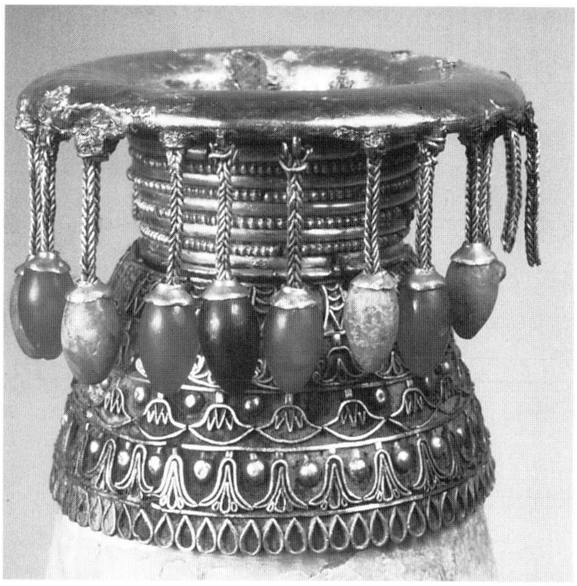

Figure 4.12 Alabastron of King Aspelta, detail. (Photograph: Susanne Gänsicke.) (See Plate 19.)

surfaces that have been embellished by chased religious designs. Other gilded pieces from Nuri include mummy trappings (Fig. 4.10), and numerous small jewellery fragments, beads and amulets (Dunham 1955: Fig. 14, pl. 120B: centre [20.1059, mummy mask]; 219, Fig. 169, pl. 113B [for 20.299 buttons, 20.300, rosettes]; 229, pl. 64A [for trapping 20.343]). A calcite perfume bottle (Figs 4.11 and 4.12) shows a bejewelled rim with intricate gilded silver cloisonné (Gänsicke 1994).

Three cemeteries at Meroë, the latest and southernmost capital of Kush, span a period from about 800BC to AD350. Here too are found gilded mummy trappings (Fig. 4.13), numerous small gilded jewels such as rings, and horse's harnesses (Fig. 4.14a,b),[19] and an imported Roman goblet shows traces of gilding (Dunham 1957: 105, pl. 53).[20]

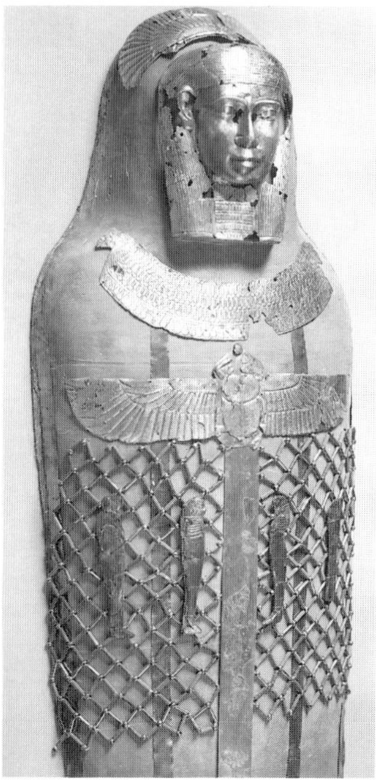

Figure 4.13 Mummy trappings of Queen Mernua (sixth century BC), from Meroë, Tomb Beg. S. 85. Gilded silver, silver, electrum, faience. (Courtesy, Museum of Fine Arts, Boston. 23.1396.)

a

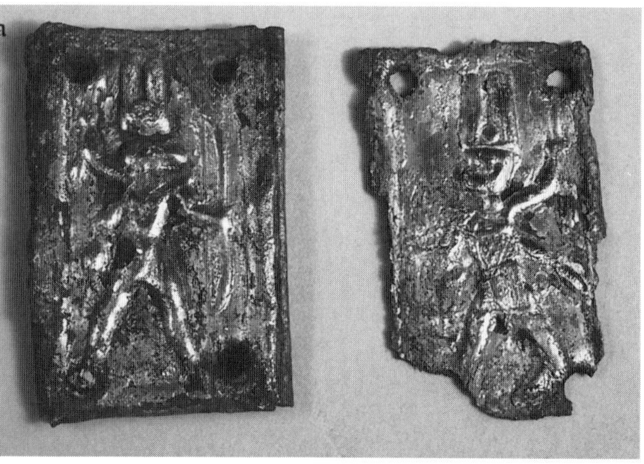

b

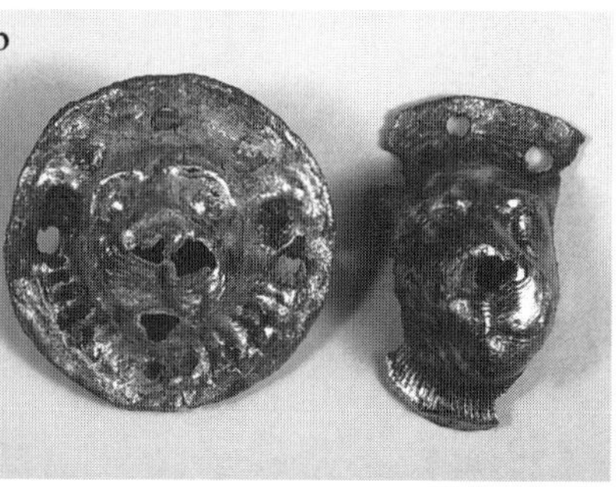

Figure 4.14a, b Plaques from a horse's harness of King Aryesbekhe (mid first century AD), from Meroë, Tomb Beg. N. 16. Gilded silver, bronze, iron. Museum of Fine Arts, Boston. Exp. no. 21-3-694d (14A); 21-3-694b,g (14B). (Photograph: Susanne Gänsicke.) (See Plate 20.)

Technical examination

Examination procedures

Twenty-six objects were sampled and forty-seven cross-sections prepared (Table 4.1). All of the objects were initially examined under a low-power stereobinocular microscope. Samples were examined by reflected light microscopy. Most were analysed in an electron beam microprobe, and some were etched for metallographic examination (see Appendix 1 for analytical details).[21] Three objects that could not be sampled were qualitatively analysed by X-ray fluorescence only.

Gilding procedures

Two gilding techniques were observed. On two groups of objects gold leaf was attached mechanically. Presumably an adhesive was utilized, although none is now visible. The first, an early group of tubular beads (Table 4.1, no. 2), consists of smooth cylinders of silver over which three layers of gold leaf (each *c*.6 μm) were applied, while the second, a much later group of plaques from horses' harnesses (Table 4.1, no. 28), is formed by plates stamped with low-relief designs over which a single layer of gold was applied (4–7 μm). The configurations of both groups of objects is such that gilding could easily be carried out after shaping.

All of the other objects examined were gilded by diffusion bonding. This procedure was the most common gilding technique for metallic objects in the ancient world before the invention of mercury amalgam gilding around the second century BC (Lechtman 1971; Lins and Oddy 1975; Oddy *et al.* 1979; Eluère 1989: 189).[22] The extensive use of diffusion gilding on these objects is not unexpected: only two date to a period later than this (ie, the plaques already mentioned, on which gilding was mechanically attached, and a ring (Table 4.1, no. 29), which was diffusion-gilded).

In diffusion bonding a layer or layers of leaf are attached to the metal substrate by hammering or perhaps with the aid of an adhesive. The gilded object is then heated, which causes diffusion across the gilding/substrate interface. If extensive enough diffusion takes place, a very strong bond is created. Unlike a mechanically gilded surface, a diffusion-gilded surface can be further worked and shaped without danger of loss of the gilding. Our criterion for identifying this technique was detection of a compositional gradient across the gilding layer in the cross-sections. Gradients such as those shown in Figure 4.15 (Table 4.1, no. 20) and in Figure 4.16 are typical on objects whose silver substrates are uncorroded. A gradient can still be seen in the gilding cross-sections of corroded samples, with the gold content decreasing and silver content increasing near the bottom of

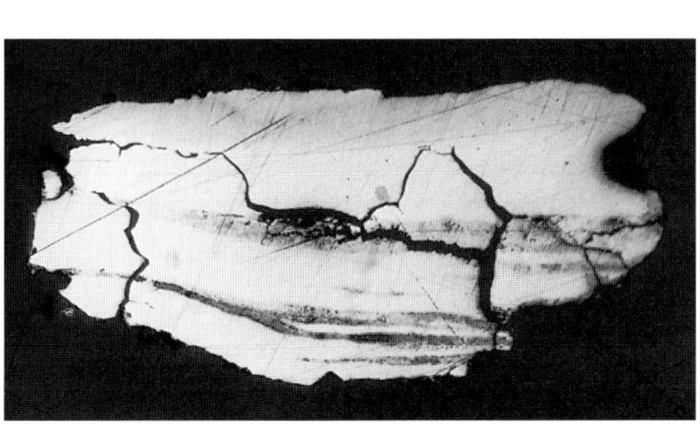

Figure 4.15 Sample of the cylinder sheath 20.275, Nuri, Pyramid Nu. 10. Thickness of gilding layer and diffusion zone is 60 μm, width of field 1.06 mm. (Photograph: Susanne Gänsicke.)

Table 4.1 List of objects examined and summary of microprobe results.

No.	Object (MFA number)	Provenance	Date	Gilding thickness[1]	Layer	Gilding composition[2] Gold	Silver	Copper	Substrate composition[3] Silver	Copper	Gold	Lead
1	Bracelets (20.1671)	Kerma X, B 49	1600BC	Sample 1: no gilding visible					92.1	4.9	2.6	0.4
				Sample 2: only small specks of gilding present		70.3	28.6	1.1	92.0	4.4	3.2	0.4
2	Gilded tubular silver	Kurru 15 (King Shabaqo)	716–702BC	3 layers, 18.2 µm total	1	98.0	1.8	0.2				
	beads (24.911, 24.912)#				2	98.5	1.5	< 0.08				
					3	97.8	2.2	<0.08				
3	Bronze mirror with gilded silver handle (21.318)*	Kurru 15 (King Shabaqo)	716–702BC	Not determined								
4	Maat figure, malachite with gilded elements (21.323)	Kurru 55 (unidentified queen)	716–702BC	3 layers, 1–2 µm each	1	91.1	8.4	0.5	97.8	1.9	0.3	< 0.09
					2	95.8	2.6	1.6				
					3	95.5	3.9	0.6				
5	Amulet, ram on inlaid column (24.972)	Kurru 55 (unidentified queen)	716–702BC	Sample 1: 4–8 µm		62.6	35.0	2.4	99.3	0.5	0.2	< 0.09
				Sample 2: 4–7 µm		62.3	36.0	1.7				
6	Amulet, ram on stone column (24.974)	Kurru 55 (unidentified queen)	716–702BC	8 µm		52.1	45.6	2.3	98.4	0.8	0.8	< 0.09
7	Amulet, ram on crystal column (24.976)	Kurru 55 (unidentified queen)	716–702BC	16 µm		54.1	44.0	1.9	92.8	2.1	5.0	~0.09
8	Hathor heads, horse's harness (23.1477)	Kurru 203 (King Shabaqo)	716–702BC	1–2 µm		96.9	2.7	0.4				
9	Plume holder, horse's harness (23.877)	Kurru 219 (King Tanwetamani)	664–653BC	Sample 1: 13–21 µm		80.7	18.2	1.1				
				Sample 2: 33–66 µm (interference from silver solder?)		41.4	45.2	11.4	97.9	1.8	0.3	< 0.09
10	Miscellaneous fragments, possibly cylinder sheaths (20.399 a + b)	Nuri 8(King Aspelta)	600–580BC	Sample 1: 20 µm		89.3	10.7	< 0.08	91.1	1.0	7.9	< 0.09
				Sample 2: 8 µm		78.8	16.9	0.6	96.9	1.2	1.9	< 0.09
				Sample 3: 5–8 µm		85.4	14.2	0.4				
11	Alabastron with bejewelled mount (20.1070)	Nuri 8(King Aspelta)	600–580BC	Sample 1: nd		68.3	28.6	3.1	93.2	1.8	5.0	< 0.09
				Sample 2: nd		59.1	38.3	2.6	91.5	2.3	6.2	< 0.09
				Sample 3: 31 µm		90.0	10.0	0.4				
				Sample 4: 16 µm		87.8	11.4	0.8				
				Sample 5: 13 µm		86.4	12.8	0.8	88.3	1.8	9.9	< 0.09
				Sample 6: 16 µm		75.5	23.3	1.2	93.8	1.9	4.3	< 0.09
12	Cylinder sheath, rosette (21.341c)	Nuri 8 (King Aspelta)	600–580BC	8 µm		86.1	9.9	4.0				
13	Cylinder sheath (21.11736)	Nuri 8 (King Aspelta)	600–580BC	Sample 1: 13 µm		84.1	15.5	0.4				
				Sample 2: 13 µm		81.0	18.5	0.5	98.0	1.4	0.6	< 0.09
				Sample 3: 26 µm		78.2	20.8	1.0	98.8	0.8	0.4	< 0.09
				Sample 4: 13 µm		66.1	32.4	1.5	98.6	1.0	0.4	< 0.09
				Sample 5: 20 µm		74.7	24.3	1.0				

No.	Object (MFA number)	Provenance	Date	Gilding thickness[1]		Gilding composition[2]			Substrate composition[3]			
					Layer	*Gold*	*Silver*	*Copper*	*Silver*	*Copper*	*Gold*	*Lead*
14	Mummy trappings (23.1396)	Meroë S 85 (Lady Mernua)	6th *c.* BC	Sample 1: 4–5 μm		54.9	44.0	1.1				
				Sample 2: 13 μm		53.0	46.0	1.0	98.8	1.2	< 0.1	< 0.09
15	Cylinder sheath fragments found between trappings (23.1396)	Meroë S 85 (Lady Mernua)	6th *c.* BC	Sample 1: 1 μm		79.0	19.1	1.9	97.3	1.7	1.0	< 0.09
				Sample 2: 2 layers,	1	92.1	7.5	0.4				
				each 10 μm	2	90.6	8.9	0.5				
				Sample 3: 13 μm		86.3	12.8	0.9	99.5	0.5	< 0.1	< 0.09
16	Cylinder sheath fragments from miscellaneous tombs (21.11744)	Nuri 3, 8, 9, 10, 26 and 27	643–513BC	10–13 μm		92.2	7.4	0.4	99.7	< 0.08	0.3	< 0.09
17	Tapering bead (20.402)	Nuri 27 (Queen Madikeñ)	593–568BC	12 μm		86.9	12.5	0.6	96.2	1.6	2.2	< 0.09
18	Miscellaneous fragments of cylinder sheath (18-3-1011)	Nuri 27 (Queen Madikeñ)	593–568BC	9 μm		85.0	14.6	0.4				
19	Ornament (20.953)	Nuri 42 (Queen Asata)	593–568BC	66 μm		37.2	59.7	3.1				
20	Cylinder sheath with inlays (20.275)	Nuri 10 (King Amani-Natake-Lebtke)	533–513BC	Sample 1: 59 μm Sample 2: 13 μm		87.6	10.9	1.5	97.0	1.7	1.3	< 0.09
21	Fragments of gold sheet, rosette (20.291)	Nuri 4 (King Si aspiqa?)	478–459BC	20 μm		83.8	15.5	0.7				
22	Buttons (20.300)	Nuri 17 (King Baskakereñ)	398–397BC	9–10 μm		70.5	28.0	1.5				
23	Small buttons (20.299)	Nuri 17 (King Baskakereñ)	398–397BC	8 μm		65.3	33.2	1.5	91.3	3.9	4.8	< 0.09
24	Tapering bead (20.301)	Nuri 17 (King Baskakereñ)	398–397BC	5 μm		71.4	28.6	1.0				
25	Bead (20.409)	Nuri 44 (Queen Batahaliye)	397–362BC	8–9 μm		75.0	23.9	1.1				
26	Mummy trapping, Qebeh-Senuf (20.343)*	Nuri 44 (Queen Batahaliye)	397–362BC	Not determined								
27	Goblet (24.971)*	Meroë Beg N 2 (King Amanikhabale)	43–26BC	Not determined								
28	Plaques, horse's harness (21-3-694 a-k)#	Meroë Beg N 16 (King Aryesbekhe)	62–78AD	None present in sample 1					97.6	0.3	2.1	< 0.09
				Sample 2: no gilding visible					97.9	0.9	1.2	< 0.09
				Sample 3: 4 μm		80.4	18.5	1.1	97.4	0.4	2.2	< 0.09
				Sample 4: 7 μm		81.3	17.8	0.9				
29	Ring (22-12-2a)	Meroë W 115	25–184AD	59 μm		75.1	23.7	1.2				

* No samples taken; gilding composition and technique could not be determined with certainty, since only surface analysis by energy-dispersive X-ray fluorescence was carried out.

Samples from these objects contained no evidence of compositional gradients around the gilding/substrate interface. All other samples showed such gradients.

1. Thicknesses were measured by reflected light microscopy. On diffusion gilded objects, these thicknesses include some of the diffusion zone, and thus they do not accurately represent the original thickness of the applied gilding. "nd" indicates that microprobe analysis revealed no traces of gilding.
2. Compositions are averages of several point analyses (usually 3–6) and have been normalized. The limit of detection for copper was 0.08%; "<0.08" indicates that the element was not detected.
3. Compositions are averages of several point analyses (usually 3–6) and have been normalized. The limit of detection for gold was 0.1% ("<0.1" indicates that the element was not detected); the limit of detection for lead was 0.09%.

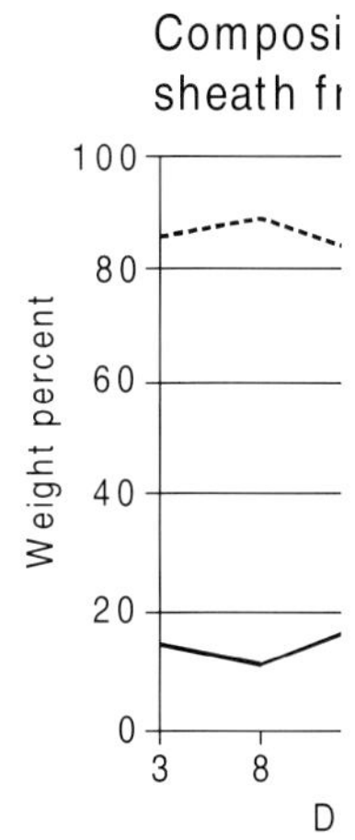

Figure 4.16 Composition of gilding on cylinder sheath fragment from Meroë, Tomb Beg. S. 85. (Graphics: Richard Newman.)

the gilding layer (see Fig. 4.23, Table 4.1, no. 16). The gradients can be readily seen by back-scattered electron imaging in the microprobe (see Fig. 4.21).

There are three main parameters that can be varied in diffusion gilding: the temperature to which the object is heated, the length of time of heating and the thickness of applied leaf or foil. Higher temperatures and longer periods of heating result in wider diffusion zones. Laboratory studies have shown that even quite low temperatures can create a strong bond by diffusion. The manner in which heating was carried out would obviously affect time and temperature. If carried out with the aid of a blowpipe, presumably quite short periods at high temperatures were used. If carried out in a furnace, much longer times were possible. Taking the diffusion profile found in uncorroded parts of a sample from Queen Mernua's cylinder sheath (Table 4.1, no. 15, sample 3), calculations show that the observed zone width could have been created by heating at 775°C for around five minutes; if the temperature were only 650°C, nearly an hour of heating would be necessary (Fig. 4.17). We can only note that a range of temperatures and heating times was undoubtedly used on the objects examined: diffusion zones range from a few micrometers to nearly sixty micrometers (Fig. 4.15). How well the heating stage was controlled during a given gilding process is not known; to some extent this is not very important for the end result. Multiple layers of leaf were applied to some objects (Fig. 4.18), while in others the gilding was laid on in single layers (Fig. 4.15). The thickness of the initially applied leaf varies considerably. Some of the leaf was as thin as about one micrometer (Berthelot 1968: 19–23, 62–4; Hatchfield and Newman 1991: 32–3). On other objects the applied single layer of leaf or foil was up to thirty micrometres thick.

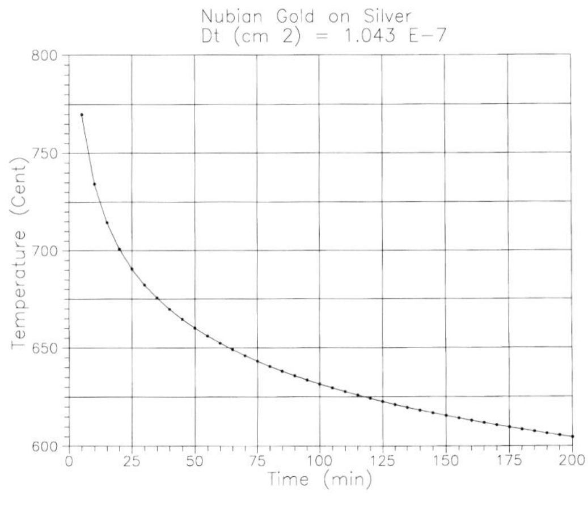

Figure 4.17 Calculated heating temperatures and times based on diffusion profile found in uncorroded parts of a sample from Queen Mernua's cylinder sheath (Table 1.15, Sample 3). (Graphics: Robert Ogilvie.)

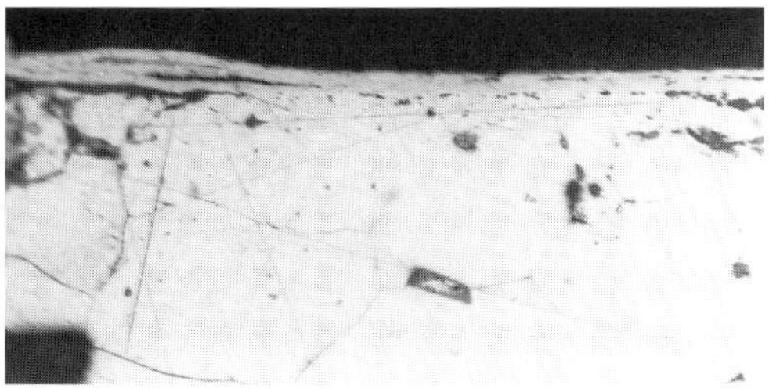

Figure 4.18 Sample taken at the base of the Maat amulet 21.323, El Kurru, Tomb Ku. 55. Three layers of gilding are visible, each approximately 1μm thick, width of field 0.21 mm. (Photograph: Susanne Gänsicke.)

The extent of diffusion would also have been affected by treatments that the gilded metal could have undergone after the gilding process. For example, sheet that was shaped by hammering may have been annealed, which of course would have produced further diffusion. Two gilded silver objects with Egyptian provenance in the Museum's collection were sampled for comparison. Both were also gilded by diffusion bonding, one with a single layer, the other with two thin layers, and the metal compositions were comparable to those of the Nubian samples.[23]

One feature nearly always noted in samples in which the substrate was substantially corroded were fingers of gilding extending from the interface of the gilding into the substrate (Figs 4.19, 4.20 and 4.21). These features have been noted on other silver objects that have been diffusion gilded, and they have been reproduced in the laboratory (Oddy *et al.*

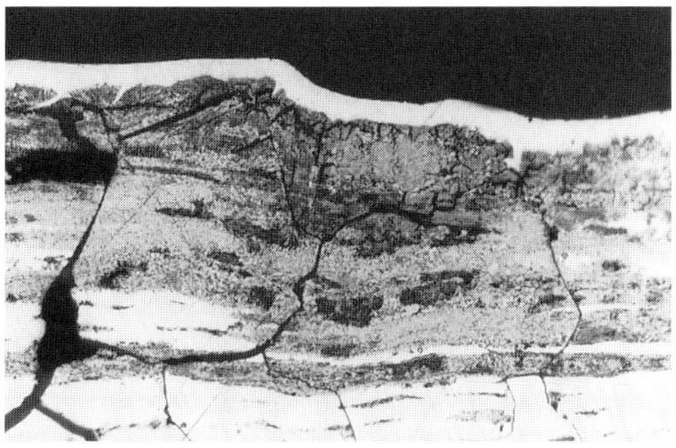

Figure 4.19 Sample of the cylinder sheath fragments of Queen Mernua, Meroë, Tomb Beg. S. 85 (Sample 3). Fingers of gilding are extending into the substrate; width of field 0.53 mm. (Photograph: Susanne Gänsicke.)

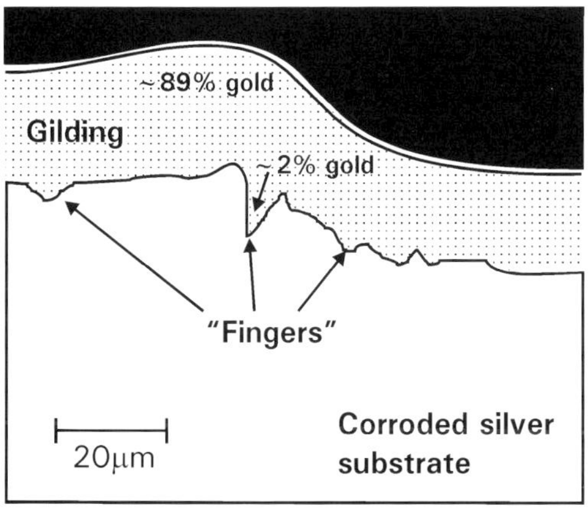

Figure 4.20 Detail drawing of Fig. 4.19. (Graphics: Richard Newman.)

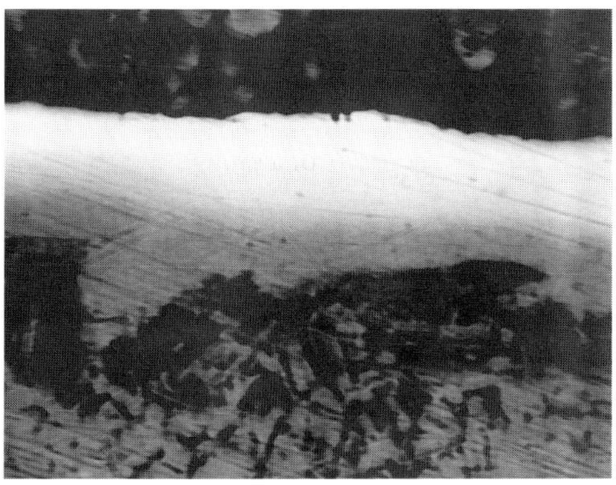

Figure 4.21 Sample of cylinder sheath fragment 21.11736, Nuri, Pyramid Nu. 8 (Sample 1). Detail of fingers in gilding layer; width of field 52 μm. (Photomicrograph: Richard Newman.)

1981). The fact that they have not corroded as rapidly as the surrounding substrate suggests that they are more gold-rich (Lechtman 1971: 10). But analyses we have carried out on a few of them indicate that their gold content is low. For example, one finger (similar to the one in Figure 4.16), which extended about 5 micrometres below the lower edge of the gilding, had a gold content of a little less than 3 per cent. These 'fingers' are probably due to preferential diffusion along grain boundaries (Kaur and Gust 1989; Kaur *et al.* 1989). They are probably also present in samples in which the substrate is not corroded, but are not readily visualized by backscattered electron imaging, presumably because their composition is quite close to that of the surrounding silver substrate.

Obviously diffusion gilding can be carried out on substrates other than silver. We found a compositional gradient in a gilding sample from a copper earring/hair ring from Kerma, an indication of the use of the diffusion gilding procedure on a copper substrate in Nubia.[24]

Composition of the applied leaf

The compositions of the applied leaf were quite variable (Fig. 4.22). Some layers were nearly pure gold, for example the thinner leafs applied in layers (Table 4.1, nos 2 and 4). Such gilding layers could have been made from refined gold. At the other extreme are gold–silver alloys with silver contents up to 60 per cent in one instance. Gold containing over about 25–30 per cent silver is quite different in colour from pure gold, and the use of a high-silver alloy would have produced an obviously white-gold coloured layer of surface coating. On some objects, such as the alabastron (Fig. 4.12), the gilded surfaces were not originally all of the same colour: this is implied by the highly variable silver contents of the gilding layers, which ranged from around 10 per cent to nearly 40 per cent.

The gilding layers that contain substantial silver could easily be native gold used as found. Other than silver and gold, the only other element present above trace levels in some of the gilding layers is copper. Copper is a common impurity in native gold and gold–silver alloys, although usually not at a level much above about 2 per cent. Somewhat higher levels than this were found in some of the samples examined, but it cannot be stated on the basis of current knowledge of native gold alloy compositions that these indicate any intentional addition of copper. Considerably higher copper levels were found in parts of a few samples, but close examination suggests that these locally elevated levels are probably due to silver solder. The relatively low amounts of copper found in the gilding layers would not have had an obvious effect on the colour of the gilding.

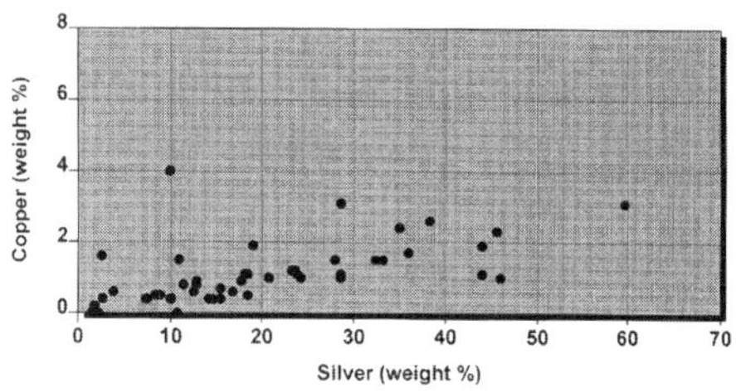

Figure 4.22 Composition of gilding layers. (Graphics: Richard Newman.)

Time of application of gilding in the overall manufacturing process

Gilding could have been applied to silver sheet before any shaping or decorative toolwork was carried out; it could have been applied after shaping

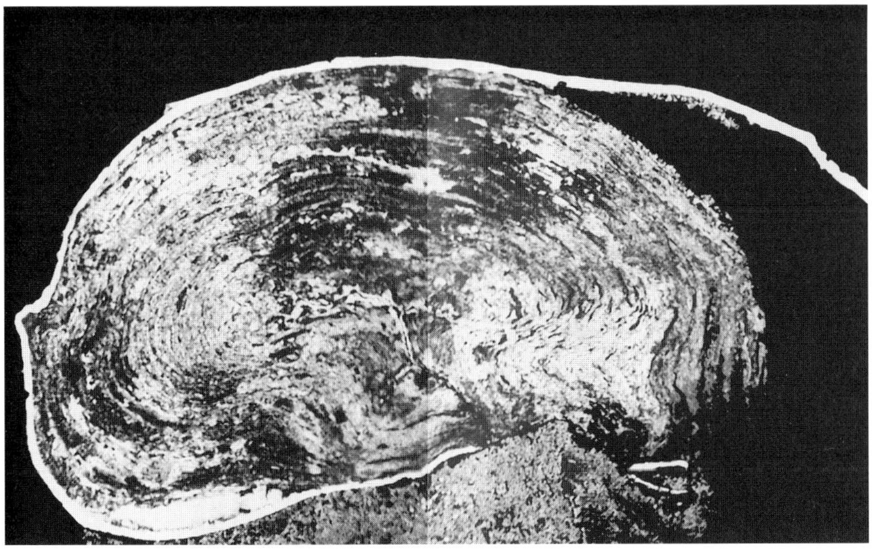

Figure 4.23 Sample of cylinder sheath fragments from Nuri (21.11744), Pyramids Nu. 3, 8, 9, 10, 26, 27. Twisting seam is clearly visible in the silver wire; width of field 1.7 mm. (Photograph: Susanne Gänsicke.)

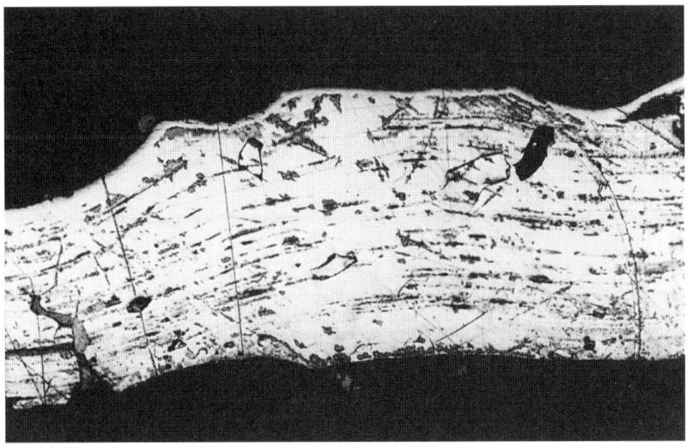

Figure 4.24 Sample from one of the cylinder sheaths of Queen Mernua, Meroë, Tomb Beg. S. 85 (Sample 1). Punched surface decoration was probably applied after diffusion gilding; width of field 1.06 mm. (Photograph: Susanne Gänsicke.)

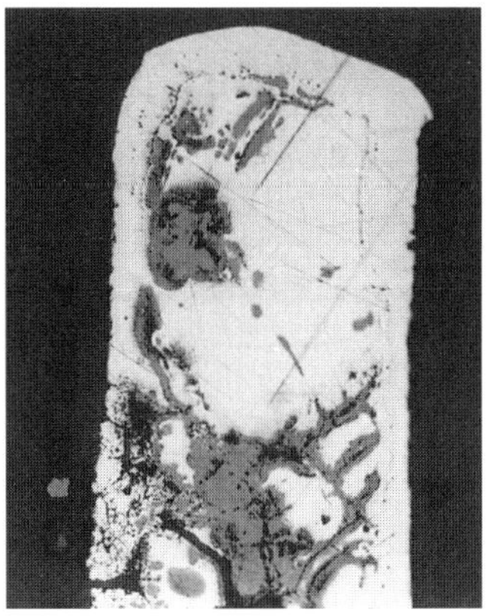

Figure 4.25 Sample from the uppermost cloisons on the cloisonné collar of alabastron 20.1070, Nuri, Pyramid 10; width of field 0.53 mm. (Photograph: Susanne Gänsicke.)

and working; or some working could have been carried out after application. On some objects, the gilding must have been carried out before shaping, for example on the half round and round wires (Fig. 4.23) that we find on the Nuri cylinder sheaths (Table 4.1, nos 10, 12, 13, 15, 16, 18 and 20) and the neck of the Aspelta alabastron (Fig. 4.12). The use of pre-gilded silver was also observed on other objects (Ogden 1982: 79; Oddy *et al.* 1983–84). Other features such as horns and ears of the El Kurru amulets (Table 4.1, nos 5, 6 and 7) also appear to have been gilded previous to assembly, as also was described for a human-headed scorpion amulet of the Twenty-Sixth Dynasty (Williams 1924: 189, pl. 28, 132 a,b). Some of the cross-sections from the cylinder sheaths included chased designs, and distortion of the grain structure of the silver substrate beneath and around the tool mark was always observed (Fig. 4.24). It seems most reasonable that gilding would have been carried out before any extensive working. However, the presence of distorted metal alone does not really answer the question, since diffusion gilding can be carried out at temperatures well below the annealing temperature of silver.[25] Another interesting object comprising numerous gilded elements is the metal mount of the alabastron (Fig. 4.12), where even the thin ribbons forming the cloisons were gilded (Fig. 4.25) (Gänsicke 1994).

On other objects gilding could have been carried out after initial shaping and assembly took place. This was presumably the case on a number of thin silver sheet elements of repetitive Hathor-headed designs (BMFA 23.1477), probably punched in molds, and on gilded beads. Both types from El Kurru show gilded surfaces formed by multiple layers of thin, superimposed gold foil that does not appear to be well bonded to the substrate. Analysis shows that diffusion gilding was utilized on the Hathor heads, while no diffusion zone can be observed on the beads as previously mentioned. The gilded solid silver mirror handle (Fig. 4.6) also shows a slightly flaking gold leaf surface that would presumably have been applied after casting.

With the exception of the mirror just mentioned, all of the objects consist of fairly thin silver (*c.*0.14–0.80 mm). Gilding could have been carried out on thicker silver sheet, which was then hammered and annealed before being shaped into objects, or gilding could have been carried out on sheet which had already been hammered to the intended final thickness.

Silver substrates

As already discussed, the sources of the silver in ancient Egypt and Nubia are not known at present. Our analyses of the uncorroded silver substrates showed that the silver was relatively pure in nearly all cases (Fig. 4.26). Copper exceeded about 2 per cent in only two objects included in this study: the bracelets from Kerma (Table 4.1, no. 1) of the mid-second millennium BC and in the small buttons from Nuri (Table 4.1, no. 23) from the fourth century BC. Gold was present above the detection limit of about 0.1 per cent in all but two samples (Table 4.1, nos 14 and 15). Although the detection limit for lead in our microprobe analyses is relatively high (about 0.09 per cent), nevertheless, we would expect to detect lead if the silver had been derived from lead ores (such as galena): lead contents in silver from such sources are typically in the range of a few tenths of a per cent. In only the Kerma bracelets just mentioned was lead present well above the detection limit. These were almost certainly made from silver that had been refined from a lead ore and then alloyed with some copper. Most likely this silver would have been imported. The gold content of such silver is usually quite low, in the range of a few tenths of a per cent. The much higher level in the bracelets (about 3 per cent) seems too high, and is unexplained at the moment, unless it is due to diffusion from gilding or to a mixture of silver from different sources.

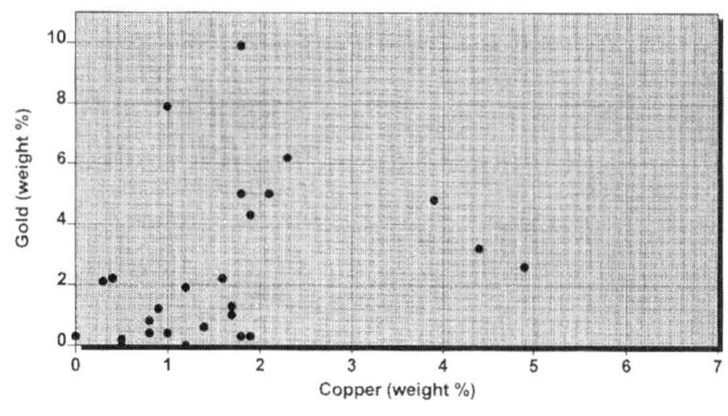

Figure 4.26 Composition of 'silver substrates'. (Graphics: Richard Newman.)

Given current knowledge, we can only say that most of the silver objects included in this study were not derived from argentiferous lead ores. The geographical and mineralogical origin of the silver metal remains uncertain. It could have been refined from silver ores, or perhaps was used as native silver metal. The gold and copper levels of most of these objects seem low enough so that they could simply be unintentional impurities.

Conclusions

Although the range of objects and the sites from which they come was limited in this study, these initial results indicate that gilding of silver by diffusion bonding, the typical technique used in antiquity previous to the invention of techniques involving mercury, appears with a certain amount of sophistication on Nubian objects from at least as early as the eighth century BC, with two examples from the first half of the second millennium BC. A variety of related techniques were used to achieve the same objective: the appearance of solid gold. Some of the gilded objects, however, were clearly meant to appear as white gold or electrum. Notably absent is the use of parcel gilding, in which only certain design elements were covered with gold to create a deliberate contrast with the silver, although using pre-gilded elements would have allowed such a deliberate contrast. In none of the samples could mercury gilding be detected, as one would expect for these early dates.

To date no significant chronological or regional developments can be ascertained, but it can be pointed out that gilding appears frequently on funeral items, especially mummy trappings. On many objects the silver substrate is quite thin while the gold foil in relation appears rather thick. As the technique of beating gold to much thinner leaf was known, such observations would not attest to the use of gilding as a means to save gold. The results of the study seem to suggest that the gilding of silver was much more varied and widespread in the Sudan, and probably in the entire Nile Valley, than previously assumed; and they appear to open avenues to further investigation of the sources for the raw materials and the relationship of silver and gold towards each other.

Acknowledgments

We wish to thank Arthur Beale, Rita Freed, Timothy Kendall and Yvonne Markowitz, at the Museum of Fine Arts Boston, and Deborah Schorsch at the Metropolitan Museum for their assistance and discussion, and Cathleen Duffy, formerly an Advanced-Level Science Intern at the Museum of Fine Arts, for help with the analysis of some samples. We also thank Robert Ogilvie, who developed a computer program to calculate heating temperatures and times based on analytically determined diffusion gradients.

Appendix 1. Analytical procedures

Cross-sections were mounted in Buehler epoxide resin and polished. For analysis, samples were carbon coated. Analyses were carried out in a Cameca MBX microprobe equipped with a Tracor-Northern 5502 energy-dispersive X-ray fluorescence system (EDS) and Tracor-Northern 1330 stage automation and wavelength-dispersive X-ray fluorescence system (WDS). Quantitative analyses were carried out by WDS, using pure metal standards and fundamental parameters (ZAF) matrix corrections. Analyses were carried out at regular intervals across the gilding/substrate interface and/or across the gilding layer itself, from top to bottom, in order to determine whether compositional gradients were present. We assume that analyses carried out at the top of a gilding layer represent the original composition of that layer when the gradient was found not to extend to the surface. In a few instances, this was not the case, and the composition we report may not be the original composition. Analyses of the silver substrate also can be assumed to be accurate reflections of the original composition when no gradient is present in the vicinity of the area analyzed. The substrates in many of the samples were severely corroded, or have been electrolytically reduced during restoration, and we cannot determine their original composition.

After analysis, some samples were repolished and etched with a chromic acid solution.

Notes

1. BMFA, 21.3091, 21.3092: silver vessels and sticks (Dunham 1950: 32, Figs. 11e, f, h, pl. 64c). BMFA, 20.334, 20.341, 20.342: gold ewer, tweezers, and lid (Dunham 1955: 81, Fig. 55, pls. 34D, 89A, B).
2. Nineteen nuggets have been excavated by Reisner in Nubia (Dunham 1950: 16, Fig. 2d, pl. 52A–B, pl. 57B6 [for Ku. Tum. 2.; # 6, natural gold nugget]; Dunham 1955: 27, Fig. 14, pl. CXV A:3/11 [for Nu. 59, 18-2-294i, natural gold nugget], 146, [for Nu. 26, 20.337, natural gold nugget]; Dunham 1957: 26 [for Bar. 13, 16-2-42, three natural gold nuggets]; Dunham 1963: 46, Fig. 31c, 3 upper [for Beg. W. 671, 23-M-372, natural gold nugget], 49, Fig. 34d, 2/1 [for Beg. W. 585, 23-M 199a, b, 2 natural gold nuggets], 54, Fig. 39c, 2 [for Beg. W. 652, 23-M-333, natural gold nugget], 63, Fig. 46f, 2 [for Beg. W. 848, 23-M-673, 674, 2 natural nuggets], 65, Fig. 48c, 1 [for Beg. W. 859, 23.311, large gold nugget], 130 [for Beg W 140, 22-2-529, 2 gold nuggets], 284, [for Beg. W 506, 23-3-223, small gold nugget], 308 [for Beg. W. 695, 23-M-409, natural gold nugget], 362, Fig. 192F, 1.d [for Beg. S. 125, 21-2-376, small gold nugget], 435, [for Beg. S. 154, 21-3-232, irregular gold nugget]).
3. BMFA, 37.606: diadem. BMFA, 04.1955: girdle fragment.
4. BMFA, Exp. nos 16-4-64, 17-1-647, 17-1-285: eyes. BMFA, Exp. no. 16-4-86: uraeus.
5. British Museum, London, BM EA 40931, 40390: bracelets. Recently, Deborah Schorsch, Metropolitan Museum of Art, and Michael Cowell, British Museum, carried out X-ray fluorescence analysis on the bracelets and could not detect evidence of gilding.
6. Catalogue Général du Musée du Caire (CG) 52372/3: earrings. BM EA 14348: earrings.
7. BMFA 1981.159: pectoral.
8. Musée du Louvre, Paris, N 767: pectoral. The analysis was carried out in 1977 by XRF. I thank Elisabeth Delange for providing me with the information.
9. Rijksmuseum van Oudheden, Leiden, G. 1: diadem. CG 52.715: amulet of Queen Khama. Ambiguity remains about the actual presence of gilding, according to Raven.
10. The object belongs to the collection of the Pelizaeus-Römer-Museum, Hildesheim; a second silver mask is present at the Staatliche Sammlung Ägyptischer Kunst, Munich.
11. Römer-Pelizaeus-Museum, Inv. Nr. 5123: Osiris.
12. CG 53.262: vessel from Bubastis.
13. Walters Art Gallery, Baltimore, 57.1416: Amun, 54.1617: Amun, 57.1419: Nefertem.
14. BM EA 60006: Amun Re, Late New Kingdom, 900 BC Musée du Louvre E 27431: King offering a figure of Maat, New Kingdom.
15. BMFA 20.1671: bracelet. BMFA, Exp. no. 16-4-1281: sheet fragments from KN 105. One item from KN 189, described as gilded silver (Reisner 1923b: 285), is in fact a copper alloy.
16. Musée du Louvre, E 6204: pendant of Osorkon.
17. BMFA 23.1477: trappings.
18. From Pyramids Nu. 2, 3, 8, 9, 10, 15, 16, 26, 27. It should here be noted that amongst the fragments of mummy beads of Queen Mernua (see note 50, 23.1396) a number of fragments of more than one cylinder sheath were discovered by Margaret Leveque. It is not clear if they were originally part of the queen's tomb of if these fragments are intrusive, but for this study they will be referred to as 23.1396.

19. Dunham 1963: 366–73, Figs 368, 369, 370 (for 23.1396 mummy trappings), 137, Fig. 102 (for Beg. W 115, 22-12-2a, signet ring), 177, Fig. 130e, 2/2 (for Beg. W 179, 24.485, ring with ram's head), 183, Fig. 130e, 2/3; Fig. 133m (for Beg. W 179, 24.515, ring in form of uraeus), 222, Fig. 155, 11 (for Beg. W 32, 23.1461, signet ring); Dunham 1957: 139, 141, Fig. 90, pl. 63A, B, 64A (for Beg. N 16, 21-3-694 a-k, horse's harness), 151, Fig. 96, pl. 64b (for Beg. N 18, 21-3-659-663, 672, 687), 194, pl. 73 (for Beg. N 24, 24.1078, rectangular plaque with enamel).
20. BMFA 24.971 (22-2-439).
21. A number of pieces were treated in past decades by electrolytic reduction, which resulted in losses and in alteration of the metal structure.
22. Eluère describes substrates of 20–200 μm and leafs of max. 5 μm on a diadem and bracelet from the 7th century BC. Lins and Oddy's article includes a comprehensive list of objects in the British Museum, Department of Western Asiatic Antiquities, some which are gilded by diffusion bonding.
23. BMFA, 87.733-745: pendants of seated gods, from Tell Nebesheh, Twenty-Fifthth Dynasty, one layer 5–10 μm gold. BMFA, 31.842: Ptolemaic trappings, from Giza, 1st century BC, gilded with two layers of 3 μm leaf.
24. Exp. no. 14-2-964.
25. Lechtman found that gold foil thicker than 1 μm tends to become springy and buckle when applied to the silver substrate without heat, and that it could not follow intricate designs.

References

Aldred, C. 1971. *Jewels of the Pharaos.* London: Thames & Hudson.

Andrews, C. 1981. *Catalogue of Egyptian Antiquities in the British Museum.* London: British Museum.

Andrews, C. 1991. *Ancient Egyptian Jewelry.* New York: Abrams.

Becker, L., Pilosi, L. and Schorsch, D. 1994. 'An Egyptian silver statuette of the Saite Period – a technical study'. *The Metropolitan Museum Journal*, **29**: 37–56.

Berthelot, M. 1968. *Archéologie et Histoire des Sciences.* Amsterdam: Philo Press.

Born, H. 1986. 'Material und Herstellungstechnik der Silberschale des Merenptah'. *Jahrbuch Preußischer Kulturbesitz*, **23**: 143–53.

Breasted, J. H. 1906. *Ancient Records of Ancient Egypt IV.* Chicago, Ill.: University of Chicago Press.

Castiglioni, A. and Castiglioni, A. 1994. 'Discovering the Berenice Panchrysos'. *Bulletin of the Egypt Exploration Society* **4**: 19–22.

Castiglioni, A., Castiglioni, A. and Vercoutter, J. 1995. *L'Elodrado dei Faraoni.* Novara: Istituto Geografico De Agostini SpA.

Cauville, S. 1987. 'Les statues cultuelles de Dendera d'après les inscription pariétales'. *Bulletin de l'Institut français d'archéologie orientale du Caire*, **87**: 71–117.

Conrad, S. 1993. 'Die silberne Mumienmaske aus Hildesheim'. *Restauro*, **99**(4): 256–9.

Craddock, P. T., Meeks, N. D., Cowell, M. R. *et al.* 1998. 'Refining of gold in the classical world'. *Proceedings of 'The Art of the Greek Goldsmith'* (forthcoming).

de Garis Davies, N. 1944. *The Tomb of Rekh-mi-Re at Thebes.* New York: Metropolitan Museum.

Drenkhahn, R. 1976. *Die Handwerker und ihre Tätigkeiten im alten Ägypten.* Wiesbaden: Harrassowitz.

Dunham, D. 1946. 'An Egyptian diadem of the Old Kingdom'. *Bulletin of the Museum of Fine Arts*, **44**, no. 255: 23–9.

Dunham, D. 1950. *Royal Cemeteries of Kush I: El Kurru.* Boston, Mass.: Museum of Fine Arts.

Dunham, D. 1955. *Royal Cemeteries of Kush II: Nuri.* Boston, Mass.: Museum of Fine Arts.

Dunham, D. 1957. *Royal Cemeteries of Kush IV: The Royal Tombs at Meroë and Barkal.* Boston, Mass.: Museum of Fine Arts.

Dunham, D. 1963. *Royal Cemeteries of Kush V: The West and South Semeteries at Meroë.* Boston, Mass.: Museum of Fine Arts.

Eggebrecht, A. (ed.) 1990. *Suche nach Unsterblichkeit.* Mainz: Zabern.

Eide, T., Hägg, T., Pierce , R. H. and Török, L. 1994. *Fontes Historiae Nubiorum,* vol. I. Bergen: John Greig SA.
Eluère, C. 1989. *Secrets of Ancient Gold*. Giun-Düderingen, Switzerland.
Forbes, R. J. 1964. *Metallurgy in Antiquity*, vol. 8. Leiden: E. J. Brill.
Fuchs, R. 1983. 'Silber'. In *Lexikon der Ägyptologie,* band V/3, W. Helck and W. Westendorf (eds), col. 362–6. Wiesbaden: Otto Harrassowitz.
Gänsicke, S. 1994. 'King Aspelta's vessel hoard from Nuri in the Sudan'. *Journal of the Museum of Fine Arts, Boston*, **6**: 14–40.
Gale, N. H. and Stos-Gale, Z. A. 1981. 'Ancient Egyptian silver'. *Journal of Egyptian Archaeology*, **67**: 103–15.
Garside, A. (ed.). 1980. *Jewelry Ancient to Modern*, 37–39. New York: Viking Press.
Gaskell, J. L. 1985. 'Reappraisal of the Gebeit gold mine, northeast Sudan: A case study'. *Prospecting in Areas of Desert Terrain*, 49–58. London: Institute for Mining and Metallurgy.
Harrell, James A., and Max Brown, V. 1992. 'The oldest surviving topographical map from ancient Egypt (Turin Papyri 1879, 1899, and 1969)'. *Journal of the American Research Center in Egypt*, **29**: 81–105.
Harris, J. R. 1961. *Lexicographical Studies in Ancient Egyptian Minerals*. Berlin: Akademie-Verlag.
Hatchfield, P. and Newman, R. 1991. 'Ancient Egyptian gilding methods'. In *Gilded Wood, Conservation and History*, D. Bigelow, E. Cornu, G. J. Landrey, C. van Horne (eds), 27–47. Madison, Conn.: Sound View Press.
Hornung, E. 1982. 'Der Ägyptische Mythos von der Himmelskuh'. *Orbis Biblicus et Orientalis* **46**: 37.
Hunt, L. B. 1974. 'The oldest metallurgical handbook'. *Gold Bulletin*, **9**: 24–31.
Jodidio, P. (ed.) 1995. Laboratoire de recherche des musées de France. Numéro spécial de *Connaissance des Arts*, **68**: 23.
Jüngst, H. 1982. 'Zur Interpretation einiger Metallarbeiterszenen auf Wandbildern Altägyptischer Gräber'. *Göttinger Miszellen*, **59**: 15–27.
Kaur, I. and Gust, W. 1989. *Fundamentals of Grain and Interphase Boundary Diffusion*. Stuttgart: Ziegler Press.
Kaur, I., Gust, W. and Kozma, L. 1989. *Handbook of Grain and Interphase Boundary Diffusion Data* vols. 1 and 2. Stuttgart: Ziegler Press.
Kendall, T. 1982. *Kush: Lost Kingdom of the Nile*. Brockton, Mass.: Brockton Art Museum/Fuller Memorial.
Kendall, T. 1994. 'Le Djebel Barkal: Le Karnak de Koush'. *La Nubie. Les Dossiers D'Archéologie*, **196**: 46–53.
Kendall, T. (in press). 'The origin of the Napatan State: El Kurru and the evidence for the royal ancestors'. *Meroitica* 15. Berlin: Humboldt Universität.
Klemm, R. and Klemm, D. 1989. 'Antike Goldgewinnung in der Ostwüste Ägyptens'. In *Proceedings of the International Symposium 'Old World Archaeometallurgy'*, A. Hauptamnn (ed.), 227–34. Bochum: Selbstverlag des Deutschen Bergbau-Museums.
Klemm, R. and Klemm, D. 1994. 'Chronologischer Abriß der antiken Goldgewinnung in der Ostwüste Ägyptens'. *Mitteilungen des Deutschen Archäologischen Instituts Kairo*, **50**: 189–222, pl. 29–35.
Knudtzon, J. A. 1907. *Die El-Amarna Tafeln*. Leipzig: J. C. Hinrichs'sche Buchhandlung.
Lacovara, P. 1990. 'An ancient Egyptian royal pectoral'. *Journal of the Museum of Fine Arts, Boston* **2**: 19–28.
Lechtman, H. 1971. 'Ancient methods of gilding silver: Examples from the old and the new world'. In *Science and Archaeology*, R. H. Brill (ed.), 2–30. Boston, Mass.: MIT Press.
Liliquist, C. 1993. 'The Boston/Lafayette jewel and other glass-inlaid ornaments'. *Varia Aegyptica*, **9**(1–2): 33–44.
Lins, P. A. and Oddy, W. A. 1975. 'The origins of mercury gilding'. *Journal of Archaeological Science*, **2**: 365–93.
Lucas, A. 1924. 'Note on the cleaning of certain objects in the Cairo museum'. *Annales Services des Antiquités de l'Égypte*, **24**: 15–16.
Lucas, A. 1928. 'Silver in ancient times'. *Journal of Egyptian Archaeology*, **14** : 313–19.

Markowitz, Y. 1993. 'Ancient amulets of Kush'. *Ornament*, **16**, 4: 46–9.
Markowitz, Y., Lacovara, P. and Hatchfield, P. 1997. 'Jewellery Fragments from the tomb of Nefertari in the Museum of Fine Arts, Boston'. In *Chief of Seers. Egyptian Studies in Memory of Cyril Aldred*, E. Goring, N. Reeves, and J. Ruffle (eds), 220–28. London: Kegan Paul.
Maystre, C. 1986. *Tabo I: Statue en bronze d'un roi méroïtique. Musée National de Khartoum, Inv. 24705*. Genève: Georg Editeur SA.
Mishara, J. and Meyers, P. 1974. Ancient Egyptian silver. *Recent Advances in Science and Technology of Materials,* **III**: 29–45.
Möller, G. 1924. *Die Metallkunst der Alten Ägypter*. Berlin: Verlag Ernst Wasmuth.
Montet, P. 1959. *La Nécropole Royale de Tanis*, vols I–III. Paris: CNRS.
Moussa, A. M. and Altenmüller. H. 1977. *Das Grab des Nianchchnum und Chnumhotep*. Mainz: Zabern.
Newman, R. 1990. 'Technical examination of an ancient Egyptian royal pectoral'. *Journal of the Museum of Fine Arts, Boston*, **2**: 30–37.
Nicholson, E. D. 1979. 'The ancient craft of gold beating'. *Gold Bulletin*, **12**(4): 161–6.
Notton, J. H. F. 1974. Ancient Egyptian gold refining. *Gold Bulletin*, **7**(2): 50–56.
Oddy, A. 1993. 'Gilding of metals in the old world'. In *Metal Plating and Patination*, S. La Niece and P. Craddock (eds), 171–81. Oxford: Butterworth-Heinemann Ltd.
Oddy, W. A., Padley, T. G. and Meeks, N. D. 1979. Some unusual techniques of gilding in antiquity. *Archaeo-Physica,* **10**: 230–42.
Oddy, W. A., La Niece, S., Curtis, J. E. and Meeks, N. D. 1981. 'Diffusion-bonding as a method of gilding in antiquity'. *MASCA* **1**(8): 239–41.
Oddy, W. A., Meeks, N. D. and Ogden, J. M. 1983–84. 'A Phoenician earring: A scientific examination'. *Jewellery Studies* **1**: 3–13.
Oddy, A., Pierce, P. and Green. L. 1988. 'An unusual gilding technique on some Egyptian bronzes'. In *Conservation of Ancient Egyptian Materials*, S. C. Watkins and C. E. Brown (eds), 35–9. London: UKIC.
Ogden, J. 1982. *Jewellery of the Ancient World*. New York: Rizzoli.
Rackham, H. (trans.) 1961. *Pliny: Natural History*, book 33. Cambridge, Mass: Harvard University Press.
Raven, M. J. 1988. 'The Antef diadem reconsidered'. *Oudheidkundige Medelingen* **68**: 77–90.
Reisner, G. A. 1923a. *Excavations at Kerma, parts 1–3.* Harvard African Studies 5. Cambridge, Mass.: Harvard University Press.
Reisner, G. A. 1923b. *Excavations at Kerma, parts 4–5.* Harvard African Studies 6. Cambridge, Mass.: Harvard University Press.
Roehrig, C. H. 1996. 'Cult figure of a falcon-headed deity'. In *Ancient Art from the Shumei Family Collection*, John P. O'Neill (ed.), 4–7. New York: Metropolitan Museum.
Sadr, K., Castiglioni, A. and Castiglioni, A. 1994. *Interim Report on the Eastern Desert Research Center (CeRDO), Archeological Activities 1994.* Varese: Centro ricerche sul deserto orientale.
Schäfer, H. 1910. *Ägyptische Goldschmiedearbeiten*. Berlin: Karl Curtius Verlag.
Scheel, B. 1985. 'Studien zum Metallhandwerk im alten Ägypten 1'. *Studien zur Altägyptischen Kultur*, **12**: 117–77.
Scheel, B. 1986. 'Studien zum Metallhandwerk im alten Ägypten 2'. *Studien zur Altägyptischen Kultur*, **13**: 181–205.
Scheel, B. 1987. 'Studien zum Metallhandwerk im alten Ägypten 3'. *Studien zur Altägyptischen Kultur*, **14**: 247–64.
Settgast, J. 1986. 'Die Silberschale des Merenptah'. *Jahrbuch Preußischer Kulturbesitz*, **23**: 125–41.
Stos-Fertner, Z. A. and Gale, N. H. 1979. 'Chemical and lead isotope analysis of ancient Egyptian Gold, silver and lead'. *Archaeo-Physica*, **10**: 299–314.
Stos-Gale, Z. A. and Gale, N. H. 1981. 'Sources of Galena, lead, and silver in predynastic Egypt'. *Revue D'Archeometrie*, **3**: 285-285.
Vercoutter, J. 1959. 'The Gold of Kush'. *Kush*, **7**: 120–53.
Vercoutter, J. 1994. 'Or et politique dans l'Égypte des origines'. *Hommages à Jean Leclant,* vol. 2. Le Caire: Institut français d'archéologie orientale.

Vernier, É. 1925. 'L'or chez les anciens Égyptiens'. *Bulletin de l'Institut français d'archéologie orientale du Caire*, **25**: 167–73.
Vernier, É. 1927. *Catalogue général des antiquités égyptiennes du musée du Caire, bijoux et orfèvreries I, II*. Cairo: l'Institut français d'archéologie orientale.
Wildung, D. 1996. *Sudan: Antike Königreiche am Nil*. Tübingen: Wasmuth Verlag GmbH&Co.
Williams, C. R. 1924. *Gold and Silver Jewelry and Related Objects. Catalogue of Egyptian Antiquities of the New York Historical Society*. New York: New York Historical Society.
Wilkinson, A. 1971. *Ancient Egyptian Jewellery*. London: Methuen.

5 Two Gilded Bronze Sculptures from the Athenian Agora

Alice Boccia Paterakis

Abstract

Fragments of a bronze equestrian statue portraying Demetrios Poliorketes, which consist of a leg, drapery, sword and other pieces, date to the end of the 4th century BC. The head from a bronze statue of Nike dates to the 5th century BC. Both statues were gilded in antiquity with sheets of gilded silver using diffusion bonding and the foil gilding method. The sheets were secured to the statues with silver wires, which were hammered into grooves cut into the surface of the bronze. Both bronze statues were analysed by atomic absorption spectroscopy (AAS) by Craddock in 1977. The leg and drapery fragments of the equestrian statue were found to be heavily leaded tin bronze, whereas the sword from the equestrian statue and the Nike head were found to be tin bronze. The gold and silver of the foil and wire were identified by Sheridan in 1976 with the scanning electron microscope (SEM). Further analysis of the foil and wire with optical metallography and the SEM with electron microprobe is called for to clarify the following points: the morphology of the gold as applied to the silver; the technique used for diffusion bonding; and the method of securing the gilded silver sheets with the silver wires.

Introduction

In 1932 a bronze head (Fig. 5.1) was found lying face down in the mud in a well in the Athenian Agora (Thompson 1940). In 1971 several fragments of a free-standing, bronze, equestrian statue were discovered in another well in the Athenian Agora: a leg, two drapery pieces, a sword (Fig. 5.2), a griffin from the helmet, several fragments representing leather straps, a bit and fragments of a palmette (Shear 1973). Both statues preserve traces of the original gilded silver with which they had been covered in antiquity. These statues are unique; the only surviving examples of bronze statuary from antiquity that were gilded with gilded silver foil using the foil gilding technique.

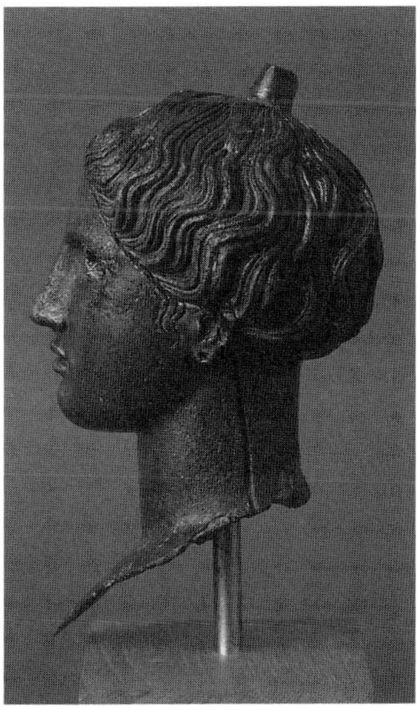

Figure 5.1 Nike head. (Photograph by Craig Mauzy, Agora Excavations, American School of Classical Studies, Athens.) (See Plate 21.)

Equestrian Statue

Description

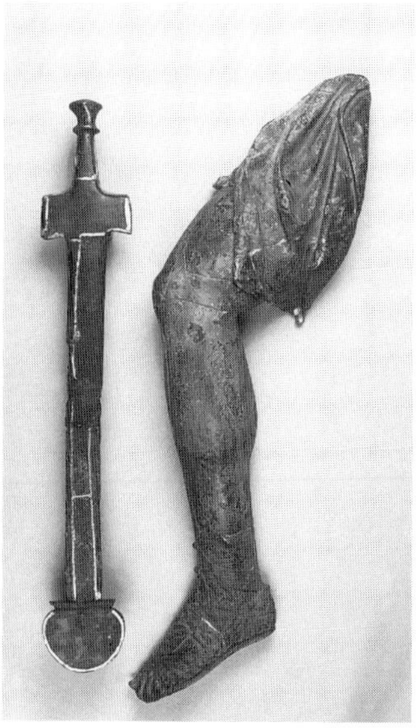

Figure 5.2 Equestrian leg, drapery and sword. (Photograph by Craig Mauzy, Agora Excavations, American School of Classical Studies, Athens.) (See Plate 22.)

The statue is of heroic size and would have stood at least 6 feet tall (Houser 1982). Much of the interpretation and identification of this sculpture is based on the work of C. Houser who is soon to publish a *Hesperia Supplement* entitled 'The Agora Horseman'. This statue demonstrates the progression from the classical to the Hellenistic style. It has been identified as the portrait of the Macedonian king, Demetrios Poliorketes, and is dated to 303–302 BC (Fig. 5.3). Demetrios was the son of Antigonos Monopthalmos (the One-eyed), an officer for Alexander the Great. Together, Demetrios and Antigonos tried to reunite Alexander's Empire. In 307 BC , they responded to a call from Athens to liberate the city. The composition is similar to that of a bronze equestrian statue probably of Alexander the Great attributed to Lysippos. By 200 BC the Athenians had grown to hate the Macedonians, which they demonstrated by flinging this statue of Demetrios down a public well (Shear 1973).

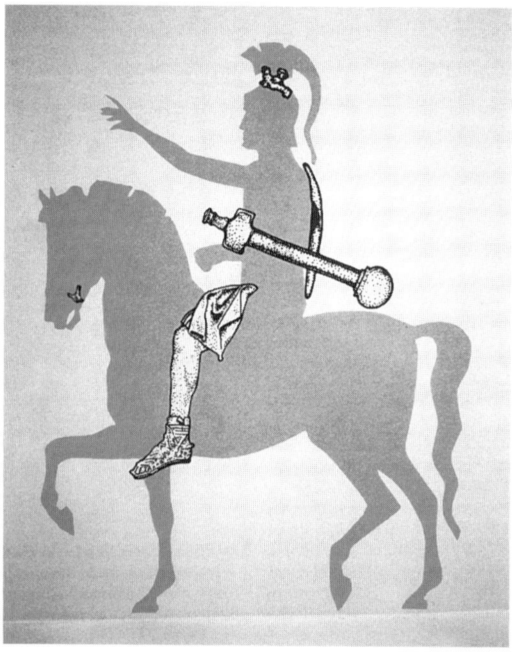

Figure 5.3 Reconstruction drawing of equestrian statue (from Houser 1982: 232, Fig. 11).

Fabrication

Fragments of the sculpture show evidence of the lost wax casting method. The drapery pieces and sword are complete castings preserved intact. The leg consists of two castings joined above the sandal. The leg was cast hollow in two pieces of bronze ranging in thickness from 0.75 cm to 1.0 cm. Both mechanical and metallurgical means were used to join the separate parts of this statue. The inner thigh is open where it conformed to the horse's body. The flange projecting on top of the thigh could have been used to anchor the leg to the adjoining statue part (Shear 1973). Lumps of solidified metal on the interior surface next to the edge that was joined to the horse may be remnants of a weld metal (Houser 1987). The drapery pieces were cast so that their edges on one side bend sharply to hook onto neighbouring pieces of the statue (Shear 1973). The sword was cast solid in one piece and was attached to its strap by rivets on the scabbard; it hung on the rider by its strap. One of the strap fragments preserves the means of attachment to the rivets of the sword. The griffin can be placed on the helmet of Demetrios using the helmet of the Athena Parthenos statue, which stood in the Parthenon of Athens, as a model (Camp 1997). It was cast of solid bronze; the wings, which were attached separately, may be of a different alloy. Rivet holes are preserved in the palmette fragments for their construction and attachment. The palmette, which is similar in style to a contemporary, Athenian stone stele, may have been attached to a bronze stele, which was part of the equestrian monument (Houser 1987), or it may have been held as a trophy of naval victory by Demetrios himself (Camp 1997).

All pieces show grooves for the attachment of the gilded silver sheets. The grooves were cut into the cold bronze after casting on all pieces except those of the palmette, whose grooves were cast (Houser 1982). They measure approximately 2 mm wide. Traces of the foil remain on all pieces with the exception of the griffin. The foil consisted of sheets of gilded silver that reached 0.5 mm in thickness in places. These sheets were hammered around the bronze and secured with the aid of a silver wire, which was hammered into the grooves locking the sheets in place. These wires measured 1.5–2.5 mm in width and 1.5 mm in thickness. Much of this wire remains in place and can be seen clearly on the leg, sword and drapery pieces. The number of separate gilded sheets used can be estimated from the arrangement of the grooves: nine for the leg, five for the smaller drapery, six for the larger drapery, and twenty-three for the sword (Shear 1973). The configuration of the areas to be gilded determined the size and form of the sheets. One can assume that the foil gilding was intended to be permanent since the statue would have been very unsightly stripped showing the irregular patterns of the grooves on the surface.

Analysis

Several samples of the bronze pieces were analyzed by P. Craddock by Atomic Absorption Spectrometry (AAS) in the Research Laboratory of the British Museum in the 1970s (Craddock 1977). All pieces of the statue, except for the sword, were found to be heavily leaded tin bronze, which is a composition typical of Late Hellenistic statuary bronzes. Three samples from the leg are similar in composition, ranging in lead content from 28 per cent to 35 per cent. One sample removed from the join between the foot and leg showed similar composition to the others but with a lower lead content of 23 per cent. This join may have been 'fusion welded' by applying molten metal similar in composition to the join. The lower lead content of the weld metal would dictate a melting temperature superior to that of the leg, a requirement of the welding metal in order to melt the underlying surfaces. The melting temperature of this copper–tin–lead alloy weld would be approximately 950°C (Lechtman and Steinberg 1970). The high lead content of the statue would have made it easier to cast, weld and surface finish. The sword is a tin bronze with a low lead content of 1.75 per cent, a composition that was also used during the Hellenistic period (Craddock 1977).

In 1976 W. Sheridan analysed a sample of the foil and a sample of a wire from the drapery piece (B 1383) covering the thigh using a scanning electron microscope (SEM) (Sheridan 1976). The three elements analysed were gold, silver and copper. Unfortunately, the areas of the foil and wire sampled are not known. The gold layer of the foil measured 15 microns (0.015 mm) and the silver layer measured 225 microns (0.225 mm), a total

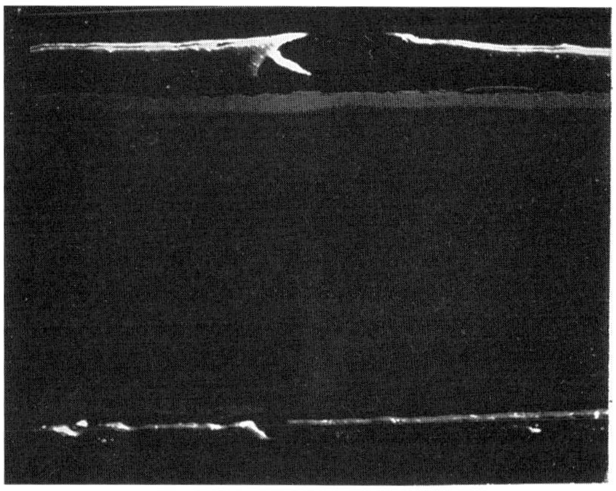

Figure 5.4 SEM photograph at ×200 magnification of the cross-section of gilded silver foil. There is a thin gold layer (lighter grey) on top of the thick silver layer (darker grey). (Photograph by W. Sheridan.)

of 240 microns in thickness. Sheridan assumed that the gold and silver are pure layers based on the absence of copper. The good coherence between the two layers can be seen from the SEM photograph in Figure 5.4. The diffusion zone appears to be approximately 1 micron thick. He attributed the diffusion to a bonding technique in which the gold was hammered on to the silver, which had been heated. The metal wire used to secure the foil to the surface of the drapery was identified as pure silver based on the absence of copper or gold. Sheridan suggested that the silver wires were pounded into the grooves, probably by means of a flattening tool hit with a mallet, so that they projected slightly above the surface of the bronze. The foil could then have been placed over the silver wires and hammered down to the silver, forming a cold weld. Gold, which is heavily diffused into the silver but which is not continuous, was identified as the lighter areas on the two vertical faces of the silver wire in Figure 5.5. Sheridan suggested that perhaps some gold may have lapped into the groove during hammering and that heavy pressure, while hammering the wire into the groove, forced diffusion. Another possibility is that the sides of the wire may have been gilded. On all the objects where foil and wires are still intact the outline of the grooves is visible in the surface of the gold, and in some places the foil is clearly seen to be trapped under the wire. This physical evidence contradicts Sheridan's theory of the foil overlying the wire.

The drawing in Figure 5.6, which is based on the microscopic examination by Paterakis of the end of a wire in a groove on the leg using ×80 magnification, reveals a gold layer on the top and sides of the wire. The width of the wire is approximately 1 mm. The wire core tested positively for lead with Plumbtesmo qualitative paper test strips. The lead may derive from the leaded bronze of the statue or it may have been alloyed with the silver of the wire. An area that is darker than the wire core on the left side of the wire may be a result of the oxidation of silver or lead. Although silver is

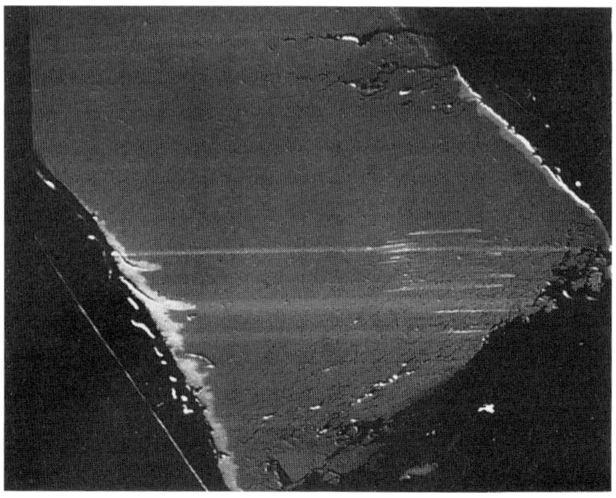

Figure 5.5 SEM photograph at ×60 magnification of the cross-section of silver wire. Gold is seen as the lighter grey areas on the two vertical faces of the silver. (Photograph by W. Sheridan.)

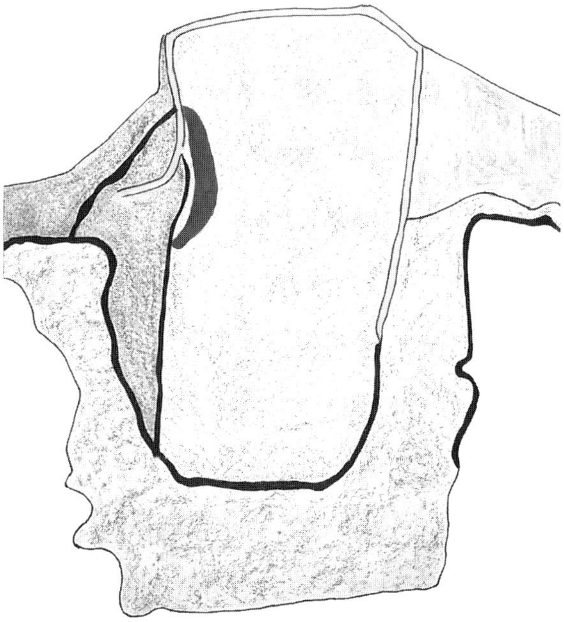

Figure 5.6 Drawing of a cross-section of silver wire in a bronze channel. (See Plate 23.)

more noble than lead, dissimilar metals in alloyed form and in contact with each other do not necessarily obey the normal rules of corrosion expressed in the order of the electrochemical series (Plenderleith and Werner 1976). There is a layer between this darker area and the black outline of the wire, which appears to be wire core metal. There is no evidence of foil in this cross-section. Perhaps the sheets did not extend to the end of this channel at the edge of the leg that would have been welded or soldered on to the horse's body.

The open structure of the channels may have necessitated a solder in order to secure the foil and wire. The cross-section in Figure 5.6 shows the wire inside the rectangular-shaped channel measuring approximately 2 mm wide, which may contain an unidentified solder with some copper corrosion products. A hard silver solder of composition 72–28 wt% silver–copper, known to have been used in antiquity, would explain the presence of these copper corrosion products (Lechtman and Steinberg 1970). This solder has a melting temperature of approximately 780°C, which is sufficiently less than that of the silver wire and of the copper alloy channel. The foil and wire would have been embedded in the molten solder in the grooves. The thin black layer outlining much of the wire may be an intermetallic compound between the silver wire and the solder (Fig. 5.6). The thin black layer outlining the right side of the channel may be an intermetallic compound between the solder and copper alloy channel.

Nike head

Description

The statue has been attributed by H. Thompson to Nike, the winged goddess of victory (Thompson 1940). She has been dated to 440–430BC based on individual features and style. From the peg on top of the head the

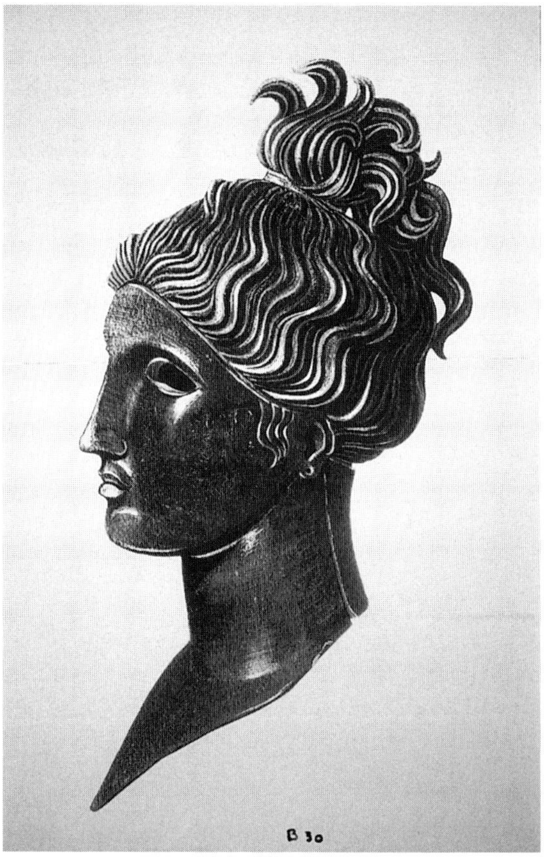

Figure 5.7 Reconstruction drawing of Nike head. (Piet de Jong, Agora Excavations, Pntg # 218.)

scheme of the hairstyle has been recovered from several ancient representations. The reconstruction drawing (Fig. 5.7) by Piet de Jong is based on the vases of Meidias and the coins of Syracuse (Shear 1933). This Nike may have been one of the Golden Nikai of the goddess Athena on which much of the gold reserves of the Athenian state was stored in the 5th and 4th centuries BC (Thompson 1940). A restoration of the statue gives a height of approximately 90 cm, or 3 feet.

Fabrication

The head was cast hollow as a separate unit by the lost wax method (Shear 1933). There is evidence that the head was removable: the finishing of the underside of the V-shaped chest and the inner surface of the neck indicate that they were visible at times; and the edges of the V-shaped chest and the neck are smoothly finished and beveled to fit into a socket (Thompson 1940).

The Nike head differs from the equestrian statue in the number and design of gilded silver sheets. The grooves were intentionally carved in inconspicuous places on the head so that the statue would be presentable without its gold (Shear 1973). The channels were cut after casting with narrow, straight-faced chisels which were first tested on the underside of the 'V' (Thompson 1940). Two periods of gilding have been distinguished by two sets of grooves. The first set of grooves include those in the hair, one across the rear of the neck from which two others descend (Fig. 5.8), and one just below the hair-line in front. It is not known what metal was used to gild the statue initially, the possibilities are silver, gilded silver and gold. Gold would be the most probable for this time period. In the crisis of 406–405BC most of the Nikai statues were stripped to contribute the gold

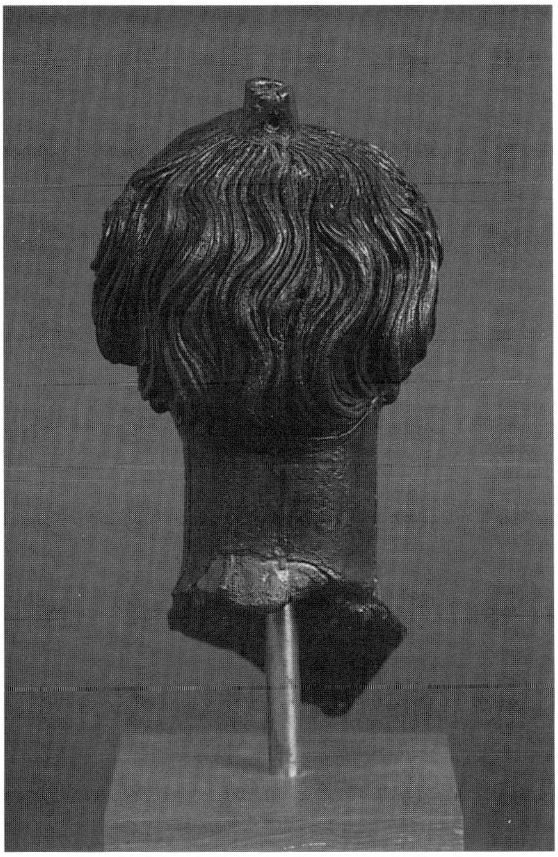

Figure 5.8 Rear of Nike head showing first set of grooves cut in neck. (Photograph by Craig Mauzy, Agora Excavations, American School of Classical Studies, Athens.) (See Plate 24.)

to the needs of the state. After stripping, most of the grooves were disguised by filling them with bronze. The second and final gilding of the Nike head may have been made possible in 336BC by gifts made to Athens by Alexander the Great. New grooves were carved in more conspicuous places except for the channel on the forehead, which was reused. This foil gilding consisted of gilded silver sheets similar in appearance to those of the equestrian statue. Remnants of gilded silver are to be seen in some of the grooves. Final removal of the foil was performed with a straight-faced chisel 8 mm wide, which left scars on the top of the head behind the peg and in front of and behind the right ear. The tops of the ears were removed in this process. In the years 300–295BC a tyrant raided the treasury and many objects of precious metal were sacrificed. The Nike head was probably stripped and discarded at this time. Evidence to support the original intention to gild the statue lies in the facts that: the head is removable; the smooth surface is necessary for physical adhesion of the metal; the most minute details in the bronze, such as the wisps of hair, would have been evident in burnished gold; and the bronze was fully modelled.

Analysis

The bronze of the Nike head was analysed by AAS in the 1970s by P. Craddock at the British Museum (Craddock 1977). It was found to be a tin bronze consistent with the other analyses made by Craddock of 5th century BC statues. A material found on the interior was identified as a mixture of calcium carbonate and bleached beeswax by chemical analysis and infrared spectroscopy (Farnsworth and Simmons 1960). This mixture could not have been used as a core material since the melting temperature of tin bronze, 1000°C, surpasses the melting temperature of the mixture,

67.5°C. It could be an adhesive that had been used to adhere the eyes to the sockets, the topknot on to the head, the head on to the body and the foil on to the statue surface or into the channels.

Discussion of gilding techniques

According to Oddy, the gilding methods that were used in antiquity up to Roman times on silver, copper and copper alloys are foil gilding, diffusion bonding, cold mercury gilding and fire gilding (Oddy 1985). Another method of gilding was the application of organic adhesives such as egg white, which would not have survived on outdoor statuary (Oddy 1993). All these methods except fire gilding fall under Lechtman's category of 'true gilding' methods in which pure gold is applied to the surface of another metal (Lechtman 1971). The fire gilding method falls under Lechtman's category of 'depletion gilding', in which the gold is applied alloyed with mercury.

The evidence to date indicates that the diffusion bonding and foil gilding methods were used on the equestrian statue. Although the gilded silver of the Nike head has not been microscopically examined or analysed with a scanning electron microscope, it may be assumed that similar diffusion bonding and foil gilding techniques using silver and gold were used for the second cladding. The gold was diffusion bonded to the silver sheets and this foil was attached to the statues with the foil gilding method. Sheridan's study shows that diffusion bonding was achieved between an overlayer containing gold and an underlayer containing silver.

According to Oddy, from at least the second millenium BC gold leaf was diffusion bonded to silver by burnishing it directly on to the silver and then heating for a short time to cause interdiffusion of the gold and silver (Oddy 1985). This technique is reported to have been common on silver objects in the Hellenistic period. Lechtman defines gold leaf as a sheet less than 1 micron thick and gold foil greater than 1 micron thick (Lechtman 1971). Gold foil has also been defined as a sheet thick enough to handle that can support its own weight (Oddy 1985). Many layers of leaf or a thickly beaten layer of gold could have been applied to the silver sheet. The 15 micron thickness of the gold on the equestrian statue requires the application of heat to render it more ductile so that it could be hammered over the contours of the statue.

The form of the gold, as leaf or foil, when applied to the silver has not yet been determined for the Athenian statues. In the scanning electron microscope photographs layers of gold leaf are not visible. According to Lechtman solid state diffusion between a gold sheet formed from layers of gold leaf and silver upon heating may result in the disappearance of the separate layers of gold leaf and may impart a thicker appearance to the gold (Kirkendall effect) (Lechtman 1971). Since the resulting gold layer on the equestrian statue exceeds 1 micron in thickness, it can be considered a foil even if it was formed from many layers of gold leaf. The silver sheet is a foil since it also exceeds 1 micron in thickness.

Sheridan did not analyse for mercury so the possibility of its use should not be ruled out. The foil could have been prepared by attaching gold leaf or foil to the silver sheets with the application of mercury followed by heating and burnishing for diffusion. In the cross-sections of the foil and wire analysed by Sheridan, the possibility of mercury in the gold layer cannot be excluded, since it would be approximately the same tone as gold in the photos (Figs 5.4 and 5.5). The first use of mercury gilding in the

Mediterranean world has not been determined. Cold mercury gilding may have preceded fire gilding (Oddy 1993). Vitruvius in the first century BC and Pliny in the first century AD mention mercury gilding (Oddy 1985). The texts of both authors have been interpreted to describe the cold mercury gilding process (Oddy 1993). According to Lechtman the application of mercury between layers of gold leaf causes the disappearance of the individual leaves (Lechtman 1971). Trace amounts of mercury identified by analysis do not necessarily confirm this technique since mercury may be present as an impurity (Lins and Oddy 1975). However, mercury should be preserved from the cold-mercury gilding and fire gilding processes, assuming the gilded layer is intact. In fire gilding some mercury is reported always to remain, which can be detected by sensitive analytical methods such as emission spectrography (Oddy 1993). Fire gilding of gold directly onto the bronze of the equestrian statue would not have been possible due to its high lead content.

Oddy describes foil gilding as the wrapping or covering of an object with foil made of solid gold (or gilded silver) (Oddy 1985). The foil is attached to the statue by physical means, that is by bending the foil around the edges of the object, by riveting it in place, or by hammering it into grooves cut into the surface. The history of foil gilding may be traced from the 3rd millenium BC, when the Egyptians used gold foil to decorate furniture and coffins. Foil gilding with gold foil was in use in Mesopotamia by at least 3,000BC. In the first century AD Pliny refers to a bronze statue of Alexander the Great imported from Greece by Nero who had it gilded. Some time later, when the gilding was removed, Pliny mentions the scars retained and the incisions in which the gold had been secured. The specific adaptation of the foil gilding technique ascribed to this statue based on Pliny's description is that of 'a l'hache' (Oddy *et al.* 1979). The 'a l'hache' method consisted of hammering and burnishing the gold foil on to the roughened surface of the bronze and securing it in place by hammering the foil into grooves that had been cut in a haphazard fashion, criss-crossing the whole surface. The foil gilding method of the Athenian pieces is distinguished from this technique by the intentional placement of the grooves in the surface of the statue. The foil gilding method using grooves in the surface in which to secure the foil does not appear to have been very common on statuary, since the Athenian statues are the only surviving examples found to date (Oddy 1985).

Conclusion

We may summarize hypothetically the processes used to gild the Athenian statues as follows.

Diffusion Bonding (gilding the silver foil)

1. Silver sheets were gilded flat by diffusion bonding of a gold foil or several layers of gold leaf under pressure and possibly with heat prior to their placement on the statues.

Foil Gilding (attaching the gilded silver foil to the statues):

2. The gilded silver sheets were cut to fit roughly areas delineated by grooves that had been cut in the surface, either before or after casting the statues.
3. The foil was hammered against the surface of the statues to conform to the contours, a process most likely aided by heating.

4. The foil was secured to the surface of the bronze pieces with the aid of silver wires that may themselves have been gilded. The cladding sheets were fixed in place by hammering the wires on top of the sheets, forcing them into the grooves. Solder may have been used to secure them in the grooves.
5. The statues were then assembled from these individually cast, foil gilded bronze pieces.

Several points regarding these techniques require clarification with the aid of further analysis, as follows:

- morphology of the gold prior to its application (ie, leaf or foil)
- application of heat during diffusion bonding of the silver and gold
- use of mercury
- composition of the gilded silver foil (alloying of the gold or silver)
- method of securing the gilded silver sheets with the wires
- composition of the wire
- determination of an embedding material (solder) in the grooves and its composition.

For the clarification of these points, sampling of the gilded silver from both statues for analysis is pending permission from the Greek Archaeological Service. D. Scott of the Getty Conservation Institute has generously offered to supervise the analysis by optical metallography, use of an electron microprobe and scanning electron microscopy. An issue related to the original gilding method that may never be clarified by analytical means, however, concerns the possibility of the simultaneous achievement of diffusion bonding and foil gilding of the statues. We don't know if diffusion of the gold and silver could have been accomplished while hammering and burnishing the sheets on to the statues with the aid of heat, thereby bypassing the initial step of bonding on a flat surface.

Acknowledgements

I wish to thank Roger Doonan, Fellow in Archaeometallurgy at the Fitch Laboratory, The British School at Athens, and Elizabeth Hendrix, student in the MA/Conservation Diploma program at New York University, for their assistance and advice during the examination of these statues. I also wish to thank Professor C. Houser for her helpful comments regarding this manuscript.

References

Camp, J. 1997 (personal communication).

Craddock, P. 1977 *Report on the Composition of Eight Fragments of Bronze Sculpture from the Athenian Agora*. London: British Museum Research Laboratory, unpublished.

Farnsworth, M. and Simmons, I. 1960. 'A unique cement from Athens'. *Hesperia* **XXIX**: 118–22.

Houser, C. 1982. 'Alexander's influence on Greek sculpture as seen in a portrait in Athens, Macedonia and Greece'. *Studies in the History of Art* **10**: 229–38.

Houser, C. 1987. *Greek Monumental Bronze Sculpture of the 5th and 4th Centuries BC.* New York, London: Garland Publishing.

Lechtman, H. 1971. 'Ancient methods of gilding silver examples from the old and the new worlds'. In *Science and Archaeology*, R. H. Brill (ed.), 2–31. Cambridge, Mass.: MIT Press.

Lechtman, H. and Steinberg, A. 1970. 'Bronze joining: A study in ancient technology'. In *Art and Technology, A Symposium on Classical Bronzes,* S. Doeringer, D. G. Mitten and A. Steinberg (eds), 5–35. Cambridge, Mass.: MIT Press.

Lins, P. A. and Oddy, W. A. 1975. 'The origins of mercury gilding'. *Journal of Archaeological Science* **2**: 365–73.

Oddy, W. A. 1985 '*Vergoldungen auf Prähistorischen und Klassischen Bronzen* [The gilding of prehistoric and classical bronzes]'. In *Archäologische Bronzen Antike Kunst Moderne Technik*, H. Born (ed.), 64–71. Berlin: Stäatliche Museen Preussischer Kulturbesitz Museum für Vor- und Frühgeschichte.

Oddy, W. A. 1993. 'Gilding of metals in the Old World'. In *Metal Plating and Patination*, S. La Niece and P. Craddock (eds), 171–81. Oxford: Butterworth-Heinemann.

Oddy, W. A., Vlad, L. B. and Meeks, N. D. 1979. 'The gilding of bronze statues in the Greek and Roman world'. In *The Horses of San Marco*, Royal Academy, London (ed.), 182–6. Milan: Olivetti.

Plenderleith, H. J. and Werner, A. E. A. 1976. *The Conservation of Antiquities and Works of Art*. Oxford: Oxford University Press.

Shear, T. L. 1933. 'Excavations in the Athenian Agora: The sculpture'. *Hesperia* **II**: 519–27.

Shear, T. L. Jr. 1973. 'The Athenian Agora: Excavations of 1971'. *Hesperia* **XLII**: 164–73.

Sheridan, W. 1976. Unpublished report on the Agora excavations.

Thompson, H. A. 1940. 'A Golden Nike from the Athenian Agora'. *Harvard Studies in Classical Archaeology,* sup. vol. **1**: 183–210.

Thompson, D. B. 1944. 'The Golden Nikai reconsidered'. *Hesperia* **XIII**: 178–81.

6 Gilding in Britain: Celtic, Roman and Saxon

Peter Northover and Kilian Anheuser

Abstract

In the history of metallurgy in the British Isles until Roman times the use of gold is highly episodic, and the practice of gilding even more so. In the Bronze Age gold sheet and foil were made from the earliest times and they were used to decorate a variety of substrates, including people; their attachment to metal was rare and inconsistent. Gold was not important again until the later part of the pre-Roman Iron Age, when in the mid-first century BC the supply of gold was temporarily interrupted, but the gold appearance of objects was maintained by plating, amalgam gilding and, possibly, depletion gilding. When gold supplies were re-established most metal went to the coinage, including gold plated forgeries, while a small amount was used to highlight the decoration of a small number of bronze and silver objects, especially in western Britain. During the Roman occupation the pattern of gilding was very much that of the Roman world; Celtic styles, confined to Ireland and Scotland, are notoriously difficult to date. The migration of Germanic tribes, first at the invitation of the Romans and then uncontrolled, brought a new range of styles, with in particular, an expansion in the use of gilding in personal ornament, especially brooches. Gilding was also a part of the most elaborate metalwork in both pagan and Christian areas, from armour to chalices. This chapter recounts this history and then describes experiments made to reproduce features of amalgam gilding observed in these objects. These experiments have defined the conditions for successful gilding in terms of surface preparation, firing temperature and finishing; variations such as the application of leaf to amalgamated surfaces are discussed. The gilding process is seen to be essentially a low temperature one but it is also demonstrated that cold amalgam gilding is not feasible.

Introduction

The British Isles lie close to the northwest coast of Europe. In antiquity they were open to influences from what may be described as Atlantic, Central and Northern Europe and the history of metalworking there is in part a record of the endless fluctuations in the relative importance of each of these. These influences rarely survived unmodified and a number of metallurgical innovations had an independent development in Britain, two such being wire drawing and the use of leaded bronze (Northover 1982,

1995). Underlying this metallurgical activity was the relative abundance of non-ferrous metal resources in the British Isles, an abundance such that in the second half of the eighteenth century AD Great Britain was the largest copper producer in the world, and even today one tin mine survives and, even more tenuously, a single gold mine. Despite these opportunities metallurgy reached the British Isles last of all in western Europe. The earliest dates for mining activity centre around 2450BC (O'Brien forthcoming), and similar dates have been obtained for the earliest metalwork in Britain – small hammered ornaments of copper and gold (Barclay and Halpin forthcoming). The consumption of gold expanded with the development of larger sheet gold ornaments, such as the *lunulae*, in the last quarter of the third millennium, exactly the time at which bronze was also introduced. The first century of bronze, after 2200BC, saw composite objects deliberately combining copper and bronze for effect, such as large, domed copper rivets in a high tin bronze dagger (Northover forthcoming). Composite objects with gold are a rather later phenomenon, of the first half of the second millennium BC, but these rarely, if ever, directly place gold against metal; more usually the substrate or connection is wood, stone, amber or jet (Eogan 1995: 22–41). It was only later in the Bronze Age, from about 1300BC onwards that gold began to come together with other metals.

Gold surfaces in the later Bronze Age

It is reasonable to suggest that gilding was applied to metal artifacts either as part of a decorative scheme, to give a golden appearance, or more emphatically, to deny the nature of the base substrate. It is only in the later Bronze Age, from the last quarter of the second millennium BC onwards, that gold began to be combined decoratively with other metals. The other exploitation can be emphatically ruled out. From about 1300BC onwards, and moving into a period of increasing climatic, social and economic stress, the scale and splendour of prestige metalwork continually expanded, for example in the production of large bronze cauldrons and repoussé sheet bronze shields. For personal and ceremonial ornaments, such as the flange-twisted gold torcs (Northover 1989), solid gold was used, with the heaviest known torc (melted down in the eighteenth century) exceeding 1.2 kg. Gold foil with impressed decoration *was* used as enhancement to other objects, for example with the Caergwrle model boat (Northover 1994). This is carved from shale and the oars, waves, keel and shields at the gunwales are formed from gold foil with fine geometrical decoration, with the gold foil wrapped around pure tin as a resilient backing. Extending the theme of non-metallic substrates for gilding, it is instructive to realise that in effect people were gilded. The gold cape from Mold in northeast Wales, and what are believed to be conical hats and crowns from Avenches, Schifferstadt and other sites in continental Europe are now interpreted as ceremonial garb.

Either side of the turn of the first millennium BC there was a hiatus in the history of gold and it almost disappeared from use. When gold use reached another climax in 900–700BC, the picture was dominated by Ireland, where numerous massive finds of Bronze Age gold metalwork have been made (Eogan 1983: esp. 69–72, 244–53), a nineteenth century find from near Mooghaun, County Clare, with 146 objects totalling 5.4 kg, being the largest. Now we also have gold being used to clad base metal in objects referred to as '*bullae*' and '*ring-money*', the core usually being bronze or lead. Ring-money offers an interesting variation in surface plating: numerous examples have a tiger-striped appearance, the stripes being a plating of electrum. The mechanism by which this was achieved has yet to be elucidated, but the choice of metals is interesting. An example from Graianog, Llanllyfni, Gwynedd in northwest Wales has been analysed. The gold substrate has 82.66 per cent gold, 11.01 per cent silver and 6.33 per cent copper: a typical Late Bronze Age gold. The stripes, however, contain 54.09 per cent gold, 4.04 per cent copper and 41.88 per cent silver (Green 1988). The regular gold is unrefined natural gold alloyed with copper, but the silver content of the electrum stripes is far above the values normal for late Bronze Age gold. Either a small amount of natural electrum had been located or, perhaps more probably, silver was incorporated in the alloy, which is itself a problem because otherwise silver is unknown in the Irish Bronze Age.

The partial gilding of bronze with gold foil also occurred very rarely during the late Bronze Age. The evidence is reviewed by Taylor (1980: 79). The objects comprise five spearheads, two swords and two discs; for example, on a spearhead from Harrogate, Yorkshire, the socket is decorated with gold foil. The means of attaching the foil is not recorded.

The first Celtic climax

The use of gold in Britain almost came to an end by or during the seventh century BC. The reasons are not clear but troughs in gold use are to be associated with other times of transition and this is one of the most important, the shift from bronze to iron and the beginning of the Iron Age. Our understanding of the metallurgical history of this process is made much harder by the fact that the survival of metalwork from the later seventh and sixth centuries BC is minimal and Britain and Ireland have the appearance of almost being without metal. There is some residual gold use in the form of simple ribbon torcs in Ireland and Scotland, but this is almost undatable.

In southern Britain, on which we now concentrate, gold was thought to re-appear in the second century BC in the form of gold coins imported from Gaul (now Belgium and France) (Cowell 1992; Northover 1992), but it is now suggested that some goldsmithing, based on a transfer of idioms from bronze could have taken place as early as the third century BC (Jope, personal communication). This resurgence of gold and goldsmithing culminated in the output of the smiths who produced the Snettisham Treasure from Norfolk in eastern England in the first half of the first century BC (Clark 1954; Stead 1991). This find is dominated by torcs in gold, silver and copper alloys, together with bracelets and other objects, and with scrap and part-processed metal. Analysis of the gold alloys shows a strong correlation with contemporary gold coinages based on imported bullion.

Gilding has a place in this hoard, since several pieces are mercury gilded (Stead 1991; Northover and Salter 1990). Here, although in some cases the gilding is very thin, it seems certain that the gilding was used to disguise the metal being gilded; for the gilt bronze button terminal torcs, for instance, gold analogues are known from other sites. The context for this lies in the history of the period. There was a major influx of gold into Britain during the Gallic War; the Gauls were buying military aid from the British tribes to defend themselves against Julius Caesar and his legions. When Gaul was overrun by the Romans, the Gallic gold coinage was suppressed and the supply to Britain interrupted. To make the gold go further, debasement was accelerated, and both amalgam gilding and surface enrichment were employed to make objects look more golden. It is also possible that depletion gilding was used, but this is not yet proven (Oddy and Cowell, 1993). This link to the Gallic War makes it likely that this gilding was carried out *c.*50BC, making the amalgam gilding some of the earliest dated examples so far west. This discovery is still rather new, and the path by which the method reached eastern England and the source of the mercury are still not known, although Iberia and Italy are the obvious candidates for the latter.

Gold plating was also extended to the coinage. There are amalgam-gilded discs in the Snettisham Treasure that have the appearance of coin blanks, but the method was not normally applied to the coinage. It appears that copper or, perhaps, base alloy blanks were wrapped in gold alloy foil or particles, which were then diffusion bonded to the core. Silver alloy was applied to coins and torcs in the same way and the technique here has been studied in more detail (Northover and Salter 1990; Oddy and Cowell 1993; Anheuser and Northover 1994).

When the southern British tribes adapted themselves to post-war conditions, the currency was stabilised around a new red-gold alloy, and the use of gold for other purposes was largely abandoned. Exceptions are gold threads from the princely burial in the Lexden Tumulus outside Colchester, Essex, and gilded mail mounts in the same find (Foster 1986), as well as non-amalgam gilding on a silver cup from a high-status burial of the late first century BC at Welwyn, Hertfordshire (Stead 1967). Gilding makes a more direct re-appearance in a 'Celtic' context in the mid-first century AD in the form of a brooch from Carmarthen, Dyfed in southwest Wales. The brooch effectively combines Roman and Celtic styles, the general brooch form being Roman but the decoration very much local in an evolved Celtic style. The brooch is silver and amalgam gilding is used to highlight the background to the design (Boon and Savory, 1972). There are several other objects with related design ideas and, often, a similar use of gilding. They are generally distributed beyond the initial Roman frontier established on the Fosse Way across the English midlands and show Celtic smiths adopting Roman materials and skills but retaining their native idiom. The same silver/gold contrast was developed with tin against copper alloys, for example the openwork brass discs against a tinned copper background in the Tal-y-llyn, Gwynedd hoard (Savory 1963). An earlier simple but striking example of this was the application of a simple, geometric thick tinned stripe down the centre line of an otherwise plain bronze sword scabbard dating to around 10BC (Northover and Salter 1990).

Despite the availability of amalgam gilding for bronze and silver substrates from the mid-first century BC, smiths working in the Celtic style were still using other gilding techniques. A hoard of harness trappings from Stanwick, Yorkshire and dating to approximately the third quarter of the first century AD still uses foil gilding (Oddy 1993). This was either soldered in place or applied with an adhesive; a scientific examination of this material has yet to be made.

Roman Britain

During the occupation of much of Britain by the Romans, during the period AD43–410, metalworking techniques gradually came to be those of the Romanised world. In southern Britain native technology and design were largely submerged by the beginning of the second century AD while they persisted for longer in the highlands of the north and west. In Scotland and Ireland, beyond the boundaries of the Empire, local styles continued to develop, but dating of finds is often problematic (MacGregor 1976; Youngs 1989; Redknap 1991; Oddy 1993). We have seen how amalgam gilding was introduced into a Celtic context in pre-Roman Britain; with the increasing dominance of Roman techniques it had probably become the standard method by the second century AD (Lins and Oddy 1975). Some of the best examples of gilding surviving from Roman Britain come from the very end of the period, for example in the treasures from Thetford, Norfolk (Johns and Potter 1983) and Hoxne, Suffolk (Johns and Bland 1994). The latter was originally contained in a wooden chest and was deposited close to AD410. Besides nearly 15,000 coins, some of gold, the chest contained golden jewellery and a large number of spoons, often gilded, with other tableware such as parcel gilt bronze pepper-pots, and a jug handle in the form of a silver tiger with niello stripes.

The early Saxon contribution

The Anglo-Saxon settlement of Britain in the fifth and sixth centuries AD still has many shadowy areas, although no longer can it be truly called a 'Dark Age', as archaeology continually increases our understanding of the period. The cemeteries of this period have yielded an abundance of gilded bronze jewellery, for example, the saucer brooches of the upper Thames valley (Dickinson 1982), where the whole visible surface when worn is gilded. Other types of brooch may be only partially gilded or mix gilding with tinning and silvering, the latter usually by close plating. The examination of many of these pieces has shown that amalgam was almost universal for simple objects (Oddy 1980). Indeed it is the gilding on brooches of these classes that provided the immediate inspiration for the experiments detailed in the second part of this paper, when what appeared optically to be a rare case of leaf or foil gilding proved to be amalgam gilding retaining over 20 per cent mercury burnished to maximum density.

Where more complex schemes of decoration were developed from the sixth century onwards to reach their peak in items such as the Tara brooch and Ardagh and Derrynaflan chalices from Ireland, panels of filigree and granulation on gold back plates were soldered in place (Youngs 1989; Whitfield 1993). By the beginning of the seventh century the Royal family of a Saxon kingdom could command the wealth to produce a great variety of elaborate armour, regalia and jewellery, of which the helmet from the Sutton Hoo ship burial can serve as an example. On an iron basis are assembled tinned bronze sheets, silver inlaid iron (the crest), silver inlaid bronze castings further enhanced with garnets (the eyebrows) and gilded bronze castings (e.g. eyebrow and crest terminals) (Bruce-Mitford 1978: 138–231). The sources of inspiration come both from the Late Roman/ Byzantine world and from Germanic and Nordic Europe, and there is much to be done still in studying the development of fine metalworking techniques outside the confines of the Roman Empire. We can add that it is in the towns of Saxon England that we first find workshop evidence of mercury gilding. The best example is a small stone mortar for mixing the amalgam from a workshop in *Hamwih,* the Saxon predecessor of the port of Southampton (Bayley 1993).

Fire-gilding in experiment and analysis.

We have noted the appearance of amalgam gilding (or fire gilding) in Britain as early as the mid-first century BC. By AD100–300 in Europe and the Middle East it had become the predominant technique until the invention of electroplating in the mid-nineteenth century. So far, fire gilding has usually been identified through its mercury content by qualitative analysis, using X-ray fluorescence or emission spectroscopy (Oddy 1993; Oddy and Cowell 1993). Amalgam gilding was carried out by grinding together small pieces of gold leaf with excess mercury, by which a paste of gold amalgam particles (Au_2Hg) in mercury was formed. Application of this paste to a copper alloy or silver object and subsequent heating produced a firmly bonded yet porous gilding layer, which by burnishing was given a smooth and reflective surface. A possible variation of the technique was the application of gold foil onto a copper or silver surface coated with mercury, where the gold amalgam formed *in situ*, with the same subsequent heating and burnishing of the object.

This section presents results of the metallographic examination and quantitative analysis of original and replication fire-gilding layers. A series of amalgam gilding experiments was conducted under controlled conditions (temperature, duration of heating), and their results are being compared with data from the examination of fire gilding on Romano-British and Anglo-Saxon jewellery to determine exactly how the gilding was carried out. For metallography and analysis small pieces of gilded copper sheet from the replication experiments were mounted vertically in epoxy resin to show a cross-section of the gilding, and polished to a mirror finish with 1 μm diamond paste. In the same manner cross-sections of fire gilding on archaeological objects were prepared from small samples (0.3–1 mm square), which were scraped with a scalpel from the gilded surface. The original and replication samples were examined by optical and scanning electron microscopy, and analysed by quantitative wavelength-dispersive electron probe microanalysis.

Characterization of original fire-gilding layers

Samples of fire gilding from 12 Roman and Anglo-Saxon copper alloy brooches from Britain were investigated metallographically and analysed by electron probe microanalysis. The residual mercury content in the gilding was found to vary between 8 per cent and 25 per cent (Table 6.1). These results were in good accordance with analyses of fire gilding from various other periods and areas undertaken in a more general study of amalgam gilding technology (Anheuser 1996). The thickness of fire gilding was always between 2 and 10 μm, much thicker than leaf gilding, which was applied cold with an adhesive (typical thickness 0.1–0.4 μm). This is explained by the fact that fire gilding is a diffusion bonding process in which a metallurgical bond through interdiffusion of gold and the substrate is formed while the object is heated. The role of mercury is to ensure a close contact of substrate and plating. Because of the diffusion of copper or silver from the substrate into the gold, plating with a thickness of less than 1–2 μm would show visible discolouration. When the object was heated during the gilding process, the mercury content of the gold amalgam decreased from 80–90 per cent to 8–25 per cent. This initially created a very porous gilding layer, which was subsequently compacted by burnishing. Depending on the extent of the burnishing, the final result could be porous and grainy, or compact and superficially resembling gold foil (Fig. 6.1).

Table 6.1 Analytical data of fire gilding.

Object	Location	Catalogue [45] and inventory no.	Hg (wt%)	No. of analysis points	Thickness of gilding (μm)
Fourth century Roman glass stud with gilded flange, from Pakenham, Suffolk	Ashmolean Museum	inv. 1927.279	7.2–8.1	3	4–6
Fourth century Roman glass stud with gilded flange, from High Beech, Loughton, Essex	Ashmolean Museum	inv. 1927.277	11.4–12.5	4	1.5–3
Anglo-Saxon wrist clasp, from Spong Hill, Norfolk	Norwich Castle Museum	Spong Hill 42/4	15.7–23.2	39	
Anglo-Saxon brooch: gilt appliqué, from Barrington, Cambridgeshire	English Heritage, Ancient Monuments Laboratory	BAEH 91, (530) <53A>	15.9–19.3	3	2.5–3
Anglo-Saxon saucer brooch, from Barrington, Cambridgeshire	English Heritage, Ancient Monuments Laboratory	BAEH 91, Tr XIV(354) <28>	20.3–26.2	7	4
Anglo-Saxon applied brooch, from Fairford, Gloucestershire	Ashmolean Museum	cat. 1.13, inv. 1961.1	12–14.8	4	4–5
Anglo-Saxon saucer brooch, from Fairford, Gloucestershire	Ashmolean Museum	cat. 2.51, inv. 1961.96	8.9–14.8	4	6–10
Anglo-Saxon square-headed brooch, from Bury St. Edmunds, Suffolk	Ashmolean Museum	cat. 13.13, inv. 1909.437	9.5–14.9	4	2–7
Anglo-Saxon square-headed brooch, from Berinsfield, Oxfordshire	Ashmolean Museum	cat. 13.12, inv. 1989.418	8–11.5	3	2–4
Anglo-Saxon saucer brooch, from Berinsfield, Oxfordshire	Ashmolean Museum	cat. 2.6, inv. 1989.299	21.2–24	6	3.5–4
Anglo-Saxon applied brooch, from Wheatley, Oxfordshire, piece of appliqué	Ashmolean Museum	cat. 1.20, inv. 1883.37	12.7–19.1	5	7
Anglo-Saxon cruciform brooch, unprovenanced	Ashmolean Museum	cat. 12.39, inv. 1992.105	16.3–18.8	5	5–7

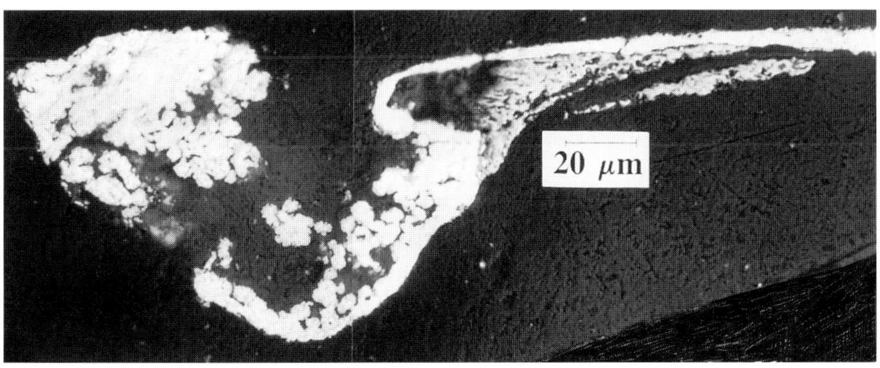

Figure 6.1 Fire-gilding on an Anglo-Saxon brooch (5–7th century) from Berinsfield, Oxfordshire. Ashmolean Museum, inv. no. 1989.418, cat. no. 13.12 [45]. The sample shows a cross-section of an edge of the object with porous unburnished gilding on the side (left) and compact burnished gilding on the face (top right). (See Plate 25.)

Replication experiments

Gold amalgam was prepared from 24 carat gold and pure mercury (99.99 per cent) by heating small pieces of gold foil with approximately five times their weight of mercury in a closed glass vessel to approximately 100°C for several days. Although the gold immediately amalgamated on its surface, even after a few weeks there were still substantial particles of undissolved gold remaining. The mass, consisting of liquid mercury, gold amalgam (Au_2Hg), and gold particles, was transferred into a linen cloth to separate the solid from the excess mercury, which was squeezed through the cloth. All undissolved gold was removed with tweezers. The remaining paste of Au_2Hg and mercury was finely ground in a porcelain mortar and was finally brought to the right consistency (smooth but keeping an imprint of the cloth fabric) by adding some mercury or removing mercury through the cloth. Gold amalgam can be conveniently handled with a wooden or steel spatula (to which it does not adhere) or an amalgamated copper spatula (to which it sticks).

Pieces of copper sheet were thoroughly cleaned with emery paper before the paste of Au_2Hg and mercury was applied. Mercury does not easily wet copper because of a thin oxide layer that always forms on the copper under atmospheric conditions. It was therefore necessary to spread the gold amalgam paste with a cotton stick dipped into dilute aqueous nitric acid, a method recommended by the Italian sixteenth-century goldsmith, Cellini (Ferrero 1971). Alternatively, before the invention of mineral acids in the late Middle Ages, a solution of alum ($KAl[SO_4]_2$) and rock salt (NaCl) in aqueous acetic acid could be used, which was already mentioned by Pliny in an amalgam gilding recipe of the first century AD (Rackham and Warmington 1952: 64–5). Modern recipes recommend treatment of the clean copper with a solution of mercuric nitrate prior to the application of gold amalgam paste (Untracht 1985: 668). Insufficient cleaning led to the formation of a patchy gilding layer. It should be mentioned that like silver and gold, copper can easily be amalgamated by rubbing mercuric sulphide (cinnabar) over the metal. This must also be considered a possible surface pretreatment for copper prior to gilding.

After application of the amalgam the samples were heated in a tube furnace for 10 minutes (plus a five-minute heating-up period) to temperatures up to 600°C. The temperature was measured with a thermocouple inside the tube, close to the sample. During the heating the temperature within the tube was kept at ±20K relative to the preset temperature. In low-temperature experiments (up to 260°C) the sample was attached to a

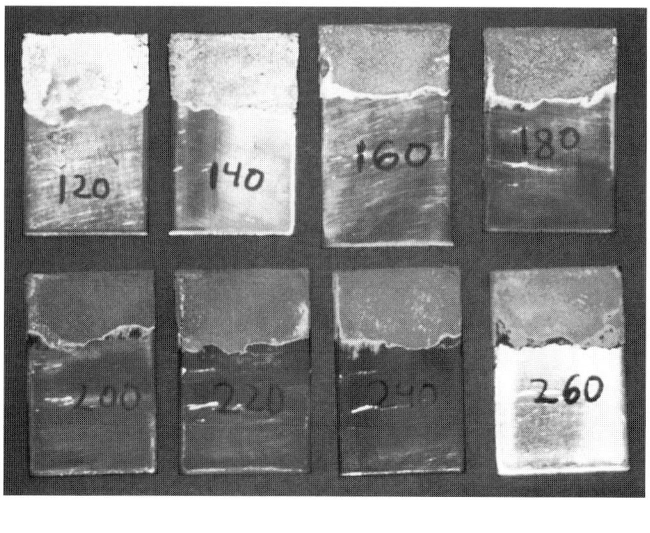

Figure 6.2 Replication fire gilding on copper: colour change of gold amalgam heated for 10 minutes at the temperature written on the sample (in °C). (See Plate 26.)

thermocouple and heated in a glass flask, so that the colour of the amalgam could be observed. At a temperature between 150°C (for very thin layers) and 250°C (thick layers) the colour of the amalgam changed suddenly from grey to dull yellow, which marked the moment when the gilding was ready for burnishing (Fig. 6.2). At this stage all metallic mercury had evaporated and the silver-coloured Au_2Hg had lost further mercury, being converted in a solid-state reaction into yellow Au_3Hg. At no point did the amalgam become fully liquid. After heating the gilding was porous, having lost more than half of its weight through the evaporation of mercury (Fig. 6.3). The samples were cooled down to room temperature and immersed in a solution of 5 per cent NaCl and 5 per cent $KAl(SO_4)_2$ in 5 per cent aqueous acetic acid for about 30 seconds, to remove copper oxide that had formed during the heating, rinsed with water, and finally burnished with an agate burnisher and washing-up liquid as a dispersant. Burnishing compressed the porous structure and created a brilliant golden surface (Fig. 6.4). If the gilding was less than about 10 μm thick, a smooth and even high-quality surface was obtained, a cross-section of which was compact, whereas thicker layers retained some porosity and tended to be uneven. Depending on its final mercury content, fire gilding either consists of a gold–mercury compound with the approximate formula Au_3Hg (20–25

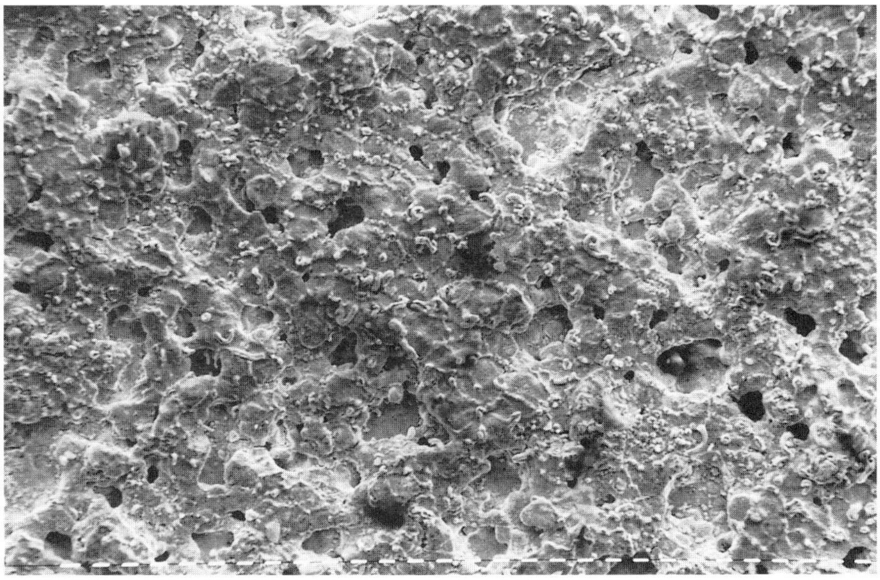

Figure 6.3 Porous surface of replication fire-gilding before burnishing. Scanning electron microscope, secondary electron image. Scale bar equals 10 mm.

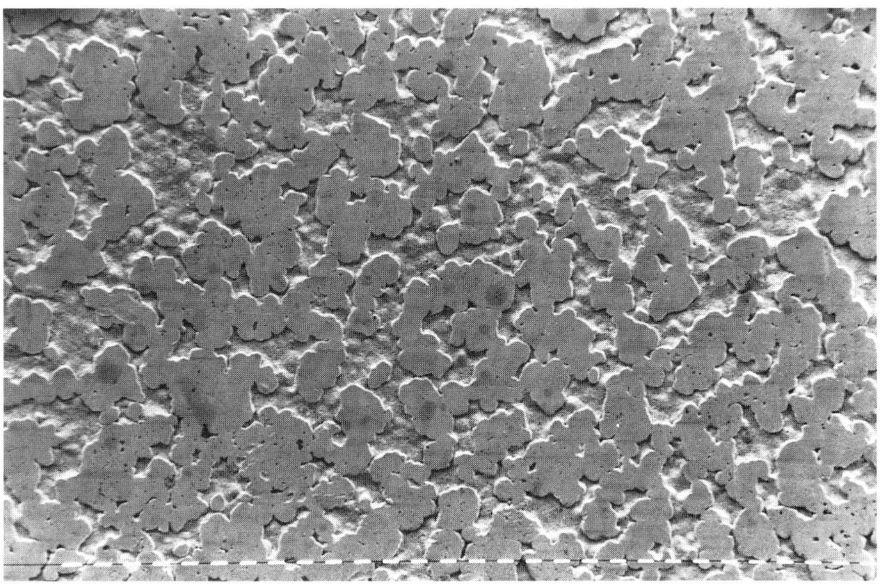

Figure 6.4 Compacted surface of replication fire-gilding after burnishing. Scanning electron microscope, secondary electron image. Scale bar equals 10 mm.

per cent Hg), or a solid solution of mercury in gold (< 15 per cent Hg), or a mixture of both. There is no visual difference between the two compounds, and both can be given a brilliant golden finish by burnishing.

To determine the influence of the heating temperature on the mercury content in the gilding, a series of samples were prepared and analysed, with temperatures being varied between 140°C and 600°C. Figure 6.5 shows the residual mercury content in the gilding as a function of temperature. An increase of the heating time from 10 to 30 or 60 minutes also had some effect on the mercury content at low temperatures, but above 300°C there was little difference. At temperatures above 350°C the copper substrate became covered with a black oxide layer, the thickness of which increased with temperature and length of time of heating. The porous structure of the gilding also led to the formation of copper oxide beneath the gilding and caused the gold to flake off. Best results were obtained at 250–350°C and with heating times not exceeding 15 minutes. Overheated gilding tended to be discoloured and adhered poorly because of oxidation of the substrate. It should be mentioned that heating was an essential step of the amalgam gilding procedure. An alternative cold amalgam gilding process carried out at room temperature, which has occasionally been suggested (Vittori 1978; Oddy 1993), turned out not to be feasible because in room temperature experiments, even after several months, the excess mercury had not evaporated and the resulting dull grey surface did not at all resemble proper gilding (Anheuser 1997).

Conclusions

In previous publications (Lechtman 1971; Oddy 1980; Bunker *et al.* 1993; Twilley 1995) it was stated that fire gilding meant heating an object beyond the boiling point of mercury (357°C), or even as high as 500°C. The results reported here show that this was not the case. Mercury evaporates at a sufficient rate already from 200–250°C, and it is at this temperature that a steep decrease in the mercury content of amalgam gilding layers was observed (Fig. 6.5). The practical temperature range at which the process was carried out would be 250–350°C. Temperatures above 350°C led to excessive oxidation of copper substrates even under the slightly reducing conditions of a charcoal fire. In case of amalgam gilding on silver the upper temperature limit was set by the rapid diffusion of

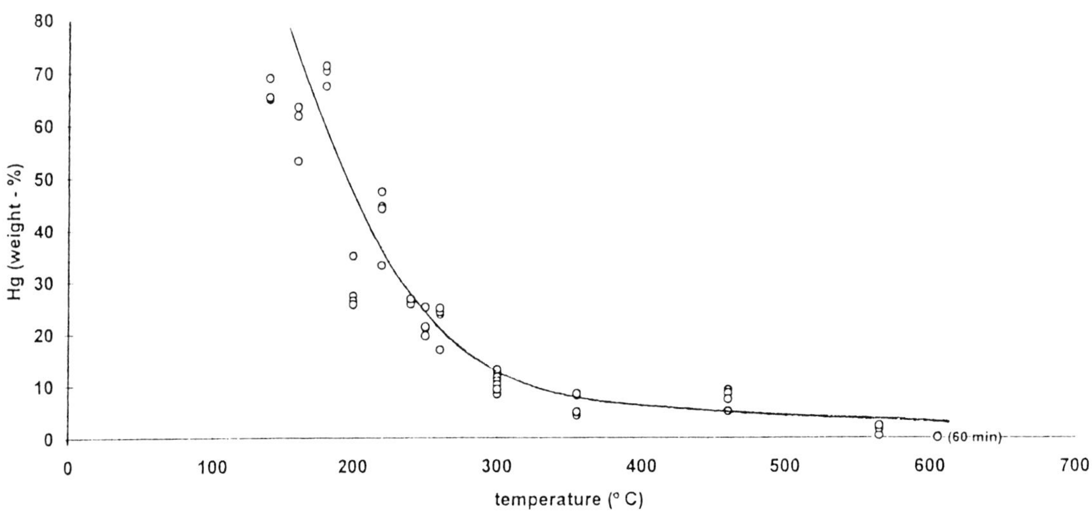

Figure 6.5 Mercury content of amalgam gilding as a function of firing temperature after 10 min heating time.

silver into the gold at high temperatures, which had to be avoided. However, a craftsman did not need to measure the precise temperature. The colour change of the amalgam from grey to yellow indicated to him that heating had been sufficient, and he was also aware of the problems caused by overheating. It is revealing in this context that Cellini in his treatise on goldsmithing (Ferrero 1971) recommended *un dolce fuoco* for the evaporation process, which he called *rasciugare* (drying). Amalgam gilding was in fact a low-temperature process, even though it could not be carried out at room temperature. This was confirmed by the high residual mercury content of 8–25 per cent in Romano-British and Anglo-Saxon fire gilding. At this level mercury should always be detectable by X-ray fluorescence analysis of the gilded surface. If mercury is only found at a much lower level with more sensitive techniques (ie, atomic absorption or emission spectrometry) its presence does not indicate fire gilding but must rather be seen as a contamination.

The thickness of fire gilding was 2–10 μm, which required more gold than the much thinner leaf gilding or electroplating, but a thicker plating was also more wear-resistant. A firm diffusion bond was formed between gold and substrate, and a minimum thickness of 1–2 μm was necessary to prevent discolouration of the gold by the diffusion process. Because the gold was applied as a paste, the technique was ideally suited for the gilding of objects with an irregular or complex shape. These properties made fire gilding an excellent method for the gilding of metalwork and explain why it eventually became the predominant gilding technique for metals, and why there was hardly any Anglo-Saxon gilded metalwork that was not amalgam gilded (e.g. MacGregor and Bolick 1993).

Acknowledgements

KA would like to acknowledge the Ashmolean Museum (Oxford) for permission to sample their objects. Funding was provided by the German Ministry for Research and Technology/German Academic Exchange Service (Hochschulsonderprogramm II).

References

Anheuser, K. 1996. 'An investigation of amalgam gilding and silvering on metalwork'. Unpublished DPhil thesis, Department of Materials, University of Oxford.

Anheuser, K. 1997. 'Cold and hot mercury gilding of metalwork in antiquity'. *Bulletin of the Metals Museum*, **26**: 27–39.

Anheuser, K. and Northover, J. P., 1994. 'Silver plating on Roman and Celtic coins from Britain: A technical study'. *British Numismatic Journal*, **64**: 22–32.

Barclay, A. and Halpin, C. (forthcoming). *Barrow Hills, Radley, Oxfordshire, Vol. 1, The Neolithic and Bronze Age monument complex*. Oxford: Oxford Archaeological Unit, Thames Valley Landscapes Monograph.

Bayley, J. 1993. 'Precious metal-working in Anglo-Saxon England'. In *Outils et Ateliers d'Orfèvres des Temps Anciens*, C. Eluère (ed.), 137–40. St-Germain-en-Laye: Antiquités Nationales, Mémoire 2.

Boon, G. C. and Savory, H. N. 1972. 'A silver trumpet brooch with relief decoration, parcel-gilt, from Carmarthen, and a note on the development of the type'. *Antiquaries Journal* **55**: 41–61.

Bruce-Mitford, R. N. 1978. *The Sutton Hoo Ship Burial*, vol. 2: *Arms, Armour and Regalia*. London: British Museum Publications.

Bunker, E. C., Chase, W. T., Northover, J. P., and Salter, C. J. 1993. 'Some early Chinese examples of mercury gilding', In *Outils et Ateliers d'Orfèvres des Temps*

Anciens, C. Eluère (ed.), 55–66. St-Germain-en-Laye: Antiquités Nationales, Mémoire 2.

Clark, R. R. 1954. 'The early Iron Age treasure from Snettisham, Norfolk'. *Proceedings of the Prehistoric Society* 20: 27–86.

Cowell, M. R. 1992. 'An analytical survey of the British gold coinage'. In *Celtic Coinage: Britain And Beyond*, Proceedings of the 11th Oxford Symposium on Coinage and Monetary History, M Mays (ed.), 202–34. Oxford: British Archaeological Reports, British Series 222.

Dickinson, T. M. 1982. 'Ornament variation in pairs of cast saucer brooches: a case study from the Upper Thames region'. In *Aspects of Production and Style in Dark Age Metalwork*, L. Webster (ed.), 21–50. London: British Museum, Occasional Paper 34.

Eogan, G. 1983. *Hoards of the Irish Later Bronze Age*. Dublin: University College Dublin.

Eogan, G., 1995. *The Accomplished Art; Gold and Gold-Working in Britain and Ireland in the Bronze Age*. Oxford: Oxbow, Monograph 42.

Ferrero, G. G. (ed.) 1971. *Opere di Benvenuto Cellini.* Torino: Unione Tipografico – Editrice Torinese.

Foster, J. 1986. *The Lexden Tumulus - a Reappraisal of an Iron Age Burial from Colchester, Essex*. Oxford: British Archaeological Reports, British Series 156.

Green, H. S. 1988. 'A find of ring-money from Graianog, Llanllyfni, Gwynedd'. *Bwletin y bwrdd Gwybodau Celtaidd,* **35**: 87–91.

Johns, C. and Bland R., 1994. 'The Hoxne late Roman treasure'. *Britannia,* **25**: 165–74.

Johns, C. and Potter, T. 1983 *The Thetford Treasure*. London: British Museum Publications.

Lechtman, H. N. 1971. 'Ancient methods of gilding silver: Examples from the Old and the New Worlds'. In *Science and Archaeology,* R. H. Brill (ed.), 2–30. Cambridge, Mass.: MIT Press.

Lins, P. A. and Oddy, W. A. 1975. 'The origins of mercury gilding'. *Journal of Archaeological Science,* **2**(4): 365–374.

MacGregor, M. 1976. *Early Celtic Art in Northern Britain*, 2 vols. Leicester: Leicester University Press.

MacGregor, A. and Bolick. E. 1993. *Ashmolean Museum, Oxford: A Summary Catalogue of the Anglo-Saxon Collections (Non-Ferrous Metals)*. Oxford: British Archaeological Reports, British Series 320.

Northover, J. P. 1982. 'The metallurgy of the Wilburton hoards'. *Oxford Journal of Archaeology,* **1**(1): 69–110.

Northover, J. P. 1989. 'The St Helier gold torc'. *Bulletin de la Société Jersiaise,* **25**(1): 112–37.

Northover, J. P. 1992. 'Materials issues in the Celtic coinage'. In *Celtic Coinage: Britain and Beyond*, Proceedings of the 11th Oxford Symposium on Coinage and Monetary History, M. Mays (ed.), 235–300. Oxford: British Archaeological Reports 222.

Northover, J. P. 1994. 'Bronze Age gold in Britain'. In *Prehistoric Gold in Europe*, G. Morteani and J. P. Northover (eds), 515–32. Dordrecht: Kluwer Academic Publishers.

Northover, J. P. 1995. 'The Isleham wire-drawing dies'. In *Trans Europam – Festschrift für Margarita Primas*, B. Schmid-Sikimic (ed.), 15–22. Bonn: Dr Rudolf Habelt.

Northover, J. P. 1999. 'The earliest metalworking in southern Britain', in *The Beginnings of Metallurgy*, A. Hauptmann, E. Pernicka, Th. Rehren and Ü Yalçin (eds), 211–26. Bochum: Deutsches Bergbaumuseum: Der Anschnitt Beiheft 9.

Northover, J. P. and Salter, C. J. 1990. 'The decorative metallurgy of the Celts'. *Materials Characterization* **25**: 109–123.

O'Brien, W. (forthcoming). 'Ross Island and the origins of Irish metallurgy'. In *Ireland in the Bronze Age*, W. O.'Brien (ed.), Proceedings of a conference at Dublin Castle, April 1995, Dublin.

Oddy, W. A. 1980. 'Gilding and tinning in Anglo-Saxon England'. In *Aspects of Early Metallurgy*, W. A. Oddy (ed.), 129–34. London: British Museum, Occasional Paper.

Oddy, W. A. 1993. 'Gilding of metals in the Old World'. In *Metal plating and*

patination, S. La Niece and P. T. Craddock (eds), 171–81. Oxford: Butterworth-Heinemann.
Oddy, W. A. and Cowell, M. R. 1993. 'The technology of gilded coin forgeries'. In *Metallurgy in Numismatics III*, M. Archibald and M. R. Cowell (eds), 199–226. London: Royal Numismatic Society, Special Publication 24.
Rackham, H. and Warmington, E. H. 1952. *Pliny: Natural History, 33–5.* Cambridge, Mass. and London: Loeb (text and English translation).
Redknap, M. 1991. *The Christian Celts*. Cardiff: National Museum of Wales.
Savory, H. N. 1963. 'A new hoard of La Tène metalwork from Merionethshire'. *Bwletin y bwrdd Gwybodau Celtaidd*, **20**: 449–75.
Stead, I. M. 1967. 'A La Tène III burial at Welwyn Garden City'. *Archaeologia,* **101**: 1–62.
Stead, I. M. 1991. 'The Snettisham Treasure: Excavations in 1990'. *Antiquity,* **65**: 47–65.
Taylor, J. J. 1980. *Bronze Age goldwork in the British Isles.* Cambridge: Cambridge University Press.
Twilley, J. 1995. 'Technical investigations of an early 15th century Chinese cloisonné offering stand'. In *The Ceramics Cultural Heritage*, P. Vicenzini (ed.), 161–73. Faenza: Techna.
Untracht, O. 1985. *Jewelry Concepts and Technology.* New York: Doubleday.
Vittori, O. 1978. 'Interpreting Pliny's gilding: Archaeological implications'. *Rivista di Archaeologia*, **2**: 71–81.
Whitfield, N. 1993. 'Some new research on gold and gold filigree from early medieval Ireland and Scotland'. In *Outils et Ateliers d'Orfèvres des Temps Anciens*, C. Eluère (ed.), 125–36. St-Germain-en-Laye: Antiquités Nationales, Mémoire 2.
Youngs, S. 1989. *The Work of Angels: Masterpieces of Celtic Metalwork, 6th–9th Centuries AD.* London: British Museum Publications.

7 A Study of the Gilding of Silver in Byzantium

Pete Dandridge

Abstract

While Byzantine goldsmiths used partial gilding extensively as a decorative embellishment on silver,[1] there are few scientific analyses or technical descriptions of the methods utilized by them to adhere the gold to the silver, and those that exist are focused exclusively on the Early Byzantine period – defined as beginning with the founding of Constantinople in AD324 and terminating with the end of Iconoclasm in AD843. Two recent acquisitions by the Metropolitan Museum of Art have provided an opportunity not only to review the pertinent literature, but to supplement the existing studies, and extend the time frame covered in the literature into the later part of the Middle Byzantine era, AD843–1261.

Introduction

Any study documenting the technical evolution of Byzantine metalworking traditions needs to consider the potential material relationship the Empire had to its cultural antecedents and to its neighbours. Fortunately, there is an extensive corpus of research devoted to the study of Roman and medieval metalwork. In addition there are several contemporary literary sources beginning with Vitruvius in the first century BC and Pliny in the first century AD and extending through the Romanesque period and Theophilus's treatise of the beginning of the twelfth century, corresponding roughly to the period under discussion here (Caley 1926; Rackham 1952; Morgan 1960; Dodwell 1971–72; Smith and Hawthorne 1974; Vittori 1978; Hawthorne and Smith 1979; Oddy 1988; Raub 1993).[2] With the aid of these documentary resources and the information garnered from the examination of the objects themselves, two methods for the gilding of metal have traditionally been considered as dominant in the West beginning in the second or third centuries AD and extending through the medieval period. The first is the application of gold leaf onto a substrate pre-amalgamated with mercury and the second, the spreading of an amalgam or paste of gold and mercury on to a cleaned or pre-amalgamated surface with the subsequent heating and burnishing of the object in either instance (Lins and Oddy 1975; Oddy 1988, 1993; Scott 1991: 683; Mango and Bennett 1994: 33). Vittori's interpretation of Pliny's text added the possibility of cold-amalgam gilding, or the overlaying of gold leaf on to a surface amalgamated with mercury and independent of the application of heat (Vittori 1978). Common to all of these processes is the use of gold leaf and not foil. Leaf and foil have been variously distinguished in recent literature. Oddy characterizes leaf as not having sufficient dimensional cohesion to maintain its form when picked up with the fingers (Oddy 1988: 10), while Lechtman quantifies the difference between leaf and foil by ascribing a thickness of less than one micron to leaf and to foil a value greater than a micron (Lechtman 1971: 2–3).

While there are no extant literary descriptions or treatises concerning Sasanian workshop practices, numerous studies focus on surviving examples of that culture's silverwork (Lefferts 1966; Chase 1968; Lechtman 1971; Gibbons *et al.* 1974, 1979; Meyers *et al.* 1974; Meyers 1981; Atil *et al.* 1985; Lang and Hughes 1990; Jett 1992). Only one gilding technique is reported in these investigations to have been in general use – the application of an amalgam of gold and mercury on to the silver surface (Lefferts 1966: 149–50; Chase 1968: 83, 91; Gibbons *et al.* 1979: 173; Meyers 1981: 150; Jett 1992: 59–60) – however, two exceptions to the rule have been noted: a third century AD horse rhyton in the Cleveland Museum of Art partially gilt with gold foil laid over a mercury amalgamated silver surface (Lechtman 1971: 29), and a sixth century AD silver bowl in the Sackler Gallery selectively gilt with leaf, but with no mercury detected by X-ray fluorescence analyses (Jett 1992: 59, 183).

Compared to Roman and Near Eastern practices, Byzantine techniques used in the gilding of silver have been infrequently studied and few objects have been examined. Lechtman's metallographic and electron microbeam probes of gilt silver samples from a sixth century book cover in the collection of Dumbarton Oaks, and her subsequent reproduction of the technique in the laboratory was the first and most comprehensive examination (Lechtman 1971: 6–10). Several different phenomena are visible in the cross-sections of the book cover. In one section, discrete layers of gold leaf are visible even as they conform roughly to the varied surface topography of the corroded silver substrate. In an adjacent area, the gold reads as a visually homogeneous layer in intimate contact with the substrate with the probe across the section indicating a two micron thick strata of pure gold at the surface and a twelve micron diffusion zone with gold, silver, and mercury detected. The largest amount of mercury is indicated just inside the point of highest gold concentration and falls off slowly through the diffusion zone. Lechtman attributed the gilding to the initial amalgamating of the silver substrate with mercury, which was then overlaid by multiple applications of gold leaf, either with interspersed warming or with finishing by a single, longer heating.

Snow and Weisser examined five partial-gilt silver objects dating from the sixth century in the Hama Treasure at the Walters Art Gallery (Snow and Weisser 1986). Two chalices, an ewer and a flask all exhibited discernible evidence that sheets of gold leaf were used in the gilding process with a granular gold layer also apparent on the neck of the ewer (Snow and Weisser 1986: 52, 55–6). X-ray fluorescence analyses of the gold on all of the objects confirmed the presence of trace amounts of mercury except on one of the chalices and on the neck of the ewer with its granular gilding, which was physically inaccessible to analysis. The fifth object, a box, showed no sign of leaf but mercury was detected and visual evidence was found of granular gold. The authors, cautiously interpreting their results, suggest that the detection of mercury indicates that all of the objects were gilded by initially amalgamating the silver with mercury and subsequently applying gold leaf. They further suggest, however, that the one chalice where no mercury was detected may have been gilded by burnishing leaf onto the surface of the silver (Snow and Weisser 1986: 52).

Emission spectroscopy and/or X-ray fluorescence were used by Oddy, Cowell and La Niece to identify mercury within the gilding on five sixth and seventh century Byzantine silver objects in the British Museum that are all described as fire gilt, which they define as pre-amalgamating the silver surface with mercury, applying either a paste of mercury and gold or sheets of gold leaf, warming the object, rinsing the surface, and polishing the gold (Oddy 1988: 11, 19). There was no visual or metallographic characterization of the gilding in their study.

Most recently, Newman examined an Early Byzantine, silver-gilt chalice in the collection of Dumbarton Oaks (Newman and Lie 1992: 81, Figs 5 and 6). A metallographic section and microprobe analysis showed there to be a 12 micron thick layer of granular gold above a 12 micron area of diffusion from the silver interface into the oxygenated zone. Unlike Lechtman's probe, Newman found that the greatest concentration of mercury occurred within the oxygenated zone of the silver substrate where there was little diffusion of gold. The gilding of the chalice was interpreted as the result of a paste of mercury and gold having been applied to the silver, followed by heating, and a final burnishing.

The evidence from these studies of Early Byzantine silver suggest that four different techniques were employed for their gilding: an amalgam of gold and mercury applied over a cleaned substrate followed by gentle warming as postulated by Newman; an amalgam of gold and mercury over a surface pre-amalgamated with mercury followed by gentle warming as indicated by Oddy; gold leaf applied over a surface pre-amalgamated with mercury followed by gentle warming as proposed by Lechtman, Oddy, and Snow and Weisser; and diffusion gilding in the instance of the Walters' chalice. These studies provide a context for the evaluation of the Metropolitan Museum's recent acquisitions.

In 1986, the Metropolitan Museum of Art acquired a group of objects consisting of ten chalices, three censers, a dove and a wine strainer known collectively as the Attarouthi Treasure and dating from the late sixth or early seventh century (Fig. 7.1).[3] The chalices vary in their size and weight, but all are similar in form with wide flaring feet with encircling borders, flaring stems, bulbous knobs with wind-blown acanthus leaves, and large bowls with out-turned, thickened lips. The bowls are embellished with three decorative schemes: sacred figures set within arcuated frames; unframed sacred images alternating with abstract symbols; and

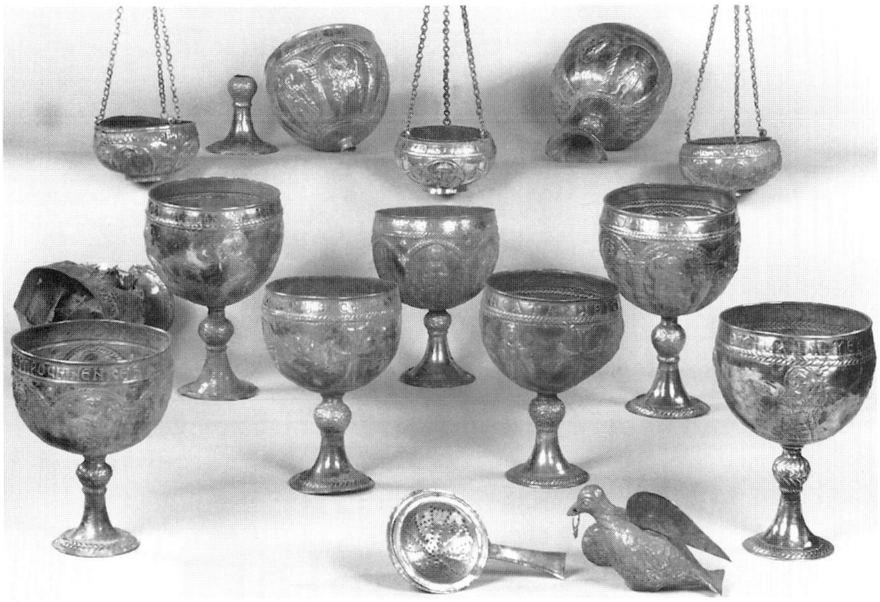

Figure 7.1 Attarouthi Treasure, before treatment. (The Metropolitan Museum of Art, Purchase, Rogers Fund and Henry J. and Drue E. Heinz Foundation, Norbert Schimmel and Lila Acheson Wallace Gifts, 1986. (1986.3.1-.15). All rights reserved, The Metropolitan Museum of Art.)

abstract symbols alone. Horizontal borders bracket the frieze-like scenes. Some of the figures represented include a youthful Christ, the Virgin *Orans* and various saints and archangels. Greek dedicatory inscriptions encircle the upper register of decoration below the chalices' and censers' rims and enabled the treasure to be localized to two churches in the town of Attarouthi, in what is now northern Syria (Frazer 1988). While the decorative and iconographic vocabulary is similar on all of the chalices, they represent the work of different goldsmiths with the exception of one identifiable pair.

What made this purchase so exciting in terms of the history of their technology was the objects' excellent state of preservation. The silver looked to have survived its burial and excavation remarkably unscathed apart from isolated instances of deformation, localized pitting and loss associated with chloride corrosion, scattered areas of copper corrosion overlaying both silver and gilded sections, calcitic deposits and soil accretion. The only apparent evidence of post-excavation intervention was in the cold reforming of distortions on two of the chalices and the reduction of surface dirt and accretion.

All of the objects in the treasure were raised and/or sunk from hammered silver sheet with embossed decorative elements raised from the interior in a repoussé technique, and chased from the front. The immediacy of the forms and the figures and the surety with which they were modelled suggest that, in most instances, only one cycle of raising and chasing would have been necessary to execute the design, although several annealings may have been required. Whether the planished surfaces of the silver were scraped prior to the execution of the repoussé designs is unclear. Certainly, the presence of centering marks on the interiors of the bowls and the feet, the horizontally incised borders, and the horizontal, parallel striations on the surfaces of many of the feet indicate that both bowls and feet were mounted on lathes and partially finished with both scrapers and burnishing tools. Unlike many of the feet, which show the parallel striations indicative of lathe finishing, the reserved areas on the chalice bowls exhibit parallel scraper marks in a loosely horizontal orientation with the tool used both to clarify the borders of the partial-gilt areas where the gilding may have flowed or spread unintentionally and to smooth the reserved fields. As a final finish, both the scraped silver and the gilt areas were compressed with a burnisher. The mechanical joining of the foot and the bowl would have been the last step in the fabrication of the chalices and would have entailed inserting the slightly flared, closed top of the foot into the soldered-on collar at the base of the bowl and gently working the lip of the collar down over the top of the stem: a 'ring joint' in Mango's typology (Mango 1986: 70).[4]

With the naked eye and with low power magnification it is apparent that gold leaf covers the gilded areas. In the recesses of the design, the leaf was poorly adhered to the substrate, and, where it is now lifting, either another layer of leaf or a matte, ocher-colored, granular gold of substantial depth is revealed (Fig. 7.2). Across the raised elements of the design, the gold leaf has a hard, seamless appearance that maintains the rich gold colour of the recesses; however, even within these well burnished areas, there have been losses to the leaf that appear to be related to lack of adhesion rather than to abrasion or wear. Where the granular gold has been exposed on either recessed or raised areas, it does not exhibit a degree of distortion or compression that would suggest that it had ever been intended as a finished surface. Further evidence for the integrity of the gold leaf are the undisturbed, scattered areas of silver chloride and copper corrosion products overlaying adjoining areas of reserved and gilt silver, the homogeneity

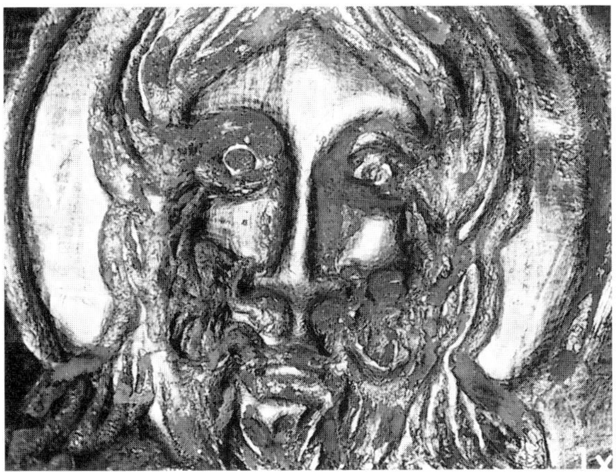

Figure 7.2 Detail of the head of Christ from chalice 1986.3.2. (The Metropolitan Museum of Art, Purchase, Rogers Fund and Henry J. and Drue E. Heinz Foundation, Norbery Schimmel and Lila Acheson Wallace Gifts, 1986 (1986.3.2). Photograph, author, Sherman Fairchild Center for Objects Conservation, The Metropolitan Museum of Art.)

of calcitic deposits on both silver and gold, and the absence of any corrosion or accretion between the leaf and the amalgam in metallographic cross-sections that would indicate the gold leaf represented a later regilding.

In 1993, the museum made another major acquisition of Byzantine art; in this instance, a processional cross from the late Middle Byzantine era, dated by Evans to the first half of the 11th century (Evans and Wixon 1997: 62–4).[5] The cross is composed of multiple pieces of hammered silver sheet, embellished with a rich array of figurative medallions, rinceaux patterns, and decorative borders that were all executed in a repoussé technique and gilt (Fig. 7.3). To support the thinly worked, deeply modelled figural busts, the cavities on the reverse were filled with gypsum and capped with soldered on iron discs.[6] The silver sheets making up each arm of the cross were then attached with the same solder to a thin iron core in the overall shape of the cross as the final step in its fabrication. Excess solder that flowed out from the interior during attachment is visible along the edges of the silver overlaying sections of the gilding. The central crossings were covered by separately modelled medallions depicting Christ on the front and the medical saint, Thalelaeous, on the reverse with each of these busts modelled in slightly larger scale than those on the termini and

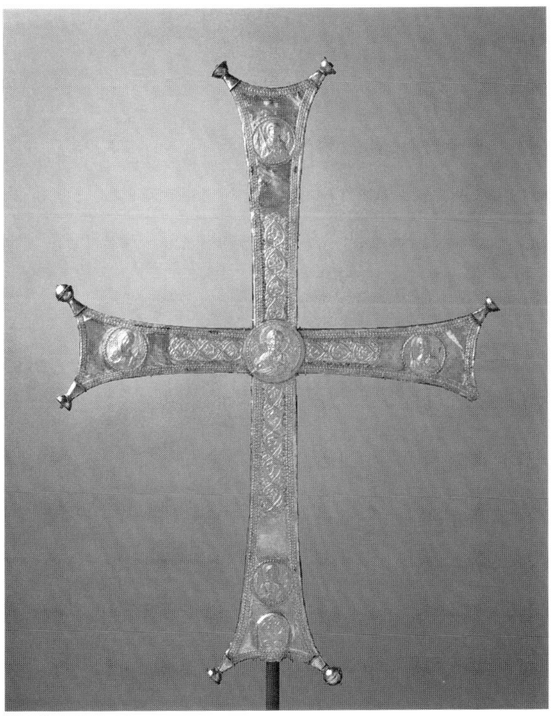

Figure 7.3 Processional Cross, after treatment. (The Metropolitan Museum of Art, Rogers Fund, 1993. (1993.163) (Photograph © 1996, The Metropolitan Museum of Art.) (See Plate 27.)

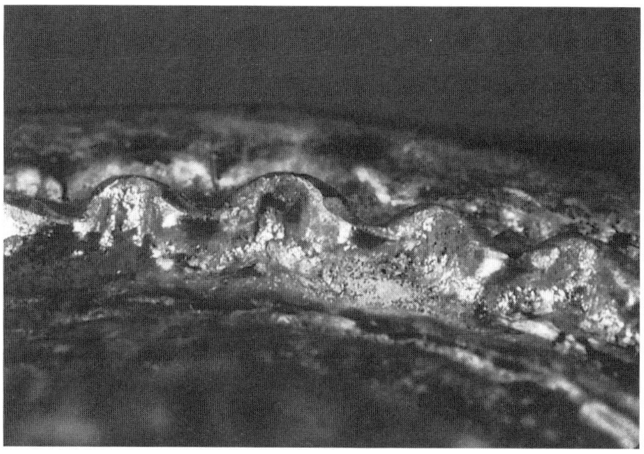

Figure 7.4 Detail of the border around the central medallion of Christ, before treatment. (The Metropolitan Museum of Art, Rogers Fund, 1993. (1993.163) Photograph, author, Sherman Fairchild Center for Objects Conservation, The Metropolitan Museum of Art.) (See Plate 28.)

supported by a similar gypsum packing and iron disc backing. The burnished silver and partial-gilt surfaces are well preserved. Under the microscope, it is possible to discern traces of leaf lifting away from the surface (Fig. 7.4), scattered outlines of a corner or an edge of leaf, and an underlying layer of granular gold. Unlike the Attarouthi chalices, the gilt surfaces were very methodically burnished, but even so, there are still discrete areas where the gold leaf has lifted away.

Analysis

Three of the chalices contained areas of loss or tearing that facilitated sampling. The extracted cross-sections were analysed by metallography, scanning electron microscopy,[7] and electron beam microprobe.[8]

The first of the chalices to be discussed bears the donor inscription, 'Of St. Stephen of the village at Attarotha' (Fig. 7.5). The sample was taken from above Christ's proper right hand in an area that had been crushed and reworked. Analysis of the silver-rich alloy in the interior of the polished section indicates that it contains 95.05 per cent silver and 4.95 per cent copper – a composition that corresponds to previously reported analyses of the major elements in Byzantine silver (Schweizer 1977; Schweizer and Meyers 1978; Carlson 1986; Craine 1980; Meyers 1992; Newman and Lie 1992: 77–8, 87). The copper represents an intentional addition and would have been added to increase the strength and hardness of the silver. Whether the silver in this chalice or in any of the objects to be discussed was freshly smelted or recycled is unclear, but in either case it would have been extracted originally from lead ores containing trace amounts of gold and iridium that would have been retained with the silver throughout the smelting process, and, depending upon the type of analyses, detected along with the silver (Meyers 1992: 170). Similarly, minor amounts of lead not removed completely during cupellation are found with the silver. The lack of gold and lead in the present analyses may be due to the fact that the limits of detection with EDS are 0.5 and 1.0 per cent respectively.

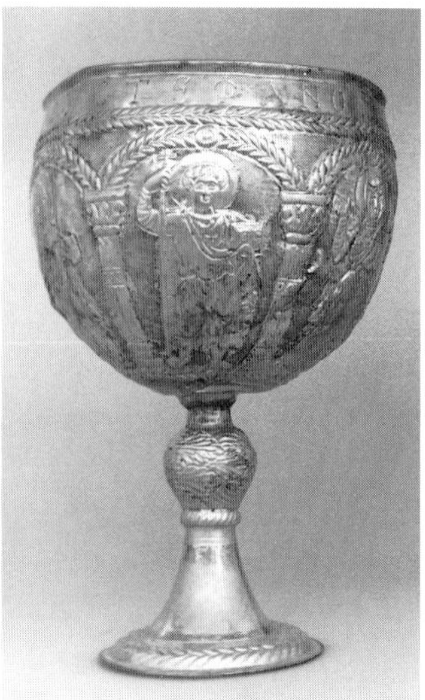

Figure 7.5 Chalice, 1986.3.2, after treatment. (The Metropolitan Museum of Art, Purchase, Rogers Fund and Henry J. and Drue E. Heinz Foundation, Norbery Schimmel and Lila Acheson Wallace Gifts, 1986 (1986.3.2). Photograph, author, Sherman Fairchild Center for Objects Conservation, The Metropolitan Museum of Art.)

The diffusion of oxygen into the silver during the cold working and annealing of the alloy creates oxygenated zones on either side of the silver that average 20 microns in depth and have well defined, sharp interfaces with the interior metal. In this sample, the zone on the chalice's inner surface has been substantially lost due to abrasion. When the section is etch-polished with a potassium dichromate reagent (Kehl 1949: 430) (Fig. 7.6)[9], the interior alloy exhibits many of the characteristics associated with ancient silver – large grain size, grain boundary displacement as evinced by

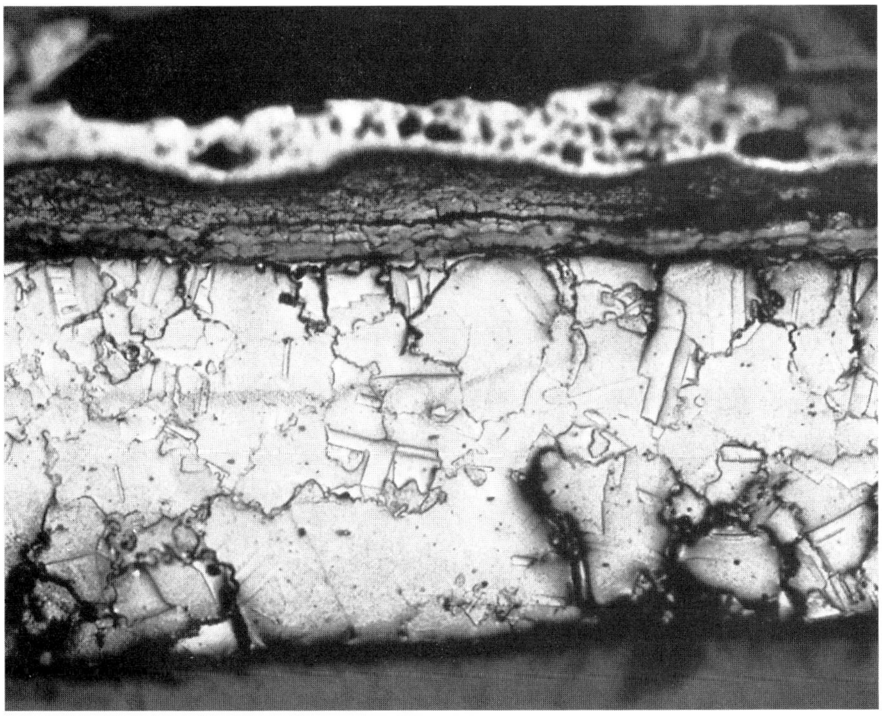

Figure 7.6 Detail of an etched cross-section from chalice 1986.3.2, ×500. (The Metropolitan Museum of Art, Purchase, Rogers Fund and Henry J. and Drue E. Heinz Foundation, Norbery Schimmel and Lila Acheson Wallace Gifts, 1986 (1986.3.2). Photograph, author, Sherman Fairchild Center for Objects Conservation, The Metropolitan Museum of Art.)

jagged contours, discontinuous and intergranular precipitation and cellular growth along the grain boundaries, and bent annealing twins and strain marks associated with cold working and annealing (Lefferts 1966: 148; Smith 1966: 35–8; Chase 1968: 88, pl. 11, Fig. 25; Schweizer and Meyers 1978; Craine 1980: 13–14; Scott 1991: 671–2, 677–83; Newman and Lie 1992: pl. 1). The more extreme intergranular corrosion and separation, the stress lines, and the deformation of the annealing twins could be associated with the chalice's crushing and post-excavation reformation in the area sampled.

Two different visual effects are apparent within each of the oxygenated zones. Abutting the interior metal is a 'reddish' strata with relatively large grains and scattered precipitation with some of the grains exhibiting a flat-

Table 7.1 Compositional analyses: EDS.

Acc. No.	Location	Ag	Au	Cu	Hg	Pb
1986.3.2	Silver substrate	95.05		4.95		
	Oxygenated zone	95.95		4.05		
	Granular gold	3.01	84.47	0.29	12.23	
	Gold leaf	2.23	97.75	0.02		
1986.3.6	Silver substrate	93.03		5.61		1.36
	Oxygenated zone	94.18		5.01		0.81
	Granular gold	7.57	79.48	0.32	12.64	
	Gold leaf	1.10	98.84	0.06		
1986.3.9	Silver substrate	95.88		4.12		
	Oxygenated zone	86.32		8.03	5.15	
	Silver/mercury interface*	76.62	6.12	4.25	13.01	
	Granular gold	31.63	54.12	0.73	13.52	
	Gold leaf	3.46	96.13	0.41		
1993.163	Silver substrate	90.91	2.32	5.06		1.70
	Oxygenated zone	90.73	2.19	4.35	2.73	
	Oxygenated zone	48.26	5.60	2.16	43.98	
	Granular gold	9.74	73.34	0.42	16.50	
	Gold zone	7.35	78.78	0.20	13.67	

* The area composed of a homogeneous, compact layer of silver-white metal ca. 1.5 microns and a porous layer ca. 1.0 microns thick.

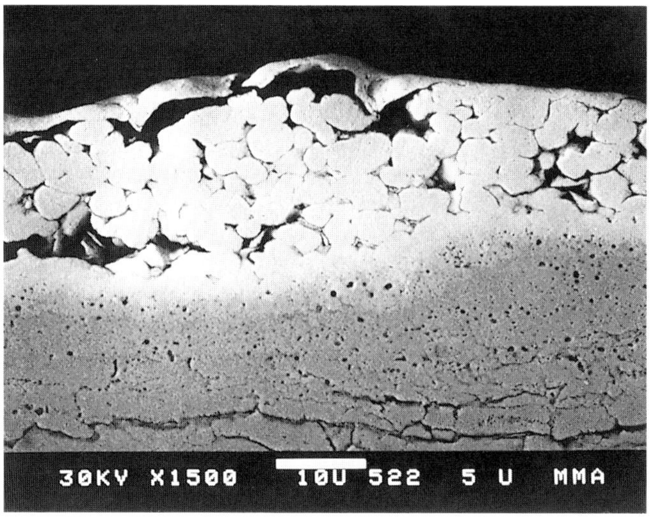

Figure 7.7 SEM photo of the polished cross-section from the gilded surface of chalice 1986.3.2, ×1500 (The Metropolitan Museum of Art, Purchase, Rogers Fund and Henry J. and Drue E. Heinz Foundation, Norbery Schimmel and Lila Acheson Wallace Gifts, 1986 (1986.3.2). Photograph, M. T. Wypyski, Sherman Fairchild Center for Objects Conservation, The Metropolitan Museum of Art.)

tened, linear orientation. Overlaying this 'reddish' zone is a pale pink, white layer with what appear to be very small grains with no visible orientation and with extensive fine precipitation around the grain boundaries that appears to be copper oxide. The overall composition of the zone is 95.95 per cent silver and 4.05 per cent copper.

In one area of the cross-section, the gilded zone is composed of a substantial layer of granular, spongy, gold alloy, pale yellow in colour, overlaid by a richly toned gold leaf with as many as four distinct layers. On the other end of the section, the same stratification is maintained, but is much more compact with the faint outline of a layer of gold leaf visible along the top surface (Fig. 7.7). The dimensions of the overall layer change depending upon the degree of mechanical compression that has occurred. The thickness of the leaf measures from 1.0 to 1.5 microns with the inadvertent overlapping and crimping of the thin gold presumably accounting for the multiple layers as opposed to there having been repeated, intentional applications (Lechtman 1971: 6). The depth of the granular layer varies from 10 to 16 microns with individual grains measuring 2–3 microns across. Just below the gold/silver interface, and distinct from the oxygenated zone, is a very homogeneous, compact layer of silver-white metal of consistent depth, 1.5 microns, and directly below that, a layer 1.0 micron thick that has a very porous structure. It may well be that the thicker, upper layer represents an amalgamated surface of silver and mercury, while the lower layer retains a depleted structure, possibly from the cleaning of the silver prior to gilding.

The analysis of the leaf, 97.25 per cent gold and 2.25 per cent silver, is consistent with the composition of the Byzantine solidus (Hendy 1969; Metcalf 1972: 429–30; Grierson 1982: 4–5, 14, 92: Oddy and La Niece 1986) – the standard gold coin in circulation at the time of the chalice's manufacture – and of reported analyses of Early Byzantine gold objects (Hockey 1989: 35; Becker *et al.* 1994: 409). While the fineness of the gold coinage fluctuated during different historical periods, it was a primary source of gold for silversmiths and their patrons (Boak 1928: 604). The granular zone has a composition of 84.47 per cent gold, 3.01 per cent silver, 12.23 per cent mercury, and 0.29 per cent copper, which was identified by X-ray diffraction as eugenite (Au_5Hg).[10]

The elemental maps for gold, silver, copper and mercury and the microprobe traces help to clarify these various zones and their relationship to one another (Fig. 7.8).[11] The gold leaf is apparent as a distinct layer as is the broad zone of granular gold and mercury. When one transposes the

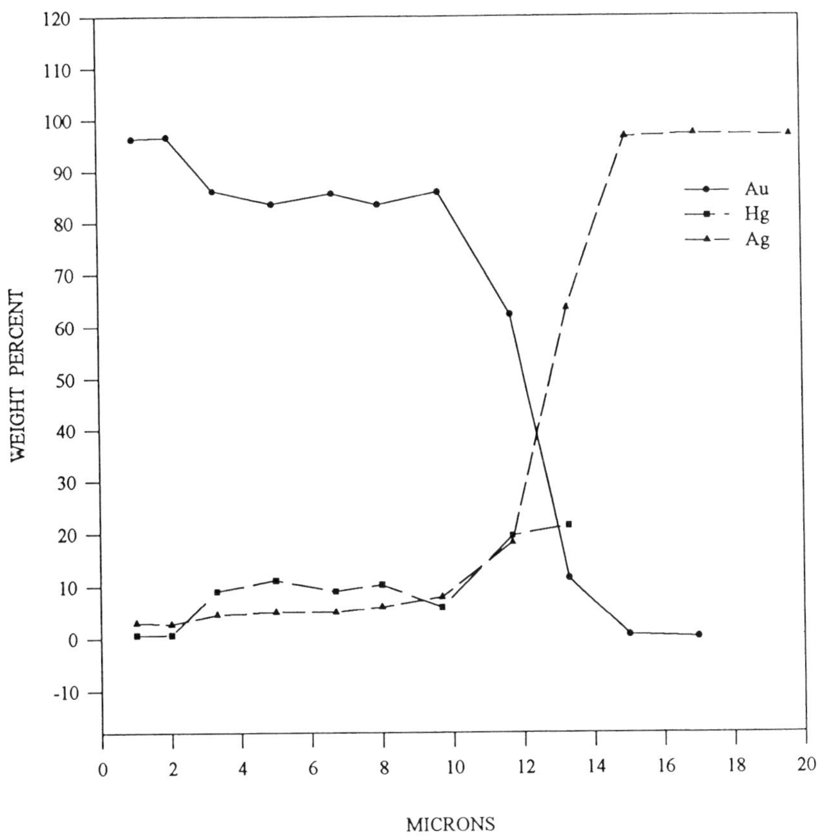

Figure 7.8 Microprobe analysis of the cross-section from chalice 1986.3.2 along a line perpendicular to the surface.

silver and mercury maps, it is evident that the mercury is in direct contact with the silver and that its highest concentration is not within the granular zone but at the silver interface where there is a 1.5 micron layer corresponding to the compact, homogeneous layer of silver-white metal observed in the etched section and the scanning electron micrograph.[12] The trace for mercury indicates that there are two distinct zones: one within the granular layer at an average concentration of 10.0 per cent and the other at the silver surface where the mercury attains its maximum concentration of 20.0 per cent.

The chalice inscribed 'For the repose of the soul of Anastasia and the salvation of John, Deacon of St. Stephen of the village of Attarathion' was

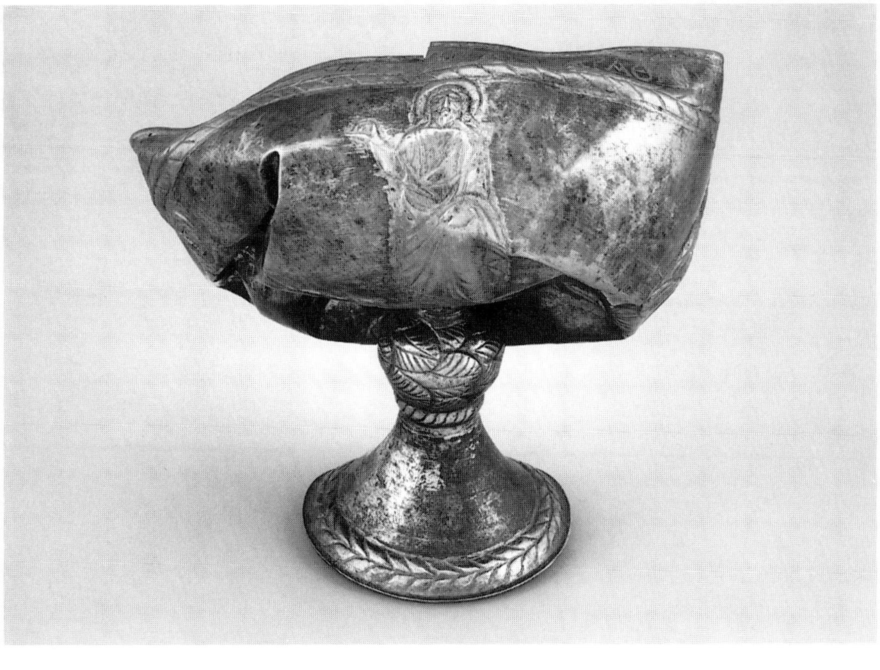

Figure 7.9 Chalice, 1986.3.6, before treatment. (The Metropolitan Museum of Art, Purchase, Rogers Fund and Henry J. and Drue E. Heinz Foundation, Norbert Schimmel and Lila Acheson Wallace Gifts, 1986. (1986.3.6) All rights reserved, The Metropolitan Museum of Art.)

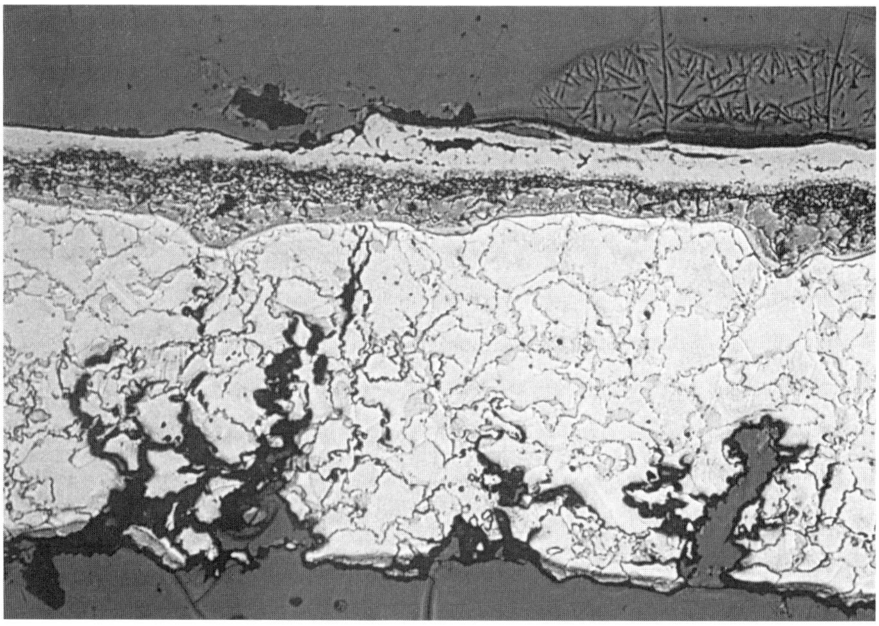

Figure 7.10 Detail of an etched cross-section from chalice 1986.3.6. (The Metropolitan Museum of Art, Purchase, Rogers Fund and Henry J. and Drue E. Heinz Foundation, Norbery Schimmel and Lila Acheson Wallace Gifts, 1986 (1986.3.6). Photograph, author, Sherman Fairchild Center for Objects Conservation, The Metropolitan Museum of Art.) (See Plate 29.)

received in an entirely crushed condition (Fig. 7.9). No treatment was undertaken on the object so that a record of the treasure's condition at the time of acquisition could be preserved. Sampling occurred at the proper right edge of the Virgin's drapery along a pre-existing tear. In its polished state, the cross-section displays many of the same physical characteristics and strata seen in the initial example. The composition of the interior silver is 93.03 per cent silver, 5.61 per cent copper, and 1.36 per cent lead with the amount of copper and lead detected decreasing to 5.01 and 0.81 wt% respectively within the oxygenated zone. The greater incidence of precipitation or porosity clustered at the grain boundaries and randomly scattered throughout the oxygenated zone is at variance with the first sample but can be accounted for possibly by the lead within the alloy. Etching discloses the displaced grain boundaries, discontinuous precipitation, cellular growth, bent annealing twins, and strain marks within the body of the silver, and shows a structure within the oxygen infused area of relatively large grains with no orientation overlaid by very small grains (Fig. 7.10). The two narrow bands directly below the silver/gold interface of a visually homogeneous alloy and of a porous alloy are very distinct in the etched sample.

The gilded zone is composed of a pale yellow granular layer overlaid by an equally deep stratum of multiple pieces of gold leaf (Fig. 7.11). The overall depth averages eight microns. The gold leaf appears to have maintained its compositional and physical integrity while in intimate contact with the granular layer evinced by the color of the leaf and the mainte-

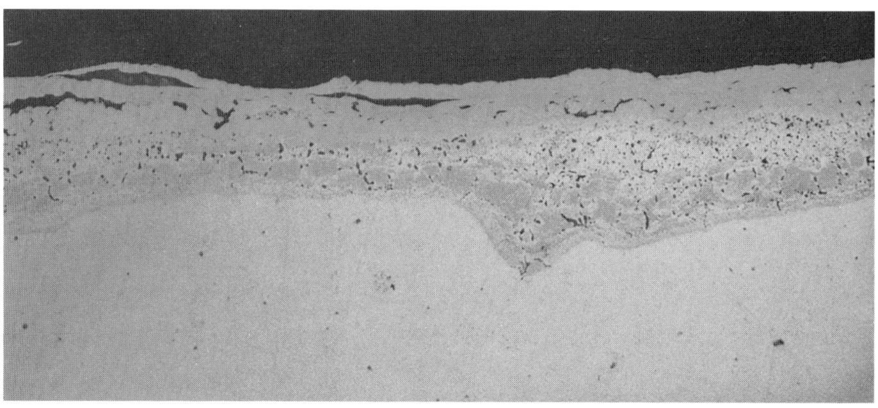

Figure 7.11 Detail of a polished cross-section from chalice 1986.3.6, ×1000 (The Metropolitan Museum of Art, Purchase, Rogers Fund and Henry J. and Drue E. Heinz Foundation, Norbery Schimmel and Lila Acheson Wallace Gifts, 1986 (1986.3.6). Photograph, author, Sherman Fairchild Center for Objects Conservation, The Metropolitan Museum of Art.) (See Plate 30.)

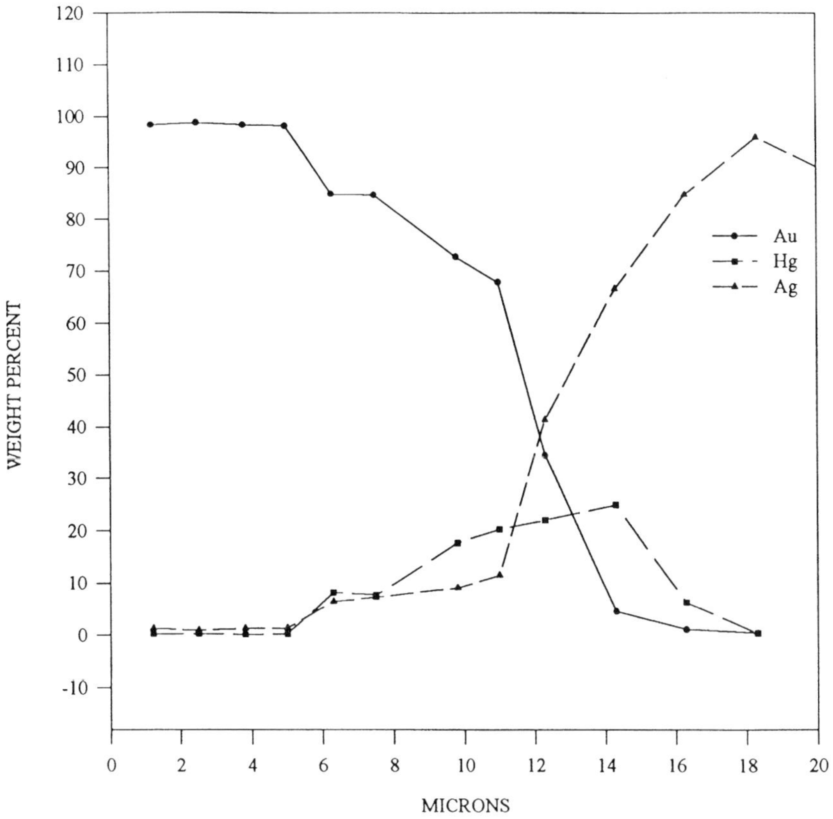

Figure 7.12 Microprobe analysis of the cross-section from chalice 1986.3.6 along a line perpendicular to the surface.

nance of its linear form. The composition of the gold leaf was analysed as 98.84 per cent gold, 1.10 per cent silver, and 0.06 per cent copper. The recorded concentrations in the granular zone are 79.48 per cent gold, 7.57 per cent silver, 12.64 per cent mercury and 0.32 per cent copper with eugenite being the only identifiable crystalline material. The elemental maps and the microprobe trace (Fig. 7.12) indicate that there is a discrete layer of nearly pure gold on the surface, that the concentration of mercury increases from the granular gold zone to the silver surface, and that there is a broad zone of silver, gold, and mercury interdiffusion.

A third sample from the Attarouthi Treasure was taken from the edge of an area of loss on the foot of the chalice embossed with alternating stars and crosses and inscribed, 'Of St. John of the village of Tarothis' (Fig. 7.13). The composition of the silver substrate is 95.88 per cent silver and

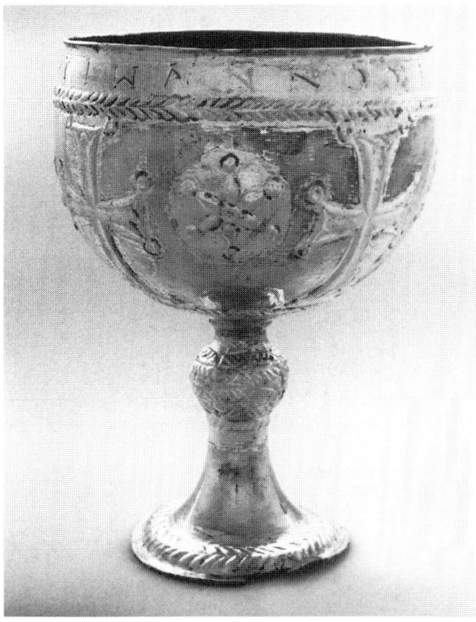

Figure 7.13 Chalice, 1986.3.9, after treatment. Areas of reserved silver exhibit runs of fluid amalgam and parallel striations from the scraper. The dark areas within the gilding represent losses in the leaf exposing the granular layer below. (The Metropolitan Museum of Art, Purchase, Rogers Fund and Henry J. and Drue E. Heinz Foundation, Norbery Schimmel and Lila Acheson Wallace Gifts, 1986 (1986.3.9). Photograph, author, Sherman Fairchild Center for Objects Conservation, The Metropolitan Museum of Art.)

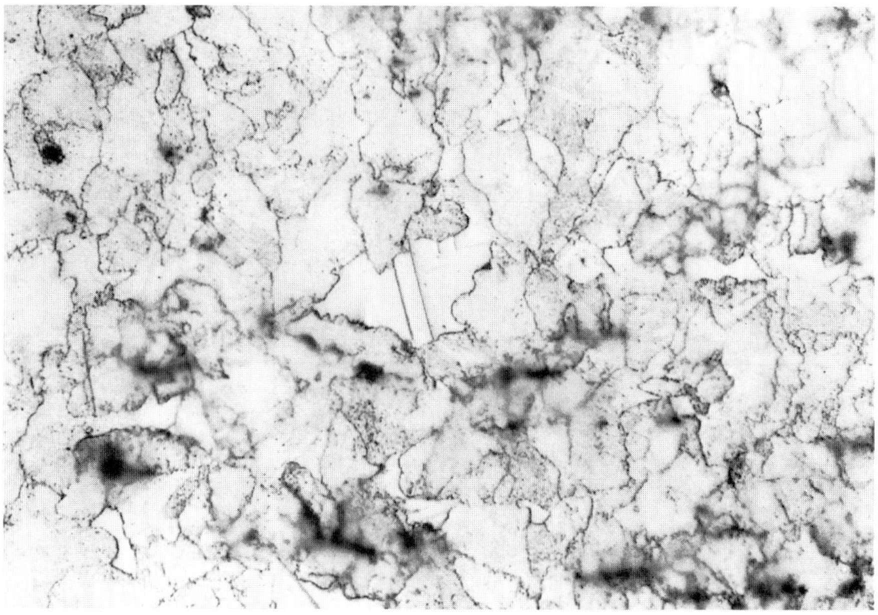

Figure 7.14 Detail of annealing twins and 'jigsaw' patterned grain boundaries of etched cross-section from chalice 1986.3.9. (The Metropolitan Museum of Art, Purchase, Rogers Fund and Henry J. and Drue E. Heinz Foundation, Norbery Schimmel and Lila Acheson Wallace Gifts, 1986 (1986.3.9). Photograph, author, Sherman Fairchild Center for Objects Conservation, The Metropolitan Museum of Art.)

4.12 per cent copper, and unlike the other samples, appears in the polished section to have inclusions or stringers with a distinctly linear orientation scattered throughout the body of the silver rich phase. Etched, the structure is characteristic of ancient silver and similar to the other examples apart from the substantially smaller grain size (Fig. 7.14). The oxygenated zone is non-existent in some portions of the section, while averaging 5–7 microns in others, with extensive porosity overall and particularly within the pale pink, silver-white, upper layer. The gilded zone is well burnished and extremely thin and compact, ranging from 1–5 microns deep, and while the leaf is still physically distinct in some areas and visually recognizable overall by its deep orange colour, in many places it has lost its linear, sheet-like character and has a granular form similar to the pale yellow grains it overlays (Fig. 7.15). The fineness of the leaf measures 96.13 per cent gold, 3.46 per cent silver, and 0.41 per cent copper.

When one observes the elemental maps and the microprobe traces (Fig. 7.16), it is apparent that there is a distinct layer of gold leaf, that the greatest concentration of mercury and copper is at the silver surface, and that there has been a substantial degree of interdiffusion of silver, gold and mercury within the thin granular zone. Compositional analysis confirms the degree of interdiffusion with copper and mercury concentrations recorded as 8.03 and 5.15 per cent respectively within the 'reddish' layer of the oxygenated zone; 6.12 and 13.01 per cent in the pale pink, silver-white,

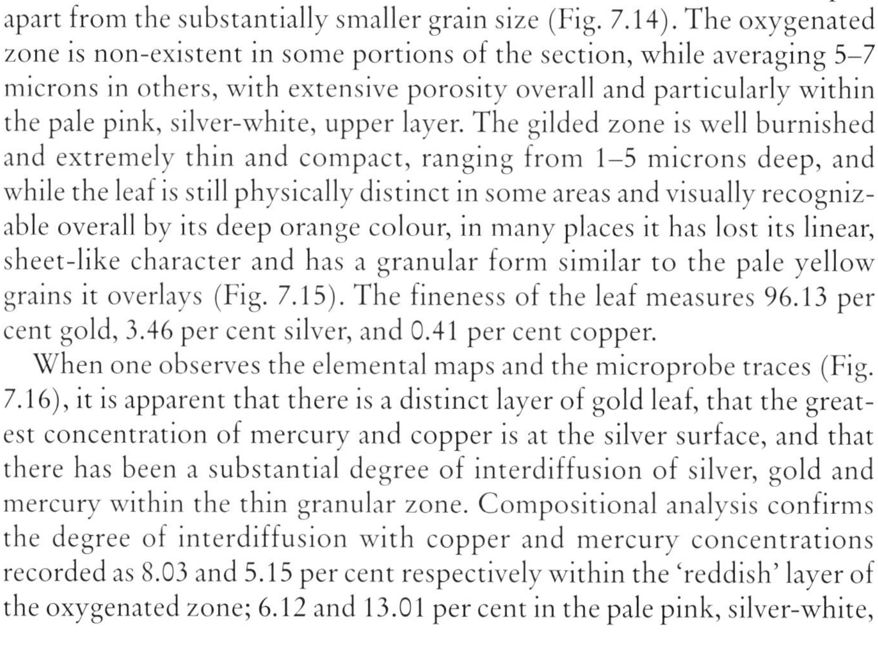

Figure 7.15 Detail of the polished cross-section from gilded surface of chalice 1986.3.9, ×1000 (The Metropolitan Museum of Art, Purchase, Rogers Fund and Henry J. and Drue E. Heinz Foundation, Norbery Schimmel and Lila Acheson Wallace Gifts, 1986 (1986.3.9). Photograph, A. Heywood, Sherman Fairchild Center for Objects Conservation, The Metropolitan Museum of Art.) (See Plate 31.)

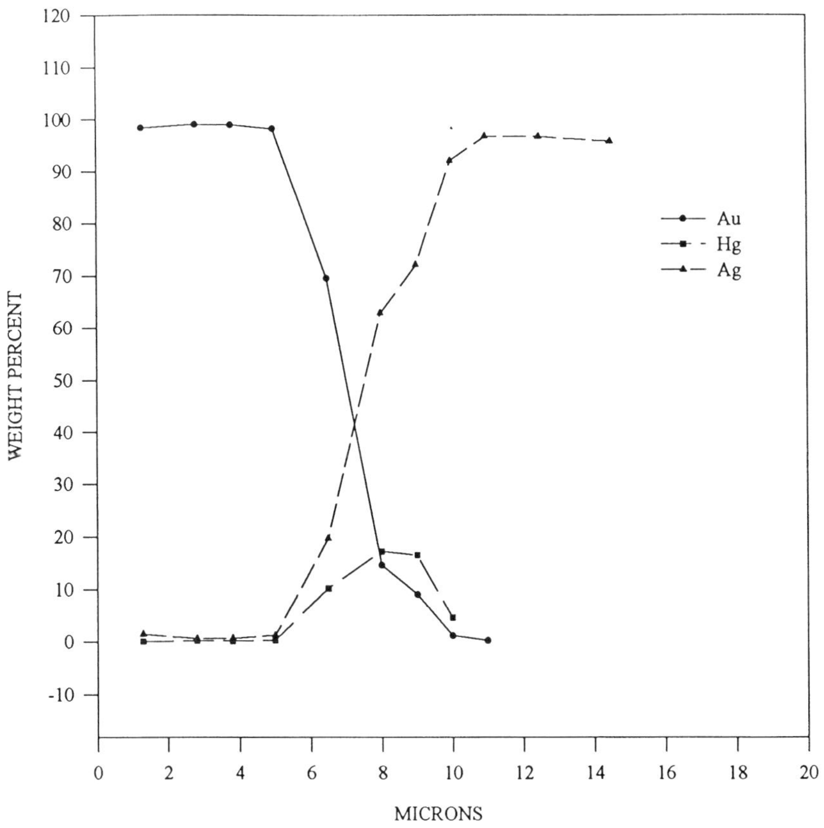

Figure 7.16 Microprobe analysis of the cross-section from chalice 1986.3.9 along a line perpendicular to the surface.

porous layer above; and in the pale yellow granular layer at the base of the gilded zone, one finds 54.12 per cent gold, 31.63 per cent silver, 13.52 per cent mercury and 0.73 per cent copper.

A fourth sample was removed for analysis from the processional cross along a tear in the outer edge of the wavy border at the end of the proper left front arm of the cross just at the edge of the gilding (Fig. 7.17). The polished section looks very familiar – possible traces of thin leaf, a thin granular zone of gilding and oxygenated zones that are preserved along both surfaces of the silver. One aspect of the oxygenated zone appears unusual with respect to the other examined sections in that there are scattered areas with a somewhat different texture and with a tint of yellow colour possibly associated with the diffusion of mercury and gold (Fig.

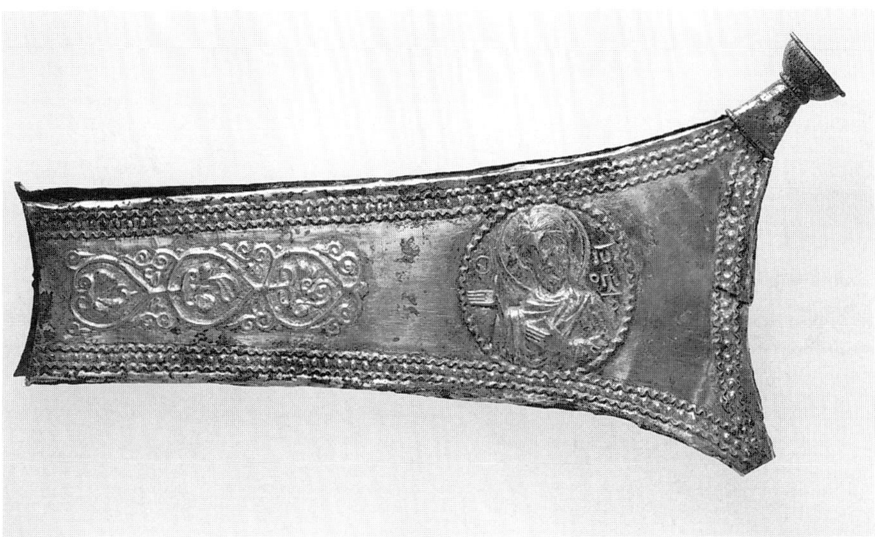

Figure 7.17 Detail of the proper left, front arm of the processional cross, 1993.163, after treatment. (The Metropolitan Museum of Art, Rogers Fund, 1993. (1993.163) Photograph, author, Sherman Fairchild Center for Objects Conservation, The Metropolitan Museum of Art.)

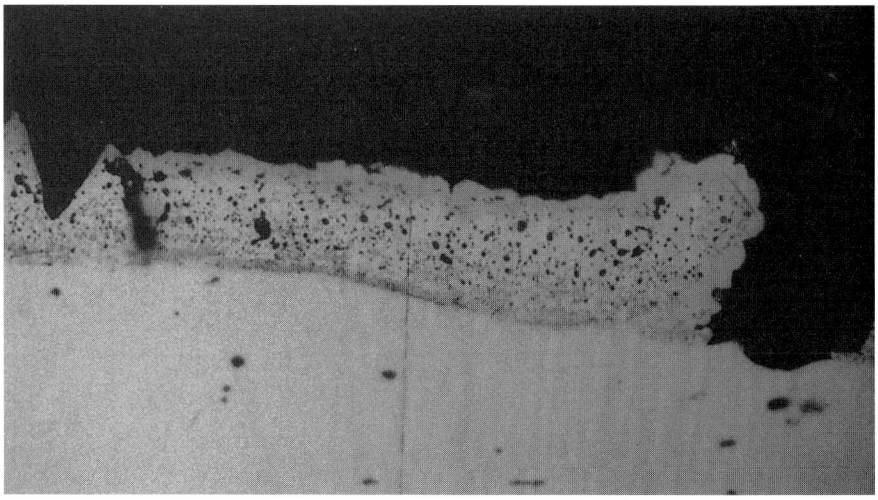

Figure 7.18 Detail of the polished cross-section of processional cross, 1993.163, ×1000. The diffusion of mercury well into the oxygenated zone is evinced by fingers of pale yellow colour. (The Metropolitan Museum of Art, Rogers Fund, 1993. (1993.163) Photograph, author, Sherman Fairchild Center for Objects Conservation, The Metropolitan Museum of Art.) (See Plate 32.)

7.18). The strata of gilding is at some variance to what we have seen as well – the leaf is very difficult to discern, and there is no colour alteration that might differentiate the leaf and the granular layer or amalgam. The absence of any discernible leaf appears to be due to the area sampled, the extreme edge of the gilded surface, with the stratification of the gold at the surface a result of the burnishing of the granular amalgam.

When etched, the large grain size, distorted grain boundaries, grain boundaries with smoothly arching sections, discontinuous precipitation along the grain boundaries, possible cellular growth, scattered large inclusions with flattened orientation and slightly bent annealing twins are all apparent (Fig. 7.19).

Analysis of the metallographic section indicates that the silver-rich interior alloy is 90.91 per cent silver, 5.06 per cent copper, 2.32 per cent gold, and 1.70 per cent lead. At one position within the oxygenated zone there is 90.73 per cent silver, 2.19 per cent gold, 4.35 per cent copper and 2.73 per cent mercury while in the area that displays the modified texture and colour, the composition is 48.26 per cent silver, 5.60 per cent gold, 2.16 per cent copper and 43.98 per cent mercury. Comparing the gold, mercury, silver, and silver–gold dot maps of the analysed area confirms the increased levels of mercury into the depth of the oxygenated zone, and identifies a distinct area of high mercury content, several microns in diameter, into which little or no gold has diffused (Fig. 7.20). The graphic display of the microprobe traces indicates a broad zone of interdiffusion between mercury, silver and gold with a concentration of mercury, 30 per cent, at

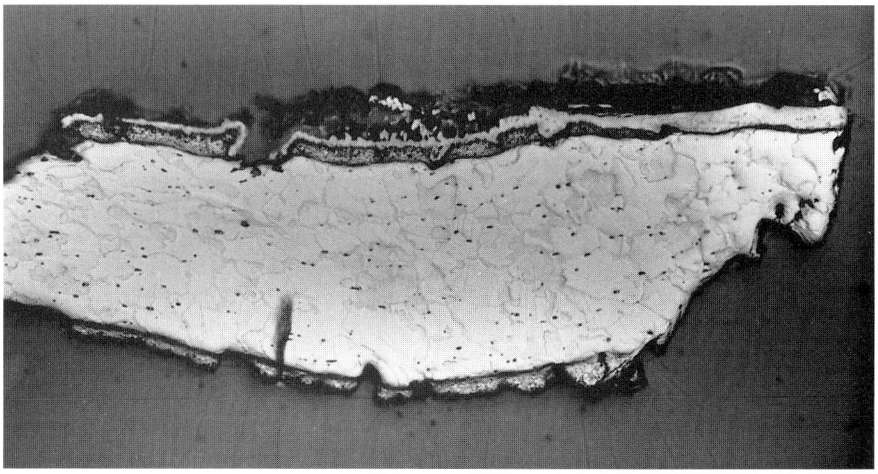

Figure 7.19 Etched cross-section from the processional cross, 1993.163, ×200 (The Metropolitan Museum of Art, Rogers Fund, 1993. (1993.163) Photograph, author, Sherman Fairchild Center for Objects Conservation, The Metropolitan Museum of Art.)

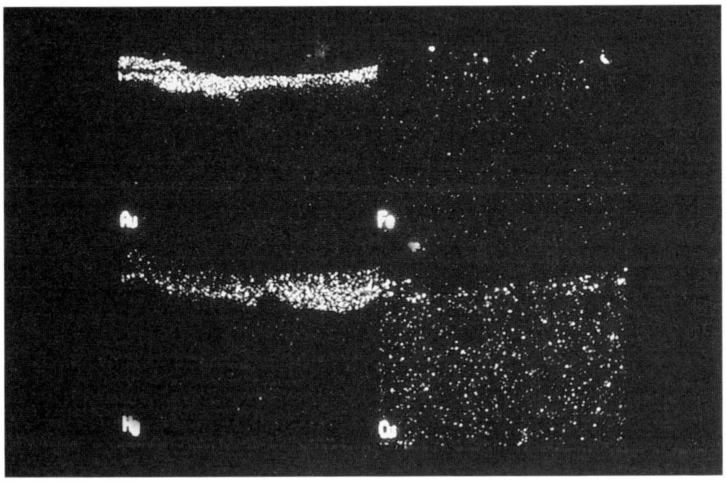

Figure 7.20 Electron probe scan for gold, mercury, iron and copper of the cross-section from the processional cross, 1993.163, ×1000. (The Metropolitan Museum of Art, Rogers Fund, 1993. (1993.163) Photograph, M. T. Wypyski, Sherman Fairchild Center for Objects Conservation, The Metropolitan Museum of Art.)

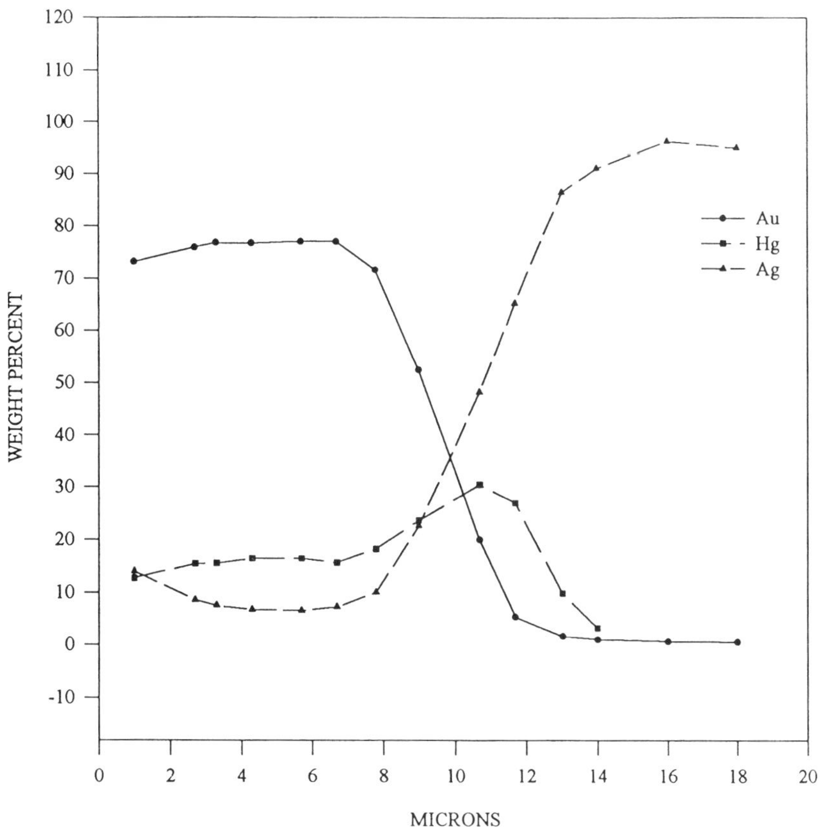

Figure 7.21 Microprobe analysis of the cross-section from the processional cross, 1993.163, along a slightly oblique line to the surface.

the silver interface (Fig. 7.21). Within the granular gold zone the composition is 78.78 per cent gold, 7.35 per cent silver, 0.20 per cent copper and 13.67 per cent mercury identified as eugenite by X-ray diffraction.

Conclusion

What then distinguishes the chalices, the cross and their cross-sections from one another, and is there any physical evidence to suggest their method of gilding? Fortunately, Anheuser's recent replication experiments and Lechtman's earlier studies provide controls against which the sections and the analyses can be compared (Lechtman 1971: 14; Anheuser 1996: 58–61).

All of the objects sampled in this study show visual and microscopic evidence of a layer of gold leaf overlaying a stratum of granular gold.

Lechtman and Anheuser have reproduced the granular morphology both with a paste of gold and mercury applied on to silver and with leaf applied over a mercury amalgamated silver substrate with no discernible inclusions of unalloyed gold particles evident in either process (Lechtman 1971: 10, 14; Anheuser 1996: 56; Anheuser and Northover 2000). Elemental mapping and microprobe analyses of the cross-sections from the chalices indicate a variable degree of interdiffusion of gold, silver, and mercury occurring within the granular zone with the concentration of mercury falling consistently between 12–16 per cent, and with the maximum concentration of mercury occurring at the silver interface below the granular gold zone across a relatively narrow band with concentrations of between 18–23 per cent. In the case of the processional cross, the mercury concentration is much higher at the silver interface, between 30–43 per cent, and is detected well into the oxygenated zone as well; moreover, areas of unalloyed mercury can be detected at the silver interface. Similarly, Anheuser and Lechtman found in their replication experiments that the concentration of mercury is greatest at the silver interface (Lechtman 1971: 14; Anheuser 1996: 59, 61), and Anheuser showed that the concentration varied depending upon the method of application (Anheuser 1996: 59–61). If an amalgam of gold and mercury was applied to the silver, the concentration of mercury would not exceed 25 per cent, whereas, if the silver surface was amalgamated initially with mercury then overlaid with gold leaf, the concentration of mercury at the silver interface would range between 25 and 50 per cent, and the silver and mercury amalgamated zone would be around 10 microns deep.

The correlation between our analytical results and those of Anheuser and Lechtman suggests that, initially, the Attarouthi chalices were gilded with an amalgam of gold and mercury while the silver surface of the processional cross was amalgamated just with mercury. Furthermore, the microprobe analyses of the chalices' granular gold zones and silver interfaces parallel those from Newman's study of the Byzantine chalice from Dumbarton Oaks (Newman and Lie 1992: Fig. 6), as well as, the Sasanian objects examined by Lechtman all of which are described as having been gilt with a paste of gold and mercury (Lechtman 1971: Figs 1.22 and 1.27).

Given the depth and concentration of design elements on the cross, it may have seemed advisable to wet the silver first with mercury to promote greater coverage. Initial gilding was then carried out either with leaf or with a paste of gold and mercury. The annealing of the chalices and the cross after this first application of gold volatilized off most of the unreacted mercury, and induced the formation of the intermetallic compound, eugenite, with a distinctly granular morphology apparent in all of the cross-sections irrespective of the gilding method.

To this point, the gilding of the chalices and the cross conforms to the descriptions in the treatises and to our knowledge of Roman, medieval, and Sasanian methodology, and, as such, represents a continuation of well established traditions; however, the presence of a visually and microscopically discrete layer of gold leaf overlaying the granular zone and maintaining its linear form, purity, and colour represents an additional step in the gilding process that is not described in the literature. It appears that the application of the gold leaf was a separate and final step, that it occurred after the final anneal of the amalgam, and that it was dependent for its adhesion on the thoroughness of the burnishing – a cold technique. Anheuser found that if there is any metallic mercury remaining in the granular gold zone, leaf applied over the surface would be solubilized (Anheuser 1996: 56). By minimizing the excess metallic mercury on the amalgamated surface prior to the application of the leaf, and by eliminating a final warming, the interdiffusion of gold, silver and mercury could be restricted to the granular gold amalgamated zone and the visually arresting impression of the deep, rich colour of the gold leaf maintained. The lack of thoroughness in the burnishing of the gold leaf within the recesses of the repoussé decoration on the chalices suggests they were not intended to be handled extensively, and perhaps, their visual effect was their primary function. In response, the goldsmiths devised a method of gilding that achieved the richness of a diffusion gilded surface with the ease and economy of amalgam gilding.

The use of gold leaf, applied cold, as a final, finished surface in a multi-stepped gilding process does not appear to be unique to the museum's chalices and cross. Microscopic examination of the Middle Byzantine, silver processional cross in the collection of the Musée National du Moyen Âge et des Thermes de l'Hôtel de Cluny dated to the late eleventh or early twelfth century (Durand 1992: 329–30) revealed visual and microscopic indications of a similar gilding technique – discrete sheets of gold leaf overlay a granular gold layer on the figural medallions and rinceaux patterns, layers of leaf adhered loosely to the granular substrate within recessed areas barely accessible to a burnisher, and leaf lifting away on well burnished portions of the design as well. The book cover from Dumbarton Oaks and its gilding may bear reinterpretation. It was understood by Lechtman to be the result of gold leaf applied over a silver substrate amalgamated with mercury (Lechtman 1971: 6–10); however, her microprobe does not indicate a broad zone of amalgamated silver and mercury, no mercury was detected within the silver, and the highest percentage of mercury is found within the granular gold zone just within the gold peak. It is possible that the layers of gold leaf apparent in her cross-section represent a final, cold application over a silver surface amalgamated with a thin paste of gold and mercury and gently annealed. Indeed, the re-evaluation of the gilding technique on Byzantine silver-gilt objects described as retaining the distinct outlines of gold sheet may reveal other comparable examples.

The occurrence of an apparently similar gilding technique on a statistically limited number of examined objects, separated chronologically by half a millennium, suggests that the application of a final layer of gold leaf over a prepared gold ground in a cold technique was potentially more widely utilized. As a variant on the commonly practised amalgam gilding techniques in the west and east, including Byzantium, it provided a deeply saturated gold colour to contrast vividly with the reserved, burnished silver in a juxtaposition that may have had aesthetic or religious significance that encouraged its selective use by generations of goldsmiths.

Acknowledgements

The opportunity to examine the processional cross in the Cluny Museum was afforded by the generosity of Vivian Huchard, Director, and Elisabeth Antoine, Curator. I would like to thank Richard Newman, Research Scientist, Department of Objects Conservation and Scientific Research, Museum of Fine Arts, Boston for undertaking the electron microprobe analyses, and thereby expanding the breadth of the study. Several colleagues in the Sherman Fairchild Center for Objects Conservation have been very supportive with their time and expertise: James H. Frantz, Conservator in Charge, Richard E. Stone, Conservator, and, in particular, Mark T. Wypyski, Associate Research Scientist, who provided all of the EDS analyses and scans, and the SEM photomicrographs.

Table 7.2 Summary of microprobe results.

Distance from surface in microns	Gold	Mercury	Silver	Copper	Total
MMA 1986.3.2					
1 (1.0)	96.3 (94.6)	0.64 (0.63)	3.0 (2.9)	0.03 (0.03)	100 (98.2)
2 (2.0)	96.6 (96.1)	0.69 (0.69)	2.7 (2.7)	ND	100 (99.5)
3 (3.3)	86.1 (82.7)	9.1 (8.6)	4.5 (4.4)	0.30 (0.30)	100 (98.2)
4 (5.0)	83.5 (82.5)	11.2 (11.1)	5.0 (5.0)	0.29 (0.29)	100 (98.8)
5 (6.7)	85.6 (84.4)	9. 1 (9.0)	5.0 (5.0)	0.27 (0.27)	100 (98.6)
6 (8.0)	83.4 (81.0)	10.3 (10.0)	6.0 (5.8)	0.25 (0.24)	100 (97.1)
7 (9.7)	85.9 (84.7)	5.9 (5.9)	7.9 (7.8)	0.20 (0.20)	100 (98.6)
8 (11.7)	62.2 (60.4)	19.5 (18.9)	18.2 (17.6)	0.15 (0.15)	100 (97.0)
9 (13.3)	11.5 (10.9)	21.4 (20.2)	63.4 (60.0)	3.8 (3.6)	100 (94.7)
10 (15.0)	0.58 (0.57)	ND	96.6 (94.9)	2.8 (2.7)	100 (98.2)
11 (17.0)	0.17 (0.17)	ND	97.2 (95.6)	2.6 (2.6)	100 (98.4)
12 (19.7)	ND	ND	96.8 (95.3)	3.2 (3.1)	100 (98.4)
substrate	ND	ND	96.0 (93.9)	4.0 (3.9)	100 (97.7)
MMA 1986.3.6					
1 (1.2)	98.4 (98.0)	0.22 (0.22)	1.3 (1.4)	0.04 (0.04)	100 (99.7)
2 (2.5)	98.8 (98.3)	0.25 (0.25)	0.91 (0.91)	0.05 (0.05)	100 (99.5)
3 (3.8)	98.4 (97.2)	0.14 (0.14)	1.3 (1.3)	0.08 (0.08)	100 (98.7)
4 (5.0)	98.2 (97.1)	0.30 (0.30)	1.4 (1.3)	0.07 (0.07)	100 (98.S)
5 (6.3)	84.9 (82.3)	8.2 (8.0)	6.5 (6.3)	0.33 (0.32)	100 (96.9)
6 (7.5)	84.7 (81.9)	7.7 (7.4)	7.3 (7.1)	0.28 (0.27)	100 (96.7)
7 (9.8)	72.8 (71.4)	17.7 (17.4)	9.1 (9.0)	0.34 (0.33)	100 (98.1)
8 (11.0)	67.9 (65.8)	20.2 (19.6)	11.5 (11.1)	0.38 (0.37)	100 (96.8)
9 (12.3)	34.1 (32.5)	22.0 (20.7)	41.4 (39.1)	2.0 (1.9)	100 (94.2)
10 (14.3)	4.6 (4.4)	24.9 (23.2)	66.6 (62.0)	3.7 (3.5)	100 (93.1)
11 (16.3)	1.1 (1.0)	6.3 (5.9)	94.8 (78.7)	7.7 (7.2)	100 (92.8)
12 (18.3)	0.39 (0.37)	0.40 (0.38)	95.9 (90.1)	3.2 (3.1)	100 (93.4)
13 (20.3)	ND	ND	89.2 (81.0)	10.8 (9.S)	100 (90.9)
MMA 1986.3.9					
1 (1.3)	98.4 (95.4)	0.13 (0.13)	1.5 (1.4)	0.03 (0.01)	100 (97.0)
2 (2.8)	99.1 (96.8)	0.20 (0.20)	0.61 (0.60)	0.07 (0.07)	100 (97.6)
3 (3.8)	99.0 (96.8)	0.18 (0.18)	0.66 (0.65)	0.08 (0.08)	100 (97.7)
4 (5.0)	98.2 (95.3)	0.38 (0.37)	1.3 (1.3)	0.14 (0.14)	100 (97.1)
5 (6.5)	69.5 (66.6)	10.2 (9.8)	19.7 (18.9)	0.62 (0.60)	100 (95.9)
6 (8.0)	14.6 (13.0)	17.2 (15.2)	62.9 (55.7)	5.2 (9.9)	100 (88.6)
7 (9.0)	9.0 (8. 1)	16.5 (14.9)	72.1 (65.2)	2.5 (2.2)	100 (90.4)
8 (10.0)	1.2 (1.1)	4.6 (4.2)	92.1 (84.6)	2.1 (2.0)	100 (91.9)
9 (11.0)	0.26 (0.25)	ND	96.8 (91.9)	2.9 (2.8)	100 (94.9)
10 (12.5)	ND	ND	96.7 (90.1)	3.3 (3.1)	100 (93.1)
11 (14.5)	ND	ND	95.8 (93.1)	4.2 (4.1)	100 (97.3)
MMA 1993.163					
2 (1)	73.1 (66.7)	12.7 (11.6)	14.0 (12.8)	0.19 (0.17)	100 (97.4)
3 (2.7)	75.9 (7 3.1)	15.4 (14.9)	8.5 (8.2)	0.19 (0.18)	100 (96.4)
4 (3.3)	76.8 (74.3)	15.5 (15.0)	7.5 (7.2)	0.18 (0.17)	100 (96.7)
5 (4.3)	76.7 (74.6)	16.4 (15.9)	6.7 (6.5)	0.20 (0.19)	100 (97.2)
6 (5.7)	77.0 (75.4)	16.4 (16.0)	6.5 (6.3)	0.21 (0.21)	100 (97.9)
7 (6.7)	77.0 (72.3)	15.6 (14.7)	7.2 (6.8)	0.22 (0.21)	100 (94.5)
8 (7.8)	71.6 (65.9)	18.2 (17.7)	10.0 (9.7)	0.32 (0.31)	100 (97.6)
9 (9. 0)	53.5 (51.3)	23.6 (22.5)	22.6 (21.6)	0.24 (0.23)	100 (95.5)
10 (10.7)	19.9 (18.9)	30.4 (28.9)	48.2 (45.8)	1.4 (1.3)	100 (95.0)
11 (11.7)	5.3 (5.0)	26.9 (25.3)	65.2 (61.3)	2.5 (2.4)	100 (94.0)
12 (13.0)	1.6 (1.5)	9.8 (9.2)	86.5 (81.5)	2.1 (2.0)	100 (94.2)
13 (14.0)	1.0 (1.0)	3.2 (3.1)	91.1 (87.7)	4.7 (4.5)	100 (96.3)
14 (16.0)	0.64 (0.61)	ND	96.3 (92.1)	3.1 (2.9)	100 (95.7)
15 (18.0)	0.47 (0.45)	ND	95.0 (91.5)	4.6 (4.4)	100 (96.4)

Notes: Normalized weight percentages are given, followed (in parentheses) by actual results from the analyses. Totals may not match the sums of the four components due to rounding. ND, not detected (limits of detection: gold, 0.11 per cent; mercury, 0.12 per cent; copper, 0.03 per cent).

Notes

1. There is no evidence from Byzantium for the segregation of metalworkers in silver and in gold, indeed, *The Book of the Prefect* suggests the opposite. The collected ordinances are thought to have been published in the 10th or 11th century, and silver-merchants and goldsmiths are both described as working in gold, silver, pearls and precious stones (Boak 1928: 604).
2. For a critical analysis of the various literary sources see Anheuser (1996: 22–33).
3. The Metropolitan Museum of Art, 1986.3.1-.15. Purchase, Rogers Fund and Henry J. and Drue Heinz Foundation, Norbert Schimmel and Lila Acheson Wallace Gifts.
4. For further discussions of the techniques used in the fabrication of Byzantine silver and silver-gilt vessels see Snow and Weisser (1986: 38–51) and Newman and Lie (1992: 77–81).
5. The Metropolitan Museum of Art, 1993.163. Purchase, The Rogers Fund. For photographs and descriptions of The Metropolitan and related crosses see Evans (1997: 60–67).
6. The solder was analysed by EDS and found to contain 85 per cent tin, 9.5 per cent silver, 3 per cent lead, 1 per cent nickel, and 1 per cent copper.
7. The SEM analyses and elemental mapping were undertaken by Mark T. Wypyski, Associate Research Scientist, The Sherman Fairchild Center for Objects Conservation, The Metropolitan Museum of Art, with a Kevex model Delta IV energy dispersive X-ray spectrometer attached to a modified Amray model 110 (1600 T) scanning electron microscope equipped with a normal beryllium detector that does not detect elements below the atomic weight of sodium.
8. Electron microprobe analyses were undertaken by Richard Newman, Research Scientist in the Department of Objects Conservation and Scientific Research, Museum of Fine Arts, Boston. The four polished metallographic samples were carbon coated and analysed by quantitative wavelength-dispersive X-ray fluorescence in an electron beam microprobe. The instrument utilized was a Cameca MBX probe, equipped with a Tracor Northern 5502 energy-dispersive X-ray fluorescence system (EDS) and a Tracor Northern 1310 wavelength-dispersive X-ray fluorescence system (WDS) and stage automation (Department of Earth and Planetary Sciences, Harvard University). Four elements were analysed; pure metal standards were used for gold, silver and copper, while cinnabar was the standard for mercury. Matrix corrections were carried out by the ZAF (fundamental parameters) method. The instrument was operated at 25 keV, with a beam current of about 15 amps. All analyses were carried out with a point beam at intervals of 1 µm or 2 µm; however, the area from which X-rays are generated during each analysis is around 3 µm in diameter, so there is considerable overlap in the actual areas analysed from one point to the next.
9. All of the silver cross-sections of MMA objects referred to in this study were etched with the same potassium dichromate reagent.
10. Samples were mounted in a Debye Scherrer Camera and exposed to X-rays generated by a Philips X-ray diffraction unit with a copper anode and nickle filter at 40 keV and 30 mA for a period of two hours. D-spacings were analysed with the Fein-Marquart Search-Match Program.
11. The elemental mapping provides relative concentrations of the elements chosen over a specified area at 1000× magnification. For the microbeam traces, every attempt was made to select areas on the metallographic cross-sections where the gold leaf, granular zone, and silver substrate were in intimate contact.
12. We did not analyse the oxygenated zones on the inner surfaces of the chalices apart from confirming under the metallographic microscope that they had the same texture and visual appearance as the similar zones beneath the gilding; however, Pieter Meyers suggested that if mercury were detected within the interior zones, it might be an indication of the conditions under which the objects were warmed during gilding.

References

Anheuser, K. 1996. 'An investigation of amalgam gilding and silvering on metalwork'. Unpublished PhD dissertation, Oxford University.

Anheuser, K. and Northover, P. 2000. 'Gilded archaeological metals: Celtic, Romano-British and Saxon'. In *Gilded Metal Surfaces*, Terry Drayman-Weisser (ed.), 109–22. London: Archetype.

Atil, E., Chase, W. T. and Jett, P. 1985. *Islamic Metalwork in the Freer Gallery of Art*. Washington: Smithsonian Institution.

Becker, L., Schorsch, D., Williams, J. L. and Wypyski, M. 1994. 'Technical and material studies'. In *Art of Late Rome and Byzantium: In the Virginia Museum of Fine Arts*, A. Gonosová and C. Kondoleon (eds), Appendix 1, 406–9. Richmond: Virginia Museum of Fine Arts.

Boak, A. E. R. (trans.) 1928. 'Notes and documents: "The Book of the Prefect"'. *Journal of Economic and Business History*, **1**: 597–619.

Caley, E. R. (trans.) 1926. 'The Leyden Papyrus'. *Journal of Chemical Education*, **3**: 1149–66.

Carlson, J. H. 1986. 'Energy dispersive X-ray fluorescence analysis for elemental composition'. In *Silver from Early Byzantium: The Kapor Koran and Related Treasures*, M. Mango (ed.), Appendix III.2, 9–19. Baltimore: Walters Art Gallery.

Chase, T. 1968. 'The technical examination of two Sasanian silver plates'. *Ars Orientalis* 7: 75–93.

Craine, C. 1980. 'The examination and treatment of a sixth-century openwork lamp from the Byzantine Church of the Holy Sion'. In *AIC Preprints*, 9-19. San Francisco: American Institute of Conservation.

Dodwell, C. R. 1971–72. 'Gold metallurgy in the twelfth century: the *De Diversis Artibus* of Theophilus the Monk'. *Gold Bulletin* **4–5**: 51–5.

Durand, J. (ed.) 1992. *Byzance: L'Art Byzantin dans les Collections Publiques Français*, exhibition catalogue. Paris: Bìbliothèque Nationale.

Evans, H. C. and Wixom, W. D. (eds) 1997. *The Glory of Byzantium: Art and Culture of the Middle Byzantine Era, AD 843–1261*, exhibition catalogue. New York: Metropolitan Museum of Art.

Frazer, M. E. 1988. 'Silver liturgical objects from Attarouthi in Syria'. In *Abstracts of Papers: The Fourteenth Annual Byzantine Conference*, Houston, Nov 1988, 13–14.

Gibbons, D. F., Ruhl, K. C. and Staikoff, L. S. 1974. 'Analysis of Sasanian silver objects: A comparison of techniques'. In *Archaeological Chemistry*, C. W. Beck (ed.) Advances in Chemistry Series 138: 11–21. Washington, DC: American Chemical Society.

Gibbons, D. F., Ruhl, K. C. and Shepherd, D. G. 1979. 'Techniques of silversmithing in the Hormizd II plate'. *Ars Orientalis*, **11**: 163–76.

Grierson, P. 1982. *Byzantine Coins*. Berkeley, Calif.: University of California Press.

Hawthorne, J. G. and Smith, C. S. (trans.) 1979. *Theophilus: On Divers Arts: The Foremost Medieval Treatise on Painting, Glassmaking and Metalwork*. New York: Dover Publications.

Hendy, M. F. 1969. *Coinage and Money in the Byzantine Empire: 1081–1261*. Washington, DC: Dumbarton Oaks.

Hockey, M. 1989. 'The composition and structure of a Byzantine torc'. *Jewellery Studies*, **3**: 33–9.

Jett, P. 1992. 'Materials and methods of manufacture'. In *Ancient Iranian Metalwork in the Arthur M. Sackler Gallery and the Freer Gallery of Art*, A. C. Gunter and P. Jett, 49–60. Washington, DC: Smithsonian Institution.

Kehl, G. L. 1949. *Metallographic Laboratory Practice*. New York: McGraw-Hill.

Lang, J. and Hughes, M. J. 1990. 'Joining techniques'. In *Aspects of Early Metallurgy*, W. A. Oddy (ed.), 169–77. London: British Museum.

Lechtman, H. N. 1971. 'Ancient methods of gilding silver: Examples from the Old and New Worlds'. In *Science and Archaeology*, R. Brill (ed.), 2–30. Cambridge, Mass.: MIT Press.

Lefferts, K. C. 1966. 'Technical notes'. *Bulletin of the Metropolitan Museum of Art*, Nov 1966: 147–51.

Lins, A.and Oddy, A. 1975. 'The origins of mercury gilding'. *Journal of Archaeological Science*, **2**: 365–73.
Mango, M. M. (ed.) 1986. *Silver from Early Byzantium: The Kaper Koran and Related Treasures.*, exhibition catalogue. Maryland: Walters Art Gallery.
Mango, M. M. and Bennett, A. 1994. 'The Sevso Treasure: Part I'. *Journal of Roman Archeology* Supplementary Series **12**: 33.
Metcalf, D. M. 1972. 'Analysis of the metal contents of medieval coins'. In *Methods of Chemical and Metallurgical Investigation of Ancient Coinage*, E. T. Hall and D. M. Metcalf (eds), 383–434. London: Royal Numismatic Society.
Meyers, P. 1992. 'Elemental compositions of the Sion Treasure and other Byzantine silver objects'. In *Ecclesiastical Silver Plate in Sixth-century Byzantium*, S. A. Boyd and M. M. Mango (eds), 169–89. Washington, DC: Dumbarton Oaks Research Library and Collection.
Meyers, P. 1981. 'Technical study'. In *Silver Vessels of the Sasanian Period: Royal Imagery* I, P. O. Harper, 144–83. New York: Metropolitan Museum of Art.
Meyers, P., van Zelst, L. and Sayer, E. 1974. 'Major and trace elements in Sasanian silver'. In *Archaeological Chemistry*, C. W. Beck (ed.), 22–33. Washington: American Chemical Society.
Morgan, M. H. (trans.) 1960. *Vitruvius: De Architectura*, book VII, Ch. 8, 215–16. New York: Dover Publications.
Newman, R. and Lie, H. 1992. 'The technical examination and conservation of objects in the Sion Treasure'. In *Ecclesiastical Silver Plate in Sixth-century Byzantium*, S. A. Boyd and M. M. Mango (eds), 77–88, Figs 1–11, pl. 1. Washington, DC: Dumbarton Oaks Research Library and Collection.
Oddy, A. 1993. 'Gilding of metals in the Old World'. In *Metal Plating and Patination: Cultural, Technical and Historical Developments*, S. La Niece and P. Craddock (eds), 171–81. Oxford: Butterworth-Heinemann.
Oddy, A. and La Niece, S. 1986. 'Byzantine gold coins and jewellery: A study of gold contents'. *Gold Bulletin* **19**(1): 19–27.
Oddy, W. A. 1988. 'The gilding of Roman silver plate'. In *Argenterie Romaine et Byzantine*, N. Duval and F. Baratte (eds), 9–21. Paris: De Boccard.
Rackham, H. (trans.) 1952. *Pliny: Natural History*. London: Heinemann.
Raub, C. 1993. 'How to coat objects with gold: Pliny, Leyden Papyrus X, Mappae Clavicula and Theophilus seen with a modern chemist's eyes'. In *Outils et Ateliers d'Orfèvres des Temps Anciens*, C. Eluère (ed.), 101–10. St Germain en Laye: Society of Friends of the Museum of National Antiquities & Chateau de St Germain en Laye.
Schweizer, F. 1977. 'Analyses et examen technique'. In *Objects Byzantins de la Collection du Musée d'Art et d'Histoire*, M. Lazovic, N. Dürr, H. Durand, *et al. Geneva*, new series, 5–62.
Schweizer, F. and Meyers, P. 1978. 'Structural changes in ancient silver alloys: The discontinuous precipitation of copper'. In *ICOM Committee for Conservation*, Zagreb, 78/23/5/1–16.
Scott, D. 1991. 'A technical and analytical study of two silver plates in the collection of the J. Paul Getty Museum'. In *Material Issues in Art and Archaeology*, vol. II, P. B. Vandiver, J. D. Wheeler and G. S. Wheeler (eds), 677–87. Pittsburgh, Pa.: MRS.
Smith, C. S. 1966. 'The interpretation of the microstructures of metallic artifacts'. In *Applications of Science in the Examination of Works of Art*, R. H. Brill (ed.), 20–52. Boston: Museum of Fine Arts.
Smith, C. S. and Hawthorne, J. G. (trans.) 1974. *Mappae Clavicula. Transactions of the American Philosophical Society*, new series, 4, 4. Philadelphia, Pa.: American Philosophical Society.
Snow, C. E. and Weisser, T. D. 1986. 'A technical study of the Hama Treasure at the Walters Art Gallery'. In *Silver from Early Byzantium: The Kaper Koran and Related Treasures*, exhibition catalogue, M. M. Mango (ed.), 38–65. Maryland: Walters Art Gallery.
Vittori, O. 1978. 'Interpreting Pliny's gilding: Archaeological implications'. *Revista di Archeologia* **2**: 71–81.

8 The Gilding of Metals in China

Paul Jett and W. T. Chase

Abstract

This chapter presents the results of a study of gilded metal objects from China in the collection of the Freer Gallery of Art. Also discussed are examples of gilded objects described in the art historical and technical literature. Metallographic samples were taken from a small number of the Freer objects where previous breaks and losses allowed for the removal of a sample with little or no visible change. In those cases where metallographic sampling was not performed because the good condition of the objects did not allow it, technical study entailed visual examination using a stereobinocular microscope, X-radiography, and surface analysis using energy-dispersive X-ray fluorescence. The compositions of the substrate alloys of a number of the objects had been previously determined by atomic absorption analysis or wet chemical analysis using drilled samples.

Introduction

A vast number of examples of gilded metal objects from China are available for study and cited in the art historical and technical literature. Given the number of objects, it is somewhat surprising that so few of them have undergone scientific examination. Previous scientific research on Chinese gilding techniques focused mainly on the question of the origin of fire gilding (Lins and Oddy 1975);[1] most other areas of inquiry have only begun to be addressed. In those cases where technical descriptions of gilding methods are given in the art historical literature, it appears that many times the method is simply assumed and not based upon technical study. Misidentifications of technique are particularly more prevalent in the older literature, which preceded the scientific study and publication that has given us much of our current understanding of the technical aspects of gilt metals. Two examples from older art historical literature whose descriptions entail this type of difficulty are noted later in the discussion on the origin of fire gilding.

While the main subjects of this study are gilded objects with either copper alloy or silver alloy substrates, at least one example of iron decorated with gold is necessary. The use of gold decoration on ironwork seems rarely to be gilding as it is usually defined, and the example will serve to illustrate how blurred the distinction between gilding and other decorative techniques can become. The example also shows the tremendous skill with which these two metals were combined in Chinese metalworking.

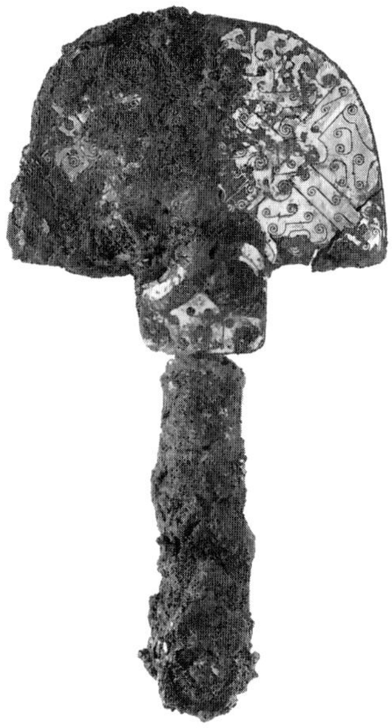

Figure 8.1 Belthook; iron with gold foil and silver inlay. (Courtesy of the Freer Gallery of Art, Smithsonian Institution, Washington, D.C., no. F1915.195.)

An iron garment hook in the Freer Gallery of Art collection is representative of a particular type dated to the Warring States Period (480–221 BC), which are described as being covered with gold foil and engraved (Fig. 8.1)(Lawton 1982: 103). This description is accurate but misleading if it is taken to mean that the appearance of the decoration was made by engraving through the gold layer to reveal the iron substrate. Microscopic examination and X-radiography reveal that the actual method of decoration is different and quite a bit more intricate. The dark lines of the decoration are the iron substrate of the garment hook, but they were not exposed by cutting through a layer of gold foil. The gold of the decoration consists of many separate small pieces of applied foil and wire. The broader areas of the gold decoration were made using foil, the edges of which were set into grooves cut into the iron, and the spiral patterns were created by inlaying short curls of gold wire. A magnified radiographic image best shows the junctures of the pieces of foil and wire, and that the wire is thicker than the foil, which give the wire its lighter (ie, denser) radiographic appearance (Fig. 8.2). The various separate pieces of foil also appear more dense along their perimeters because the edges of the foil, folded perpendicular to the surface and set into grooves in the substrate, are thicker given the perpendicular angle of view of the radiographic image. The dark linear design results from exposing narrow bands of the surface of the iron substrate between the pieces of gold foil and wire.[2] Although the decoration entailed dozens of separate pieces of gold, the overall effect is that of a single sheet of gold with an engraved decoration. Work with this appearance is sometimes referred to as gilded, but it may be more correct to describe it as a combination of inlay and overlay. The use of a similar type of decorative work on Chinese bronzes of the same period has been called 'strand inlay' or 'strand-plate inlay' (Jacobsen 1984: 83–4).

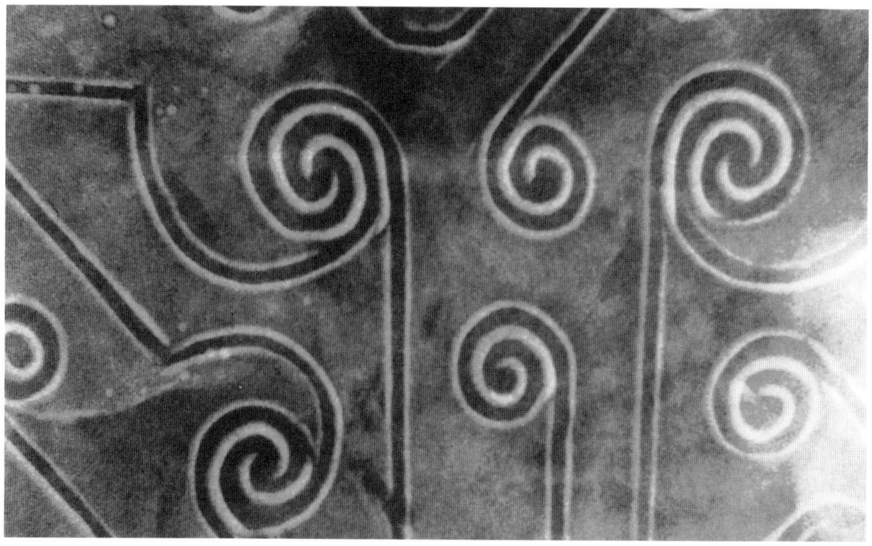

Figure 8.2 Belthook (no. F1915.195). Magnified radiographic image of gold decoration set in iron.

Objects such as the garment hook just described demonstrate how sophisticated and highly developed the use of iron for decorative purposes was in China, but these achievements will always be a minor story in comparison with that concerning the uses of gold combined with silver or copper alloys. Among these, copper alloys were by far the most important substrates used for gilding.

Gilded copper alloys

The oldest gilded objects from China are gold-foil wrapped articles composed of wood, cowrie shell and bronze – a number of objects of this type have been found in the course of archaeological excavations and are dated to the Shang Dynasty (*c*.1500–*c*.1050BC) (Singer 1971: 7; Uldry 1994: 56). Small bits of gold foil were also found in the most famous of Shang sites, Anyang. From southern China, and also dated to the period of the Shang Dynasty, was a find of spectacular bronze figures and other objects; among the finds were forty-one bronze heads, a number of which were partially covered with gold (Bagley 1990: 57–8). Photographs of the heads reveal the type of gilding; wrinkles and sharply defined edges where the gilding stops clearly indicate the use of foil (Uldry 1994: 57; Hirmer Verlag 1995: 255–8).

A good example of foil gilding is a plaque dated to the sixth century BC from the collection of the Freer Gallery of Art (Fig. 8.3). The plaque consists of a slightly convex and rather thin substrate covered by gold foil on one side. Surface analysis of the substrate using X-ray fluorescence shows the main constituents to be copper, tin and lead, but what was once probably bronze has been completely mineralized by corrosion. X-ray fluorescence analysis of the gold foil indicates an alloy containing 93 per cent gold and 7 per cent silver. None of the common methods of attaching foil to a metal substrate are obvious. The foil is not set in grooves in the support, and while the foil does wrap slightly around the edges of the support, it does not seem adequate to hold the gold in place. The use of such a technique also would not explain the foil's attachment to concave surface areas along the edge. On the back of the plaque, the reverse side of the gold foil can be seen in a few areas where holes have developed through the completely corroded substrate. The foil in these areas appears quite smooth and slightly specular. X-ray fluorescence analysis of the exposed spots of the back surface of the foil does not yield results different from those obtained by analysis of the front. Mercury is not found and, given the appearance of the foil, the use of any mercury-dependent gilding methods seems out of the question. Another feature of the gilding layer is a decoration consisting of raised lines. The lines do not appear to correspond to convex details on the substrate surface, but rather, the lines were probably created by repoussé work on the back of the foil. If this were the

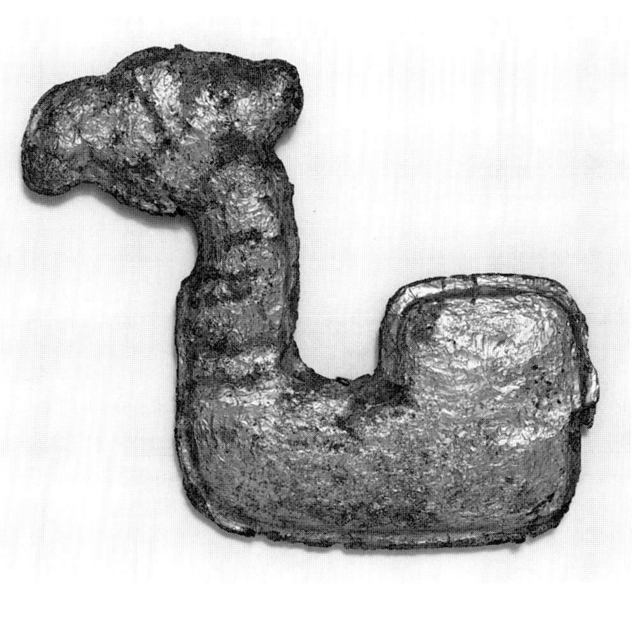

Figure 8.3 Plaque; gold foil over copper alloy support. (Courtesy of the Freer Gallery of Art, Smithsonian Institution, Washington, D.C., no. F1880.100.)

case, the foil would then need to have been applied to the bronze substrate using some method that would not distort the fragile raised decoration. All the features noted suggest, albeit inconclusively, that the foil was attached to the substrate using an adhesive.

This conclusion is supported by the examination of another object in the Freer collection. The object (Accession no. 80.101) is a piece of gold foil in the shape of a dragon, which has a metal composition and style of decoration identical with the gold on the plaque. The raised decoration on the foil bears tool marks on the reverse, or back side, proving that the decoration was executed by repoussé. If the foil once had a substrate like the plaque described previously, the joining of the foil to the support must have entailed a method that did not disturb the surface decoration of the foil. The use of an adhesive is thereby indicated.

Numerous archaeological finds demonstrate that foil gilding on copper alloys was widely used throughout the Spring and Autumn period (770–481BC)(Li 1985). However, during this era and the succeeding Warring States period (480–221BC), great changes came to Chinese metalworking in general and gilding in particular. Research by Lins and Oddy twenty years ago indicated that the earliest examples of fire gilding were from China and dated to the Warring States period (Lins and Oddy 1975). The vast majority of evidence gathered in the intervening years has only supported this theory. The wealth of fire-gilded objects with highly sophisticated workmanship that appears in the Warring States period has led some to argue that this must indicate an earlier invention of fire gilding (Wang 1984), but as Oddy has rightly pointed out (1993), what is needed is proof – objects with clear provenance that have undergone technical examination – for any claim to be made for an earlier use of fire gilding. There are as yet no examples of gilded objects dating prior to the Warring States period for which scientific study indicates the use of fire gilding.

A few older publications describe certain Shang Dynasty bronzes as gilded and further speculate that fire gilding was used. A 1935 publication on gold in ancient China includes a description of a Shang Dynasty bronze axe that contains a lengthy and persuasive discussion about finding traces of gilding on the axe; the discussion concludes with the speculation that the axe was fire-gilded (Andersson 1935: 5–6). Unfortunately, the examination of the axe did not include analysis or metallographic study. Another Shang Dynasty bronze axe, previously on loan to the Brooklyn Museum, has been reported to bear gilding and it was also suggested that the method of gilding used was fire gilding (Singer 1971: 6–7). This instance is also one of visual examination without supporting chemical analysis or microscopic examination. To the best of our knowledge, the locations of the axes today are unknown. These two bronzes have been problematic for a long time and the authenticity of the first axe has been questioned (Gyllensvard 1953: 11; Jacobsen 1984: 44, 63). While these bronzes may bear gilding (which would be, in itself, anomalous for this type of object), the use of fire gilding at so early a date is too contradictory to all other evidence to be accepted.

Ancillary evidence for the introduction of fire gilding during the Warring States period also exists in writings of the time. Literary references to liquid (or metallic) mercury and the distillation of mercury from cinnabar begin in the third century BC and occur in abundance thereafter (Needham 1974). Although it is not known when the production of mercury from cinnabar (mercury sulphide) began in China, texts dating to the first century AD refer to significant amounts of mercury being in existence as early as the fifth century BC (Needham 1974).

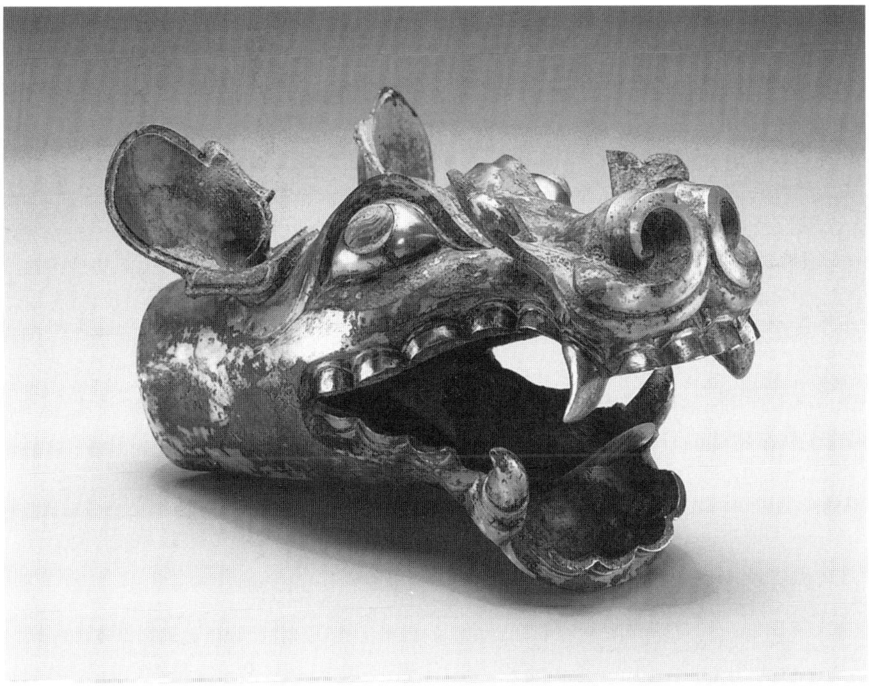

Figure 8.4 Terminal; bronze with gilding, silvering and inlay. (Courtesy of the Freer Gallery of Art, Smithsonian Institution, Washington, D.C., no. F1932.14.) (See Plate 33.)

Fire gilding was just one of many decorative metalworking techniques used during the Warring States period, and it was common for a number of techniques to be used in combination. A magnificent example of this trend is a dragon head in the Freer collection which is dated to 350–300BC[3] (Fig. 8.4). The head was probably a terminal of some sort, perhaps for a chariot pole (Lawton 1982: 64); it measures about 25 cm in length. The substrate is a copper alloy that X-ray fluorescence surface analysis indicates to contain small percentages of lead and tin. Most of the surface was fire-gilded, after which it was incised. Eyes composed of silver and glass were set into openings, and the backs of the inlay can be seen on the interior of the head. The silver, analysed by X-ray fluorescence, was found to contain about 97 per cent silver, small percentages of copper and gold, and nothing else in significant amounts. Another added part is the tongue, also made from silver. It has a plain polished surface on the top while the underside, which can hardly be seen, was fire-gilded and incised. The decoration of the terminal included one other technique. The teeth, though silver in color, are not made from silver like the eyes and tongue: they are composed of the same copper alloy as the head and were silver plated using silver amalgam. Also, the teeth were first fire-gilded and then silvered. Quite a number of Chinese objects have been shown to have amalgam silvering over amalgam gilding, but why this technique was used has not been determined (Oddy *et al.* 1979; La Niece 1990; Bunker *et al.* 1993). It may be that the gold amalgam layer intervenes to improve the bonding of the silver amalgam to the copper alloy substrate (although there are ample examples of silvering applied directly to copper alloy substrates from other cultures and time periods).

There are a number of reported examples of bronzes bearing silvering and gilding in a range of hues, and it has been speculated that the variation in colour was made by varying the silver content of a gold amalgam or by mixing gold and silver amalgams (Jacobsen 1984: 99–100; Bunker 1994: 75). But X-ray fluorescence analysis of four silvered and gilded objects in the Freer collection, the dragon head terminal and three garment hooks, clearly indicates the use of two distinctly different amalgams for silvering and gilding and not simply a gold amalgam containing varying amounts of silver. Surface analyses of the silver-toned areas on the garment hooks, where the silver layer appeared thickest, detected mostly silver and mercury and very little gold. Also, like the dragon head, the garment hooks bear areas of amalgam silvering applied to fire-gilded substrates.

Most published examples of mercury-silvered objects from China and those found in the Freer collection date only to the Warring States period and the Han Dynasty, so it seems the technique, or at least its greatest popularity, was short-lived. For a long time following the Han Dynasty, the decorative techniques employed in combination with gilt copper alloys appear generally less elaborate than those used in the earlier periods. Buddhist bronze sculptures exemplify this trend – they are typically gilded, and while other added decorative materials are used, they are used with restraint. Inlaid stones are sometimes found, particularly on works from the later periods. More rare is painted decoration applied over gilding; one example of this feature can be seen on the halo of a Buddha figure in the Freer collection (Fig. 8.5). The inscription of the figure dates it to AD451.

The use of fire gilding on copper alloys continued unchanged in spite of changes occurring in the constituents of the copper alloys over time. The substrates of the earliest fire-gilded works generally consist of copper that contains only small percentages of lead and tin (on the order of two per cent).[4] However, fire-gilded bronzes come into use some time soon after.

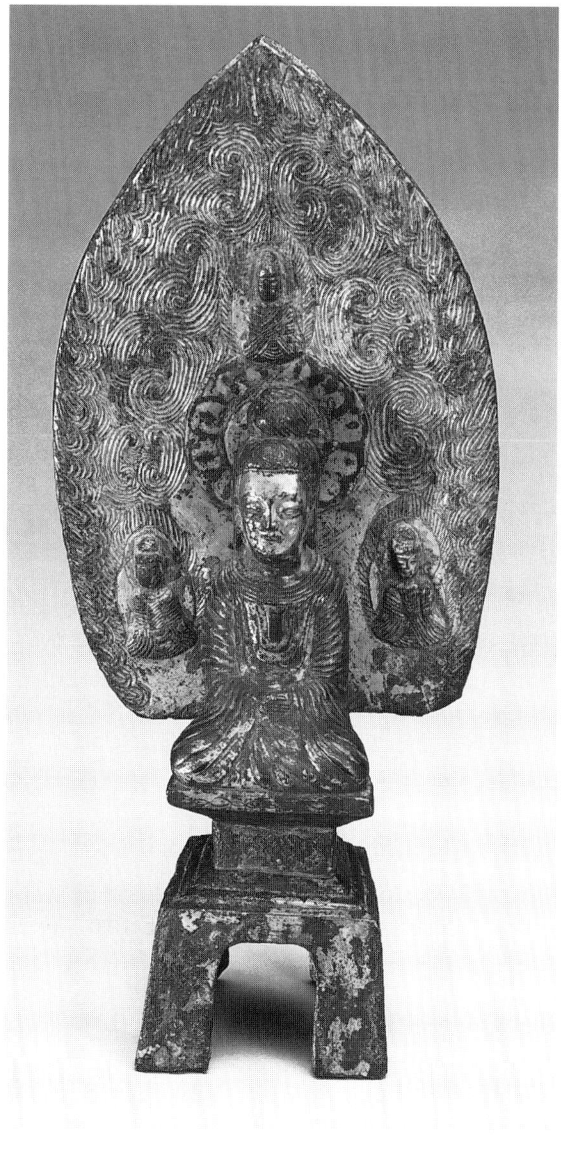

Figure 8.5 Buddha figure; bronze with gilding and paint. (Courtesy of the Freer Gallery of Art, Smithsonian Institution, Washington, D.C., no. F1911.121.) (See Plate 34.)

A mirror in the Freer collection (Accession no. 17.194) that is dated to the first century BC has been selectively gilded to highlight the design (Bulling 1960: 55, pl. 38); surface analysis using X-ray fluorescence indicates that the metal substrate is a typical mirror alloy of leaded high-tin bronze. Analyses of more than one hundred Buddhist bronzes, many of which are gilded, show that the use of bronze substrates with relatively high percentages of lead and tin (10–20 per cent) was common by the fifth century AD (Jett and Douglas 1992; Jett 1993). This use of materials continued into and through the Qing Dynasty (AD1644–1911). The date of the first general use of brass in Chinese metalwork is unclear, but it does not seem to greatly precede, if at all, the early sixteenth century, when brass appears in coinage (Bowman *et al.* 1989; Dai and Zhou 1992). From that time on the uses of brass and bronze paralleled each other, and the advent of brass seems to have simply added another copper alloy to the list of metals used as substrates for fire gilding. Fire-gilded animal figures in the Freer collection dated to the Qing Dynasty have been analysed using atomic absorption spectroscopy and found to contain as much as 27 per cent zinc in the substrate alloy (Jett and Douglas 1992).

Although fire gilding is just one of many decorative metalworking techniques used in China, the longevity of its use is remarkable. From its beginning nearly two and a half millennia ago, the use of fire gilding has continued up to the present day.

Gilded silver alloys

The history of gilded silver in China followed a different course than that of gilded copper alloys. Whereas bronze was the pre-eminent metal in early China, silver was rare, or rarely used. It has been stated that prior to the Warring States period, there was essentially no silver used in central China at all (Bunker 1994: 74), and the authors are unaware of any examples of gilded silver prior to the introduction of fire gilding. In fact, there are very few published examples of gilded silver that date earlier than the Tang Dynasty (AD618–907). One is the tongue of the dragon head terminal just described; another is a partially gilded pouring vessel, or *Yi*, which is considered to date to the Western Han period (third to second century BC) (Bunker 1994: 74–5; Uldry 1994: 112). There is also a small group of heart-shaped cups with bird head handles that are dated roughly to the Warring States period and that are partially gilded (Gyllensvard 1953: 133, pls 76a,b; So and Bunker 1995: 151–2). The few examples of gilded silver objects dated earlier than the Tang Dynasty allow for little study or comment about the subject. It is only with the coming of the Tang Dynasty that there are gilded silver objects in enough abundance that one can begin to determine techniques and trends in gilding. Two silver objects in the Freer collection are considered good examples of Tang metalwork; both are dated to about AD700–750.

The first example, a bowl that measures just under 6 cm in height and about 15 cm in diameter at the rim, is comparable to a number of works in other collections (Fig. 8.6) (Kelley 1984: 47). The interior of the bowl is undecorated but mirrors the relief forms of the exterior. Chased decorations consisting of floral and faunal motifs were selectively gilded and set in contrast to the ungilded, ring-matted background. Surface analysis of the gilding using X-ray fluorescence shows the presence of mercury, and microscopic examination reveals a granular, pasty structure in the recesses of the surface, typical of fire gilding. But the gilding also has a sharp-edged

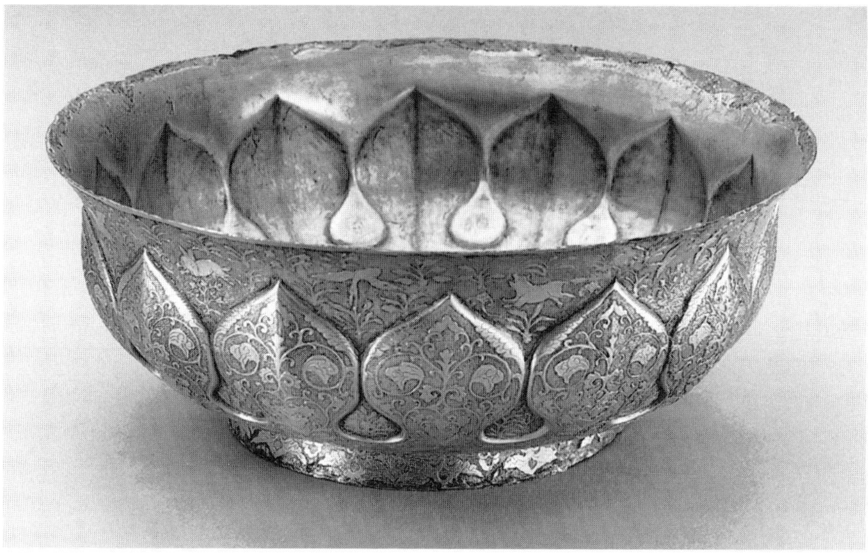

Figure 8.6 Bowl; gilt silver. (Courtesy of the Freer Gallery of Art, Smithsonian Institution, Washington, D.C., no. F1931.8.)

appearance with strong definition between the gilded and ungilded areas; this is untypical for fire gilding. In the process of fire gilding the amalgam can flow as it is heated to vaporize the mercury it contains; on pieces that have been selectively gilded, the gilding often spreads beyond the area that it was intended to cover. This is a feature often seen on the fire-gilded silver of some cultures, such as that of Iranian origin from the Sasanian period. While there is some run-over of the gilding on the Freer bowl, it was found to be minimal. This appearance is undoubtedly due in part to the severe etching that corrosion has caused on the ungilded surface areas, but it also appears to reflect a highly controlled use of fire gilding that was observed on all of the fire-gilded Tang silver in the Freer collection.

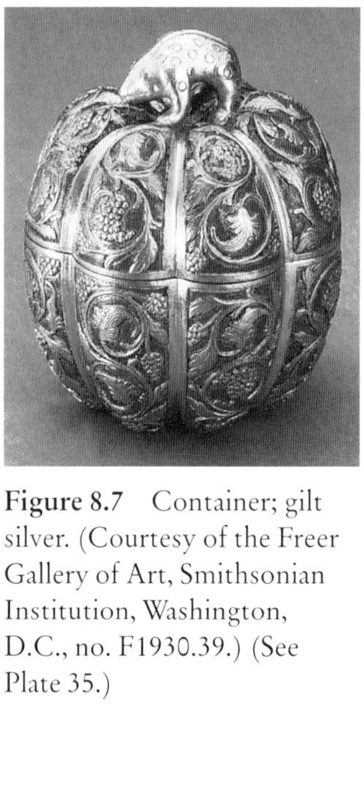

Figure 8.7 Container; gilt silver. (Courtesy of the Freer Gallery of Art, Smithsonian Institution, Washington, D.C., no. F1930.39.) (See Plate 35.)

The second example from the Freer collection is the only piece of gilded silver attributed to the Tang Dynasty that is not fire-gilded. The object is a small lidded container in the shape of a melon; a tiny bear figure serves as the handle of the lid (Fig. 8.7). The bear, the raised vine motif that covers the surface and the vertical bands that separate the lobes of the melon have been gilded; the background of the exterior surface and the interior of the container were ungilded. Microscopic examination reveals visual evidence of leaf gilding. The gilding has sharp, straight edges that, at some points, are unattached to the substrate and stand free of the surface. Another telltale visual feature of leaf gilding is the gilding's appearance over areas of relief decoration. In some places, the leaf was wrinkled and flattened to fill in the recesses of the relief; conversely, in other spots the leaf was stretched over recesses in the surface created by tooling. In both cases, the leaf was very precisely cut to fit the outline of the areas of decoration to be gilded, which gave the sharp, straight edges noted previously. X-ray fluorescence analysis of the gilding did not detect mercury, and the method of attaching the leaf was not ascertained.

Leaf gilding is also found on a pair of identical silver bowls in the Freer collection (Fig. 8.8); although recorded as being from the Tang Dynasty, the bowls are now believed to be of later date, perhaps as late as the Ming Dynasty (AD1368–1644).[5] A simple floral decoration was tooled on the insides of the bowls and somewhat crudely gilded using leaf. The straight-cut edges of the leaf do not conform well to the silhouettes of the design and overlap surrounding background areas. It is clear that the gold lies on the surface and is not inset. As with the melon-shaped container, X-ray fluorescence does not detect any mercury, and the method used to attach

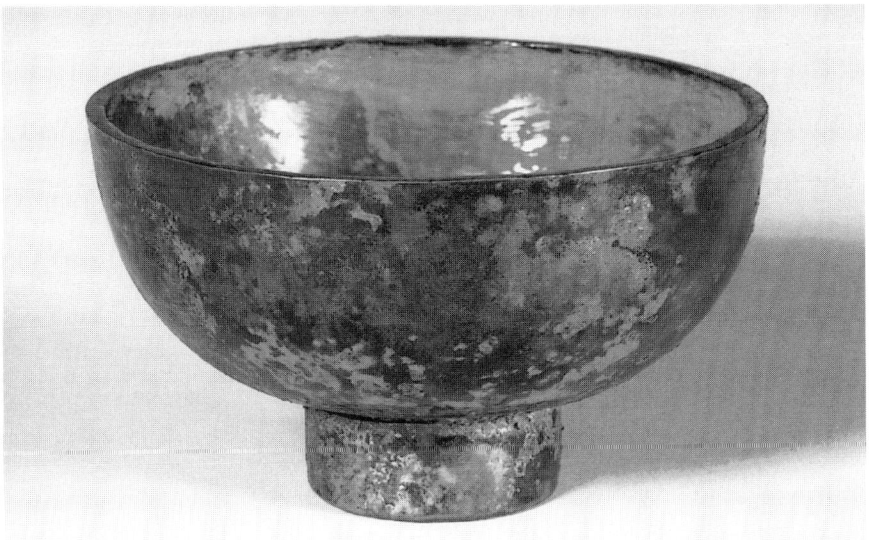

Figure 8.8 Bowl; gilt silver. (Courtesy of the Freer Gallery of Art, Smithsonian Institution, Washington, D.C., no. F1916.718.)

the leaf to the substrate was not discernible. Metallographic examination of the bowls or the melon-shaped container might reveal the method used to affix the leaf, but the good condition of these objects does not permit sampling to be considered.

Further examples of silver objects with leaf gilding similar to that seen on the bowls are found among excavated pieces dated to the Song Dynasty (AD960–1279). Published photographs show that a number of them have roughly cut gold leaf decorations (Cultural Relics Publishing House 1992: pl. 132, 319). The presence of leaf gilding on the melon-shaped container seems anomalous given the fire gilding found on most Tang Dynasty silver. In spite of this, the container in combination with the other examples of leaf gilding just noted indicate that at some time between the Tang and Song periods leaf gilding for silver came into use. But leaf gilding did not replace fire gilding; the two techniques were used coincidentally. It is unclear why or to what extent this occurred. The roles and relationship of fire gilding and leaf gilding on silver in the later Chinese dynasties await further study. We do know, however, that the use of fire gilding continued into modern times.

Conclusion

The ideal situation for the technical study of gilded metal objects is to have large groups of objects that are both well dated and of certain provenance. If it is then possible to take metallographic samples from a number of the objects, then definitive statements can be made about the materials and techniques employed. The cross-sectional view that a metallographic section provides combined with the analytical information that can be obtained from such a sample can yield a wealth of information about the nature of the gilding. Conversely, without sampling, the range of examination techniques one can employ is limited and so what can be determined about technique is also limited.

Even a bare outline of the history of gilded metals in China must remain incomplete at this time – progress to fill in the many gaps in our knowledge will be best served by the analysis of metallographic samples of objects, particularly objects recovered through excavation. As this technological history unfolds there will doubtless be many surprises. We may come to learn that, as in so many other areas of metalworking technology, the Chinese forged their own unique path.

Notes

1. Fire gilding, also called mercury gilding and amalgam gilding, entails the use of a mixture (or amalgam) of mercury and gold. The amalgam is applied to the surface to be gilded, heated to vaporize the mercury, and finally burnished, to give a smooth gold layer. Excellent studies and reviews that describe the various types of gilding used on metals in antiquity have been published (such as Oddy 1993), and the reader is referred to those for basic information about gilding techniques.
2. While the iron is now dark in colour, its original colour is unknown. The garment hook also bears some silver inlay, and the question remains as to how the two light-coloured metals were differentiated from each other. Some iron garment hooks have been found with a blue-black surface colour that appeared to be representative of the original appearance (Chase 1967: 89–90).
3. So 1994: personal communication.
4. Freer Gallery Laboratory Records, Unpublished analyses.
5. So 1994: personal communication.

References

Andersson, J. G. 1935. 'The Goldsmiths in Ancient China'. *Bulletin of the Museum of Far Eastern Art*, 7: 1–38, pls I–XXI.

Bagley, R. W. 1990. 'A Shang city in Sichuan Province'. *Orientations*, **21**(11): 52–67.

Bowman, S. G. E., Cowell, M. R. and Cribb, J. 1989. 'Two thousand years of coinage in China: An analytical survey'. *Historical Metallurgy*, **23**(1): 25–30.

Bulling, A. 1960. *The Decoration of Mirrors of the Han Period*. Ascona: Artibus Asiae.

Bunker, E. C. 1994. 'The enigmatic role of silver in China'. *Orientations*, **125**(11): 73–8.

Bunker, E. C., Chase, T., Northover, P. and Salter, C. 1993. 'Some early Chinese examples of mercury gilding and silvering'. In *Outils et Ateliers d'Orfevres des Temps Anciens*, C. Eluere (ed.), 55–66. St-Germain-en-Laye: Societé des Amis du Musee des Antiquites Nationales.

Chase, W. T. 1967. *Chinese Belt-Hooks in the Freer Gallery of Art*. Master's thesis. New York University.

Dai, Zhigiang and Zhou, Weirong. 1992. 'Studies of the alloy composition of more than two thousand years of Chinese coins'. *Historical Metallurgy*, **26**: 45–55.

Cultural Relics Publishing House 1992. *Gems of China's Cultural Relics: 1992.* Beijing: Cultural Relics Publishing House.

Gyllensvard, B. 1953. *Chinese Gold and Silver in the Carl Kempe Collection*. Stockholm: Nordisk Rotogravyr.

Hirmer Verlag 1995. *Das Alte China: Menschen und Götter im Reich der Mitte 5000v. Crh.-220 nach Chr.* München: Hirmer Verlag.

Jacobsen, R. D. 1984. *Inlaid Bronzes of Pre-Imperial China: A Classic Tradition and Its Later Revivals*. PhD dissertation. University of Minnesota.

Jett, P. 1993. 'A study of the gilding of Chinese Buddhist bronzes'. In *Metal Plating and Patination*, S. La Niece and P. Craddock (eds), 193–200. Oxford: Butterworth-Heinemann.

Jett, P. and Douglas, J. G. 1992. Chinese Buddhist bronzes in the Freer Gallery of Art: Physical features and elemental composition. In *Material Research Society Symposium Proceedings*, vol. 267, P. B. Vandiver, J. R. Druzik, G. S. Wheeler and I. C. Freestone (eds), 205–23. Pittsburgh: Material Research Society.

Kelley, C. W. 1984. *Chinese Gold and Silver in American Collections*. Dayton, Ohio: The Dayton Art Institute.

La Niece, S. 1990. 'Silver plating on copper, bronze and brass'. *Antiquaries Journal* **70**: 102–14.

Lawton, T. 1982. *Chinese Art of the Warring States Period*. Washington, DC: Smithsonian Institution Press.

Li, Xueqin. 1985. *Eastern Zhou and Qin Civilizations*. New Haven, Conn.: Yale University Press.

Lins, P. A. and Oddy, W. A. 1975. 'The origins of mercury gilding'. *Journal of Archaeological Science* **2**: 365–73.
Needham, J. 1974. *Science and Civilization in China*, vol. 5, part 2. Cambridge: Cambridge University Press.
Oddy, A. 1993. 'Gilding of metals in the Old World'. In *Metal Plating and Patination*, S. La Niece and P. Craddock (eds), 171–81. Oxford: Butterworth-Heinemann.
Oddy, W. A., Padley, T. G. and Meeks, N. D. 1979. 'Some unusual techniques of gilding in antiquity'. In *Eighteenth Symposium on Archaeometry and Archaeological Prospection*, Bonn, 1978 (Archaeo-Physika, no. 10), 230–42. Bonn: Rheinland-Verlag.
Singer, P. 1971. *Early Chinese Gold and Silver*. New York: China Institute in America.
So, J. F. and Bunker, E. C. 1995. *Traders and Raiders on China's Northern Frontier*. Washington DC: Arthur M. Sackler Gallery.
Uldry, P. 1994. *Chinesisches Gold und Silber*. Zurich: Museum Rietberg.
Wang, Haiwen. 1984. 'A study of gold plating technology'. *Gugong Bowuyuan Yuankan (Palace Museum Journal)* **2**: 50–58, 84.

9 Archaeological Gilded Metals Excavated in Japan

Ryu Murakami

Abstract

It is no exaggeration to say that in ancient Japan many metalworking techniques were brought from the Asian continent; gilding was no exception. A review of ancient excavated objects from Japan shows that most of the metal objects with gold surfaces were gilded, that is, a layer of gold was applied over a base metal, giving the appearance of a solid gold object. Indeed, in Japan ancient metal artifacts in pure gold were rather rare.

Most of the ancient gilded metal objects in Japan were fire-gilded, a technique, also called mercury amalgam gilding, by which a pasty amalgam (mixture) of mercury and gold is applied to a clean base metal surface, then heated to drive off the mercury (Oddy 1981, 1993). To study and characterize this gilding technique on ancient Japanese objects, newly fire-gilded samples were prepared. Using a scanning electron microscope (SEM), a comparison was made between the excavated artifacts and control samples replicated with traditional methods. The micromorphological features of fire-gilded surfaces were clearly revealed. Other analytical tools such as X-ray fluorescence analysis (XRF) and X-ray diffraction (XRD) were also used.

Among the metal artifacts excavated in Japan some examples with foil or leaf-gilding, techniques whereby varying thicknesses of gold sheet are attached to the surfaces of base metals, have been found. Artifacts gilded using these technique were also examined with SEM.

Historical aspects of gilding in Japan

Throughout Japanese history, the location of the Japanese archipelago at the eastern end of the Asian continent has been a very important factor for technological innovation. Since, in ancient Japan, natural resources were poor and mining technology was not well developed, most raw materials for metalwork, such as tin, lead and iron, had to be imported. Skilled craftsmen from the Asian continent, and especially from the Korean peninsula, probably visited the Japanese archipelago and introduced advanced techniques for smelting ores and manufacturing metal objects. This was particularly the case in the early stage of Japanese metalworking during the Yayoi period (fourth century BC to third century AD). Gilding techniques, including fire, foil and leaf techniques, were adopted by the end of the fifth century AD from the continent, as evidenced by the

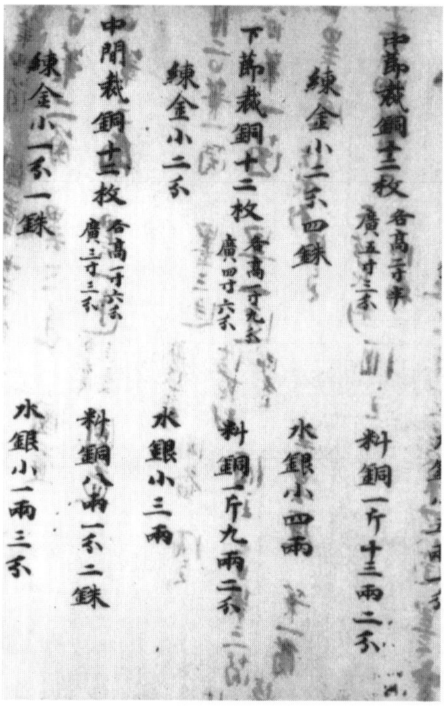

Figure 9.1 Record of construction of the *Saikondo* (West Main Hall) of *Kofuku-ji* Temple (AD742), which was handed down in *Shoso-in* (*Todai-ji* Temple Treasure House). It mentions the old recipe of fire gilding.

ornaments on horse trappings, the cores of which were made of iron, covered with gilded thin copper plates (Kobayashi 1962). With the introduction of Buddhism in AD538 through Korea, it did not take long before gilding technology was also applied to Buddhist architectural ornaments and Buddha images in Japan. The archives (Todai-ji Temple Treasure House AD742) concerning the Buddha image workshop in *Kofuku-ji* temple in Nara mention the materials for fire gilding in AD734 (Fig. 9.1). The huge bronze Buddha image (originally 16 m in height), with fire-gilded surface, of Todai-ji temple, consecrated in AD752, is representative of fire-gilding at that time. Afterwards, fire gilding became one of the most popular techniques in Japanese metalworking, and continues until the present. Other techniques of gilding, both foil gilding and leaf gilding (like foil gilding, but with thinner sheets of gold), have been used and were highly developed in modern times, especially in the Edo period (seventeenth century to nineteenth century).

Surface characterization of gilded metals

Preparation of fire-gilding samples

To examine the micromorphological features of the fire-gilded surface, some prepared samples were made by a method commonly adopted during the Edo period in Tokyo and still maintained in Tokyo National University of Fine Arts and Music today (Murakami and Niiyama 1993). The experimental procedure for fire gilding follows. First, pasty gold–mercury amalgam is made with an initial mixing ratio of 1:10 by weight. The gold easily dissolves into the mercury, resulting in a silver-coloured amalgam. To adjust the fluidity of the amalgam paste, excessive mercury is squeezed out with Japanese paper (in ancient times, deerskin must have been used (Murakami and Niiyama 1993)); Niiyama explains that the feeling of the fluidity 'may be compared with that of an earlobe'. A pure copper plate is used as a substrate, because most ancient gilding was done on copper plates. The copper surface is carefully rubbed with charcoal of paulownia,

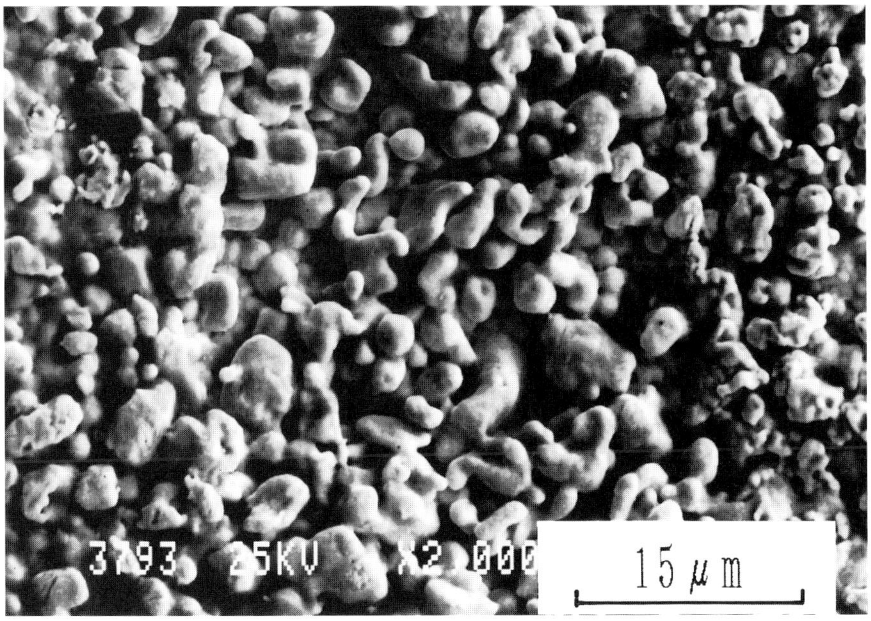

Figure 9.2 The gilded surface just after mercury was burnt off (SEM observation).

a very fine polishing material. After polishing, the wetability of the surface is increased by the application of mercuric nitrate. Plum vinegar was traditionally used for this purpose (Murakami and Niiyama 1993). Pasty gold amalgam is uniformly spread on the copper plate. In this stage, the surface still has a silvery lustre. Once the back of the copper plate is heated over a gas burner at about 350°C, the brilliant silver disappears very quickly, and the surface turns to matte gold. This phenomenon is caused by vaporization of the mercury.

Figure 9.2 shows an SEM magnified view of the surface condition just after heating. It is easy to recognize numerous gold–mercury amalgam particles that have precipitated on the surface. Each particle is 5–6 μm in diameter (Fig. 9.3). The fine irregularity of the particles creates a matte image. To get a brilliant gold surface, a final burnishing process is essential; an iron spatula is used for burnishing (Murakami and Niiyama 1993). It is an important step in the fire gilding process. Figure 9.4 shows the results of burnishing; the upper half of the plate was not burnished, and the lower half was burnished.

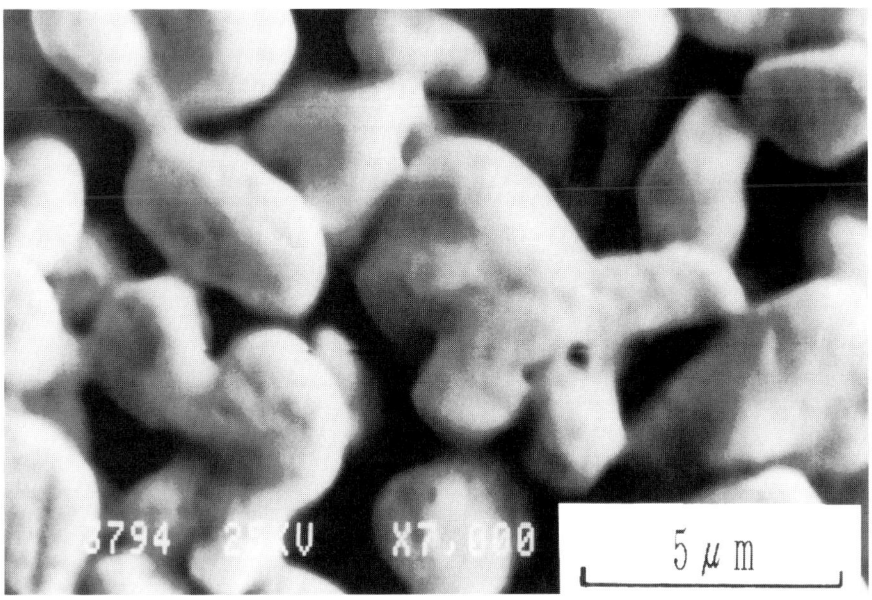

Figure 9.3 Gold–mercury amalgam particles (SEM observation).

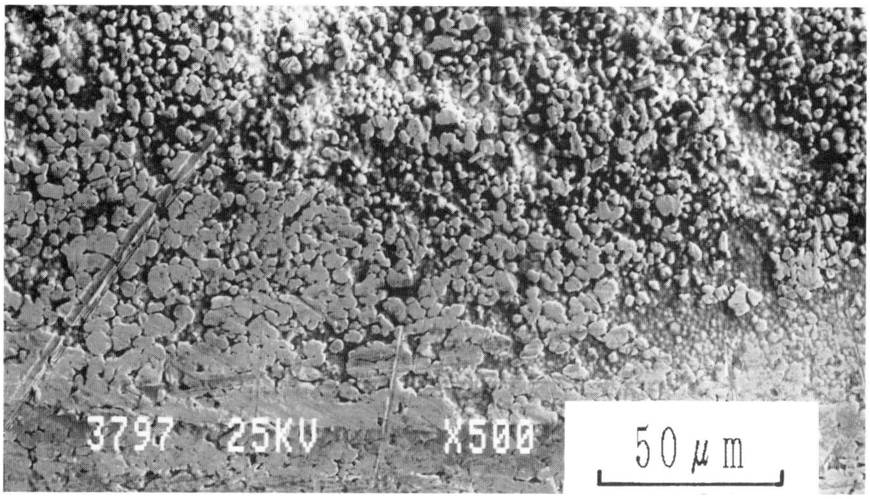

Figure 9.4 The result of burnishing can be seen in the power part of the plate (SEM observation).

Analysis of surface of prepared sample

SEM observation of the prepared sample revealed that some gold–mercury amalgam particles remained even after careful burnishing (Fig. 9.5). It is important to note, however, that these particles have been 'levelled' in the process of burnishing. Accordingly, the particles seem 'flat'. In addition, fine traces of burnishing with the spatula are evidenced by parallel striations (Fig. 9.6). If these features are detected on the surface of an excavated artifact, it would be evidence for fire gilding with gold–mercury amalgam.

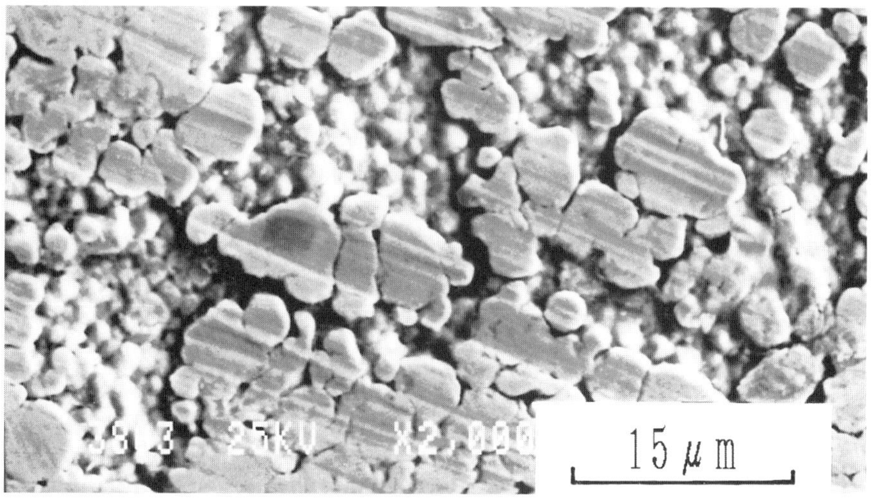

Figure 9.5 The gold–mercury amalgam particles remaining after burnishing (SEM observation).

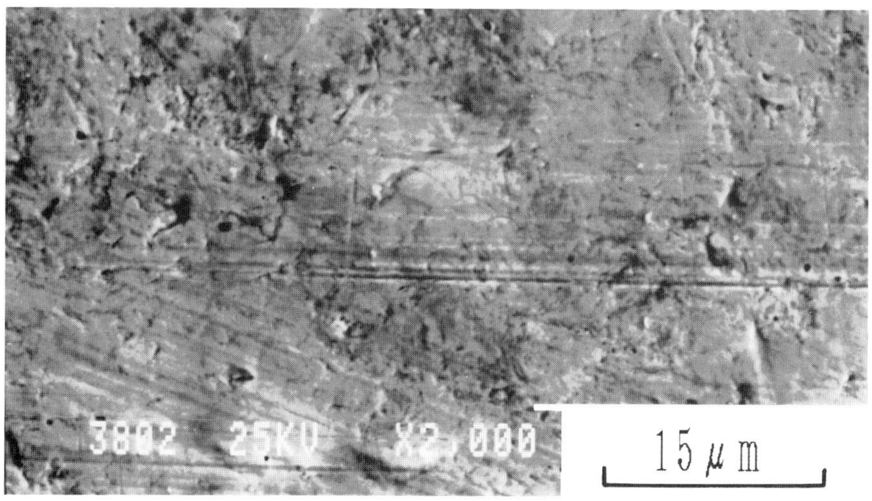

Figure 9.6 The traces of burnishing with a spatula (SEM observation).

Reappearance of the brilliance of a gilded surface by removing the patina

Usually, the surface of an excavated gilded copper or copper alloy object is covered with thick patina that has developed during long-term burial. In order to compare the gilded surface of an excavated object with modern samples, it is necessary to remove the patina without altering the underlying surface, so that traces of the original metalworking and finishing can be closely observed.

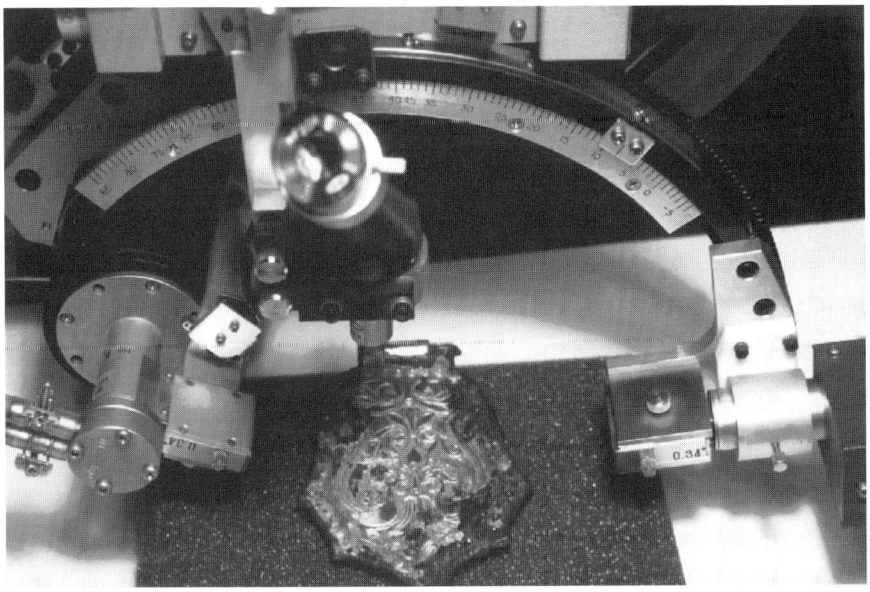

Figure 9.7 Non-destructive XRD apparatus. Here it is used for analysis on a horse ornament excavated in Fujinoki Tumulus. (See Plate 36.)

The following procedures were followed to clean the gilding on excavated objects. First, non-destructive XRD (Fig. 9.7) was used to identify the patina, which consisted of malachite and cuprite (Fig. 9.8). This analytical method is unusual because it is not necessary to remove a sample from the object and the directions at which X-ray beams are applied can be changed while holding a sample or the object in a fixed position. Second, Sawada's method (Sawada 1993) was adopted to remove the patina. Using this technique, highly water-absorbent resin mixed with EDTA or dilute formic acid is applied to the patina on the surface. A chemical reaction between the EDTA and the patina only takes place at the contact between

Figure 9.8 The result of XRD analysis of patina on the gilded surface of a horse trapping excavated in Daitsu-ji Tumulus.

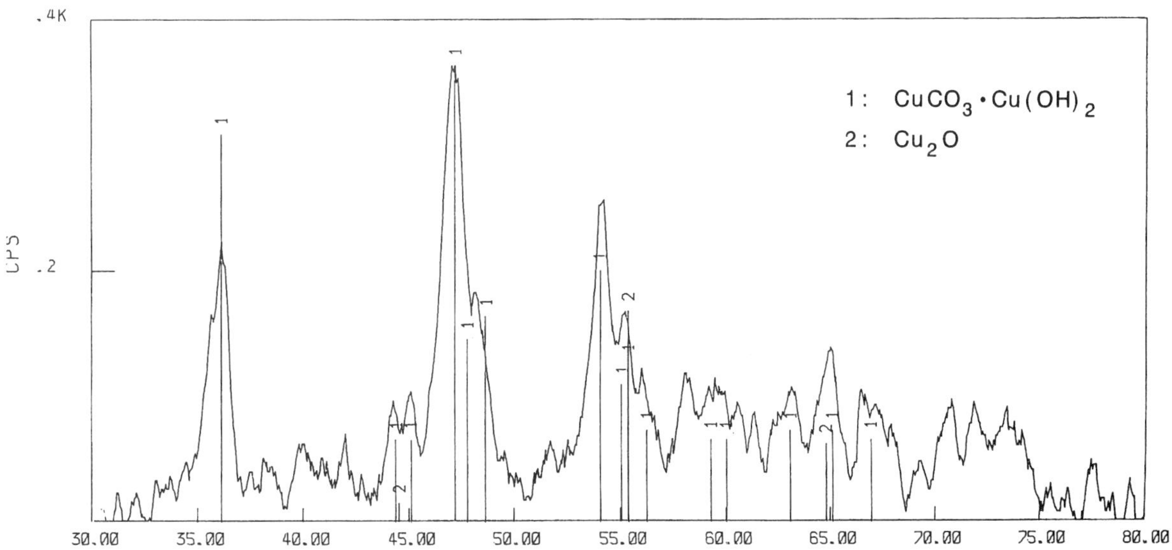

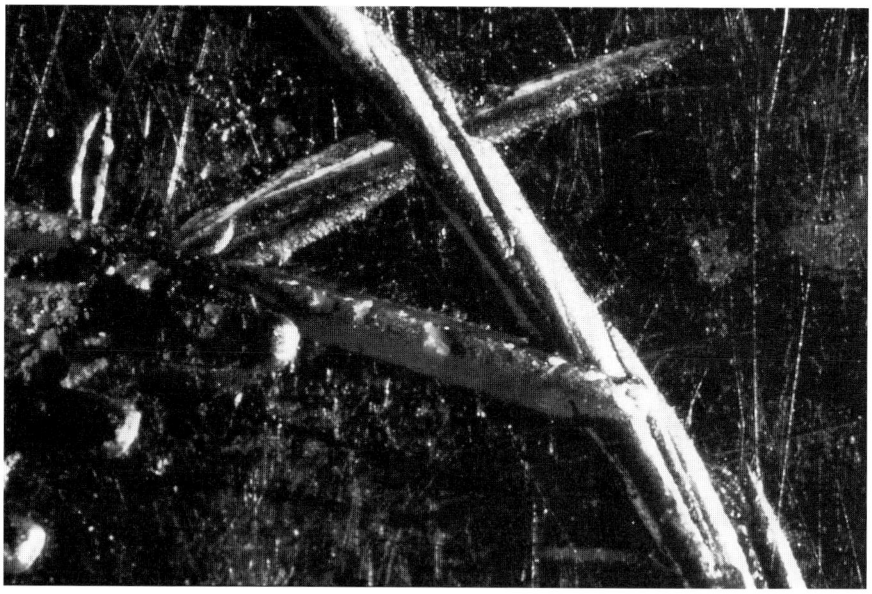

Figure 9.9 The detail of the gilded surface after the patina was removed. (See Plate 37.)

the metal surface and resin. Moreover, dissolved patina is absorbed by the resin. This process is repeated until the metal surface is exposed. Then the metal object is washed with distilled water, dehydrated with alcohol, and dried in an oven. This method is advantageous because, unlike mechanical methods, the original surface is not touched. Moreover, the chemical effect on the body of the metal object is minimized because the resin holds the chemicals. Microscopic analysis after this treatment revealed that burial has no or little effect on the gilded layer (Fig. 9.9).

Characterization of ancient fire-gilding

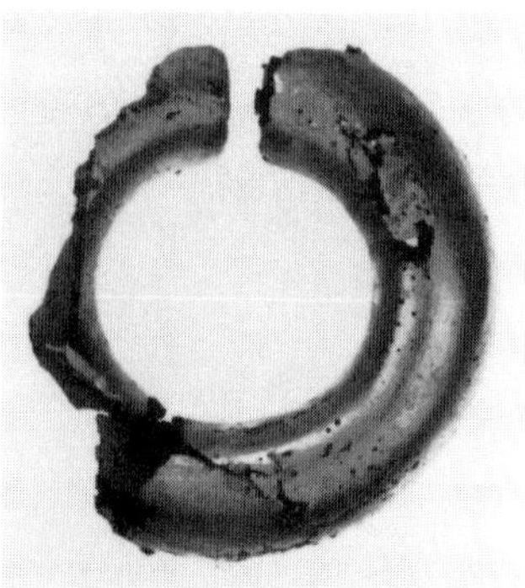

Figure 9.10 The earring excavated in *Takagawa* Tumulus, *Hyogo* prefecture. (See Plate 38.)

Figure 9.10 shows an earring from the late sixth century AD excavated in Takagawa Tumulus in Hyogo prefecture (Murakami 1991). Because this earring was broken into three parts, it was easy to observe its details. It was hollow, and made of a thin copper tube. The overlapped part of the copper tube (Fig. 9.11) was joined with silver brazing (Fig. 9.12), which was a silver–copper eutectic alloy. This silver brazing is the oldest one that has been recognized in Japan (Murakami 1995a). The outer gold surface was examined with SEM. Gold–mercury amalgam particles that were levelled and parallel striations resulting from burnishing were recognized (Fig. 9.13). Residual mercury in the gilded layer was confirmed by X-ray analysis in the SEM (Fig. 9.14). These features indicate that the surface of

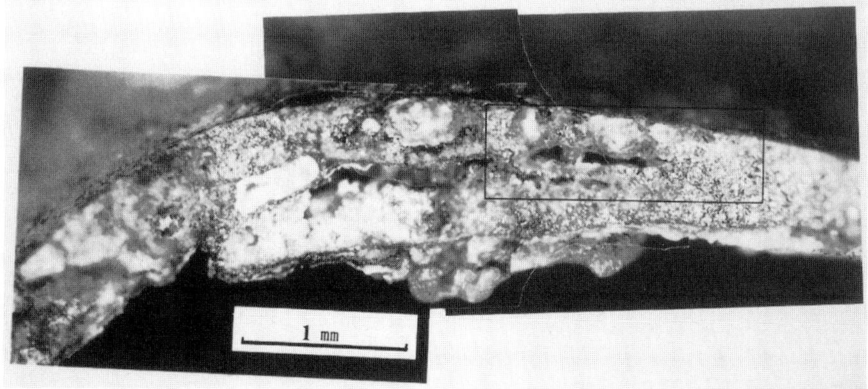

Figure 9.11 The area of overlapped copper on the earring is indicated by the rectangle. (See Plate 39.)

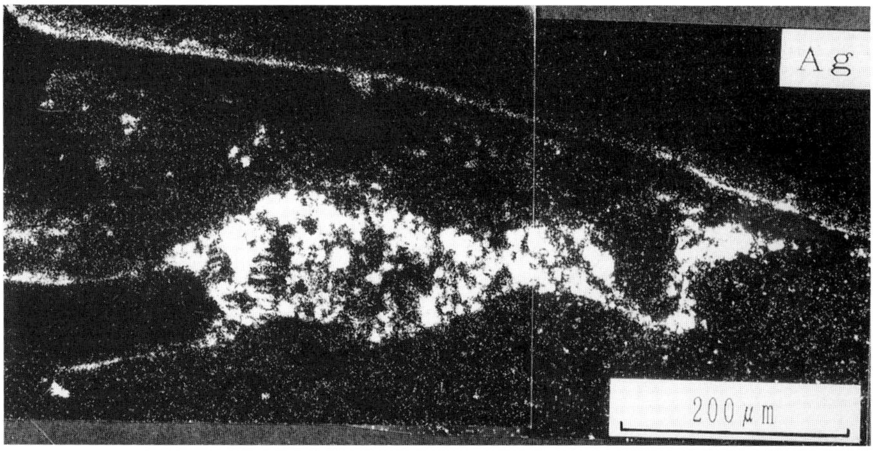

Figure 9.12 A map of the silver-rich area at the join of the overlapped coper on the earring (see area of rectangle Fig. 9.11).

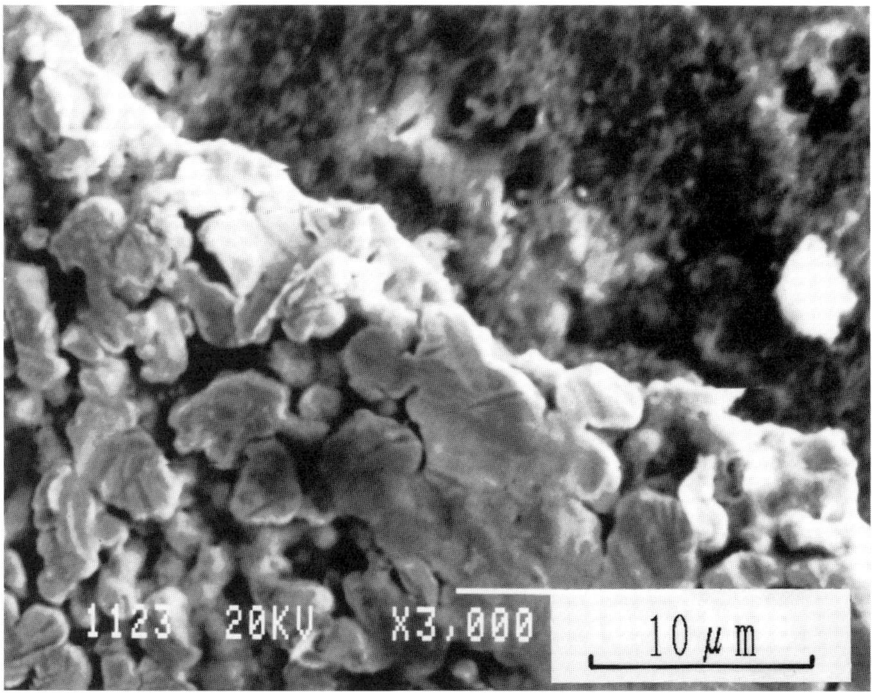

Figure 9.13 The surface of the gilded layer of the earring excavated in *Takagawa* Tumulus (SEM observation).

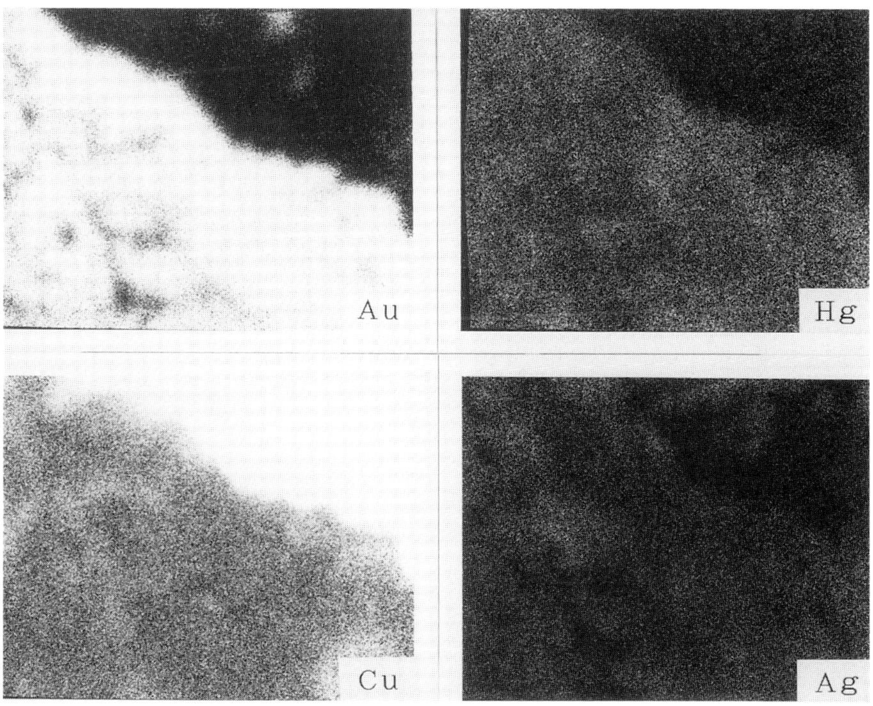

Figure 9.14 The analytical result of the gilded layer of the earring excavated in *Takagawa* Tumulus (X-ray analysis with SEM).

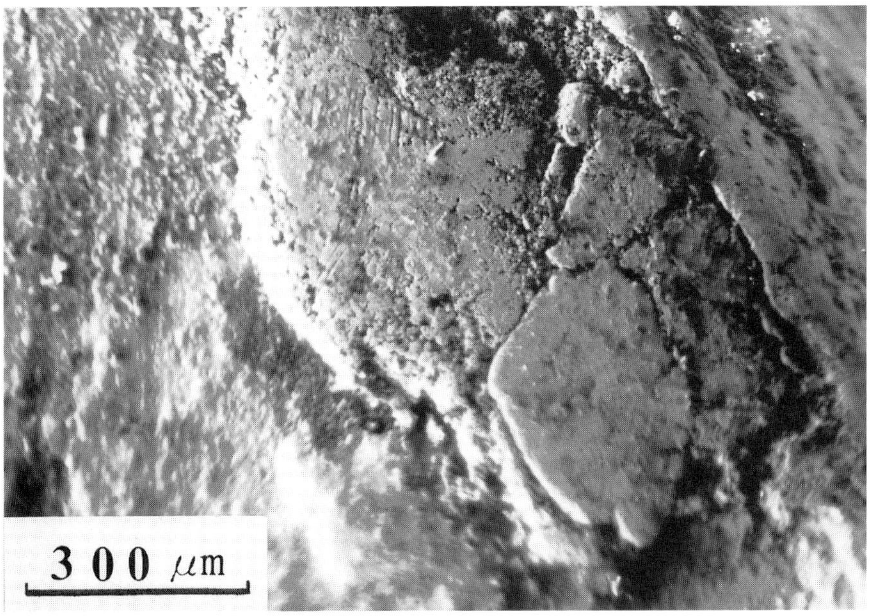

Figure 9.15 The lump of gold–mercury amalgam particles at the end of the earring excavated in *Takagawa* Tumulus (SEM observation).

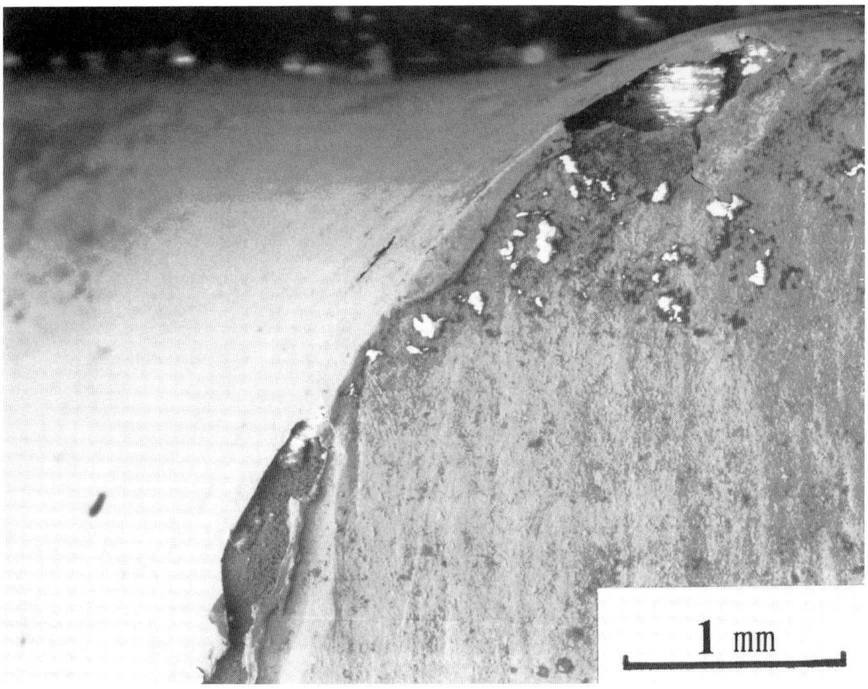

Figure 9.16 The surface of the end of earring excavated in *Takagawa* Tumulus (SEM observation).

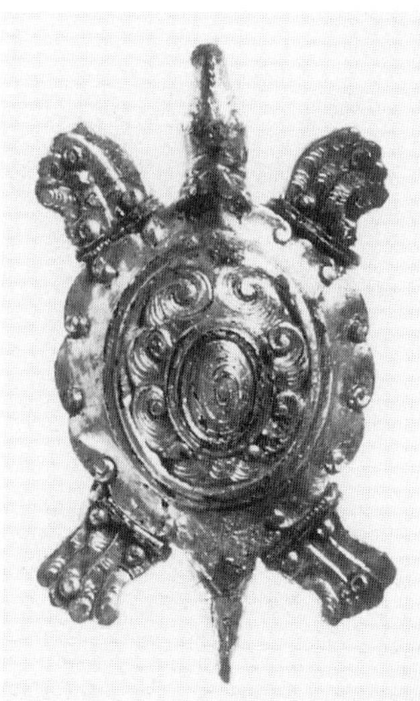

Figure 9.17 The turtle figure ornament excavated in *Sasatsuka* Tumulus, *Nagasaki* prefecture. (See Plate 40.)

this earring was finished by fire gilding. Figure 9.15 shows the lump of gold–mercury amalgam that remained at the edge of the earring's end. Neither side of the ends was burnished (Fig. 9.16); the gap between the ends was too narrow to burnish them with something like a spatula.

Observation of a cross-section also tells us the features of a gilded layer (Murakami 1994a). A gilded copper turtle figure ornament (Fig. 9.17) was excavated in Sasazuka Tumulus in Iki Island, Nagasaki Prefecture, and was a part of horse trappings (excavated by Katsumoto Town Board of Education). Iki Island was one of the key points in ancient Japan, because it is located between the Japanese archipelago and the Korean peninsula. SEM observation of the turtle ornament gave us considerable information about its gilded layer. It was about 15 μm in thickness and was composed of gold–mercury amalgam particles as mentioned above (Fig. 9.18). The flat and brilliant surface was produced by burnishing the topmost of the aggregation of amalgam particles (Fig. 9.19) Due to long-term burial, copper in the body was totally changed into fine cubic cuprite (Fig. 9.20).

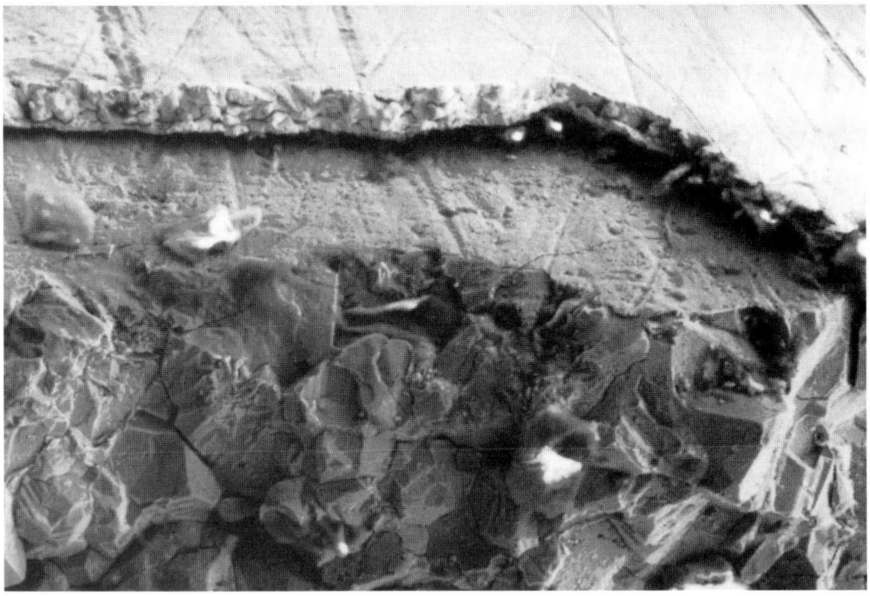

Figure 9.18 The gilded layer of the turtle figure ornament excavated in *Sasatsuka* Tumulus (SEM observation).

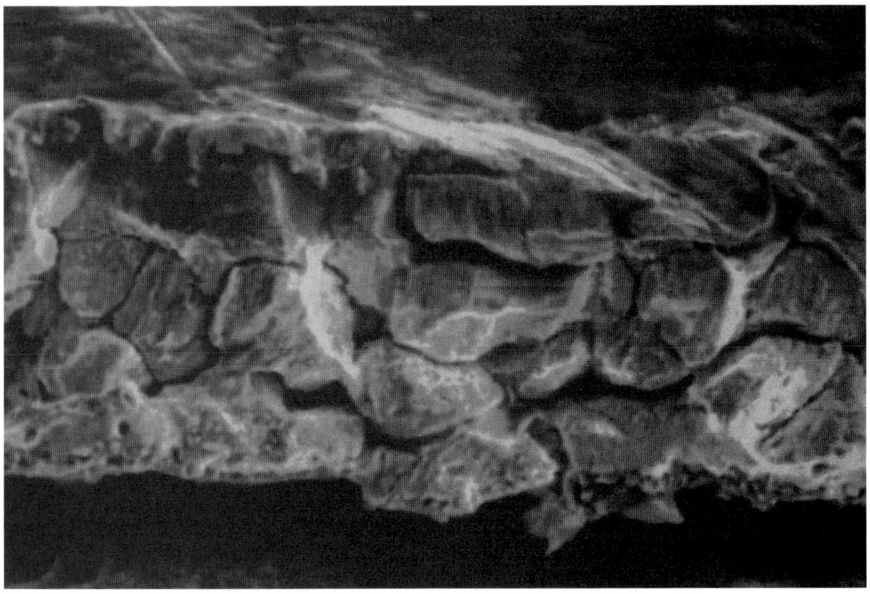

Figure 9.19 The cross-section of the gilded layer, which is composed of gold–mercury amalgam particles (SEM observation).

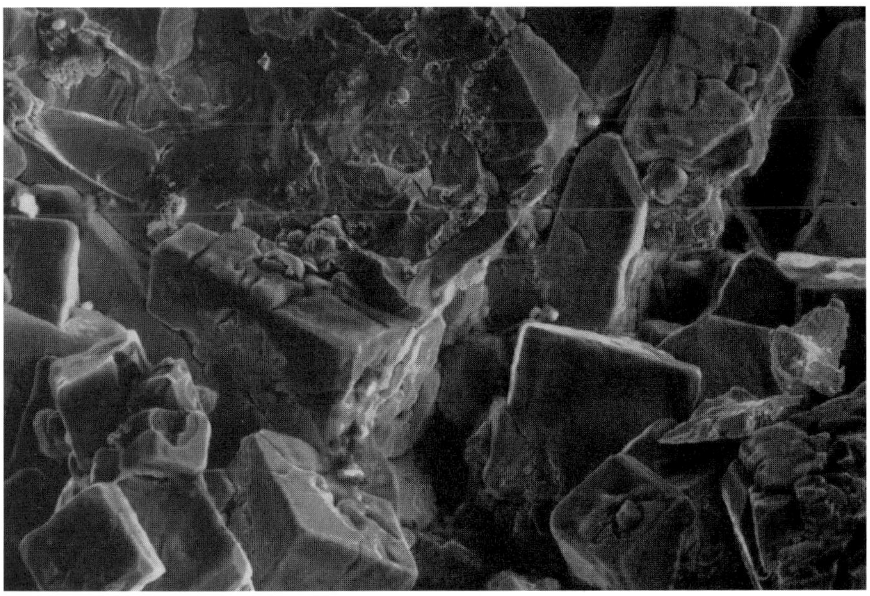

Figure 9.20 The copper of the body of the turtle was totally changed to cuprite (SEM observation).

As mentioned above, every fire-gilded layer is composed of fine gold–mercury amalgam particles; but, unfortunately, it is difficult to determine provenance by observing the micromorphological features of the gilded layer alone (Kang Daiill and Sugisita 1991; Jett 1993). For example, the copper turtle of Sasazuka Tumulus may have been brought from Korea, because Iki Island is located between the Japanese archipelago and the Korean Peninsula. Therefore the origin of the turtle ornament still remains to be investigated.

Foil gilding: another gilding technique

Among the gilded objects subjected to scientific analyses, most were fire-gilded. However, some objects with other gilding techniques, such as foil gilding or leaf gilding, have also been excavated in Japan. A pair of *gyoyo* (harness pendant of horse trapping) was excavated in Daitsu-ji Tumulus, Shiga prefecture (Murakami 1995b) dating from the beginning of the seventh century AD (Fig. 9.21). The body was made of iron covered with fire-gilded thin copper plate. Eighty-four silver rivets were arranged in the hem. The structure of the silver rivets was revealed by cross-section observation with SEM. Figure 9.22 shows that the bodies of the rivets were made of iron and only their heads were covered with silver foil (La Niece 1993), which was 20–50 μm in thickness. This *gyoyo* was finished by both fire gilding and foil silvering. It is clear that although precious metals, gold and silver, were being conserved, they were being used effectively, contrasting the silver rivets against the golden background.

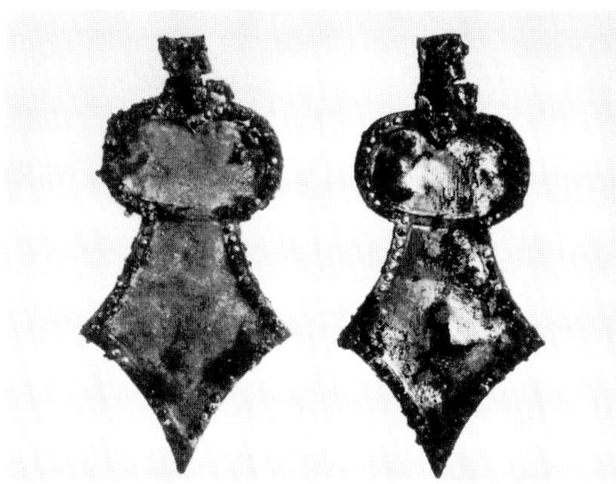

Figure 9.21 A pair of the horse trappings excavated in Daitsu-ji Tumulus, *Shiga* prefecture, before treatment (left) and after treatment (right). (See Plate 41.)

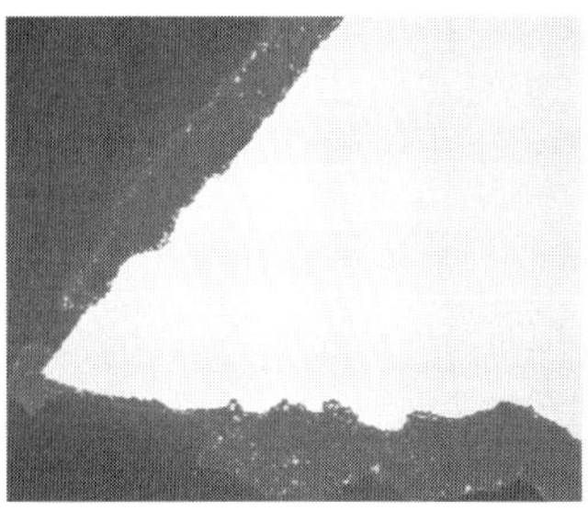

Figure 9.22 The analytical result of the cross-section of the rivet: yellow areas are iron and blue areas are silver. (See Plate 42.)

An earring (Fig. 9.23) excavated in Oohara Tumulus in Shimane Prefecture (Murakami 1994b), dating from the end of the sixth century AD, was found to be foil-gilded. Its core was made of copper. Mercury was not detected by XRF, and gold–mercury amalgam particles were not observed even with SEM. Overlapping of thin gold foil was discovered

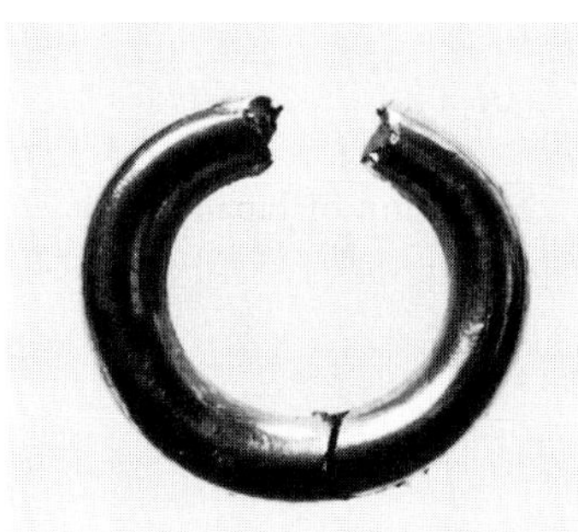

Figure 9.23 The earring excavated in *Oohara* Tumulus, *Shimane* prefecture. (See Plate 43.)

(Fig. 9.24). It follows that this earring was foil-gilded. If this gold foil had been thinner, it would be called leaf gilding. It remains to be determined whether an organic adhesive was used under the silver or gold leaf or foils. Due to the poor condition of these metal objects, it would be very difficult to detect the evidence, even if an adhesive had been used.

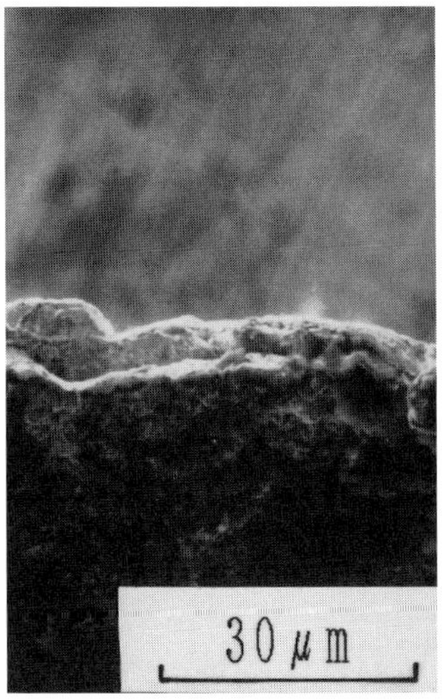

Figure 9.24 The overlapped part of the thin gold foil (SEM observation).

Conclusion

The information in this chapter is based on observations of micromorphological features. Evidence of fire and foil gilding in metal objects excavated in Japan which are dated between the fifth and seventh centuries AD have been identified. The evidence for fire gilding includes gold–mercury amalgam particles that have been levelled, and parallel striations that presumably resulted from burnishing. Foil-gilded surfaces lack gold–mercury amalgam particles, and overlapping thin gold sheets over a base metal can be observed.

In Japan numerous ancient gilded metals are excavated annually. It is not yet possible, only by observations of micromorphological features, to determine which of these ancient gilded metal objects dating between the fifth and seventh centuries were made in Japan and which were imported from Korea or China. Additional studies of surface micromorphology on provenanced objects may allow us to differentiate these objects in the future.

References

Jett, P. 1993. 'A study of the gilding of Chinese Buddhist bronzes'. In *Metal Plating and Patination: Cultural, Technical and Historical Developments*, S. La Niece and P. Craddock (eds), 171–81. Oxford: Butterworth-Heinemann.

Kang Daiill, K. and Sugisita, R. 1991. 'An analytical study of gilt objects in ancient Korean Peninsula'. *Kobunnkazai no kagaku*, **36**: 26–34.

Katsumoto Town Board of Education. Unpublished.

Kobayashi, Y. 1962. *Ancient Technology*. Tokyo: Hanawa-shobo.

La Niece, S. 1993. 'Silvering'. In *Metal Plating and Patination: Cultural, Technical and Historical Developments*, S. La Niece and P. Craddock (eds), 171–81. Oxford: Butterworth-Heinemann.

Murakami, R. 1991. 'Materials and manufacturing technique of earring excavated at Takagawa tumulus'. Archaeological report of Takagawa tumulus.

Murakami, R. 1994a. 'The realm of metalworks technique in Japan researched from microscopic Viewpoint'. *Ars Buddhica*, **213**: 117–28.

Murakami, R. 1994b. 'Materials and manufacturing techniques of earring excavated in Oohara Tumulus, Shimane'. Archaeological report of Usukokuri and Oohara Tumulus, 282–3.

Murakami, R. 1995a. 'Ancient metal jointing techniques: focusing on silver brazing'. In *Advances In The Study of Cultural Properties*, (published in commemoration of the 40th anniversary of the Nara National Cultural Properties Research Institute), 969–88.

Murakami, R. 1995b. 'Scientific research on the horse trapping excavated in Daitsuji Tumulus'. Archaeological report of Daitsuji Tumulus, 178–81.

Murakami, R. and Niiyama, S. 1993. 'Surface characterization of Japanese ancient copper alloys gilded amalgam method'. *Kobunnkazai no Kagaku*, **38**: 45–54.

Oddy, A. 1981. 'Gilding through the ages'. *Gold Bulletin*, **14**(2): 75–9.

Oddy, A. 1993. 'Gilding of metals in the Old World' In *Metal Plating and Patination: Cultural, Technical and Historical Developments*, Susan La Niece and Paul Craddock (eds), 171–81. Oxford: Butterworth-Heinemann.

Sawada, M. 1993. *A New Technique for Removal of Corrosion Products on Gilded Copper Alloy Artifacts, Current Problems in the Conservation of Metal Antiquities.* Tokyo: Tokyo National Research Institute of Cutural Properties.

Todai-ji Temple Treasure House. Record of construction of the Saikondo (West Main Hall) of Kofukuji-ji Temple, which was handed down in Shoso-in. Todai-ji: Todai-ji Temple Treasure House.

10 Three Gilded Tibetan Vessels

Bruce Christman

Abstract

Three silver vessels from seventh century Tibet, now in the Cleveland Museum of Art (CMA), have been gilded in an unusual way and are the subject of this study. Examination of the vessels shows that all three have been foil gilt. The foil at first inspection appears to be gilt silver. Further examination shows the foil to be made of a gold–silver alloy with a gold-rich surface that has been burnished. Examples of depletion gilding are rare in the Old World. This unusual combination of depletion gilt foil applied to these vessels sets them apart from gilding done on silver in other parts of the world such as in Tang China and in the Sasanian Empire. The fabrication methods used to create the Tibetan vessels are presented with an emphasis on the gilding technique. A summary of Sasanian and Tang silversmithing techniques is also presented for comparison with the Tibetan silver.

Introduction

In 1986, the Cleveland Museum of Art was fortunate to acquire three (at the time) unique partial gilt silver vessels (Figs 10.1 and 10.2) with an unknown provenance, but thought to be from Tibet and date to the seventh century.[1] The three vessels – a beaker, a rhyton and a vase[2] –

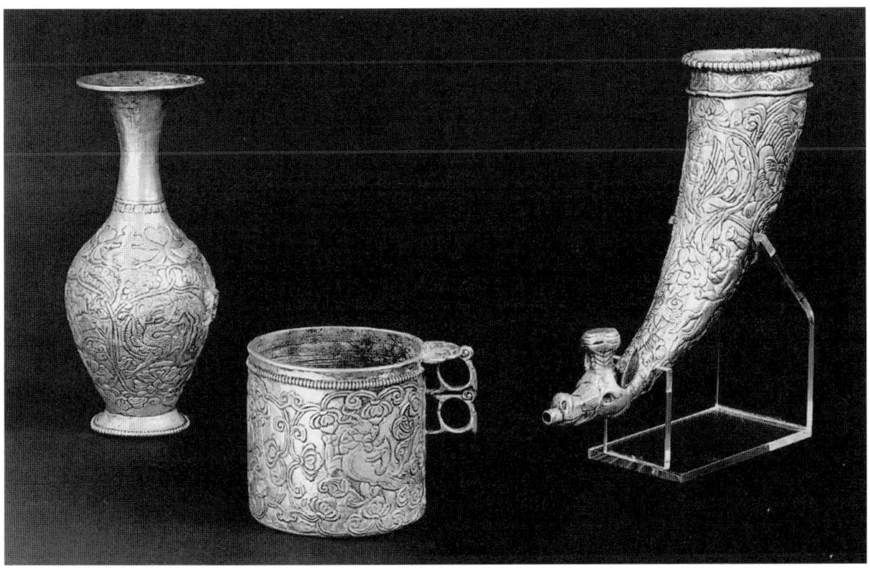

Figure 10.1 Three gilt silver vessels from Tibet dating to the seventh century. Starting from the left is the vase (1988.67), the beaker (1988.68) and the rhyton (1988.69). (© Cleveland Museum of Art.)

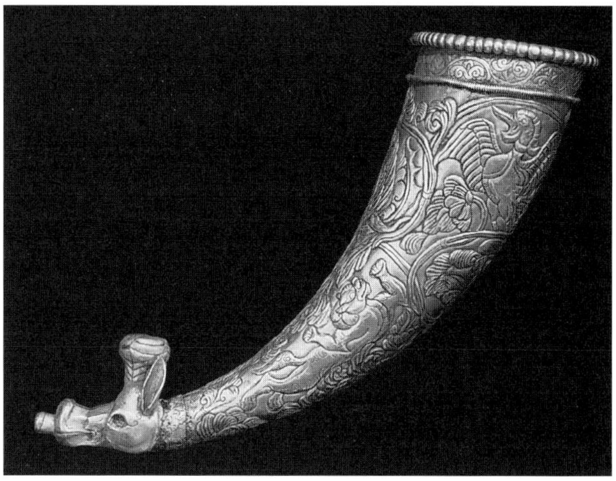

Figure 10.2 A gold–silver alloy foil coloured by depletion gilding was set into the recessed areas of the decoration to create the partial gilt appearance on the Tibetan rhyton (1988.67). The foil is held in place by driving the foil into the silver substrate using chasing tools around the perimeter of the gilt area. (© Cleveland Museum of Art.) (See Plate 44.)

combine shapes and decorative motifs that derive from both Chinese and Sasanian sources. Each of the three pieces were constructed in a somewhat different manner using standard silversmithing techniques such as seaming and raising. Both repoussé and chiselling work were used to create the relief decoration. The most notable construction feature is the method of gilding. All three vessels were gilded with the same unusual technique, which at first examination was thought to be the mechanical attachment of gilt silver foil to the vessels. Additional analyses showed that the foil is not simply gilt silver but is a debased gold alloy containing appreciable silver, which was chemically treated to create a gold-coloured surface by depletion gilding. More than anything else it is the gilding technique that brings the vessels together as a group. Other than coins and possibly jewellery, the use of depletion gilding in the Old World is very rare.

In order to better understand how and where these three vessels were made, several topics must be explored. First, a brief stylistic analysis will be given in order to demonstrate that the vessels were used in Tibet. This will include a linguistic analysis of the inscription on the beaker. Next the fabrication technique for each of the vessels will be presented, including the method of gilding. Finally, Chinese Tang Dynasty and Sasanian silversmithing techniques will be briefly discussed allowing comparisons to be made with the three Tibetan vessels.

Stylistic and linguistic analyses

As a group, the three vessels display features that come from both ends of the Silk Route – the principal route between Eastern Asia and Western Asia/Europe in which goods and ideas were exchanged. The road extended between China to the east and the Sasanian Empire to the west. During the seventh century, the Sasanian Empire occupied what is present day Iran and Iraq while China was ruled by the Tang Dynasty.

Each of the vessels is richly decorated with raised plant and floral patterns and gilding in the recessed areas. Lions are surrounded by foliage on the beaker. A winged lion and bird in the upper half and two lions in the lower half are framed by foliage on the vase. Alternating fish and birds incised by chasing decorate the interior of the mouth of the vase. Images of a dragon, a bird, a man–bird creature and a lion are found on the rhyton. The designs on these vessels blend both eastern and western elements. Such motifs as the lions and the dragon are clearly an influence from China, while the bird and man–bird motifs show the influence of Sasanian sources (Czuma 1993). The rhyton form itself derives from the west, while the vase shape is eastern, and is similar to the Kundika (an Indian ritual water container) or the Chinese ceramic mei-ping shape (Czuma 1993). The blending of eastern and western influences suggests that the vessels were made somewhere along the Silk Route.

The bottom of the beaker bears an inscription that was chased into the metal. The other two vessels were not inscribed, although three circles similar to those beneath the inscription on the beaker are found on the bottom of the bulls head on the rhyton. Two experts in Tibetan language – Heather Karmey-Staddard and Hugh Richardson – have determined that the beaker bears the earliest known Tibetan inscription, which when translated identifies the vessels as the 'personal possessions of the Chinese person of high birth' (Czuma 1993). This rather ambiguous inscription may refer to the Chinese princess, Wen-cheng, who was one of two wives of Songtsen Gampo (AD627–50). Songtsen Gampo can be considered the father of Tibet as he not only unified the country politically, but introduced Buddhism. His two marriages appear to have been diplomatically arranged with royal families of nearby states and were used to solidify his political hold on the area. His other wife, Princess Bhrikuti, cemented ties to Nepal. The inscription on the beaker would seem to indicate that the silver objects belonged to Songtsen Gampo's Chinese wife.

Fabrication

As a group, the three Tibetan vessels display many of the standard techniques used by silversmiths of many cultures and time periods to fabricate hollow ware. The beaker was probably first sunk, then raised; the vase was entirely raised; the rhyton body was seamed from a single sheet. For each of these methods the silversmith must begin with a sheet of silver. An ingot would be cast and then hammered into sheet metal. This was probably done to fabricate the Tibetan vessels, but all evidence of how the sheet silver was produced has been obliterated by subsequent hammering, decorating and finishing work. Sinking is the process of stretching the metal as it is formed into a vessel. Sinking is done by hitting the metal with a hammer on a flat anvil, which begins to force the metal into a bowl shape. Alternatively the metal can be hammered into a depression in wood or a sand bag. Raising is the process of compressing the metal as it is hammered over an iron stake. The rims of raised vessels are thicker than the original thickness of metal. Vessels formed by seaming begin by creating a cylinder or cone shape from sheet metal and soldering the edges together.

Each of the vessels has sections that were joined using a silver solder. Analysis has not been undertaken on the solder, but visually the solder appears to be a silver alloy. It is slightly yellow in comparison to the adjacent silver, and bubbles are often seen in the solder joints, which is typical of silver solders.

X-ray fluorescence spectroscopy analysis shows that the three vessels were made from a silver–copper alloy. Copper was historically used by silversmiths to strengthen the alloy and, in turn to strengthen the vessels made. Several analyses of the silver surfaces were made on each of the vessels.[3] The beaker has an average composition of 99 per cent silver with 0.3 per cent copper; the vase has an average composition of 98 per cent silver and 1.7 per cent copper; the rhyton has an average composition of 99 per cent silver and 0.6 per cent copper. Small quantities of gold (0.05–0.2 per cent), lead (0.05–0.08 per cent) and traces of bismuth and iron were

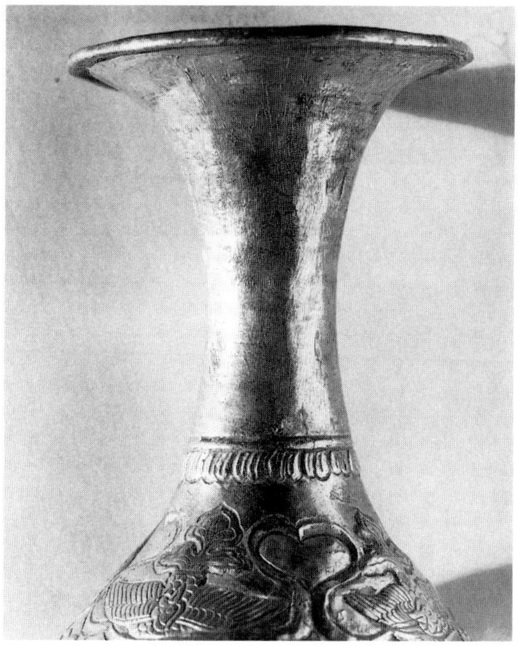

Figure 10.3 Detail of the neck of the vase showing raised enriched silver regions resulting from pickling acids entering cracks that formed in the metal during cold working. These regions of enriched silver have become raised with subsequent slow corrosion of the surrounding silver–copper alloy. (© Cleveland Museum of Art.)

detected in all samples. The three vessels appear to have an unusually high silver content, which is probably due in part to a combination of surface enrichment and the analytical method used. The X-rays penetrate approximately 10–1,000 Å into the sample, so that the analyses are primarily of the surface and not the gross composition. Surface enrichment could result from pickling the silver after annealing, enrichment of the foil *in situ*, or preferential leaching of the copper from the alloy during burial. Evidence for surface enrichment from pickling is found on the vase in the form of fine raised lines on the neck (Fig. 10.3). These lines result from the manufacturing process and subsequent corrosion of the metal. Small cracks opened in the metal during the raising process. During repeated annealing and pickling procedures, an enriched silver layer was formed in the cracks. Over time, the silver–copper alloy has slowly corroded while the pure silver surrounding the cracks did not.[4] It is interesting to note that the only vessel to display this corrosion pattern has the highest copper content of the three vessels.

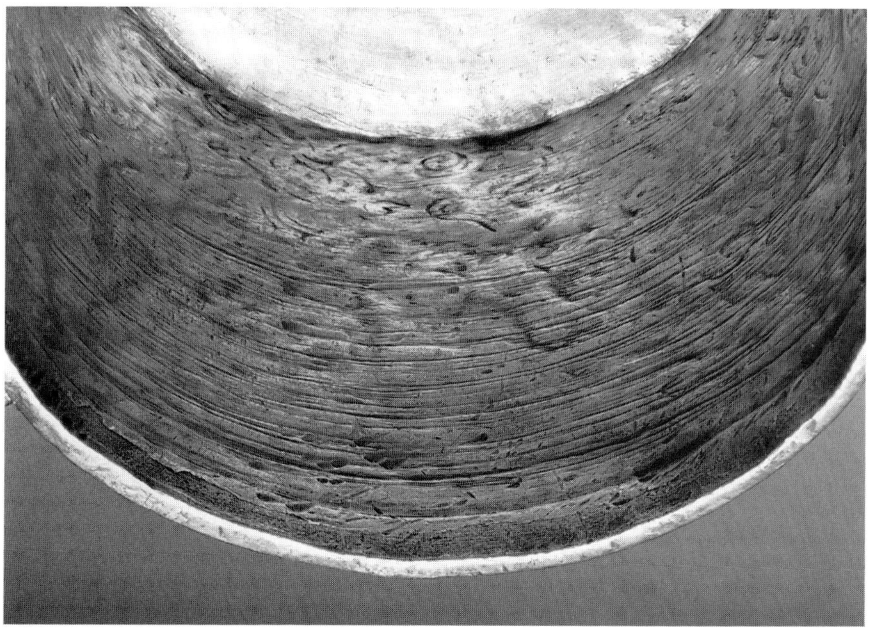

Figure 10.4 Interior of the beaker showing the lines from the silversmith's stake catching in the silver during the raising process. The outline of the gilt decoration on the exterior can also be observed. The outline is a result of anchoring the foil to the vessel with chasing tools. (© Cleveland Museum of Art.)

Beaker

The body of the beaker was made from an unusually thick sheet of silver – between 3 to 4 mm at the thickened rim. The thickness of the metal suggests that the vessel was probably formed by a combination of sinking and raising. The thick silver would first have been sunk into a depression or hammered with a peening hammer to form a 'cup' shape. The sinking process would result in some stretching or thinning of the metal. Final formation of the cup was then accomplished by raising. Numerous concentric gouge marks on the interior are evidence of raising (Fig. 10.4). These marks result from the silversmith constantly catching the end of the stake in the metal. The handle was cast and soldered to the side of the vessel. The design in relief on the outside of the beaker was created by cutting away the silver with a sharp chisel (Fig. 10.5). At the edges of the relief decoration additional depth was created with chasing. Chasing tool marks are seen in areas where the gilded foil is missing. Also in these areas remnants of tool marks from the cutting and abrading process are seen even though many of the marks have been obliterated by subsequent chasing work. The beaded decoration at the top of the beaker was created by chiselling and chasing.

Figure 10.5 Detail of the beaker showing carved and chased decoration. The broken edge of the gilt foil is apparent. Note the chisel and chasing tool marks in the areas where the foil is missing. (© Cleveland Museum of Art.)

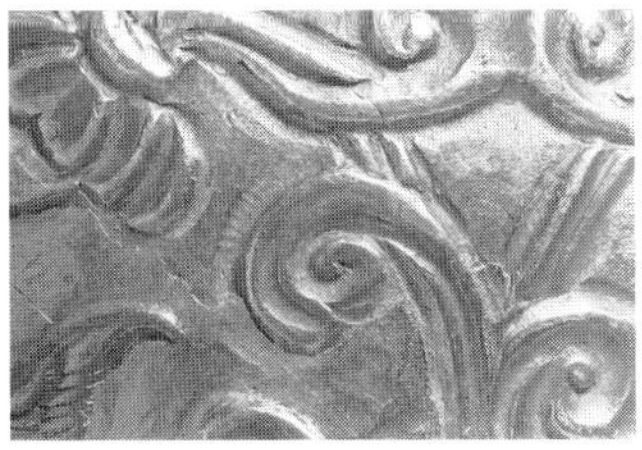

Rhyton

The rhyton body was made from one sheet of silver approximately 1 mm thick. The silver sheet was bent to a cone shape and soldered on the interior curve to form the body of the rhyton (Fig. 10.6).A lap seam was used to make the seam stronger and is very visible despite subsequent tooling. The spout of the rhyton is a bull's head, which was soldered on. Including the bull's head, the spout was made of at least five pieces of silver. The topknot, the two ears and cylindrical spout inserted in the bull's mouth were soldered to the bull's head. Both head and topknot were created by the repoussé technique. The beaded band at the top of the vessel is hollow, indicating that it was fabricated from tubing. Each bead was carefully tooled into the tubing before the band was soldered to the rhyton. The lower beaded band was made using solid wire. The raised decoration was created with repoussé and chasing work. Repoussé is the process of hammering out the metal from the back or interior of a vessel to create a raised area. The process is often used in conjunction with chasing on the front or exterior to give further definition to the raised area.

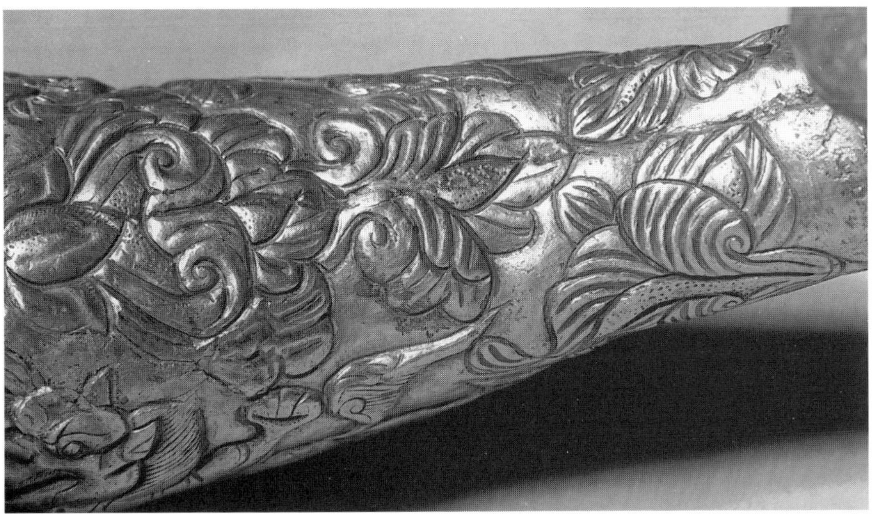

Figure 10.6 Detail of the rhyton showing the seam on the interior curve. (© Cleveland Museum of Art.)

Vase

The body of the vase was raised from a single sheet of silver approximately 1 mm thick. The foot was fabricated from sheet silver and soldered to the bottom of the body. Solid wire was formed into beads and soldered to the bottom of the foot. Like the rhyton, the decoration was created with repoussé and chasing techniques. The small rectangular silver appliqué with a gilt lion design has a different composition[5] and appears to have been used at a later date to repair a crack in the metal. It has been soldered in place using silver solder.

The gilding technique

All three vessels were gilded in the same manner using the well known technique of foil gilding (Oddy 1993). The areas that are gold in colour have gilt foil attached to the surface of the silver (Fig. 10.5). Visually the foil appears to be silver with a pale gold colour on the top surface, which has been burnished. The foil is approximately thirty microns thick and most often laid into the low areas of the relief decoration. The perimeter of the gilt metal was chased to drive the foil into the silver substrate. Numerous losses of the foil make it possible to determine that no solder, adhesives or other foreign material are present beneath the foil. The silver foil was attached to the vessels simply with metallic bonds formed by tooling. The force used to drive the foil into the silver vessel can be seen by the impressions left inside the beaker by the tools used to attach the foil (Fig. 10.4). The perimeters of the gilt areas are clearly seen despite all the marks left by the silversmith's stake. It is not clear whether additional burnishing or hammering was done to join the metal foil to the substrate more securely. It seems possible, as there are areas where the foil was adhered to non-recessed areas. The numerous losses of the gilt foil indicate that the foil is not well adhered to the vessels.

X-ray fluorescence analyses of the gilt silver foil showed that only gold and silver are present. Mercury could not be detected in any of the gilt areas, confirming that the gold had not been applied either as an amalgam or as leaf using mercury. A small sample of foil removed from the neck of the vase was examined using scanning electron microscopy with energy dispersive spectroscopy.[6] No mercury was detected, confirming the initial X-ray fluorescence results. In addition, the energy dispersive spectroscopy analysis showed gold and silver to be present throughout the sample in varying concentrations. The highest concentration of gold (75 per cent) is found on the top surface. Both elements are found in varying but roughly equal concentrations in the central region. The side nearest the vessel has regions that are richer in gold (57 per cent gold to 43 per cent silver) and other regions that are richer in silver (30 per cent gold to 70 per cent silver).

Scanning electron microscope photographs of the cross-section show a thin, compact, dense surface on the gold-rich side, which was created by burnishing (Fig. 10.7) Otherwise, the entire cross-section of the foil has a porous and spongy structure (Figs 10.8 and 10.9). No boundary or other demarcation between the gold surface and the substrate is observed, which, if present, would indicate that the gold surface was gold leaf attached to a baser foil. There are no traces of an adhesive or solder observed in the cross-section. If the gold had been attached by diffusion bonding or by burnishing, there would be areas where the bond was

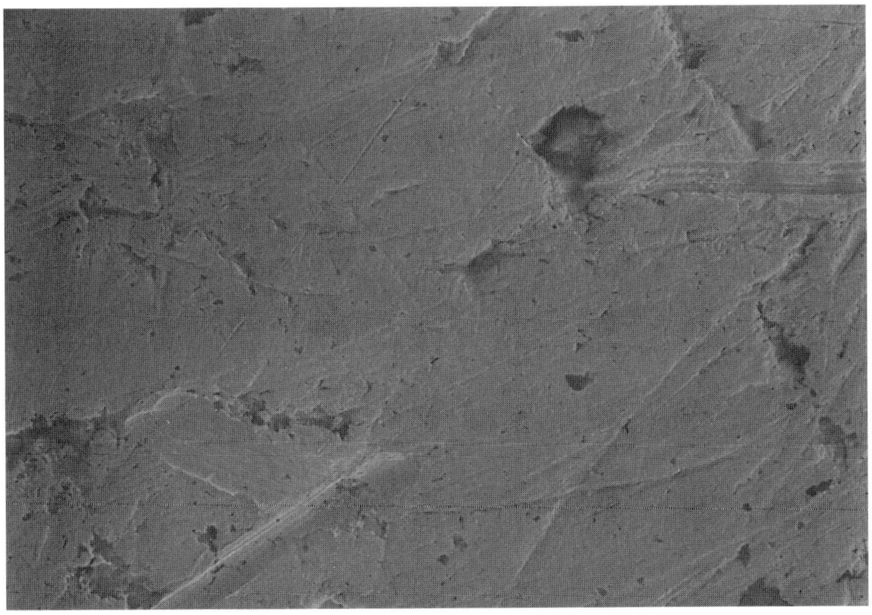

Figure 10.7 Scanning electron microscope image of the burnished gilt surface (500×).

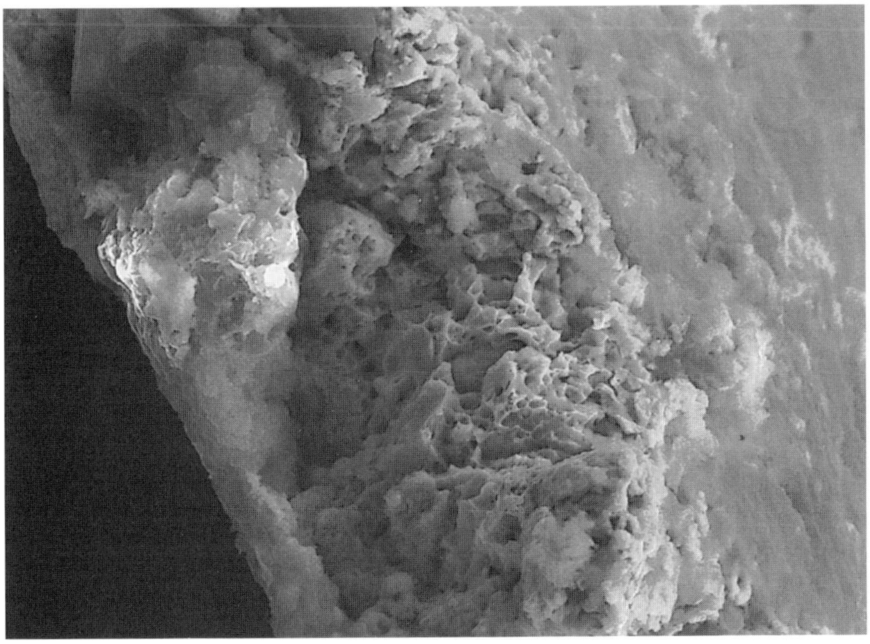

Figure 10.8 Scanning electron microscope image of the cross-section of the gold–silver alloy foil (2,000×). The burnished gold rich side is on the viewers left. The remainder of the cross-section shows a spongy structure.

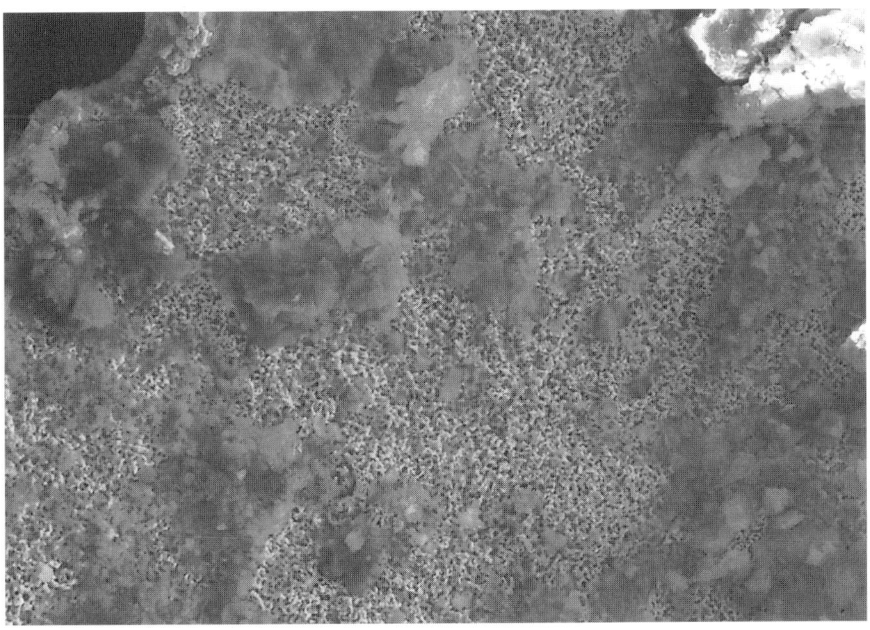

Figure 10.9 Scanning electron microscope image of silver-rich side of the foil (500×).

incomplete or a sharp change in composition would be noted. Instead, only a porous spongy structure is observed, except for the very thin compact surface created by burnishing. Both Lechtman (1971) and Oddy *et al.* (1990) have noted that fire-gilt surfaces have a spongy appearance. However, mercury should be detected by either X-ray fluorescence spectroscopy or scanning electron microscope-energy dispersive spectroscopy (SEM-EDS) analyses, as some diffusion of the mercury into the silver and gold layers would be expected. Without the presence of mercury, it is highly doubtful that any gilding using mercury was used. Other methods of applying gold can also be eliminated, as the layer of gold would remain visible by scanning microscopic examination.

The only other logical gilding technique that would produce a spongy structure is depletion gilding. The presence of both gold and silver throughout the sample provides additional evidence for depletion gilding. A gold–silver alloy was first beaten into foil, followed by a chemical process that made the surface gold-rich. The exact process for creating the gold colour is not known.

There are two possible ways to remove silver from gold: cementation and pickling. In the cementation process the gold–silver alloy is packed in an absorbent material such as clay or brick dust with a corrosive material such as sodium chloride or alum (potassium aluminum sulphate) and heated. The silver at the surface is corroded by the salt and absorbed into the clay or brick dust. The heat causes more of the silver in the alloy to migrate to the surface, where it undergoes the corrosion process. This process takes many hours or even days to create a gold-rich surface. The prolonged times result in a composition gradient between the surface and the alloy. The second method, the pickling of gold–silver alloys to create a gold-rich surface is relatively easy today with the use of nitric or sulphuric acids. These distilled acids were not yet known in the Old World during the seventh century. Lechtman (1970) in her study of pre-Columbian depletion gilding methods suggested that strong corrosive materials such as ferric sulphate and salt or ferric sulphate, salt and iron oxide would produce quite strong acids, which at room temperature would produce a thin gold-rich surface. It seems likely that some variation of pickling was used to enrich the surface in gold, as only a thin burnished layer of gold is observed at the surface.

The fact that both sides of the foil are not gold rich raises questions about the way in which the depletion gilding was done for these three vessels. Was the surface enrichment of the foil done before application to the vessels or was it done after application? Unfortunately, the evidence is not conclusive. The surfaces of the three vessels have an extremely high silver content and it is tantalizing to assume that the gold–silver foil was first applied to the vessels and then the entire vessel subjected to the corrosive chemicals. The corrosive chemicals would have easily leached the copper from the silver alloy and much of the reverse of the foil would have been protected. The surface enrichment of the silver noted could also be a result of more routine pickling after annealing, or perhaps the leaching of the copper from the surface during burial. However, on the back of the foil sample, there is variation in the concentrations of gold and silver. It is possible that these differences in gold concentration might be a result of seepage of the chemicals to the back side of the foil, either before application or after application to the vessels.

The use of depletion gilding is well documented in pre-Columbian Central and South America, where it was the principal method of colouring metals (Lechtman 1970), but its use in the Old World is much rarer and less well documented. In a survey of gilding techniques in the Old World

(Oddy 1993), depletion gilding is not even mentioned. Although it was not a common technique in the Old World, there is evidence that the process was known and used. The earliest literary evidence of enriching gold alloys comes from an Egyptian document known as the Leyden Papyrus, which dates to the late third century AD (Caley 1926) The papyrus contains 111 recipes for working with various metals. Two of the recipes would produce quite strong acids, which would readily leach silver from gold. The first recipe (number 13) entitled 'The colouration of Gold' is a mixture of *misy* plus salt and vinegar. *Misy* appears to be a sulphide or sulphate, probably iron pyrites. The second recipe, (number 25) entitled 'Gold Polish' is a mixture of *misy*, alum and salt.

Very few objects coloured by depletion gilding have come from the Old World. The earliest apparent example of depletion gilding, anywhere in the world comes from Ur (La Niece 1995). Three chisels now in the British Museum were excavated by Wooley in the 1920s from a site known as the 'Queen's grave' (PG800). The chisels date to *c.*2600BC, and have all the features of depletion gilding; being cast and hammered in a base gold alloy, the surfaces having a higher proportion of gold than the interior metal and the surfaces having been heavily burnished. The use of depletion gilding on gold jewellery is not clear (Ogden 1993). Odgen notes that there is evidence of surface enrichment on some ancient jewellery but cannot determine whether it is intentional or a result of burial. The depletion gilding technique could also be used on silver–copper alloys to create a silver surface. Colouring of coins in the Old World seems to have been the most common use of the depletion gilding technology. Silver–copper alloyed Roman coins dating from AD63 to AD260 were 'silvered' by pickling in citrus fruit acids or vinegar. The copper alloy with as little as 12–18 per cent silver produced a continuous silver surface after being pickled.[7] A second use of depletion gilding of coins occurs in sixteenth-century Japan, where gold and silver alloyed coins were made more golden in colour. A mixture of iron sulphate, copper sulphate, potassium nitrate, sodium chloride and resin was applied to the surface and heated. This process works well for alloys with a relatively high concentration of gold. In the case of the Japanese coins, the gold content was 50–75 per cent.[8] A fourth example of intentional surface enrichment occurs much closer in geography and time to the three Tibetan vessels. A plate (Fig. 10.10) in the Cleveland Museum of Art was made in India during the Gupta Period (AD300–500).[9] The plate was made from a silver–copper alloy (55:45) with subsequent preferential removal of the copper to create a silver-coloured surface (Gibbons and Ruhl 1979). The relief decoration on the bowl was created by cutting away silver-rich metal exposing the more coppery coloured metal beneath. These recessed areas were then covered with gold, resulting in silver-coloured figures on a gold background. While depletion gilding was rarely used in the Old World, it is clear the craftsmen had the knowledge and materials needed.

The analytical evidence has confirmed that mercury was not used in the gilding process either as an amalgam or to bond gold leaf to the silver. It also shows that a foil made from gold–silver alloy was used with subsequent depletion of silver from the top surface to create the gold colour. The reason for use of the depletion gilding is unclear. Perhaps mercury was unavailable or the amalgam gilding process was unknown to local silversmiths. It is also possible that the technique was developed to conserve precious gold or to take advantage of local electrum deposits; an alloy of gold and silver that occurs naturally. For whatever reason this gilding process was used, the three vessel appear to be an anomaly in the history of gilding in the Old World, as nearly all gilding at this time was being done

Figure 10.10 Gupta *Plate with Figures* (1972.71) was made from a silver–copper alloy treated to create a silver-rich surface. The figures were created by removing the spongy silver-rich areas leaving the raised figures in silver. The recessed areas around the figures were gilded. (© Cleveland Museum of Art.)

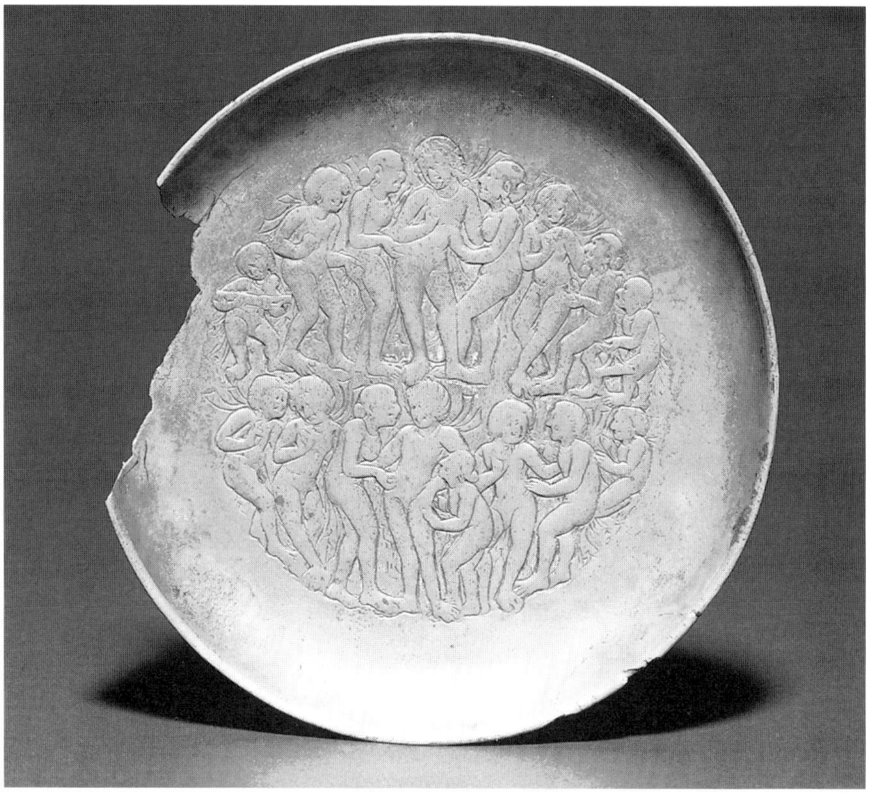

by mercury amalgamation. The vessels have been gilded with a remarkable combination of a relatively common method of foil gilding with the very unusual process of depletion gilding.

Summary of Tang and Sasanian Silversmithing Techniques

The craftsmanship of the three Tibetan vessels differs dramatically from that of silver vessels being produced by Chinese or Sasanian silversmiths of this period. While there are a limited number of ways the silversmith has to create hollow ware, there are very real differences in the ways these techniques can be used.

Analysis of Sasanian silver show that 4–8 per cent copper was added to strengthen the silver (Meyers and Harper 1981). In general, the Sasanian silver ware was created by raising the metal; some vessels have cast handles and feet soldered in place.[10] Four techniques were used for the decoration: repoussé; chasing; carving; and adding sheet metal to the surface. These methods can be used singly or in combination. Cleveland's rich collection of Sasanian silver contains several objects typical of Sasanian craftsmanship. The plate bearing King Hormizd II Hunting Lions (Fig. 10.11) is typical of the royal plates.[11] The body of the king was shaped by hammering the metal into a three dimensional shape. Decoration was chased onto this piece before it was added to the plate to create raised design areas. The piece was held in place by cutting a channel in the metal of the plate surface and folding one side of the channel over the added piece. Other relief decoration was done by cutting metal away with chisels. A comb-like chisel used by the Sasanians left very fine grooves, much like fingerprints, which are difficult to discern in the surface of the silver.[12] The decoration on a bowl in the CMA collection was done entirely by this method.[13] On other vessels the decoration was created with repoussé and chasing. Often repoussé and chiselling would be combined as was done on the vase with

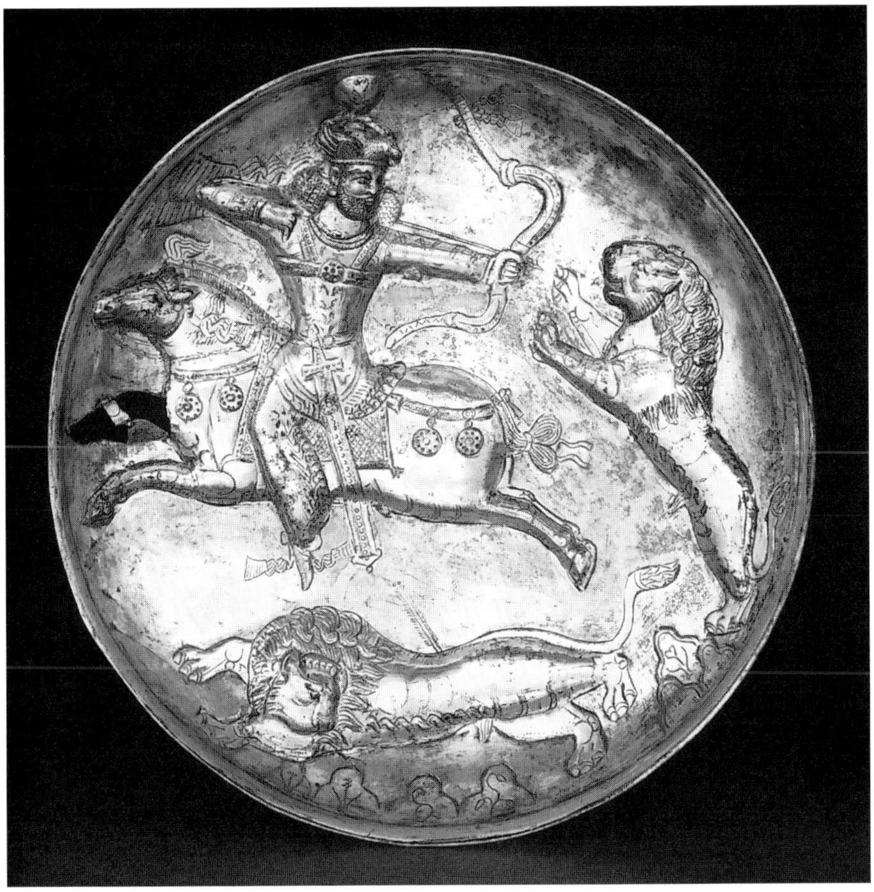

Figure 10.11 *Dish with King Hormizd II Hunting Lions* (1962.150) is typical of much Sasanian silver. The raised areas of the figure were added sheets of metal held in place by cutting a channel in the plate and then crimping one side of the channel over the added metal. Shallower areas of raised decoration were created by removing metal with a chisel. The vessel was partial gilt by the direct application of a mercury amalgam. (© Cleveland Museum of Art.)

four dancing figures (Fig. 10.12).[14] The vessels were usually partial gilt by applying a gold amalgam directly to the surface, firing it, and then burnishing and polishing. Handles and feet were generally attached to the vessels

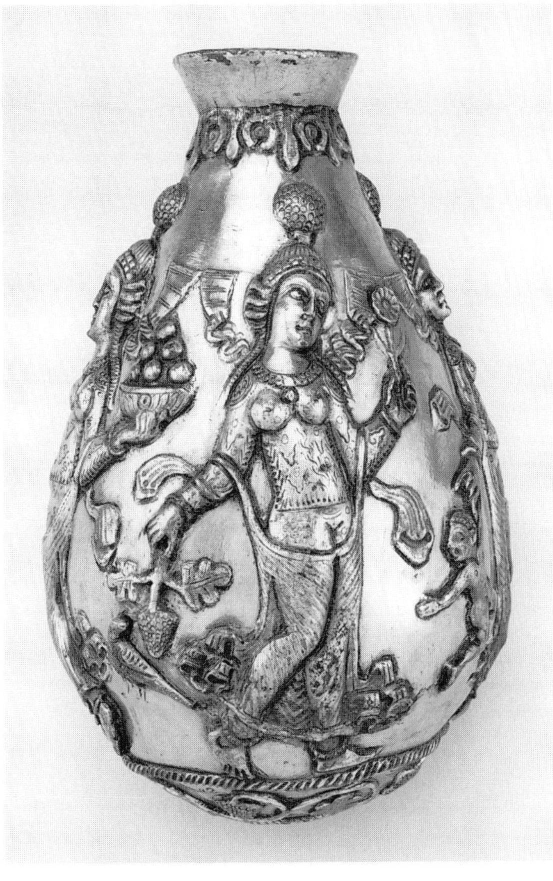

Figure 10.12 The raised decoration on the *Vase with Four Dancing Figures* (1962.294) was done with repoussé work combined with chasing work (the female figures) and by the removal of metal with a chisel (the grapes, birds and vines). The vessel was partial gilt by the direct application of mercury amalgam. (© Cleveland Museum of Art.)

using a silver solder. Visually the solder used to hold the foot onto the ewer appears to be a silver alloy, as it is a slightly different colour and air bubbles are occasionally seen. Again no analysis has been done to the solder.

Tang Dynasty silversmiths also alloyed copper with silver. No analyses of Tang silver have been published. One vessel in the Cleveland Museum of Art, a tray[15], was analysed by X-ray fluorescence spectroscopy. The tray (Fig. 10.13) has an average composition of 96.6 per cent silver, 2.6 per cent copper with 0.1 per cent gold and 0.04 per cent lead. Quantitative analyses on a stem cup[16] were not possible, as gold and mercury were present on all the exterior surfaces. Tang silver ware was formed by raising or seaming sheet metal with solder. Decoration was created by chasing and punching. Occasionally, some shallow repoussé work was done. The vessels were usually partial gilt by applying a gold amalgam to the surface. The two objects mentioned above, the tray and stem cup, are typical of Tang silver. The stem cup (Fig. 10.14)was fabricated in a number of sections. The body of the cup was seamed on one side and the bottom soldered on. The seams are butted together very precisely and are visible only as dark grey lines on the interior. The composition of the solder is not known. The tooled decoration completely hides the joins on the exterior of the cup. The stem was fabricated from several pieces soldered together. The salver was raised from a single sheet of silver with three feet soldered on the bottom. Each foot was formed from two pieces of sheet metal hammered into a three-dimensional shape. The two halves of the foot were then soldered together. In both instances the decoration is flat without any repoussé work. The decoration, which covers every available space, was executed with very fine and delicate lines made with chasing and punching tools. Most of the background texture was executed with a fine circular punch.

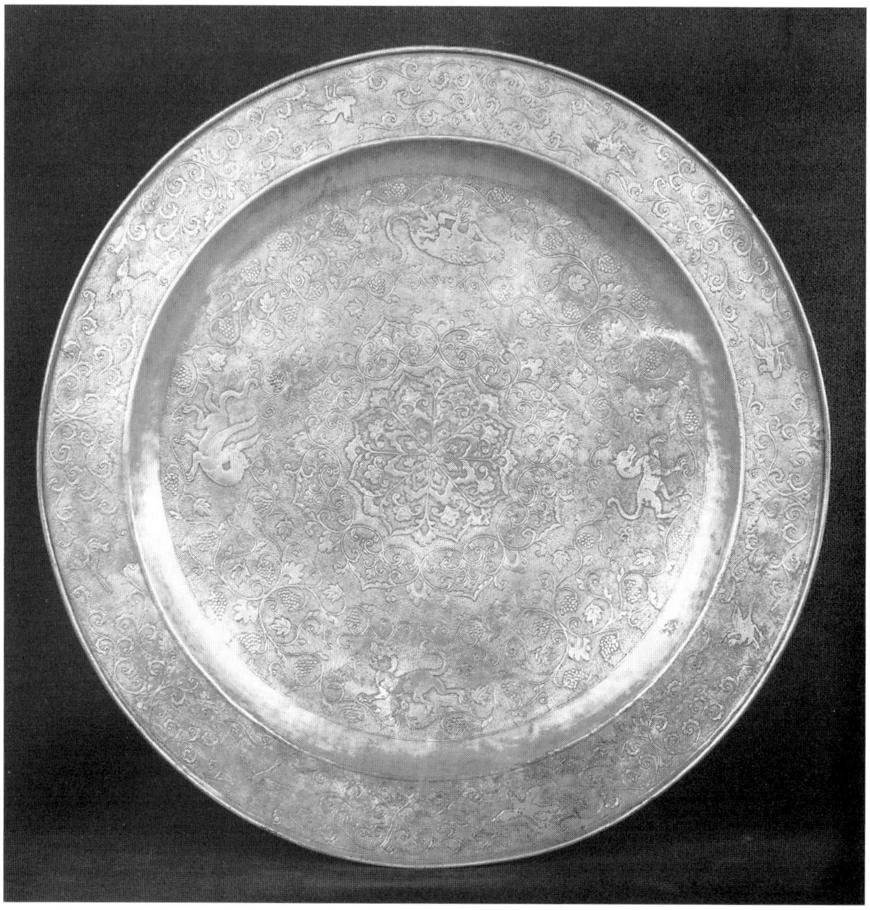

Figure 10.13 Tang Dynasty *Salver* (1972.34) was decorated by chasing and punch work. The top surface was gilt with the application of a mercury amalgam. (© Cleveland Museum of Art.)

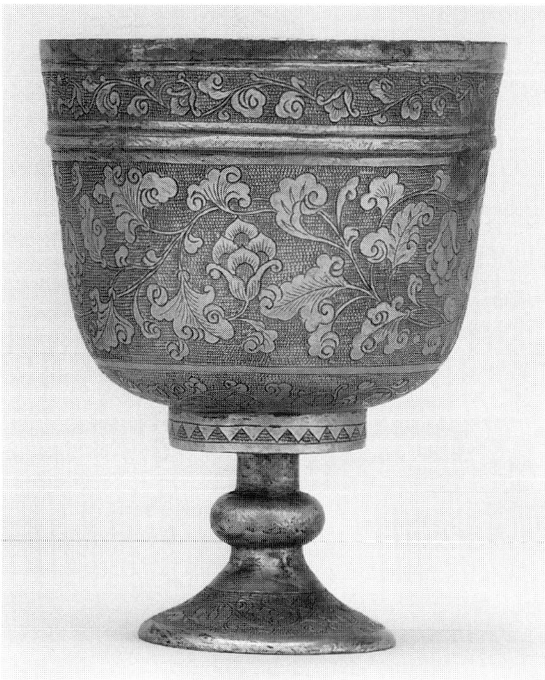

Figure 10.14 Tang Dynasty *Stem Cup* (1951.396) was made by seaming the body of the cup on one side and soldering the bottom in place. The dark coloured solder seams are visible inside the cup but are hidden by the extensive decoration on the exterior. The exterior was gilded by the application of a mercury amalgam. (© Cleveland Museum of Art.)

The author has never observed any engraving on Tang Dynasty vessels. Tang Dynasty vessels were gilded using a mercury amalgam that was applied directly to the silver.

In general the Tibetan objects lack the refinement of execution found in either Sasanian or Tang silver. The creators of the Tibetan vessels seem less sure of how to form the vessels and execute the decoration. The stake marks left inside the beaker indicate shaping this vessel was difficult. The seam on the rhyton remains visible despite subsequent tooling, while the seam on the Tang stem cup is invisible on the exterior. The chisel marks left by Sasanian silversmiths are very different from those on the Tibetan beaker. The Sasanian chisels left fine grooves in the surface, which resemble fingerprints, while the Tibetan chisel marks were coarse and more gouge-like. The difference in chiselling may indicate a different craft tradition; or perhaps the Tibetan vessel could have been more crudely executed because the tool marks would be hidden by the applied gilt-silver foil. The tooled decoration on the Tibetan vessels is not as delicate and fine as that found on either Tang or Sasanian silver. The silver solders used by the Sasanian and Tibetan silversmiths are similar, probably being a silver alloy. The composition of the Tang solder is not known but it has a very different appearance from that used on the Sasanian and Tibetan silver. While the silver solder used by the Tibetan and Sasanian smiths is slightly different in colour than the silver vessels and often has air bubbles, the solder used on the Tang objects is dark grey in colour and is usually free of porosity. Of most importance, the three Tibetan vessels were gilded by adding gilt-silver foil to the surface. In contrast, Tang and Sasanian silver were mercury gilded directly onto the fabricated vessels. The use of depletion gilded foil applied to the three Tibetan vessels makes them unique in the Old World.

Conclusion

The Tibetan inscription on the beaker indicates that the vessels were used in Tibet during the seventh century and probably belonged to the Chinese princess Wen-cheng who married Tibet's first ruler, Songtsen Gampo. The blend of eastern and western motifs and vessel shapes indicates that the three objects were made somewhere along the Silk Route. The craftsmanship – especially the gilding technique – is quite different from either Chinese or Sasanian silver ware. The differences in craftsmanship strongly indicate that neither the vessels nor the silversmiths who made them were imported from China or the Sasanian Empire. The most obvious place of origin of these vessels would be Tibet. In addition to being very rare, these vessels are the only ones in the Old World known to be gilded by a combination of depletion gilt metal attached as a foil.

Aknowledgements

This paper would not have been possible without the acquisition endowment of the Cleveland Museum of Art and the wisdom and insight shown by the Curator, Stan Czuma, and the Director, Evan Turner, in purchasing these objects. This study emerged as a result of the initial examintion of these objects for purchase and was further enhanced by the examination of related objects in the rich and varied collections. At the museum, I would like to thank my colleagues in the Conservation Department for their patience while completing the manuscript. Particular thanks goes to Patricia Griffin for generously sharing her knowledge of gilding and acting as 'devil's' advocate and to Judith DeVere for editing the manuscript. At Case Western Reserve University, I would like to thank John Sears at the Center for Surface Analyses of Materials. I particularly would like to thank Terry Drayman-Weisser for her interest and ideas but most of all for her support and encouragement.

Notes

1. Since acquisition, two other Tibetan silver vessels are now known: one is in Jokhang in Lhasa (Czuma 1993) and the other is in the Miho Museum, Japan. (Miho Museum 1997) The bowl in Japan dates approximately one hundred years later and is foil gilt. For a more complete discussion on the three vessels in the Cleveland Museum of Art see Czuma (1993: pp. 131–5, 1991, 1992/93).
2. *Beaker* 1988.68, (H. 30.5 cm.), The Severance and Greta Millikin Purchase Fund; *Rhyton* 1988.69, (H. 30.5 cm.), Gift of Mrs. Clara Taplin Rankin, and *Vase* 1988.67 (H. 22.9 cm.), Purchase from the J. H. Wade Fund.
3. Each vessel was analysed using a Kevex 0700 computer system coupled with a modified 075A front end, with the X-ray tube and detector mounted on a carriage so that analyses were done in the open air. Quantitative calculation of the various elements was performed using *Exact*, Kevex's software package, a fundamental parameters program used for calculating unknown elemental concentrations based on net peak intensities and matrix effects. The results given have not been normalized – that is, they have not been adjusted to total 100 per cent. Several factors affect the mass balance, such as minor fluctuations in the X-ray tube output. The *Exact* equations assume a smooth flat surface, which is often not the case. Finally, there are statistical errors attributable to variations in the fluorescence reaction. Experimental error is ±2 per cent for silver and copper and ±50 per cent for the trace elements of gold and lead. Experimental conditions were 50kV, 3.3 mA, using a 2 mm collimator on the tube and a 6 mm collimator on the detector. A mixed barium carbonate and strontium carbonate secondary target was used to reduce the time of analysis. Data was collected for 180 seconds. Analyses was done by the author at the Cleveland Museum of Art.

Beaker

	Copper	Silver	Gold	Lead
Bottom of beaker	0.3	100.9	0.04	0.04
Front of beaker	0.2	100.3	0.05	0.04
Proper left side	0.4	99.1	0.04	0.05
Proper right side	0.3	101.4	0.05	0.05

*Traces of iron and bismuth were noted in all samples.

Rhyton

	Copper	Silver	Gold	Lead
Bull's head, left cheek	0.6	100.9	0.06	0.08
Bull's head, right cheek	0.6	99.2	0.07	0.07
Top knot, right side	0.6	99.5	0.3	0.09
Top knot, right side	0.6	99.7	0.2	0.08
Bird at top, p. left side	0.6	99.1	0.2	0.08
Feline creature, p. left side	0.7	98.1	0.3	0.1
Feline creature, p. right side	0.5	99.2	0.3	0.08

*Traces of iron and bismuth were noted in all samples.

Vase

	Copper	Silver	Gold	Lead
Center of lion	2.0	98.7	0.2	0.07
Figure w/bird's feet	1.7	98.3	0.2	0.05
Dragon's head	1.4	99.8	0.3	0.04
Chest of phoenix	1.9	99.6	0.2	0.06

*Traces of iron and bismuth were noted in all samples.

4. Pieter Meyers, Head of Conservation, Los Angeles County Museum of Art: Personal communication while in Cleveland examining the vessels, 1993. See also Meyers 1981.
5. The composition of the appliqué on the vase was determined as above and was found to be:

	Copper	Silver	Gold	Lead
Top of appliqué	0.2	96.9	2.8	0.1
Bottom of appliqué	0.2	97.6	3.3	0.1

   Analyses are picking up some gold from the center of the appliqué. The copper content of the appliqué is a tenth of that found on the rest of the vessel (see footnote 3).
6. SEM/EDS analyses was done by John Sears at the Center for Surface Analyses of Materials at Case Western Reserve University using a Hitachi S-4500 Field Emission SEM and a 25 kV beam.
7. Cope, L. H. 1972. 'Surface-silvered ancient coins'. *Methods of Chemical and Metallurgical Investigation of Ancient Coinage,* E. T. Hall and D. M. Metcalf (eds). London: Royal Numismatic Society, Special Publication 8. Cited in La Niece (1993: 206).
8. Gowland, W. 1915. 'Metals and metal-working in Old Japan'. *Japan Society* **13** and W. Gowland 1896. 'Japanese metallurgy', part 1. *Journal for the Society of Chemical Industry* **15**, cited in Lechtman (1970: 50).
9. *Plate with Figures*, 1972.71 (diameter 18.7 cm.), John L. Severance Fund.
10. For a more complete discussion on Sasanian silver techniques and compositions see Meyers and Harper (1981) and for more detailed descriptions and analyses of Sasanian Silver at the Cleveland Museum of Art see Lechtman (1971).
11. *Dish with King Hormizd II Hunting Lions*, 1962.150 (diameter 20.8 cm), John L. Severance Fund.
12. Pieter Meyers, Head of Conservation, Los Angeles County Museum of Art: Personal communication while in Cleveland examining the vessels, 1993.

13. *Silver Bowl*, 1963.478 (h. 6.4 cm). John L. Severance Fund.
14. *Vase with Four Dancing Figures*, 1962.294, (h. 18.5 cm). Gift of Katherine Holden Thayer.
15. *Salver* 1972.34 (diameter 30.5 cm). Purchase, Leonard C. Hanna, Jr. Bequest.
16. *Stem Cup* 1951.396 (h. 8.9 cm). Purchased from the Edward L. Whittemore Fund.

References

Caley, E. R. 1926. 'The Leyden Papyrus X'. *Journal of Chemical Education*, **3**: 1149–66.

Czuma, S. 1991. 'Museum acquistions'. *Bulletin of the Asian Institute*, New Series **5**: 190.

Czuma, S. 1992/93. 'Some Tibetan and Tibet-related acquisitions of the Cleveland Museum of Art'. *Oriental Art*, **38**: 231–2.

Czuma, S. 1993. 'Tibetan silver vessels'. *The Bulletin of the Cleveland Museum of Art*, **80**: 131–5.

Gibbons, D., Ruhl, K. and Shepherd, D. 1979. 'Techniques of silversmithing in the Hormizd II Plate'. *Ars Orientalis* **11**: 163–76.

Gibbons, D. and Ruhl, K. 1979. 'The metallurgical technique of the silver "Plate with figures", Gupta Period'. *Ars Orientalis* **11**: 177–82.

La Niece, S. 1993. 'Silvering'. In *Metal Plating and Patination*, S. La Niece and P. Craddock (eds), 201–10. Oxford: Butterworth-Heinemann.

La Niece, S. 1995. 'Depletion gilding from third millennium BC Ur'. *Iraq*, **LVII**: 41–7.

Lechtman, H. 1970. 'The gilding of metals in pre-Columbian Peru'. In *Application of Science in Examination of Works of Art*, Proceedings of the seminar 15–19 June 1970, 1973, William J. Young (ed.), 38–52. Boston: Museum of Fine Arts.

Lechtman, H. 1971. 'Ancient methods of gilding silver: Examples from the Old and the New Worlds'. *Science and Archaeology*. Cambridge, Mass.: MIT Press.

Meyers, P. and Harper, P. 1981. *Silver Vessels of the Sasanian Period, Volume I: Royal Imagery*. New York: Metropolitan Museum of Art (Princeton University Press).

Meyers, P. 1981. 'Technical studies of Sasanian silver'. *Museum Applied Science Center for Archaeology Journal*, **1**: 244–5.

Miho Museum 1997. *Miho Museum Catalog*, no. 59. Kyoto: Miho Museum.

Oddy, A. 1993. 'Gilding of metals in the Old World'. In *Metal Plating and Patination*, S. La Niece and P. Craddock (eds), 170–81. Oxford: Butterworth-Heinemann.

Oddy, W. A., Cowell, M. R., Craddock, P. T. and Hook, D. R. 1990. 'The gilding of bronze sculpture in the classical world'. *Small Bronze Sculpture from the Ancient World*. Malibu, Calif.: J. Paul Getty Museum.

Ogden, J. 1993. 'Aesthetic and technical considerations regarding the colour and texture of ancient goldwork'. In *Metal Plating and Patination*, S. La Niece and P. Craddock (eds), 39–49. Oxford: Butterworth-Heinemann.

11 Lacquer as an Adhesive for Gilding on Copper Alloy Sculpture in Southeast Asia

Donna K. Strahan and Christopher A. Maines

Abstract

Asian lacquer has been used to attach gilding to copper alloy sculptures in Southeast Asia for centuries. This paper provides a survey of Asian lacquer tree species and a brief history of their geographic presence. Descriptions of the use of lacquer gilding in Thailand and on Thai Buddhist copper alloy sculptures in the Walters Art Gallery are provided. Analyses of lacquer samples from the sculptures were performed using optical microscopy, scanning electron microscopy (SEM) and pyrolysis-gas chromatography/mass spectrometry (Py-GC/MS). These analytical methods identified the tree species, the number of lacquer layers and the inclusions in the various layers. Deterioration and conservation problems of lacquer gilding are also discussed.

Introduction

Asian lacquer has been used as an adhesive for gilding on metal surfaces for over two millennia. There is little discussion of this topic in the literature, mainly because the adhesives are often overlooked or they have disintegrated leaving little trace. The opportunity to study lacquer-gilded copper alloys specifically on Southeast Asian sculpture was presented when the Walters Art Gallery was given the Griswold collection of art. The collection consists of nearly 300 objects, roughly 150 of which are copper alloy Buddha images. Many of these images have traces of lacquer and gilding on the surface. Before concentrating on the Southeast Asian sculpture, it will be worthwhile reviewing the use of adhesives for gilding copper alloys.

One of the earliest methods of applying gold foil to copper alloy objects world-wide was through the use of an adhesive. This technique goes back roughly 6000 years in China and continued into at least the middle of the first millennium BC in Egypt and the ancient Near East. Unfortunately the adhesives used during ancient times readily decay during burial and have not been adequately identified. The ancient adhesives are probably similar to the natural adhesives used today to adhere gilding to wood, gesso, plaster, stone and leather, in both the East and the West. Most probably these adhesives were fish, animal or vegetable glues, gums, or albumin obtained from blood, eggs and milk. Gilded areas on ancient Greek terracottas were adhered in place with egg whites and this method could easily have been used on metal objects. These organic adhesives were not strong enough to hold the foil for extended periods of time; they either lost their adhesion, were washed away during burial, or became brittle and flaked off, taking the gold foil with them (Oddy 1984; Oddy and Cowell 1993).

Asian lacquer originates from various tree species and has been used to attach gilding to copper alloys, predominately in Southeast Asia. To obtain an understanding of lacquer use, a brief history and its characteristics are provided. Analyses of samples of Buddhist sculpture from Thailand were performed by optical microscopy, scanning electron microscopy with energy dispersive X-ray spectroscopy, and pyrolysis-gas chromatography-mass spectrometry. Lastly, lacquer deterioration and conservation problems are discussed.

Asian lacquer and its characteristics

In addition to the aforementioned adhesives, lacquers are used throughout Asia as both coatings and adhesives for various purposes, including attaching gilding to metal substrates. Asian lacquer should be distinguished from resin lacquer. Resin lacquer, also known as shellac or lac, is a gummy resin residue produced on certain Indian and Southeast Asian trees by specific insects (*Laccifer lacca*, *Coccus lacca* or *Carteria lacca*). The dried insect residue is collected from twigs and dissolved in a suitable solvent, such as alcohol, in preparation for application. Shellac is used predominantly in India, Southeast Asia and Persia as a coating on wood, fabric, metal, plaster or even on leather. It has also been used to adhere gilding to various substrates (Harry N. Abrams 1984)

Asian lacquer, also known by its Japanese name, *urushi*, is found throughout Asia. It has been used since ancient times and continues to be made today. Lacquer is produced from the sap of various species of the *Rhus* and *Melanorrhoea* trees. The genus *Rhus* contains many different species, which include North American poison ivy (*Rhus* or *Toxicodendron radicans*), poison oak and poison sumac. Historical records indicate that lacquer trees have been cultivated at least since the Han Dynasty in China (206BC–AD220). The thrust of this paper will be the use of Asian lacquer as an adhesive for gilding, and from here on the term lacquer will refer only to Asian lacquer.

Rhus verniciflua is the most common tree tapped by the Chinese, Japanese and Koreans for lacquer sap. The Japanese term *urushi* is generally applied only to the lacquer of this species. In former times it grew over most of China south of the Yellow River, extending from the east coast to Gansu, Sichuan and Yunnan in the southwest. In Japan lacquer is also produced from *Rhus ambigua* (or *Rhus orientalis*), known as *tsuta-urushi*, and *Rhus trichocarpa*, known as *yama-urushi*. However, these lacquers are generally not used for items of the highest quality, but rather for everyday objects. *Rhus succedanea* grows in Southeast Asia as well as in Japan and the Ryukyu Islands. Lacquerwares made from *Rhus succedanea* may have been produced in South China on the border with Vietnam. *Rhus ambigua* grows in Taiwan, the Ryukyu Islands and Japan. Lacquerwares and their technology may have been introduced to other countries in Southeast Asia from the southern Chinese provinces (Garner 1979).

Melanorrhoea laccifera and *Melanorrhoea usitata* produce similar sap to *Rhus* species. However, there are additional components present, which result in a less durable film. In addition, the enzyme laccase, which is responsible for cross-linking reactions that lead to a cured lacquer film, is less active in *Melanorrhoea* than in *Rhus*. *M. laccifera* grows in Cambodia and is used for making lacquer. *M. usitata*, known as *thitsi*, grows in Burma and Thailand and also provides the standard material for the manufacture of Vietnamese lacquerwares today (Garner 1979).

Regardless of the type of tree sap used, the refining and fabrication techniques of lacquer are similar. Special care must be taken when processing lacquer because it can cause severe, and even fatal, contact dermatitis. Similar to rubber trees, the *Rhus* trees produce a gummy sap, which is tapped seasonally as it exudes from long gashes made in the bark. The quality varies according to the area in which it was grown, the type of tree and the season. Since the sap begins to harden upon exposure to oxygen, it is collected in containers and covered, usually by paper, to control this. The sap is then filtered and the excess water removed by heating or allowing it to evaporate away slowly.

The main constituents of raw lacquer are various long-chain alkylated catechols, water, gummy substances and a small amount of proteinaceous material, which includes the enzyme laccase. In the case of *Rhus* species, the catechols are known as urushiols and have pentadecyl, pentadecenyl or pentadecynyl side chains. The major components of the sap from *Melanorrhoea* species are 10-phenyldecyl and 12-phenyldodecyl catechols with urushiols being relatively minor components.

Once purified and applied to a surface, lacquer requires a temperature between 20°C and 28°C and high relative humidity (over 55 per cent is best) to harden into an impervious film (Kenjo 1988). The enzyme laccase promotes the dimerization of the catechol nuclei as well as the cross-linking of the catechol nuclei to unsaturated sites on the various side chains. Because *Melanorrhoea* lacquers contain large proportions of the 10-phenyldecyl and 12-phenyldodecyl catechols, which have no sites of unsaturation, these molecules cannot cross-link. Therefore, the films of *Melanorrhoea* lacquers are not as strong or durable as those from *Rhus* lacquers.

Asian lacquer gilding history

Asian lacquer has been in use in China since about 4000BC. Lacquerware (made of a wood base covered with lacquer layers) was already a highly developed craft during the Shang Dynasty (c.1500–c.1050BC). Although commonly used as a coating for wood, during the Zhou Dynasty (c.1050–221BC) a red and black lacquer-like substance was also used as a decorative inlay material on ancient bronzes. In antiquity dark pitch-like adhesives were also used to secure inlay materials such as turquoise. Although to date little scientific investigation has determined the type of adhesive used, lacquer is a likely candidate (Gettens 1969; Jacobsen 1984). Studies are presently underway into the use of adhesive materials used to attach gold leaf and inlay on early Chinese metal objects. Until more research is completed, the actual use of lacquer as an adhesive for gold leaf on Chinese artifacts can only be speculated upon.

Mercury gilding (or amalgam gilding) on copper alloy objects was introduced in the Warring States Period in the late fourth century BC. Afterwards, mercury gilding appears to be the predominant method used on both copper alloys and silver (see Jett and Chase, this volume, Chapter 8). Previous research has shown that setting down gold foil with a lacquer adhesive to a bronze or iron base dates from at least the Six Dynasties/T'ang Dynasty (Jacobsen 1984).

A number of gilded copper alloy objects in Japan have been gilded by the mercury amalgam method. Most of these date from after the fourth century BC. However, as more analysis is carried out on Japanese copper alloys, gilding attached by lacquer may also be identified. Certainly over the last 1000 years gilded lacquer on wood and other organic substrates was quite common and highly developed in Japan. Similar to China and Japan, Korean copper alloy objects have been mercury gilded. Once again no adhesives have been scientifically identified, and few objects have actually been analysed (see Murakami, this volume, Chapter 9).

Lacquer trees are not indigenous to Tibet, Nepal or most of India. There is no published evidence for the historical use of Asian lacquer to attach gold leaf to metals in any of these countries. Copper alloy objects from these regions are generally gilded using the mercury gilding method.

Unlike those from northern Asian countries, Southeast Asian copper alloy objects are commonly gilded with lacquer as the adhesive. Lacquer gilding has been practised at least since the fifteenth century (Wangchareontrakul 1990; Strahan 1997). This is especially true of copper alloy images of the Buddha (Fig. 11.1). Burma, East Bengal, Thailand and Laos all historically use lacquer to adhere gilding to copper alloys.[1] While both shellac and Asian lacquer are used in Burma and East Bengal, shellac is used predominantly as a medium for a colorant rather than as an adhesive (Fraser-Lu 1994). The *Melanorrhoea usitata* tree grows wild up to elevations of nearly 1000 metres in the drier forests of the Shan States and the Kathe and Bhamo districts of Burma. Although Burma and China have had relations for centuries, the dates and routes for the transmission of the art of making lacquer are not completely understood. Current evidence suggests that Burma acquired the art of lacquer either directly from China or through various trading and tributary contacts with neighbouring states

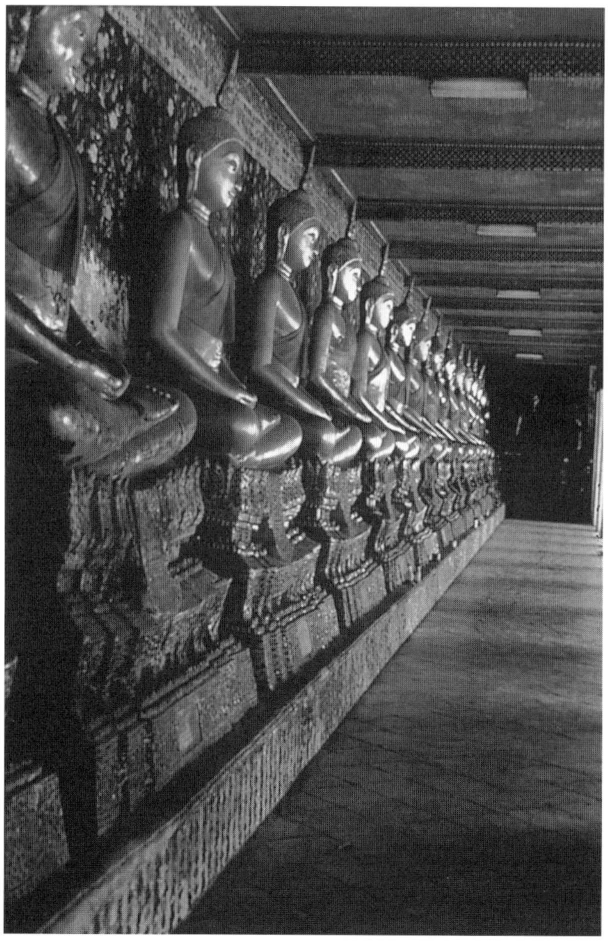

Figure 11.1 A gallery of copper alloy Buddhas coated with gilded lacquer at Wat Suthat in Bangkok, Thailand. (D. Strahan.) (See Plate 45.)

such as Thailand. Current opinion indicates that the Burmese lacquer industry became fully developed in the fourteenth or fifteenth century. This also seems to be the time period when lacquer was highly developed in Thailand and Laos. At that time craftspeople moved around freely in these areas (Fraser-Lu 1989).

While present evidence does not provide many examples of lacquer to adhere gilding to copper alloys from Cambodia and Vietnam, in all probability it was a common practice. Although a few examples of Cambodian objects in the Walters Art Gallery support this (54.2722, 54.2689, 54.2715), it is possible that many of these images were gilded later in Thailand where they were purchased.[2] It is unclear from the literature if lacquer gilding was used on Vietnamese metals, but one Walters porcelain Buddha is covered with a thick lacquer gilding layer, supporting the theory that this technique was known (Fig. 11.2). Because of the Vietnamese strong lacquer tradition, it is likely that they applied lacquer gilding to copper alloys, although there are no examples in the Walters (Smithsonian Institution 1961).

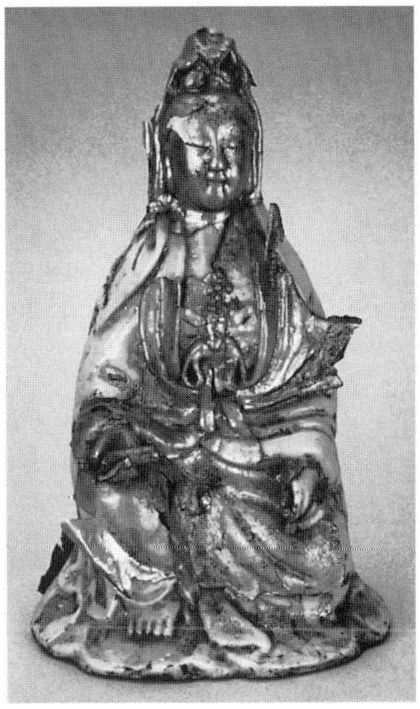

Figure 11.2 Vietnamese porcelain lacquer-gilded Buddha (48.2638). The smooth porcelain surface does not provide enough texture for the lacquer. It is readily lifting and curling off the entire surface. (D. Strahan, The Walters Art Gallery, Baltimore.) (See Plate 46.)

Lacquer gilded copper alloys in Thailand

The Buddhist sculpture from Thailand in the Walters Art Gallery provides close to 150 examples of the use of Asian lacquer to adhere gilding to copper alloy surfaces (Woodward 1997). They range in time period from the eighth through the nineteenth century and thus provide a good overview of Thai techniques. Other types of copper alloy objects are also covered with lacquer gilding, such as a gong in the Walters collection; however, discussion here will be focused on images of the Buddha (Fig. 11.3). Nearly all of the copper alloy images in the collection have traces of one or more layers of lacquer and gilding. According to tradition, Buddha images must shine like gold (Griswold 1954). While brightly polished bronze and brass alloys give the appearance of gold, many images were coated with lacquer and completely gilded with thin squares of gold leaf giving them the appearance of solid gold. The most common colour of lacquer used on Thai images is its naturally oxidized black; however, sometimes cinnabar (mercuric sulphide) is added to create a red lacquer.[3] The red colour may be used as an enhancement for the gold, similar to coloured boles common in western gilding traditions.

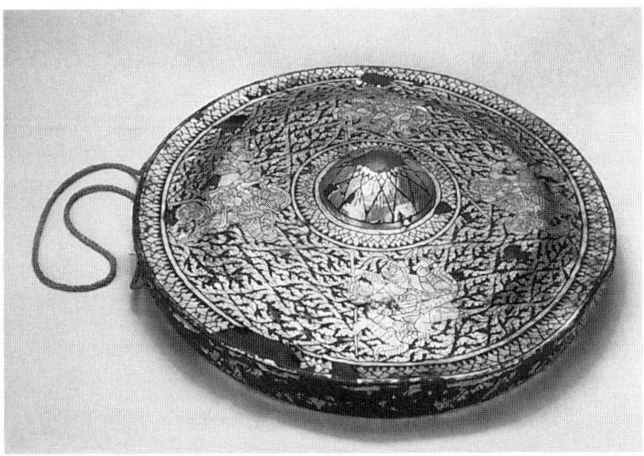

Figure 11.3 Nineteenth century Thai copper alloy gong (54.2836) with lifting black lacquer and gilding. (D. Strahan, The Walters Art Gallery, Baltimore.) (See Plate 47.)

During its lifetime an image is often subjected to incense soot and to further embellishment by the application of small squares of gold leaf by devotees (Fig. 11.4). Many layers of leaf can be applied to a single image and may over time, because of a lack of adhesion, begin to flake off,

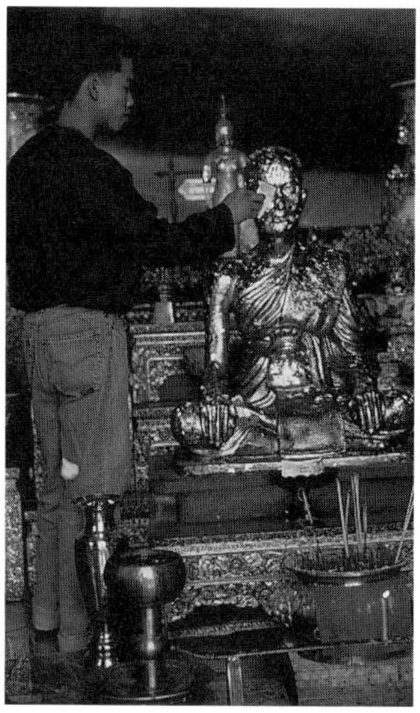

Figure 11.4 A devotee applying 'daily gilding' to an image in a temple in Bangkok. (D. Strahan.) (See Plate 48.)

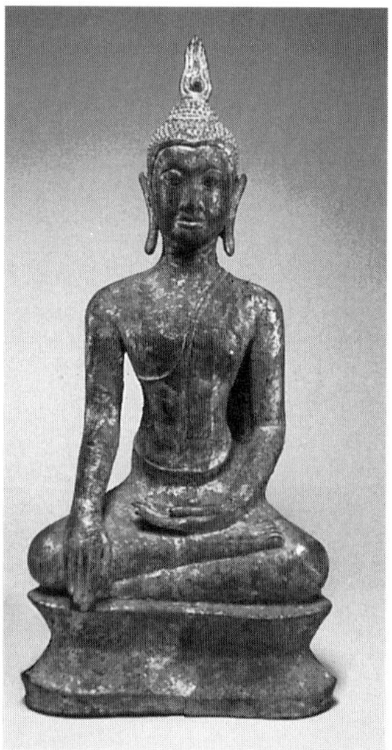

Figure 11.5 The earliest identified lacquer-gilded Buddha in the Walters (54.2802). (The Walters Art Gallery, Baltimore.) (See Plate 49.)

sometimes taking sections of a lacquer layer with them. Cracking of the lacquer can occur with the slight expansion and contraction of the substrate metal. Therefore, it is common for an image to be completely relacquered and regilded many times. It is also common for early images, which may not have been gilded, to receive gilding centuries later. This makes the dating of the gilding layer difficult as it may have been carried out centuries after the image was cast.

Images that have been buried within temple ruins for long periods of time are often renewed upon excavation with lacquer and gilding (Griswold 1955; Griswold 1974). More than half of the Walters images were previously buried and restored after excavation. In many cases more than one layer of lacquer and gilding was found. Of the Walters Thai images examined the largest number of lacquer gilding layers was four (54.2775).

The earliest extant examples of lacquer on Thai sculpture in the Walters date from the latter part of the Early Ayutthaya period (fourteenth and fifteenth centuries) (Fig. 11.5). It is difficult to date the earliest use of lacquer, but several of the Early Ayutthaya images very similar to those from a cache buried in AD1424 at Wat Ratchaburana contain a thin lacquer layer with gold leaf (Woodward 1997). Therefore, gilded lacquer was in use in Thailand at least by the early fifteenth century in the region of Ayutthaya. Lacquer gilding was found on earlier Dvaravati period images (approximately AD600–900). However, unlike in Figure 11.5, there is no proof that the lacquer and gilding were contemporary with manufacture or at least prior to burial. Even with the dominance of bright brass alloys in the later fifteenth century, lacquer gilding has continued to be practised up to the present day.

There is no evidence of mercury gilding in Thai metal objects until the Bangkok Period (after 1782). Mercury gilding appears to have been used primarily on silver objects. One example in the Walters is a silver gilt and niello bowl (Fig. 11.6). Even after the introduction of mercury gilding, lacquer gilding continued to be used. One reason for this may be because brasses and highly leaded bronzes are commonly used as alloys for Thai sculpture; the presence of zinc or high lead makes using mercury gilding difficult. Lacquer, however, can be applied easily to any alloy (Hawthorne and Smith 1963; Lins and Oddy 1975).

Thai casting methods often produce highly porous alloys with multiple flaws on the metal surfaces. These require hours of finishing work to repair. The syrupy consistency of fresh lacquer gives it the ability to fill in holes and joins on metal sculptures, thus saving time while producing a smooth surface. Before the lacquer has cured completely, it also has good adhesive qualities making it ideal for attaching gold leaf to metal surfaces.

Because Thai Buddhas were gilded over the centuries it is possible by extrapolation from modern lacquer fabrication methods to make assumptions about techniques used on early lacquered objects. During the manu-

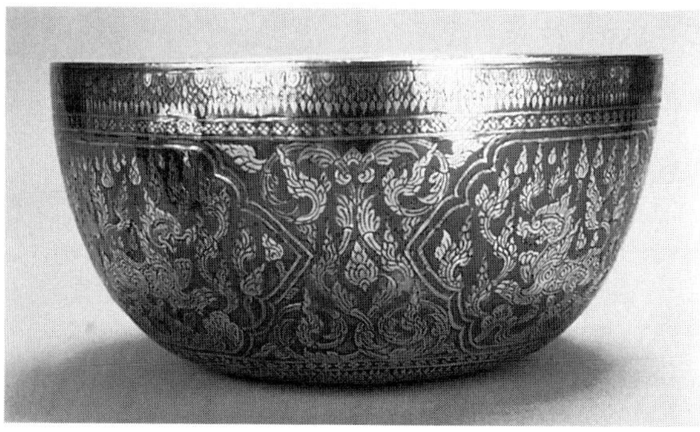

Figure 11.6 This silver partial gilt and niello bowl (45.21) has been gilded using mercury gilding. (D. Strahan, The Walters Art Gallery, Baltimore.) (See Plate 50.)

facture of modern lacquer it is common to add bulking agents to create a paste for a more viscous material. This allows a more even application to three-dimensional objects as well as filling voids and producing a smooth surface. The type of material used on older objects probably varied from location to location along with its availability. Generally today the following finely ground, and/or burnt particles are added as bulking agents: bone, sawdust, rice husks and clay. Common organic materials may be added as extenders or drying retardants. These include cashew oil, honey and tung oil. Colorants are also sometimes added. The natural oxidized colour of lacquer is dark brown to black. The most common coloured lacquer is red produced by mixing cinnabar (mercuric sulphide) or iron oxide in with the liquid lacerate. It should be noted that the palette for lacquer objects is quite limited. Many commonly used pigments, especially inorganic oxides, react with catechols and result in discoloured and poorly cross-linked films. Many pigments such as carbonates and vegetable colorants cause an adverse chemical reaction with raw lacquer, either giving it a muddy appearance or hindering it from drying. The colour of lacquer is confined to a small palette usually red, black or yellow (the pigments cinnabar, carbon black or iron oxide and orpiment) (Burmester 1988: 178).

Depending on the object, successive layers of lacquer are applied, which continually become thinner and finer with fewer inclusions. Abrasives sometimes mixed with oils may be used to polish the final surface. The gold leaf itself is usually applied over a thin pure layer of lacquer in the tacky state.

Because the lacquer is not very adherent to the metal surface and can flake off, it must be renewed repeatedly over the centuries. The repair and regilding of Buddha images are commonly carried out. Today the disappearance of the lacquer trees through deforestation and increased population have caused many foundries to seek alternative adhesives, such as auto body putty and house paint of varying shades. These new materials are applied to older excavated images to restore them for temple use. Older images may have incomplete lacquer coatings. These are often scraped off even if they are not restored, to give a more continuous appearance.

Further complicating the study of the original gilding technique is the fact that additional gold leaf is continuously applied by devotees in temples by placing individual squares of gold leaf on an image by pressure or with an oil, which is supplied with the leaf. Some images can be thickly covered with hundreds of sheets of gold leaf (Fig. 11.7).

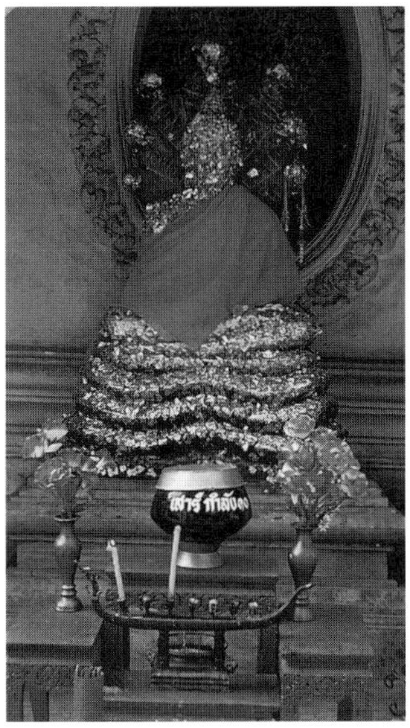

Figure 11.7 Image in Nakhon Pathom with multiple layers of gold leaf. (D. Strahan.) (See Plate 51.)

Bulked lacquer has also been used to change the stylistic appearance of Buddha images. Examples of this survive from around the seventeenth century, such as an image at Wat Pho in Bangkok. The cast image under the lacquer layers is an earlier period than the Early Bangkok Period appearance on the exterior.[4]

Microscopic analysis of lacquer from Thai Buddha images

A preliminary study of the multiple lacquer layers and types of inclusions was carried out on a number of Thai Buddha images from different periods in the Walters Art Gallery collection. The lacquer layers were compared to one another and to the earliest known lacquered image in the collection, the Early Ayutthaya image mentioned previously (Fig. 11.5; 54.2802). Microscopic samples of lacquer were removed from several of the Buddhas for analysis. Methods used for this study were optical microscopy (UV fluorescence microscopy) and scanning electron microscopy with

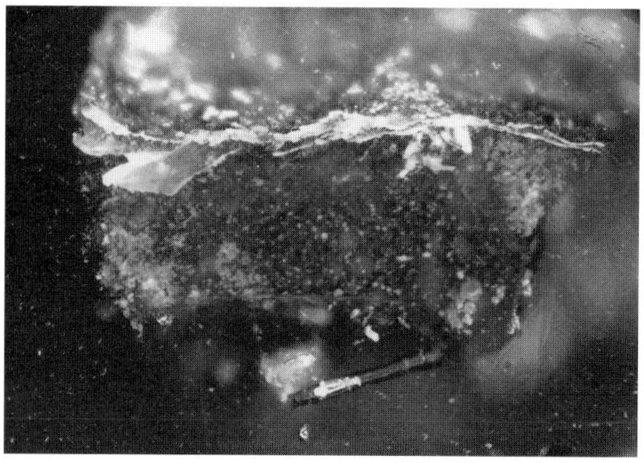

Figure 11.8 Optical microscopic cross-section under reflected light of Early Ayutthaya image 54.2802 (20×). Layers of bulked lacquer with gold leaf on top. (D. Strahan.) (See Plate 52.)

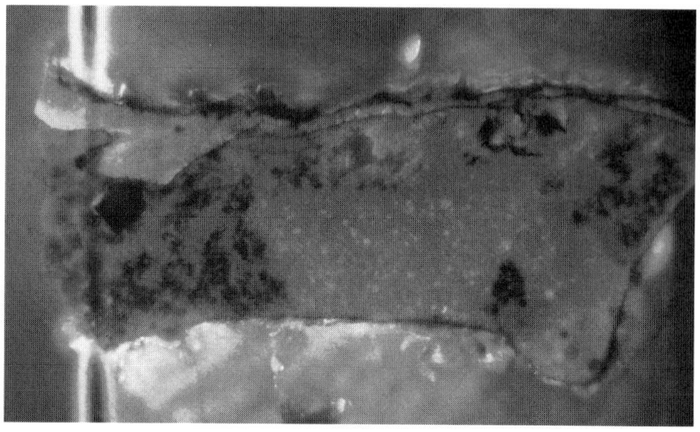

Figure 11.9 Ultraviolet light microscopic cross-section of the same area in Figure 11.8 (54.2802). While difficult to see in visible light, the inclusions in the lacquer layer are clearly seen in ultraviolet light (20×). (D. Strahan.) (See Plate 53.)

energy dispersive X-ray spectroscopy (SEM–EDS) analysis of lacquer cross-sections.[5] Some samples show a series of layers of both lacquer and gilding. Others have a series of lacquer layers before the application of gilding.

One Early Ayutthaya image (54.2802) has copper alloy burial corrosion products on the surface covering a lacquer gilded surface. A single layer of thin lacquer 100 microns thick is present under the corrosion in contact with the metal surface. The lacquer was very pure with no inorganic constituents added. SEM–EDS analysis showed that copper and lead had migrated through cracks in the lacquer to the surface both under and over the gold indicating that the lacquer was present before this corrosion was formed; ie, before burial. Two 2–3 micron thick layers of gold leaf were situated on top of the lacquer. A dot map of potassium showed the entire lacquer layer contained ashes (probably from plants). Under ultraviolet light on the optical microscope, organic constituents were identified as having been mixed with the lacquer. They appear to be burnt plant material at a rough ratio of 1:5. Iron was present only on the surface and was therefore not used as a lacquer colorant (Figs 11.8 and 11.9).

The Buddha image from the Sukhothai Period (fourteenth to fifteenth century) (54.2775) is another cross-section example, which reveals three layers of lacquer (a total of 325 microns thick), each of which contain plant material. Silica, probably in the form of sand, has been mixed with the bottom or first lacquer layer. The topmost lacquer layer (100 microns thick) contains both calcium and silica, possibly from ground stone. No phosphorus or aluminium was detected, therefore neither clay nor bone was added to the mix (Figs 11.10, 11.11 and 11.12).

The thickest and most intricate lacquer and gold layering can be seen on an eighteenth century Buddha (54.2553), which contains four layers (Fig. 11.13). The bottom layer, which was in contact with the bronze, was bulked with plant fibres and ashes (400 microns thick). Next a layer of lacquer bulked with ashes and/or plant material was applied (100 microns thick). Again the ashes were identified by the presence and location of potassium in the dot map. Burnt plant fibres were large enough to be seen by the optical microscope (Fig. 11.14). Finally, a layer of pure lacquer with few inclusions was added (100 microns thick). The inclusions were visible under ultraviolet light. This layer was unusual in that it had fine particles of gold throughout. The gilding layers were placed on top, producing a

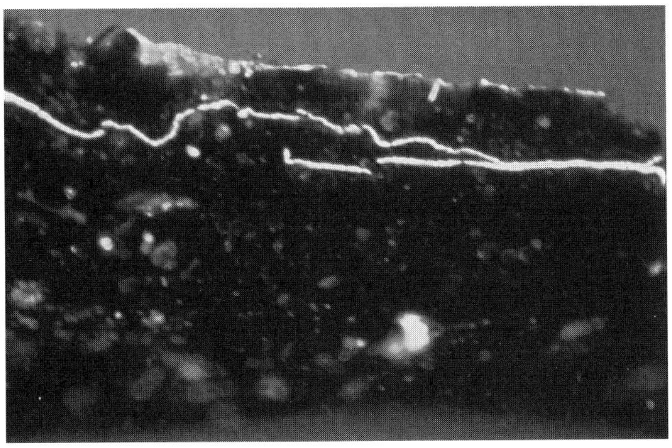

Figure 11.10 Optical microscopic cross-section of a Sukhothai Period image (54.2775) showing three distinct lacquer layers (20×). (D. Strahan.) (See Plate 54.)

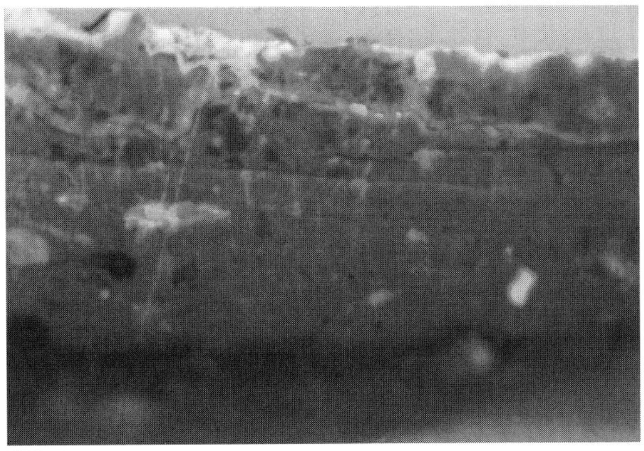

Figure 11.11 Ultraviolet light microscopic cross-section of the same area in Figure 11.10 (54.2775). Once again burnt plant material mixed with the lacquer becomes clearer under ultraviolet light (20×). (D. Strahan.) (See Plate 55.)

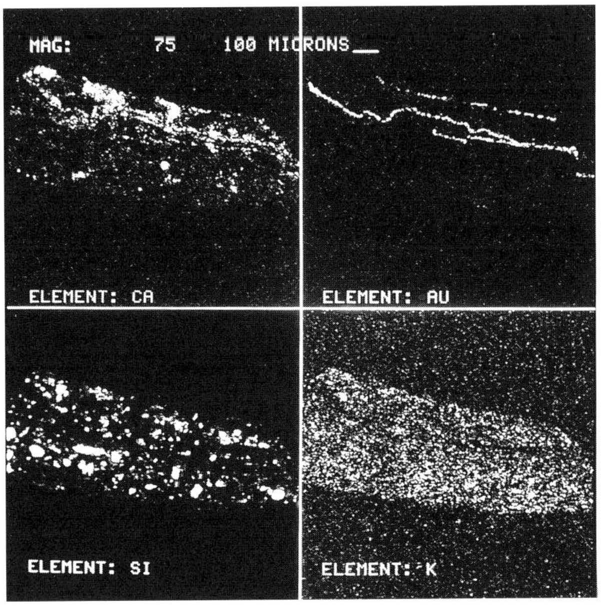

Figure 11.12 SEM dot map of the cross-section in Figures 11.10 and 11.11 (54.2775). The dot map provides elemental clues to the identity of the inclusions added to the lacquer: calcium (Ca), silica (Si), gold (Au), potassium (K). (D. Strahan.)

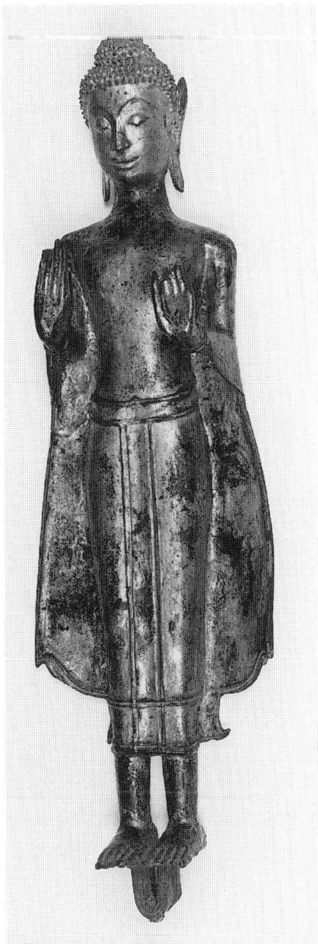

Figure 11.13 This Late Ayutthaya period Buddha (54.2553) is covered with thick lacquer and gilding layers, which are nearly intact. (D. Strahan, The Walters Art Gallery, Baltimore.)

layer roughly 2–3 microns thick. This gilding thickness is typical for all images studied (Fig. 11.15).

Two campaigns of lacquer (150 microns thick) were identified on 54.2815, a standing Buddha (Central Thailand, sixteenth century). The lacquer next to the metal is very pure with no inclusions. The upper lacquer layer was applied in two levels. The first level contains large inclusions, which contain calcium, aluminium, silica and potassium. They may be plant ash, clay and sand particles. The final level contains the same type of inclusions but in much finer quantities.

Lacquer gilding on a number of other images was analysed. Nearly all contained multiple layers of lacquer within a similar range of thicknesses, and had comparable inclusions such as plant ash, sand and clay. The thickness of the gold leaf was also similar with little change over time. Thus far no ground bone has been found, as was reported in use for both ancient Chinese lacquer Buddhas and modern Burmese images (Strahan 1993). Aside from the addition of bone, all other microscopic analyses and

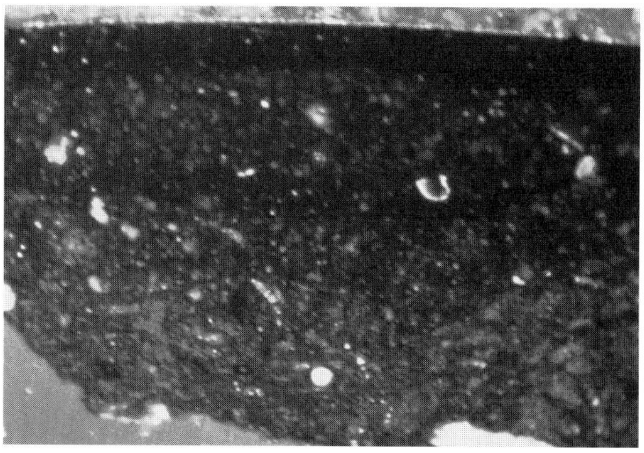

Figure 11.14 Optical microscopic cross-section of Late Ayutthaya image (54.2553) (20×). (D. Strahan.) (See Plate 56.)

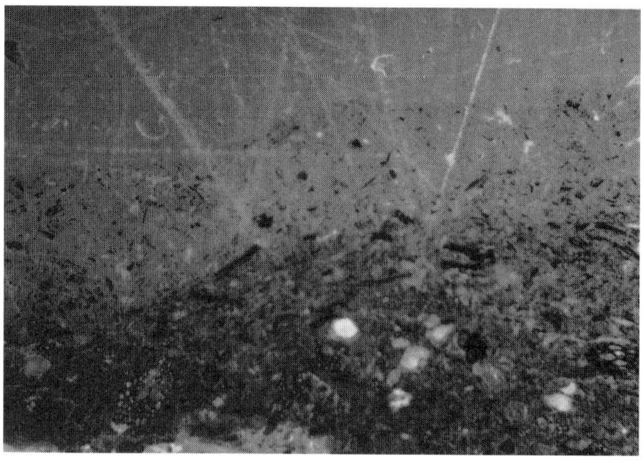

Figure 11.15 Ultraviolet light microscopic cross-section of the same area in Figure 11.14 (54.2553). The inclusions and gold flakes mixed in three of the four lacquer layers are clearly seen under ultraviolet light. (D. Strahan.) (See Plate 57.)

published data confirm that the method used in neighbouring Burma today is similar to that found on historical Thai sculpture. Lacquer gilding is a traditional process at least several hundred years old. Today rough metal surfaces to be lacquered are made smooth by the application of a paste. The paste consists of ash from burnt sawdust, rice husks, or bone, mixed with lacquer and is applied in one or more layers (Fraser-Lu 1994).

Pyrolysis-gas chromatography/mass spectrometry analysis

Four samples from Thai Buddhas in the collection of the Walters Art Gallery were analysed. The samples consisted of one to several layers of gold leaf and what was presumed to be Asian lacquer (Table 11.1; Fig 11.16).

The layers were not separated prior to analysis because of their small size (0.25–0.5 mm on a side). The lacquer layers for samples 2 through 4 contained inclusions reflected in optical microscopy and identified by SEM–EDS to consist of inorganic particles, such as quartz, and organic particles, such as fibres and bits of wood. None of the inclusions were

Table 11.1 Analysis of the layers on four samples from Thai Buddhas in the collection of the Walters Art Gallery.

Sample	Number	Period	Century	Layers
1	54.2802	Early Ayutthaya	15	one lacquer layer and gold leaf
2	54.2775	Sukhothai	14–15	multiple layers (3–4) of lacquer and gold leaf
3	54.2553	Ayutthaya/Bangkok	18–19	multiple layers (4) of lacquer and gold leaf
4	54.2815	Ayutthaya	16	multiple layers (2–3) of lacquer and gold leaf

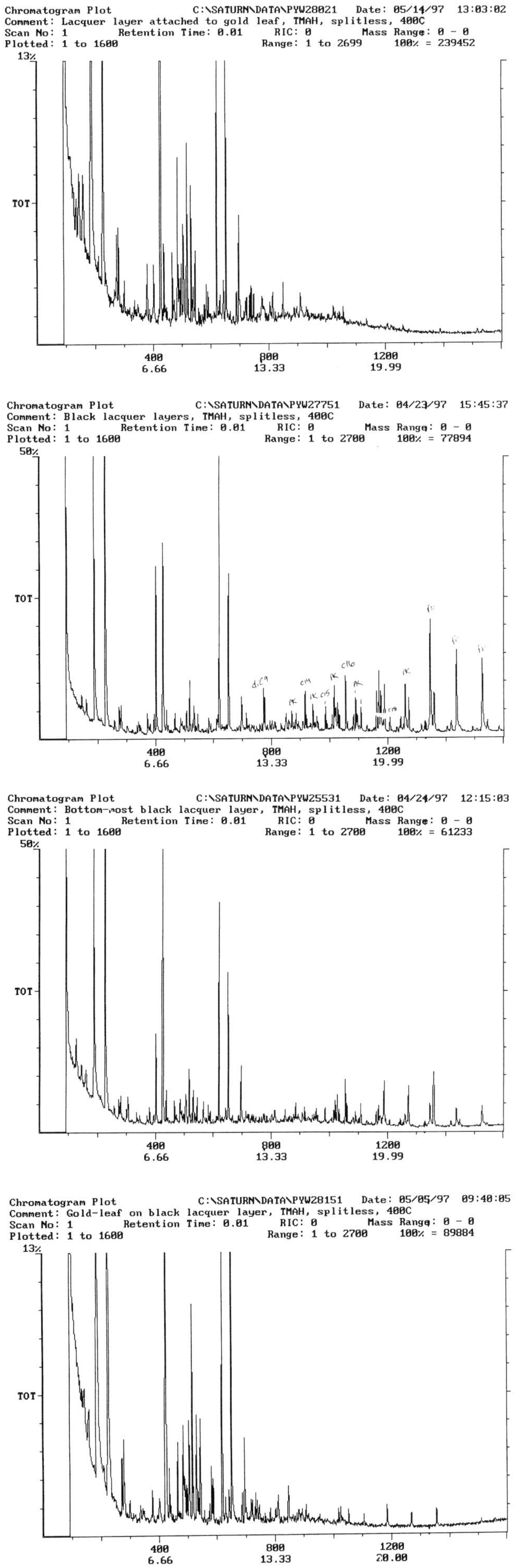

Figure 11.16 Pyrolysis charts of four samples.(16. 54.2802, 17. 54.2775, 18. 54.2553, 19. 54.2815).

apparent in the chromatograms, as would be expected from the experimental conditions described below.[6]

The results of the Py-GC/MS analyses of the four samples clearly demonstrate that Asian lacquer obtained from the *Melanorrhoea usitata* is the primary organic material present. As noted earlier this species grows predominately in Thailand and Burma. No evidence was seen of other organic media such as modern synthetic polymers, alkyd paints, or terpenoid resins, like dammar or rosin. However drying oils, such as tung or linseed oils, could be present in small amounts. Since palmitic and stearic acids are the primary components of drying oils that can be seen in Py-GC/MS analyses, and since both of these fatty acids are present in *M. usitata* lacquer, one cannot completely rule out the possibility of a small amount of drying oil being present in any sample identified as *M. usitata*. However the methyl ester of azelaic acid, another fatty acid present in drying oils, was only seen in sample 2; azelaic acid is a degradation product of oleic acid and is always seen in aged films (Mills and White 1994).

The characteristic components of *M. usitata* lacquer are 3-(10-phenyldecyl)-catechol and 3-(12-phenyldodecyl)-catechol, which together comprise about 40 per cent of the sap. An additional 3 per cent of the sap is composed of resorcinols and phenols with long hydrocarbon side-chains terminating in a phenyl group (Du and Oshima 1985; Du *et al.* 1986). *Rhus* and *Anacardium* (cashew oil) species do not produce any phenyl-terminating molecules. These molecules cannot enter into cross-linking reactions since all of the long hydrocarbon segments of the side-chains are completely saturated. Therefore, the side-chains remain intact, even after the film is dried. During pyrolysis of the film, the phenyl-terminating groups are detached from the catechol nucleus and the hydrocarbon portions undergo random scission, resulting in phenalalkyl, phenylalkanol, and phenylalkanone side-chains of varying length.

The pyrograms of the Thai lacquer samples have only three types of molecules present: phenylalkanes, phenylalkanones and fatty acid methyl esters. All four samples show at least the presence of phenylalkanes, which indicates the presence of *M. usitata* lacquer. Sample 1 was at the detection limits of Py-GC/MS, and so only very small peaks associated with alkanols could be differentiated from the background. No other species were observable in this sample. A conclusive identity of the organic material present could not be established.

Sample 2 contained phenylalkanes, phenylalkanones and fatty acid methyl esters. The pyrograms show that phenylalkanones and fatty acid methyl esters are more abundant than the phenylalkanes, indicating that this sample was a mixture of *M. usitata* lacquer and another material that was most likely a drying oil. However, since the three or four individual lacquer layers could not be separated, it is not clear whether each layer was a mixture or whether one or more layers could have been composed of pure *M. usitata* lacquer.

Sample 3 contained primarily phenylalkanes, although there were some phenylalkanones and fatty acid methyl esters present. Azelaic acid methyl ester was noticeably absent, which indicates that there was no drying oil present in the sample. Therefore, this sample was probably exclusively composed of *M. usitata* lacquer.

Sample 4 contained only phenylalkanes and palmitic acid methyl ester. This sample was exclusively composed of *M. usitata* lacquer.

Three lacquer samples from eighteenth to nineteenth century copper alloy objects were analysed solely by Py-GC/MS. All three contained both *M. usitata* lacquer with a drying oil.[7]

Deterioration and conservation problems

Generally speaking, when cured, lacquer is unaffected by most acids and alkalis, most organic solvents and water at room temperature, and is impervious to insect attack. Exposure to ultraviolet radiation and to extremes of temperature can initiate deterioration of lacquer and may cause irreversible discoloration and the dissolution of some constituents (Araki and Hisayoshi 1978). However, for lacquered objects protected with a gold layer this is not a problem, of course gold was missing in many cases (Araki and Hisayoshi 1978; Carriveau 1984; Webb 1987).

Heavily abraded gilded surfaces from use and handling are typical of many Buddha images. A more serious problem is lacquer that is no longer adherent to the metal substrate. It often cracks and flakes off, taking the gilding with it (see Fig. 11.3). Different coefficients of expansion and contraction between the lacquer and metal also cause the loss of attachment. To complicate matters, once lacquer has cured it is very brittle. If the adhesion is lost in areas, minute holes in the lacquer layer open a pathway for moisture to enter, which encourages corrosion beneath the lacquer. The presence of moisture in small areas sets up an electrolytic cell, which will lead to localized corrosion in these locations. Fresh lacquer is acidic by nature, but there is no evidence for an acidic nature in cured and aged lacquer.

The lacquer-gilded surfaces of images that have been buried in the ground for extended periods of time are generally covered with copper alloy corrosion. Metal ions can migrate from the corroded metal surface up through the lacquer layer and cover the lacquer gilding. The SEM dot map shows the migrating copper and lead ions detected within the lacquer cross-section of image 54.2802 (Fig. 11.17). During burial, corrosion also develops between the metal and lacquer layer as can be seen in 54.2684 (Fig. 11.18). Corrosion has separated the lacquer gilding from the metal surface. The location of corrosion and migrating metal ions can be seen in the SEM dot maps (Fig. 11.19). Corrosion has also been seen between the lacquer and metal on Buddha images that have not been buried. Regardless of where it is located, expanding corrosion causes a loss of adhesion and can also push lacquer off of a metal surface.

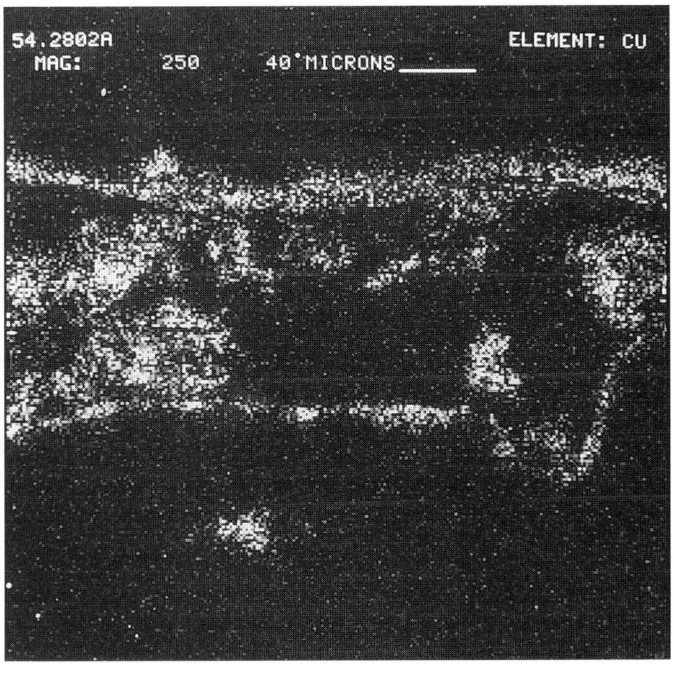

Figure 11.17 Dot map of 54.2802 cross-section. Compare with Figures 11.8 and 11.9. (D. Strahan.)

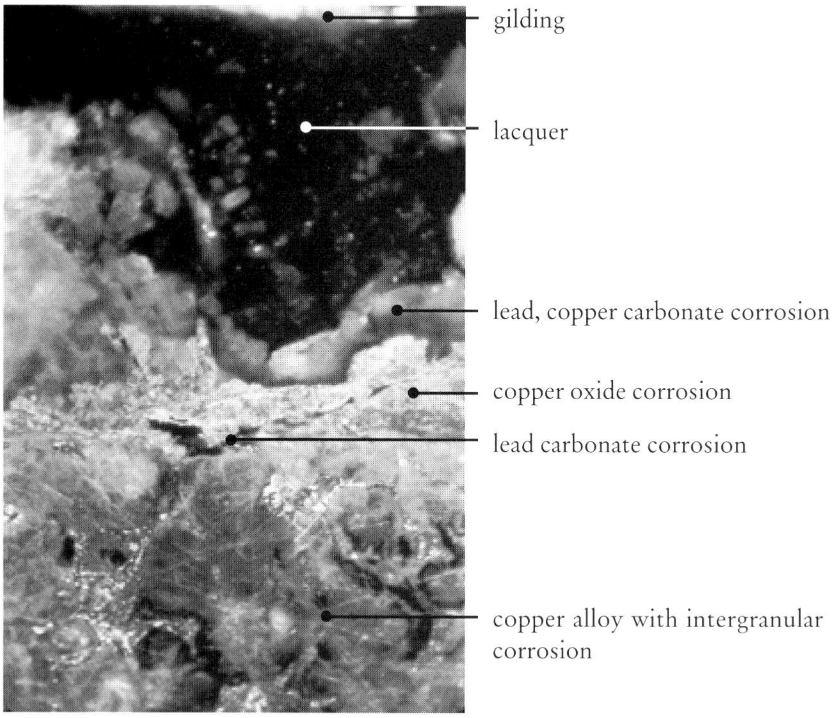

Figure 11.18 Dark field microscopic cross-section of 54.2684 showing metal alloy, corrosion and lacquer gilding layers (20×). (D. Strahan.) (See Plate 58.)

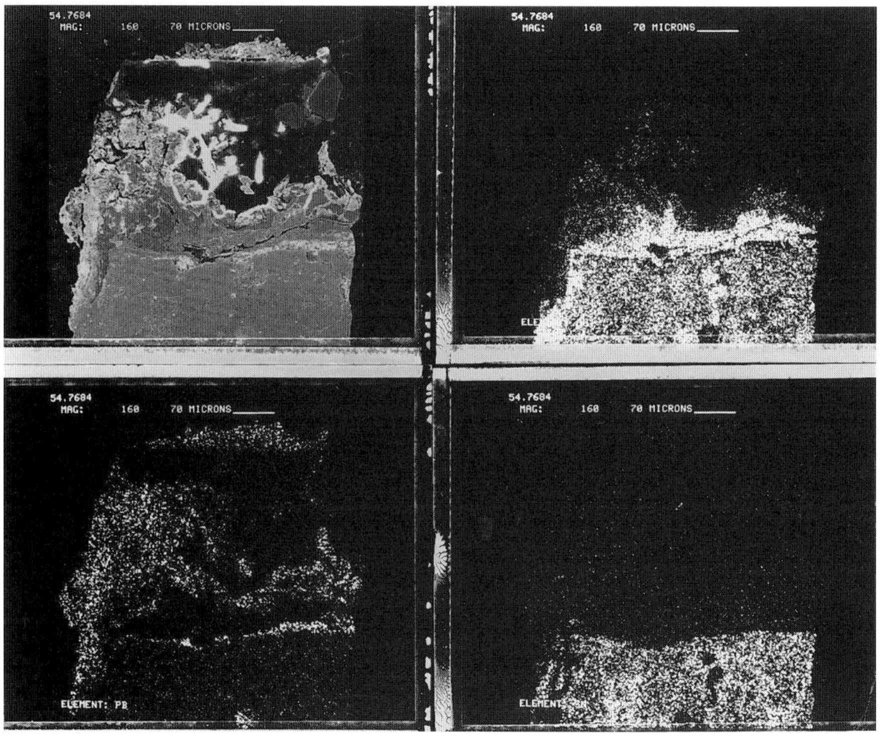

Figure 11.19 SEM photograph of 54.2684 (upper left) with dot maps of the sample showing the location of copper (upper right), lead (lower left) and tin (lower right) in the corrosion and lacquer layers. The lead has migrated up through the lacquer as the corrosion developed. (D. Strahan.)

Since little work has been done in this area, it is important to take samples before conservation treatments are performed. Any cleaning or repair methods may alter future analysis. Conservation treatments for lifting or flaking lacquer gilding can be difficult but may be achieved with a solvented acrylic resin. However, if the lacquer gilding has curled up, its brittleness often causes new fractures or losses when attempts are made to press it into place. As lacquer deteriorates or experiences long exposure to light, it begins to break down, discolouring it and making some components soluble.

If removal of copper corrosion products over the gilded lacquer is necessary, it is best to use dry techniques or wet methods that will not penetrate the gilding and lacquer layers. A method used successfully at the Walters for many years is to use gels to hold in place reagents that dissolve the corrosion products overlying the gilding. The gels are removed and replaced repeatedly until the gilding has been cleaned to the desired level.

Conclusion

Asian lacquer has been used to adhere gold leaf to copper alloy objects since ancient times. The gilded surfaces of copper alloy Buddhist sculpture in Southeast Asia have been applied with Asian lacquer as the adhesive at least since the fifteenth century. The lacquer was applied to the metal surface in a fluid, liquid form with or without bulking agents. Sheets of thin gold leaf were laid on top of the lacquer while it was still tacky. Over time, the gilding on many of the sculptures was restored using the same technique. Analysis of a number of images from Thailand revealed that multiple campaigns of lacquering were carried out using lacquer from *Melanorrhoea usitata* as the primary organic material present. Ashes, sand and plant fibres were used as bulking agents. Unlike Chinese and modern Burmese methods, no crushed bone bulking agents were found.

This study on the use of lacquer gilding on copper alloys in Southeast Asia is not complete. In depth future studies combined with statistical analysis may find identifiable technical changes over specific time periods or from different geographical areas. More research on chemical properties of lacquer and how it ages should be carried out. The study of lacquer is particularly pertinent because both the lacquer trees and the technology are disappearing and being replaced with modern materials.

Notes

1. Personal communication with Sylvia Fraser-Lu and Sarah Bekker, 22 June 1995.
2. Of Cambodian sculptures with apparent mercury gilding, the author is aware of only one that has received laboratory analysis. Elemental analysis has identified mercury gilding on the skirt, crown, necklace and bracelets of a Cambodian eleventh- to twelfth-century bronze image of a female deity, now in the Cleveland Museum of Art (82.51). It has been published in the *Cleveland Museum of Art Bulletin* 70:1 (Jan. 1983), 35 and 55. Patricia Griffin is gratefully thanked for information concerning her ongoing analysis.
3. This is particularly common on Burmese lacquer-gilded metal objects.
4. A lacquer-gilded stucco face, which was removed from a large bronze Buddha during restoration is in the collection of the Mary Washington College Galleries, Fredericksburg, Virginia. See also Griswold (1955).
5. Samples for microscopy were mounted in Bioplastic polyester resin, then ground and polished to expose the layers in cross-sections. The samples were examined using reflected, polarized and ultraviolet light microscopy. Elemental analysis of cross-sections was performed by SEM–EDS using a Jeol electron microprobe at the facilities of the Central Facility for Microanalysis of the University of Maryland in College Park.
6. *Experimental setup*
   The samples were analysed via Py-GC/MS using a Pyroprobe 1000 and Pyroprobe 1500 GC interface (CDS Analytical) attached to a Saturn 3 GC/MS (Varian Chromatography Systems) consisting of a 3400 Star CX gas chromatograph and an ultratrace ion trap mass spectrometer. Each sample was placed in a quartz boat and treated with 2 µl of tetramethylammonium hydroxide (TMAH) at 25 wt% in methanol (Aldrich Chemical Company). The quartz boat was then inserted into the resistively heated platinum-coil probe, which in turn was

inserted into the heated 1500 GC interface. The 1500 interface was flushed with helium before, during and for ten seconds after insertion to remove atmospheric oxygen and thus prevent the formation of combustion products. The interface chamber containing the probe and sample were put inline with the flowpath of the GC via a valve on the interface, and the probe coil was heated to 400°C. In the presence of the methylating reagent, the pyrolysis probe functions primarily as a heating device to assist in volatilizing the sample, and does not cause a thermal decomposition of the sample as is the case in true pyrolysis. TMAH methylates carboxylic acid groups via trans-esterification under these conditions. The pyrolysis chamber and GC injector, used in splitless mode, were kept at 320°C. The DB-5 J&W Scientific capillary column (30 m × 0.32 mm) was ramped from 40°C to 200°C at 5°C min^{-1}, and then from 200°C to 305°C at 10°C min^{-1} and held at 305°C for 10 minutes. The transfer line and ion trap manifold were maintained at 260°C. The helium flow rate was approximately 2 ml min^{-1}. The chromatograms for the samples were plotted only to 1800 seconds (30 minutes) because there were no peaks beyond this point.

7. 54.2836, nineteenth century; Gong contained fatty acid methyl esters and a few phenylalkanes indicating that this sample was a mixture of *M. usitata* lacquer and a drying oil.

   54.2694, eighteenth to nineteenth century; Burmese seated Buddha contained fatty acid methyl esters and a few phenylalkanones indicating that this sample was a mixture of *M.usitata* lacquer and a drying oil.

   54.2742, eighteenth to nineteenth century; Lan Na seated Buddha contained large amounts of phenylalkanes, moderate amounts of fatty acid methyl esters, and a small amount of phenylalkanones indicating that this sample was composed primarily of *M. usitata* lacquer with some drying oil present.

References

Araki, T. and Hisayoshi, S. 1978. 'Exhibition lighting and lacquered ware discoloration,'. *Kobunkazi no Kagaku (Science of Old Cultural Properties)*, **23**(11): 1–24.

Burmester, A. 1988. 'Technical studies of Chinese lacquer'. *Urushi: Proceedings of the 1985 Urushi Study Group.* Tokyo: Getty Conservation Institute.

Carriveau, G. W. 1984. 'Technical examination of oriental lacquer'. In *Archaeological Chemistry III*, Joseph B. Lambert (ed.), 395–402. Washington: American Chemical Society.

Du, Y. M., Oshima, R., Yamauchi, Y., Kumanotani, J. and Miyakoshi, T. 1986. 'Long chain phenols from the Burmese lac tree, *Melanorrhoea* usitata'. *Phytochemistry*, **25**: 2211–18.

Du, Y. M. and Oshima, R. 1985. 'Analysis of long-chain phenols in the sap of the Burmese lac tree, *Melanorrhoea usitate*, by capillary gas chromatography'. *Journal of Chromatography*, **318**: 378–83.

Fraser-Lu, S. 1994. *Burmese Crafts: Past and Present*. Kuala Lumpur: Oxford University Press.

Fraser-Lu, S. 1989. *Burmese Lacquerware*. Bangkok: Tamarind Press.

Garner, H. 1979. *Chinese Lacquer*. London: Faber and Faber.

Gettens, R. J. 1969. *The Freer Bronzes*, vol. II , 'Technical studies'. Washington, DC: Smithsonian Institution.

Griswold, A. B. 1974. 'Three inscribed Sukhodayan images in the Bo Tree Monastery'. *Artibus Asiae*, **36**: 65–74.

Griswold, A. B. 1955. 'Medieval Siamese images in the Bo Tree Monastery'. *Artibus Asiae*, **18**: 46–60.

Griswold, A. B. 1954. 'Bronze-casting in Siam'. *Bulletin de l'Ecole Francaise d'Extreme-Orient*, **46**: 639.

Hawthorne, J. G. and Smith C. S. 1963. *On Divers Arts: The Treatise of Theophilus*, book. III. Chicago, Ill.: University of Chicago Press.

Jacobsen, R. D. 1984. 'Inlaid bronzes of pre-imperial China: A classic tradition and its later revivals'. PhD thesis, University of Minnesota.

Kenjo, T. 1988. 'Scientific approach to traditional lacquer art'. *Urushi: Proceedings of the 1985 Urushi Study Group*, Tokyo: Getty Conservation Institute.
Lins, P. A. and Oddy, W. A. 1975. 'The origins of mercury gilding'. *Journal of Archaeological Science* **2**: 365–73.
Mills, J. S. and White, R. 1994. *The Organic Chemistry of Museum Objects*, 2nd edn. Oxford: Butterworth-Heinemann.
Oddy, A. 1984 . 'Gilding through the ages: An outline history of the process in the Old World'. *Gold Bulletin*, **14**: 2, 75–7.
Oddy, W. A. and Cowell, R. 1993. 'The technology of gilded coin forgeries'. In *Metallurgy in Numismatics*, M. M. Archibald and M. R. Crowell (eds), 199–226. London: Royal Numismatic Society, Special Publication 24.
Smithsonian Institution 1961. *Art and Archeology of Vietnam: Asian Crossroad of Cultures*. Washington: Smithsonian Institution.
Harry N. Abrams 1984. *Lacquer: An International History and Illustrated Survey*. New York: Harry N. Abrams.
Strahan, D. K. 1997. 'Bronze casting in Thailand,'. *The Sacred Sculpture of Thailand*. Baltimore: Walters Art Gallery.
Strahan, D. K. 1993. 'The Walters wood-and-lacquer Buddha: A technical study'. *The Journal of the Walters Art Gallery*, **51**: 102–20.
Wangchareontrakul, S.1990. 'Oriental lacquer and its care'. *SPAFA Digest*, **11**(1): 54–7.
Webb, M. 1987. 'Light degradation and care of oriental lacquerware', *AIC Meeting Preprints 1987*, 205–6. Washington: American Institute for Conservation.
Woodward, H. W. 1997. *The Sacred Sculpture of Thailand*. Baltimore: Walters Art Gallery.

12 A Review of Gilding Techniques in Ancient South America

David A. Scott

Abstract

The use of gold for gilding objects in ancient South America is reviewed in detail. The principal methods of gilding included foil gilding, fusion gilding, electrochemical replacement gilding and depletion gilding. Apart from the use of gold foil, most of the technologies that evolved in pre-Hispanic South America have no Old World equivalent. Examples are given of each technique and the evidence that has been accumulated to date for regional and cultural preferences is briefly described. The examination of ancient gilt surfaces and associated materials such as pigments, textile impressions, or burial deposits on the surface are important and should be studied carefully before conservation treatment. The difficulties of interpretation of gilding technology in the absence of metallographic examination are evident from the variety of methods used in the pre-Hispanic period.

Introduction

The surface colour of decorative and votive art was an important aspect of the appearance of an object in both the Old World and the New, despite the very different cultural values placed on gold and gilding. In the New World, gold could be traded for other commodities such as salt or fish, but it never assumed the monetary significance that it acquired in the Old World and, in many different cultural settings it fulfilled an important religious, symbolic or shamanistic purpose. Ancient South American cultures did not exploit the extraction of mercury from its ores, so there is an absence of mercury amalgam gilding throughout the continent. Consequently the development of gilding technology followed a very different path and many unusual developments occurred in the gilding of metals that were unique to South America, or that showed a sophistication in their application that did not occur in the Old World.

The earliest evidence for gold working in the New World is that discovered by Grossman (1972) who describes the find of hammered pieces of gold foil and a gold worker's toolkit, dated to the second millenium BC from the site of Waywaka in the southern Andes of Peru. In terms of archaeological chronology, this places the find in the Initial Period (Table 12.1). This is the only evidence that has come to light from such a relatively early time, and nothing more is known until artistically developed goldwork from the Chavín cultural area in northern Peru was discovered

Table 12.1 Chronology of some cultural areas mentioned.

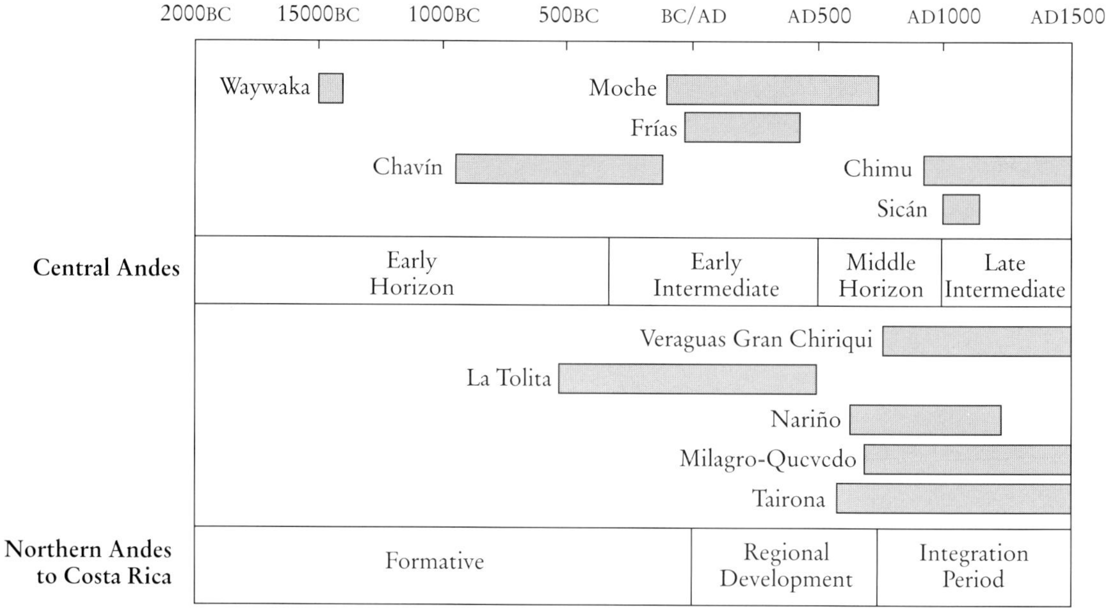

in burials in the Lambayeque valley dating to the early centuries BC, corresponding to the Early Horizon (Lothrop 1942).

An early find of exceptionally delicate gold wire from Colombia, dating to the fourth century BC, (Bouchard 1979, Scott 1991), indicates that knowledge of the skilful working of precious metals had travelled northwards from Peru and had already reached Colombia by at least this period. By the early centuries AD, knowledge of lost-wax casting had spread to Panama and Costa Rica, and by AD1000, Mixtec craftsmen were producing their own gold objects.

Within the context of gilding technologies, our knowledge of metallurgical practice and cultural choice is severely limited by the paucity of goldwork from datable excavations, and we still have a long way to go before more definitive statements can be made regarding some of these developments.

Foil gilding

Once the ability of native gold nuggets to be hammered into thin sheets has been mastered, the most obvious method of gilding is simply to cover the surface of the object in sheet gold, a technique that could be used to cover almost any substrate, the most common being wood, stone, leather or another metal. The attachment of gold foil can be achieved either by mechanical methods, usually involving roughening of the surface of the object to be gilded, followed by burnishing to adhere the gold to the substrate, or by metallurgical methods, such as soldering, sweat welding or diffusion bonding. In the Old World, the technique developed from the earliest known form of gilding, namely wrapping an object in gold foil, which can be dated from about 3000BC (Oddy 1982).

In some of their accounts of the Conquest of Peru during the fifteenth century AD, the Conquistadors noted that entire walls of buildings were covered in gold, showing that very large areas could be covered in gold foil. Very few objects made using foil gilding have been described in detail, but enough work has been done to show that foil gilding was used from Peru to Costa Rica: use was made of a complete gold cladding in one sheet, and

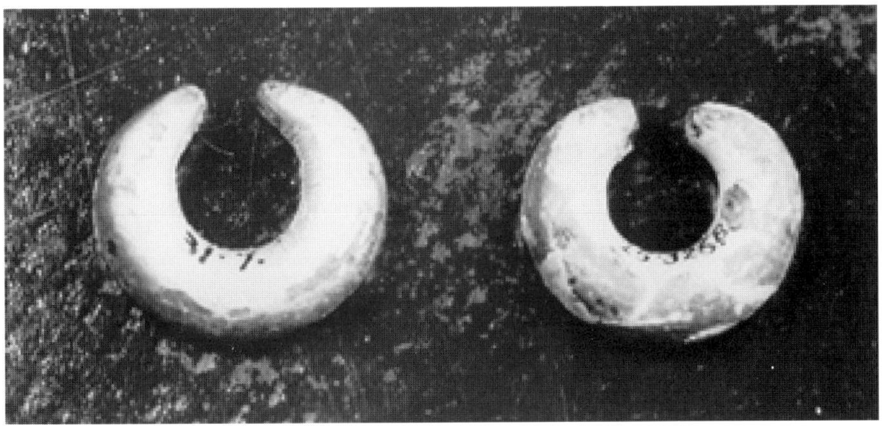

Figure 12.1 Two penannular nose rings from the area of Chanduy, Provincia de Guayas, Ecuador. The gilding, partially worn, has been made by wrapping gold foil over the copper. (Cambridge University Museum of Archaeology and Ethnology. 19 mm and 17.5 mm across.)
(See Plate 59.)

of small squares of individual gold sheet that could be partially overlapped. Macroscopic evidence for foil gilding in Ecuador has been mentioned by Saville (1907–1910), Bushnell (1951), Christensen (1954) and Ubelaker (1981), while foil-gilded pins with elaborate heads are known from Pashash, Peru dating to the early centuries AD (Grieder 1978).

In the collections of the Cambridge University Museum of Archaeology and Anthropology are two small penannular nose rings of circular cross-section, one from Punta Mambra and the other from Real, both near Chanduy, Provincía de Guayas (Bushnell 1951, Scott 1986b). Both of the nose rings appear to have been gilt by wrapping in a continuous sheet of foil, and these are shown in Figure 12.1, where some of the gilding can be seen to have worn away. Evans and Meggers (1955) excavated a rich multiple urn burial of the Milagro-Quevedo phase. The site is called La Compãnía ('Cacique Guayas'), in Los Ríos Province, Guayas basin, Ecuador. It is from the Integration Period (*c.* AD850–1532), although Hosler *et al.* (1990), note that the presence of thin copper alloy socketed feathers may indicate a date between AD850 and AD1200 for the burial itself. Several gilded objects were excavated, including a pennanular nose ring. Clear evidence for foil gilding of this nose ring can be seen in the metallographic section, reproduced in Figure 12.2. In some areas of this section, three distinct layers of foil can be observed as well as some folding of individual sheets; gold foil thickness varies in the range 0.09–0.035 mm (90–35 microns), and the composition of the foil employed is shown in Table 12.2. In the gold layer contiguous with the copper some arsenic was detected and analyses showed that the copper nose ring contains about 0.5 per cent arsenic, proving that it was made from an arsenical copper. The

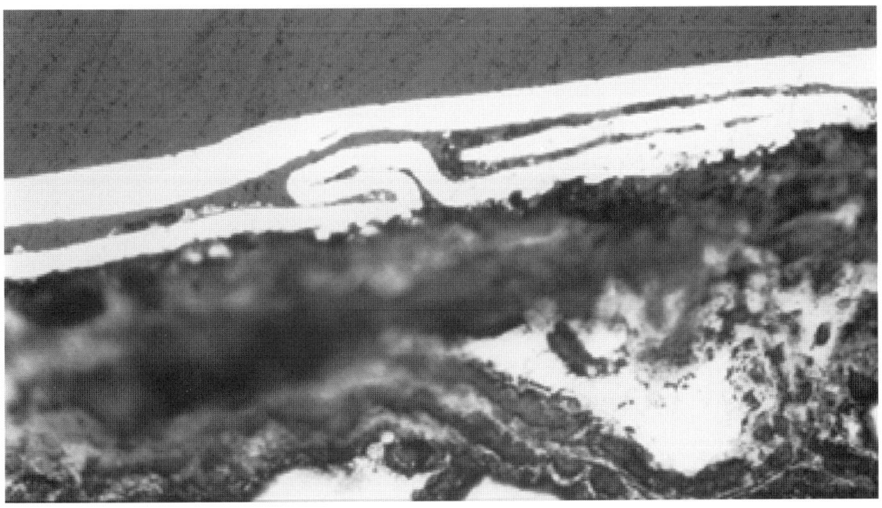

Figure 12.2 Foil-gilded layers over a pennanular nose ornament from La Compania, Los Rio Province, Milagro phase, Ecuador. The nose ring has been gilded by diffusion bonding of gold foil. Overlapping layers of gilding can be seen in this photomicrograph, together with a fold in the lower layer. In some areas interdiffusion is extensive and only one, apparently thicker layer of gilding can be seen. (Unetched, ×210.)
(See Plate 60.)

Table 12.2 Analytical data for objects discussed (wt %)

Object	Culture	Area analysed	Gold	Copper	Silver	Arsenic	Gilding method
Monkey figurine	Moche culture	outer gold surface	82	3	14.4	nd	native gold alloy: not gilded
Monkey figurine	Moche culture	surface of teeth	61	32	7	nd	fusion gilded surface
Monkey figurine	Moche culture	teeth substrate	nd	98	nd	nd	copper substrate
Decorativc sheet	Nariño area	surface (section)	59.3	32.7	8.2	nd	fusion gilded surface
Decorative sheet	Nariño area	interior (section)	nd	98.3	nd	nd	copper substrate
Nose ornament	Frias area	gilded surface	51.1	41.9	5.9	nd	fusion gilded surface
Nose ornament	frias area	silver substrate	5.5	5.5	89.0	nd	silver alloy substrate
Nose ring	Milagro-Quevedo	gilt surface	85.7	5.4	10.5	nd	foil gilding
Nose ring	Milagro-Quevedo	interior alloy	nd	96.8	nd	0.5	substrate alloy
Plaque	Moche culture	surface zone	86	4	8	nd	electrochemically plated
Plaque	Moche culture	alloy interior	2.4	95	0.1	0.2	copper substrate
Male figurine	Veraguas-Chiriqui	surface zone	78	18	3.5	nd	depletion gilded surface
Male figurine	Veraguas-Chiriqui	interior	54	30	2.5	nd	alloy interior
Disc fragment	Nariño area	surface zone	78.4	16.5	5.8	nd	depletion gilded surface
Disc fragment	Nariño area	interior	24.9	65.6	4.1	nd	alloy interior

nd: not detected. Analytical data is quoted to different significant figures since analysed on different instruments, either by electron microprobe analysis or enery dispersive secondary target x-ray fluorescence spectrometry.
Monkey figurine: private collection. Decorative sheet: lab. no. AI89 (Scott 1983a), Museo del Oro, Bogotá. Nose ornament: private collection. Nose ring: lab. no. CXXI (Scott 1986a), La Compañía, Los Ríos Province. Plaque: private collection. Male figurine: private collection. Disc fragment: lab. no. AI149 (Scott 1983a), Museo del Oro, Bogotá.

diffusion of arsenic into the gold foil, and of gold into the copper substrate, shows that diffusion bonded gold foil was used as a covering (Scott 1986b). The earliest example of foil gilding so far examined by the author is a silver and gold nose ornament, Frías in style, from Peru, dated to AD200–500, embellished with eight silver repoussé bird-heads with the eye of each inlaid with a shaped fragment of turquoise. The ornament was originally described as bimetallic, and on a cursory inspection it is impossible to decide how it has been made because it is so skilfully finished: the gold surface on both sides is highly burnished and appears continuous until the junction with the silver bird-heads is reached. Detailed surface examination found a few small areas of gold on the reverse, which could be carefully lifted with a scalpel, and under the microscope the substrate could be seen to be silver, showing that the nose ornament was made in hammered silver, finished by foil gilding with a gold layer about 100–120 microns thick. The nose ornament is illustrated in Figure 12.3.

Analytical data are given in Table 12.2, and show that the gold foil contains about 80 wt % gold, 17 wt % silver and 1 wt % copper, while the silver

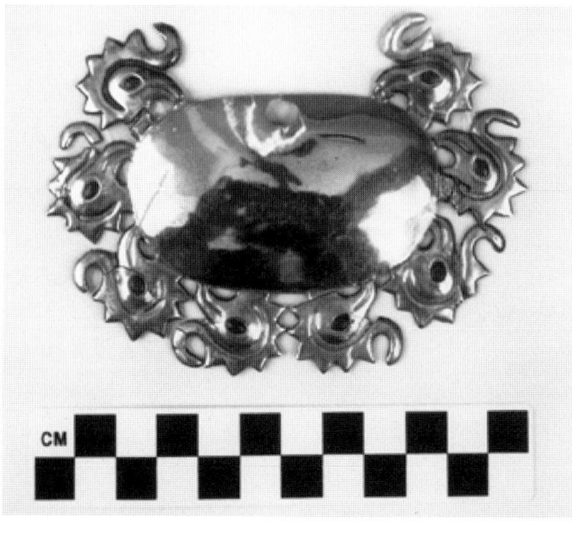

Figure 12.3 Bicoloured nose ornament, Frías in style, Peru, dated to AD200–500. Eight silver crested bird-heads are arranged symmetrically around a gilt ovoid nose ornament, which has been made by foil gilding the silver substrate of which the whole ornament is made. The eye of each bird is inlaid with turquoise. (Private collection.) (See Plate 61.)

alloy has a composition of 89 wt % silver, 5.5 wt % copper and 5.5 wt % gold. The unusual composition of the silver indicates that some alloying with copper and gold was employed to create a ternary silver–copper–gold alloy, which possesses a silvery colour. This alloy is harder than pure silver and there may have been some practical advantages to using a harder alloy for the nose ornament; alternatively, there may have been symbolic or shamanistic reasons for including some gold in the alloy, as well as applying foil gilding. In the examination of Sicán metalwork excavated from a twelve metre deep shaft tomb at the site of Huaca Loro, in the La Leche valley of northern Peru, Merkel *et al.* (1995), mention the existence of metallic wrappings applied over ceramic vessels dated to AD900–1100, corresponding to the Middle Sicán period. Merkel shows that some of these gold sheet wrappings are very thin, about 10 microns across, and are composed of gold silver copper alloys of composition 17–26 wt % gold, 15–16 wt % silver and 16–43 wt % copper, with a surface enriched in gold over about one micron, probably through working and annealing rather than depletion gilding as such. The use of these gold foils at Huaca Loro also extended to gilding wooden objects, such as darts.

Foil gilding over copper was occasionally used in Colombia, sometimes for small nose-ornaments, often in the form of twisted nails. In Panama and Costa Rica, penannular nose rings, made in carved stone or copper, have been found with gold foil coverings (Bray 1993). Foil gilding does not appear to be restricted in a geographical or temporal sense; why this should be, when other gilding techniques were also in use, remains unknown. From the very limited work that has been carried out it is presently impossible to link material choices within cultural areas to the use of foil gilding.

Depletion gilding

In one of his rare misunderstandings, Rutherford Gettens described some of the gold–copper alloy objects from Sitio Conte, Panama, as being foil-gilded (Gettens and Mooradian 1937), but they are not. They are examples of the most pervasive of South American gilding systems, which we call surface enrichment, mise-en-coleur, or depletion gilding. The latter term (Lechtman 1973) is logical since it describes a deliberate attempt to change the surface colour of an object, while surface enrichment may be associated with leaching of elemental components during burial. Alloys of gold and copper, or of gold, copper and silver, may exhibit a whole range of colours from red through green, white and yellow, as Figure 12.4 illustrates. The range of colours can be compared with compositional data shown in Figure 12.5.

If these gold–copper–silver alloys are treated chemically to remove some of the copper, or copper and silver, from the surface, the alloy may appear to be made in relatively pure gold, while the interior colour and composition of the bulk alloy will be very different from the specially treated surface. These chemical techniques of surface manipulation were certainly used by the Moche, who lived along the north coast of Peru from about 100BC–AD800 (Lechtman *et al.* 1982), and were employed by many other cultural groups such as the Chimu and Sicán in Peru, and throughout the many cultures of Colombia, Panama and Costa Rica, which shared a common metallurgical tradition before the Spanish Conquest; all made deliberate use of alloys of gold and copper. These alloys are frequently referred to as tumbaga alloys.

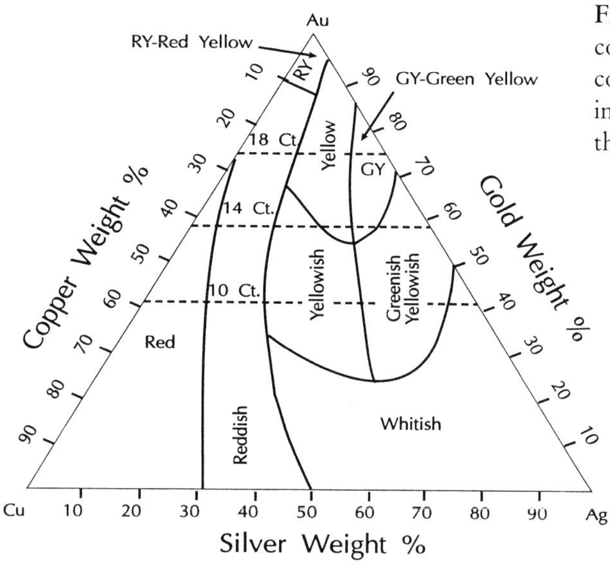

Figure 12.4 Ternary gold–copper–silver diagram showing colour fields of varying compositions. Most of the gold–copper alloys are of reddish hue; surface treatment resulting in copper or silver removal shifts the colour of the alloy to the gold region of the diagram.

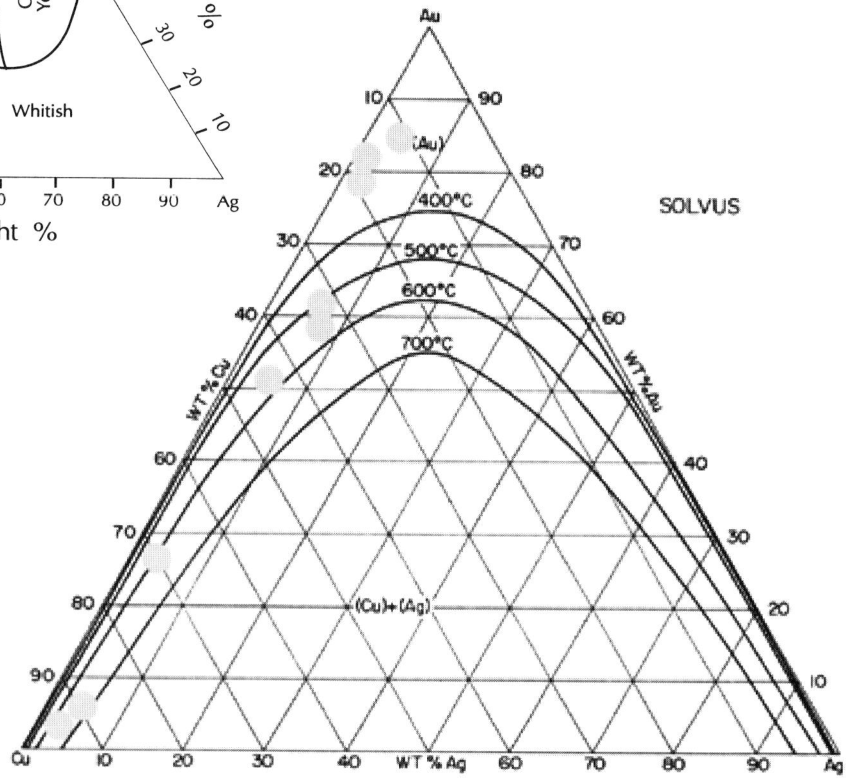

Figure 12.5 Ternary gold–copper–silver solidus diagram showing location of analysed compositions tabulated in Table 12.2.

Because of the extraction of silver and the extensive use of silver–copper alloys in ancient Peru, many of the gold objects are made in ternary alloys of gold–copper–silver, with deliberate additions of both copper and silver. This is a different situation to that found in the ancient Colombian area, where smelted silver was not available, and the only silver content of the tumbaga objects is derived from the native gold of the area (Lechtman 1973, Scott 1982, Rovira 1994). Generally the copper content of native gold in the region is low, usually less than 2.5 per cent, while the percentage of silver may vary widely from less than 1 per cent to about 30 per cent.

Lechtman (1973) examined some Peruvian objects from the Chimu and Moche cultures, which were finished by a depletion gilding process, in which both copper and silver were removed from the first few microns of the outer surfaces of the finished objects. Smith (1977) and Lechtman (1973) have suggested that corrosive mineral preparations employing basic ferric sulphates, such as copiapite, $Fe_3(SO_4)_4(OH).13H_2O$ and salt, NaCl, could well have been used to make a cementation mixture capable of removing both silver and copper from the surface, and Lechtman has demonstrated that this mixture works very effectively on gold–silver–copper alloys of compositions used by the Chimu culture. In the experimental work Lechtman showed that placing the object to be gilt in the mixture of salts with an inert filler such as brick dust and heating to temperatures in the range of 600–750°C, would result in removal of both copper and silver from the surface zone. After burnishing, the treated surface would look like very high grade gold, over an alloy that could contain substantial amounts of both copper and silver. Analogous processes for the treatment of gold alloys were known in the Old World, for example Pliny the Elder, writing of Roman practice states that 'the gold is heated with twice its weight of salt and thrice its weight of misy, and again with two portions of salt and one of schist' (Rackham 1968).

Similar recipes are known from the Leiden papyrus no. X, referring to Hellenistic practices of the third century AD (Caley 1926), but despite this Ogden (1993) notes that having examined thousands of ancient gold objects microscopically, no certain indications of depletion gilding have been found. The technique may have been used to purify gold in the Old World before use, but was perhaps seldom used on finished objects. The situation in ancient South America is completely the opposite: depletion gilding can be demonstrated to have been used often, and in some cultures, universally. Incidentally there appears to be a reluctance among Old World scholars to make use of the term 'depletion gilding', and many references refer to natural or artificial enrichment.

This is unfortunate, since there is a surprising similarity in the chemical techniques used in the Old World and New World to deplete the copper and silver contents of gold alloys: the technique is chemically the same, but different nomenclatures continue to be used. Indeed, Percy (1880), even identifies the Pliny recipe as referring to the same substance suggested by Lechtman, namely native yellow copperas, in European terms, or copiapite. Important historical information relating to the New World has been preserved in the writings of some of the early Spanish explorers or Conquistadors, for example Bernardino de Sahagun, writing in 1565 of Aztec technology (Sahagun 1565), given in translation in Bray (1978) recited that:

> . . . it was in addition treated with alum; the alum with which the gold was washed was ground. A second time it entered the fire; it was heated over it. And when it came forth once more, for the second time, it was once washed, rubbed with what was called 'gold medicine'. It was just like yellow earth mixed with a little salt; with this the gold was perfected; with this it became very yellow.

The use of natural mineral resources in the surface treatment of gold alloys is not the only method for achieving removal of copper from the surface of tumbaga objects. In areas of Colombia, where there is a profusion of natural plant materials, the Indians made use of acidic plant extracts to pickle the surface of their tumbaga alloys and to produce very effective depletion gilded layers.

One of the most interesting accounts of surface treatment of gold alloys employing plant materials, is from a document describing indigenous working of metals in Tamalameque, lowland Colombia, (Tamalameque 1555: quoted in Bray (1985)):

> . . . and in this way, placing it (a bracelet) in the fire, taking it out and putting it in water and hammering it on an anvil with the stones described, they worked until they had increased its size many times until it was finished and then the herb they brought to give it colour was crushed on a stone and, once crushed in this way, they placed it in a small pot which they brought in and added water and ground white salt and stirred all together (then they polished, heated and quenched

it in the solution several times) . . . and in this way it attained the colour and finish it should have . . . and they cleaned the said manylla (bracelet) with small quantities of fine sand that they brought in a maize husk, with their hands and water.

Plants of the oxalis family may have been used for these surface treatments, the common active constituents being 'salts of lemon', tartaric acid (HOOC·CH(OH)·CH(OH)·COOH), acidic potassium oxalate (KHC_2O_4), potassium quadroxalate ($K_3HC_4O_8 \cdot 2H_2O$), oxalic acid ($HOOC \cdot COOH \cdot 2H_2O$), and tannic acids, which are complex combinations of glucose and gallic acids. The Tamalameque document mentions that the object to be gilded was heated and quenched in the solution several times. Not only would heating result in the formation of copper oxide scale, but such scale is formed naturally in the working and annealing of tumbaga objects. The weak plant acids would be much more effective in complexing with copper corrosion products than copper metal itself. In order to study the efficacy of some simple organic acids in producing a gilded layer, a ternary alloy of 40 per cent gold, 55 per cent copper and 5 per cent silver (Scott 1983a), was cast and worked, followed by surface oxidation and pickling in a variety of organic acid solutions. The results of this work showed that a mixture of sodium chloride and oxalic acid was very effective in creating a depletion gilded surface, resulting in a gold enriched layer 3–5 microns thick. If we assume that weak plant acids were the active ingredients of the herbs mentioned in the Tamalameque document, then the admixture of these with common salt is indeed a very workable recipe. The kind of depletion gilded surface produced by this method closely resembles that found on many pre-Columbian objects from Ecuador, Colombia, Panama and Costa Rica, in which there is a natural abundance of plant material: ethnographic evidence for the use of *Oxalis decipiens* and *Oxalis martiana* in Brazil has been mentioned by Peckholt (1899), while Rivet & Arsandaux (1946) cite the use of *Oxalis pubescens* (Chulco) in Ecuador.

Surface oxidation of copper followed by leaching with plant juices, results in a surface zone composed of an island and channel morphology (Forty & Durkin 1980), which is characteristic for many depletion gilded tumbaga alloy surfaces: if unburnished the fine structure can be clearly seen in the polished microstructure or in the SEM. Figure 12.6 illustrates a depletion gilded male figurine from Costa Rica, dated to the Veraguas-Gran Chiriqui period, approximately AD700–1500. The figure may be earlier, since Bray (1992) dates some of these male figures with flexed knees to the period AD400–700. The alloy pendant is cast by the lost-wax process and the duality of the figure is shown by the animal snout and spiral ear ornaments and head-dress. The cast structure of some of these pendants is represented by the cross-section through part of a tumbaga alloy frog pendant from the Tairona region of ancient Colombia, which shows the coring in the segregated dendritic zones resulting in gold-rich and copper-rich areas. The depletion gilded surface is towards the top of the photomicrograph and represents a relatively modest enrichment in gold content of about 15 wt %, in this particular case. Figure 12.7 shows one view of the gold–copper–silver ternary solidus, which is useful in describing the compositional range, and solidus temperatures for the alloys discussed in this paper.

A representative electron microprobe line scan through a depletion gilded surface layer into the alloy substrate is shown for a Nariño example

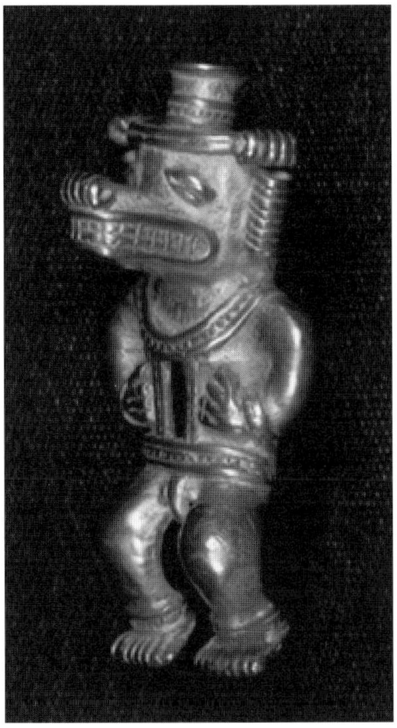

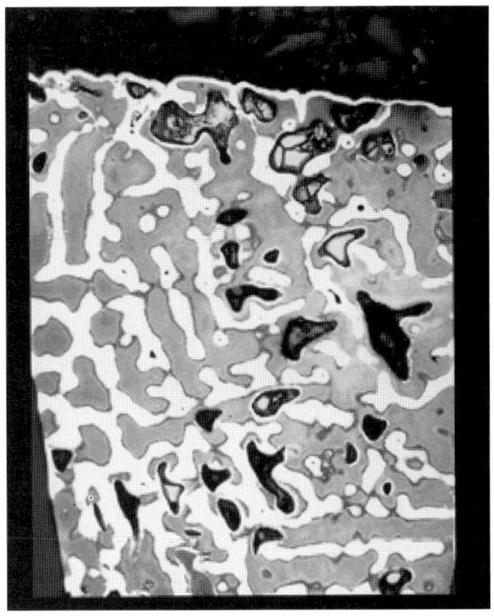

Figure 12.7 Cast tumbaga alloy from the Tairona area of pre-Hispanic Colombia, showing internal corrosion of copper-rich areas with lighter regions preserving gold-rich dendritic regions where the heavily cored structure preserves considerable segregation from casting. The upper surface is depletion gilded. (Unetched, ×85.) (See Plate 63.)

Figure 12.6 Tumbaga alloy pendant in the form of a human figurine with animal duality, from Costa Rica. The figurine is from the Veraguas-Gran Chiriqui area, and is dated to AD700–1500. Cast in a gold–copper alloy with 54 wt % gold, the surfaces are finished by depletion gilding. (Height 9.6 cm.) (See Plate 62.)

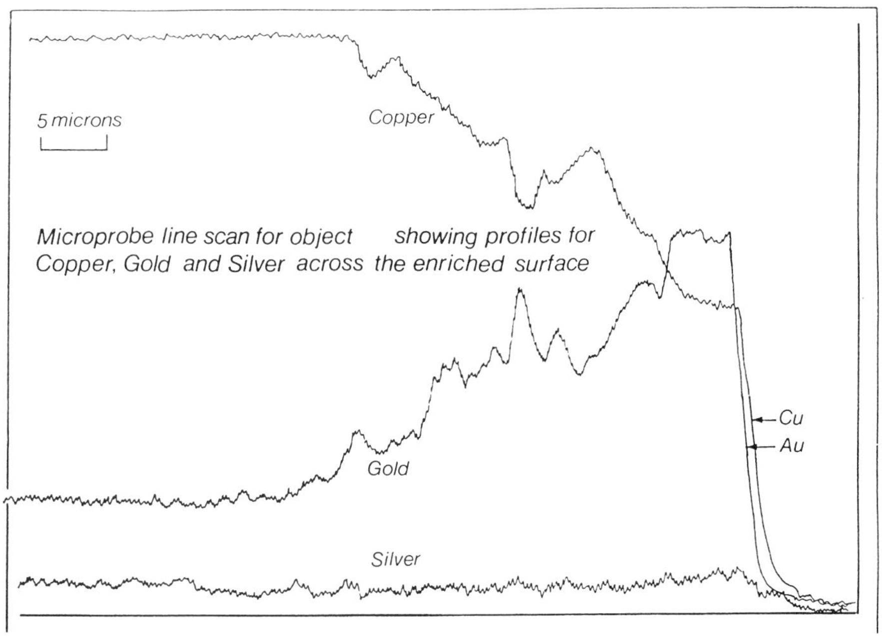

Figure 12.8 Electron microprobe line scan across a polished section of a depletion gilded sheet from the Nariño area of Colombia, Piartal Period, dated to AD800–1200. Notice that the depletion gilded surface is about five microns thick, but that in this example there is a disturbance in the gold concentration for about 30 microns until the interior composition is reached. The gold concentration in the interior is about 25 wt %, which rises to about 75 wt % at the depletion gilded surface.

in Figure 12.8. The disturbance in gold and copper concentrations extends for a depth of about 30 microns into the alloy substrate, while the actual depletion gilded surface is five microns across.

One of the remarkable uses of depletion gilding for contrasting surface colour design occurred in the Nariño area of ancient Colombia around the period from AD800–1200. An example is illustrated in Figure 12.9, a multicoloured disc, 15.3 cm in diameter from the Municipio of El Tambo, Provincia de Nariño, weighing 68.00 g (Museo del Oro number 21521: Plazas de Nieto 1979). Originally there may have been three different surface finishes visible on this disc: a burnished central gold disc is surrounded by a copper-coloured zone and an outer border of copper and

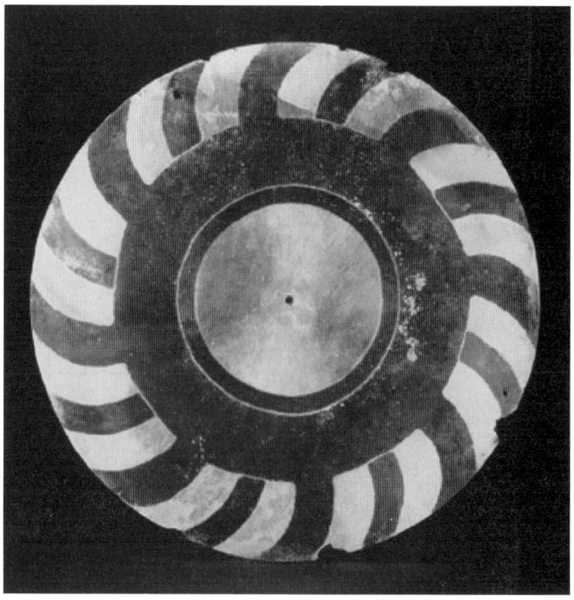

Figure 12.9 Bicoloured flat tumbaga disc from the Municipio of El Tambo, Department of Nariño, Colombia. The disc is 15.3 cm in diameter, and weighs 68 g. It is from the Piartal Period, dated AD800–1200. The tumbaga alloy, containing about 25 wt % gold, has been carefully smoothed and selectively finished by depletion gilding to create skilfully designed geometric colour contrast on both sides of the disc. These discs are typically 0.30–0.35 mm thick. Internal corrosion makes them very brittle. (Museo del Oro: Bogotá: 21,521.) (See Plate 64.)

gold panels, all produced on a hammered tumbaga disc smoothed very skillfully to remove hammer marks, producing an even, highly reflective surface. A number of techniques were used in ancient Colombia to achieve surface colour contrast through depletion gilding:

- selective abrasion of the depletion gilded surface to expose the tumbaga alloy beneath – if the gold content of this tumbaga is low, it appears coppery in colour;
- selective depletion gilding over part of the surface only;
- selective burnishing of the depletion gilded surface, producing a polished gold contrasted with matte gold finish;
- selective etching of the burnished surface to produce matte gold regions and burnished gold surround.

We do not have space to discuss all of these potential variations on the depletion gilding technique, although they are discussed in Scott (1983a), and as Figure 12.9 illustrates they were very skilfully employed in the Nariño area of Colombia. Corrosion during burial results in extensive oxidation of the exposed copper-red surfaces of these tumbaga alloys, especially at gold concentrations from 5 wt % to 30 wt %, making the colour almost black. The depletion gilded surface is selectively preserved, because the gold content is much higher, typically 70–85 wt % gold, and this part of the design retains its original finish.

The gold content of these Nariño tumbaga alloys varies from 15 wt % to 25 wt %, which is quite sufficient to allow the manipulation of surface colour by these subtle and attractive depletion gilding techniques. The de-

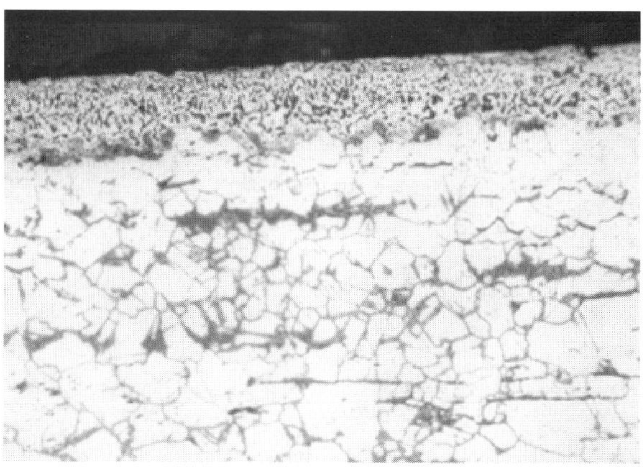

Figure 12.10 Photomicrograph showing depletion gilded layer on a Nariño sheet tumbaga. Note the development of island-and-channel morphology in the disturbed surface zone where copper removal has occurred during surface depletion. Some oxide stringers can be seen in the section, along with recrystallised grains of the gold–copper alloy sheet. (Magnification ×125, etched in potassium cyanide and ammonium persulphate for 15 s. Bright field illumination.) (See Plate 65.)

velopment of an extended island-and-channel morphology in the depletion gilded surface can be seen in Figure 12.10, where a Nariño sheetmetal plaque is shown in cross-section. The development of the island and channel morphology results from the movement of gold atoms to form stable zones in the solid, while the acidic solutions can continue to attack the bulk alloy: this morphology is very commonly observed in depletion gilded surfaces. Some conversion to cuprite is occurring in intergranular locations in the recrystallised and twinned tumbaga alloy, whilst the depletion gilded surface is well developed and clearly visible.

Fusion gilding

A few of these Nariño gold sheets proved to have been gilded by quite a different method, which proved to be something of a surprise. Many years earlier, Paul Bergsøe (1937) had drawn attention to gilded copper objects from Ecuador, and his research showed that these were indeed made of copper, with no gold content, which were gilt by dipping in a gold–copper alloy.

I prefer to call this method fusion gilding (Scott 1986a), and examples of this technology have been identified from Peru, Ecuador and Colombia. The fusion gilding process involves the application of a molten alloy to the surface of the shaped copper substrate. The fused alloy may be a binary gold–copper or silver–copper alloy with a lower melting point than pure copper. If the ornament is to be coated on all surfaces it could be dipped into a crucible of molten gold alloy, or if only one surface is to be gilt, the coating could be flushed over the surface and finished by hammering and annealing to smooth it out. The method was probably invented in Peru, during the Moche period, perhaps around the period AD100–300, and was certainly in use in the La Tolita area of northern Ecuador some time between 300BC and AD800. The examples from the Nariño area show that the technique lasted until at least AD1200. The reasons for the use of fusion gilding, as opposed to other techniques, over a period of 1,500 years remain unknown. Until many more examples from datable archaeological contexts are found, this situation is unlikely to change.

Perhaps fusion gilding was associated with certain symbolic or shamanistic concepts that were linked to immersion or transformation. Certainly the copper object is transformed during the process, and emerges dripping in gold. Laboratory studies of a gold sheet from Nariño, dating from some time between the 9th and 12th century AD, showed that it consisted of a worked and annealed copper base coated on both sides with a gold alloy of composition 51.1 wt % gold, 5.9 wt % silver and 41.9 wt % copper, about 100 microns thick. The photomicrograph of Figure 12.11 shows a cross-section through the gilt copper sheet. An electron microprobe scan

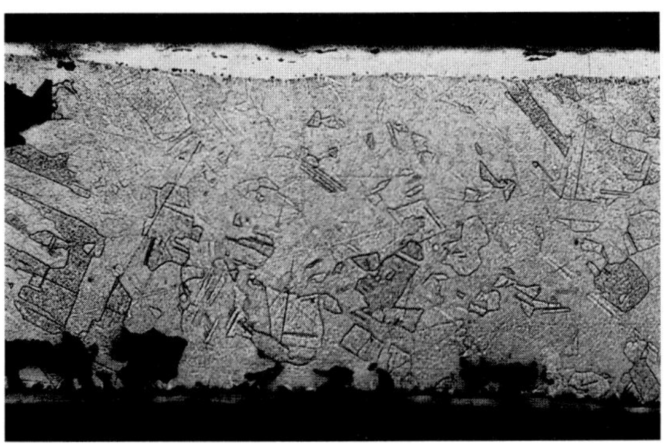

Figure 12.11 Fusion gilded gold alloy surfaces over copper sheet from Nariño, Colombia, dated to the Piartal Period, AD800–1200. The sheet is about 0.75–0.90 mm thick, and the etched section shows that both sides of the copper sheet have been coated with gold. Note the recrystallised grains of the copper substrate and the presence of slightly flattened holes between the gold layer and the copper. Before etching, these were small cuprited inclusions. The fusion gilded coating has been made from a prepared tumbaga alloy of 59 wt % gold, 8 wt % silver and 33 wt % copper. (Etched in potassium cyanide and ammonium persulphate. Magnification ×75.)

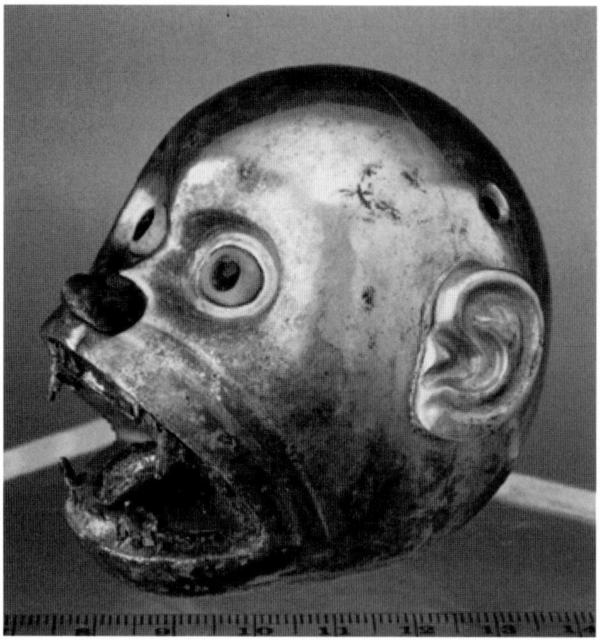

Figure 12.12 Monkey figurine, Moche Period, Peru. Sipan in style, probably dating to the early centuries AD. The monkey is made in a gold alloy, whose composition is given in Table 12.2, but the long teeth are made in copper that has been covered with gold by the fusion gilding technique. The nose and eyes are made in stones that have not been analysed. The monkey has been carefully made in a number of pieces that are joined together with soldering. A join can just be seen in the head, passing across the top of the skull. (Private collection.) (See Plate 66.)

showed the interior to be almost pure copper, with a zone of interdiffusion between the copper substrate and the fusion gilding coating of about 15 microns. This fusion gilded surface had itself been finished by depletion gilding, removing some of the copper from the gold alloy and producing a composition of about 72 wt % gold, 8 wt % silver and 20 wt % copper at the outer surface. In fact, many fusion gilded surfaces are themselves finished by depletion gilding, creating a series of compositional changes through the thickness of the object of some complexity. A Moche gold alloy monkey figurine from Peru, in Sipán style (Donnan 1995) is shown in Figure 12.12. The monkey itself is made in a hammered gold alloy, while the inset teeth are made of copper which has been gilt by the fusion gilding process. Even during the Moche period, it appears that many different gilding technologies were already in use and that the knowledge of their application probably spread from Peru northwards.

Study of a fusion-silvered copper nose ornament from Ecuador illustrates that silver, as well as gold, could be applied by the fusion technique, and it was possible by metallographic study to show that the silver coating had been applied in a molten state and smoothed by working and annealing, followed by depletion silvering as a final finishing operation (Scott 1986b).

Electrochemical replacement plating

Undoubtedly the most extraordinary discovery of the ancient Indian metalsmiths in their search for surface alteration and colouration, was the development of plating techniques, long before the advent of the European battery. Heather Lechtman (1973, 1979; Lechtman *et al.* 1982) has conclusively demonstrated that it is possible, without using mineral acids or external electric current, that gold alloys or silver alloys can be dissolved and plated onto clean copper by immersion of the copper into the plating solution, or by applying a paste to one surface of the copper object. The initial work was carried out on Moche-Vicus masks and pectorals from the site of Loma Negra on the north coast of Peru, near the Ecuadorian border. Since then other objects from Peru and Ecuador have been shown to be gilt by this technique. The techniques of electrochemical replacement and fusion gilding are fundamentally different, although both may produce gold–silver alloy surfaces. In the case of the fusion gilding technique, the microstructure can clearly be studied in cross-section and features typical of cast dendritic structures can be seen, even if the coating has subsequently been deformed by some working and annealing to smooth the surface.

The characteristics of these plated coatings are: they are very thin, typically 0.5–2 microns; they coat all surfaces and edges of the object; no mercury is detected anywhere; small surface declivities are coated with the same thin finish; and there is a diffusion zone between the plating and the substrate.

It is not possible to achieve these surface features without postulating that the gold or silver was applied from solution, which, inter alia, means that dissolution of gold must be achieved with the resources available to pre-Hispanic cultures. Lechtman (1979) was able to show that gold can be obtained in solution by heating gently in an aqueous solution of sodium chloride, potassium nitrate and potassium aluminum sulphate, using a 1:1:1 ratio and 10 g of each salt dissolved in 100 ml of water. To simulate the dissolution of a native gold, 1 g of gold and 0.15 g of silver were added to the solution. After the gold enters solution there is a practical problem: the pH of the resulting solution is so low that a copper surface will be attacked during immersion and partially dissolved, and no plating will occur. Lechtman used sodium bicarbonate additions to raise the pH of the slurry to about 9. Following this essential step, sheet copper can be gilt by immersion in the simmering slurry for about 15 min, or alternatively the slurry can be swabbed over the surface, diluted 1:3 with distilled water and maintained at pH 9. By keeping the surface wet and applying some 20–30 coats of the solution, a film of gold alloy can be deposited on one side of the object. The film is not adherent however, and Lechtman completed the process by furnace annealing in air at 850°C for 5–7 s. Any copper oxide scale formed during the annealing operation was dissolved by pickling, utilising the same salt mixture. This is plating by electrochemical replacement: copper represents a less noble end of the electromotive series compared with more noble elements, such as gold or silver, and if copper metal is dipped into the gold solution, a simple replacement occurs. Some of the copper is taken into solution and gold is simultaneously deposited onto the copper surface. In practice this deposition is dependent on different parts of the copper surface providing both anodic and cathodic zones, to enable the electrochemical exchange to occur. Anodic pits surrounded by larger cathodic deposition regions should characterise the microstructural features of replacement plated gold objects, and Lechtman provides one or two examples of these anodic pore structures (Lechtman 1979: Fig. 15, Fig. 26). It is not easy to study these electrochemical replacement gilded objects, since most of them have suffered corrosion of the copper substrate, resulting in loss of interfacial information between the very thin gold coating and the copper substrate.

There are, of course, other possibilities for the dissolution of gold, without using strong mineral acids, and Butler *et al.* (1987) show that ancient Chinese recipes may involve the production of gold halide complexes by reaction of gold or silver with different minerals contained in sealed bamboo tubes. The difficult logistical step in the plating process as reconstructed by Lechtman, is the necessity to raise the pH before plating will occur on the cleaned copper surfaces. Nevertheless, there have been no further suggestions to account for the observed features of this

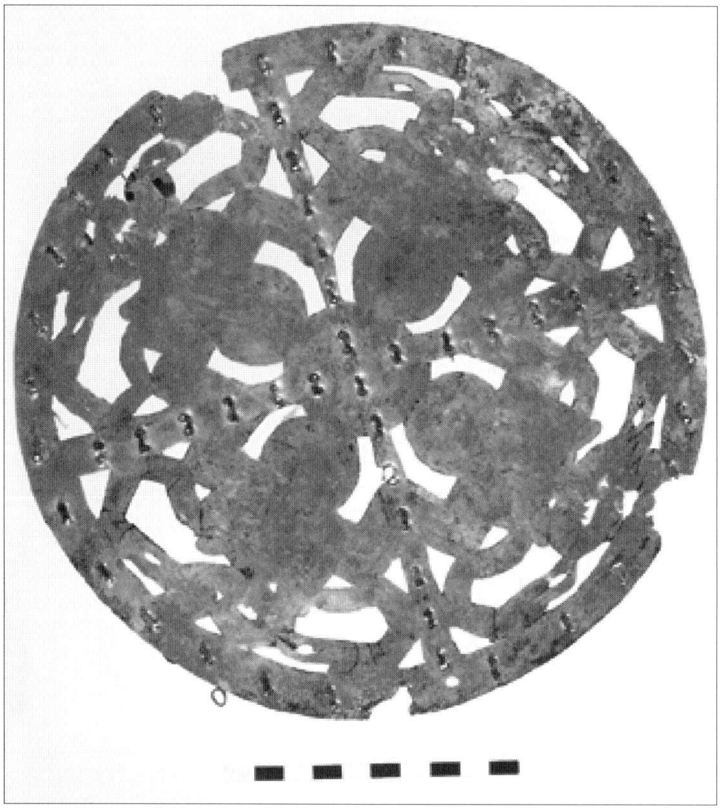

Figure 12.13 Large profile ornamental disc, Vicus in style, from the north coast of Peru. The disc represents four crab motifs. The flat sheet, about 1 mm thick, has been worked and annealed to shape, followed by the cutting out of the design, and gilding by the electrochemical replacement plating process. (Diameter 23.4 cm; gilded layer about 1 micron thick.) (See Plate 67.)

remarkable technology published since Lechtmans' seminal papers. An example of a circular profile disc, gilt by electrochemical replacement plating is shown in Figure 12.13. The worked and annealed sheet was hammered out to 1 mm in thickness, before cutting the design, and insertion of small staples designed to hold gilded dangles suspended close to the surface, followed by surface cleaning and replacement gilding. These discs with crab and scorpion motifs are common in the Vicus style of northern Peru. The composition of the underlying copper alloy used to make the disc was determined to be 93 wt % copper, 2.5 wt % gold and 0.3 wt % silver (Erlij 1974).

A photomicrograph of a small dangle from the Vicus period is shown in Figure 12.14. The fragment is extensively corroded and the gold layer survives in a pseudomorphic mass of corrosion products. The photomicrograph shows that the gold does indeed coat all surfaces, including the edge of the dangle, which is about 0.5 mm thick. The use of this extraordinary knowledge of electrochemical replacement plating seems confined to Peru and Ecuador, from the early centuries AD: no examples have yet been found further north. It is possible that the use of the technique was limited to regions where the necessary corrosive mineral salts were available: the

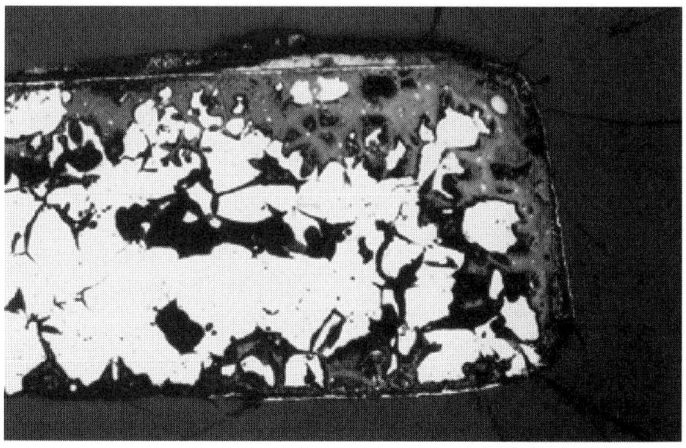

Figure 12.14 Photomicrograph of part of a cross-section through a Vicus dangle, showing the preservation of the electrochemical replacement plating over corroded copper interior. Note that the thin gold layer passes round the dangle and coats all surfaces of the object. (Magnification ×100. Unetched.) (See Plate 68.)

Peruvian region has many desert climatic zones where such minerals could be found. This hypothesis is corroborated by the absence of electrochemical replacement plating in Colombia, where more humid and lush tropical plant growth may limit the availability of suitable minerals.

The sophistication of gilding techniques employed in ancient South America means that it is often difficult to decide on exactly how a particular object has been made. For example, it is not possible to ascertain if an object is made from depletion gilded tumbaga, or from fusion gilded copper, finished by depletion gilding, without removing a complete cross-section from the piece concerned. In many cases, this cannot be justified or carried out under suitable laboratory conditions, consequently there is a danger that the more unusual techniques of fusion gilding or electrochemical replacement plating will be misdescribed or misinterpreted. It is essential to examine any pre-Columbian gilded artifact carefully, especially if it is from Peru, Ecuador or Colombia, since different methods of surface treatment may have important cultural assignations or meaning.

The corrosion that many of these gold alloys undergo further complicates the issue of identification of gilding technologies. It is often assumed that pure gold, or high grade gold, will not corrode during burial: this is not necessarily true. Examination of the surface morphology of an extraordinarily fine rectangular gold wire from the site of La Tolita, in Ecuador, very similar to an excavated example radiocarbon dated to 325BC±85 (Ny-642: CNRS: France), was examined by electron probe microanalysis (Bouchard 1979, Scott 1991). The wire is rectangular in cross-section and measures only slightly more than a human hair, at 12 microns by 40 microns. The composition of the wire is 98.5 wt % gold, 1 wt % silver and 0.15 wt % copper, showing it to be made in a native gold alloy without any addition of copper. The wire was probably made by hammering and very skilful cutting of thin sheet, since it is not drawn to shape (Scott 1991). Examination of the surface of the wire at high magnification of about 1,900×, shown in the photomicrograph of Figure 12.15, clearly reveals surface dissolution features in the form of pits and crevices

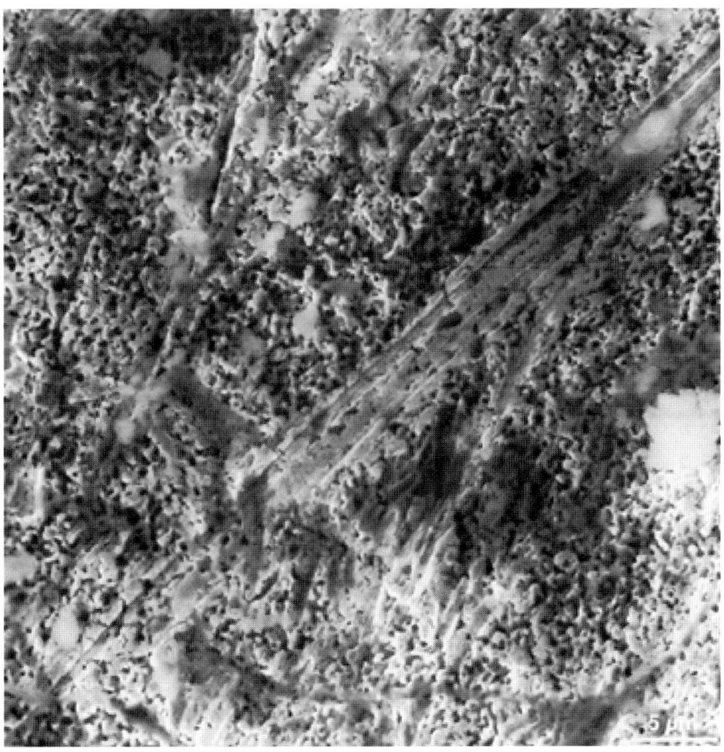

Figure 12.15 Surface morphology of gold wire from La Tolita, showing surface dissolution of this high gold content wire from burial. The scale bar at the bottom right is 5 microns, indicating the extensive dissolution of the surface that has occurred. (See Plate 69.)

in the gold surface. There is no doubt that this surface corrosion must have occurred during the 2,000 years of burial of the wire. It is well known that plants can extract gold from the ground and enhanced levels of gold in plant ash can act as a mechanism of remote prospection, showing that there is some mobility of gold in biological systems (Rapson 1982). Tarnishing, anodic dissolution and embrittlement are all important aspects of the deterioration of gold alloys. Work on general corrosion was carried out many years ago by Tammann (1919), Tammann and Wilson (1928) and Graf and Ata (1973). Addition of even a few per cent of silver or gold to copper alloys makes them susceptible to stress-corrosion cracking. In general, alloys with 10–30 at % gold (corresponding to 25–60 wt %) are very vulnerable; severity should then decrease once the gold content gets above 70 wt %. Besides general corrosion, there are changes in gold alloys that may be due to ordering, age-hardening or spinodal decomposition. Some of these are discussed in Scott (1983b). The most obvious effect of anodic dissolution is the detachment of the gilded surface from the substrate and the intrusion of corrosion products of silver or copper. The detachment of these gilt layers may imply that foil gilding was used: equally, in South America, it may not, so any visual conclusion based on this observation has to be confirmed by metallographic analysis.

Corrosion and the preservation of evidence

It is crucial that all archaeological gold or gilt objects be carefully examined and documented prior to surface cleaning; if surface cleaning can be justified at all, since it may cause irreplaceable loss of important information. It is the responsibility of the conservator to preserve, not solely to reveal a glittering gold surface by the application of chemical or mechanical cleaning techniques, but rather to evaluate and carefully observe the existing surface of the object as excavated or received.

From electrolytic reduction to plasma reduction, a period of activity spanning almost a hundred years, there have been far too many instances where the only objective has been to produce clean metal surfaces, free of corrosion, tarnish or associated materials.

Gilt surfaces may preserve textile, feather, wood, pigment or organic residues, besides associated mineral deposits or corrosion products. Peruvian gold objects may preserve traces of paint, with which gold masks were decorated: even Greek goldwork may be found with surfaces painted with cinnabar (Ogden 1993), but the variety of materials found with South American gold objects is striking, since many of them form part of elite burials that may contain a complex array of other materials (Alva and Donnan 1993). Merkel *et al.* (1995) mention that cinnabar was applied over most of the surface of a ceremonial mask excavated at Huaca Loro (Middle Sicán). On the reverse of a gold disc, with cut-out design, identified (Bray 1994) as an example of the San Agustin style of goldworking from ancient Colombia, for example, lipids have recently been identified (Cave 1995), which show a close relationship to those of human sweat, indicating the preservation of associated degradation products from a tomb burial. Most of the exceptional gold objects found in ancient Colombia come from tomb burials, so this discovery fits the expected burial environment for these plaques from San Agustin. Pyrolysis gas chromatography has been used to identify salts of long-chain fatty acids on Egyptian gold (Shedrinsky *et al.* 1991), and sampleite (Fabrizi *et al.* 1989), an apparently rare chloro-phosphate of copper, may not be as rare as previously assumed,

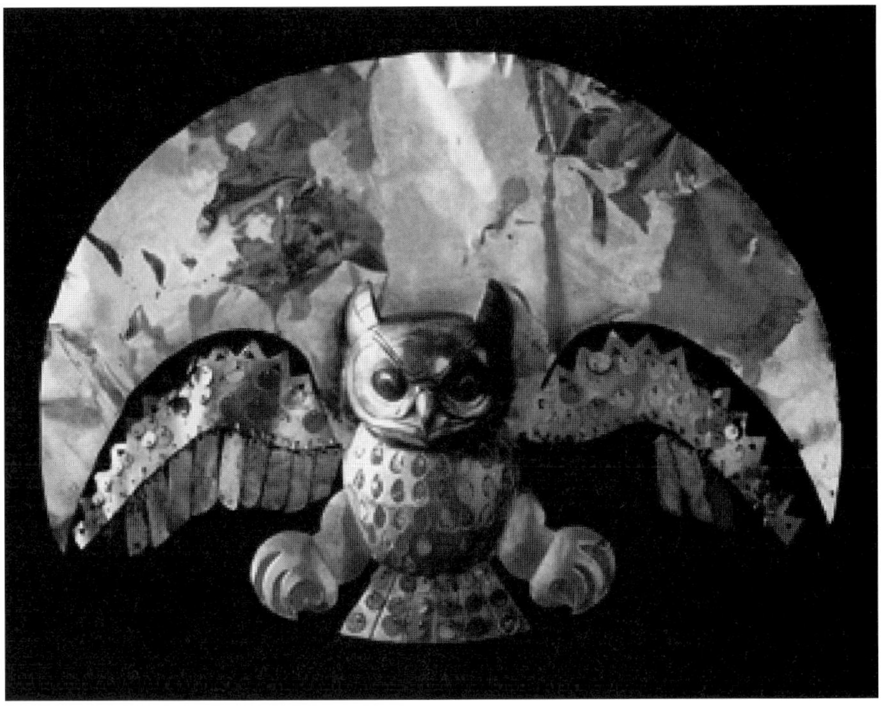

Figure 12.16 Spectacular Moche gold owl pectoral, site of La Mina, dated to 1st century AD. Not only does the golden owl have inlaid eyes and numerous dangles, some of which look like gilded copper, probably electrochemical replacement gilded, but the repousséed teardrop shaped feathers preserve red and green paint in alternating colours. The essentially intaglio design has helped preserve applied pigment. (Length 45 cm.) (Professor Christopher Donnan, UCLA, Fowler Museum of Cultural History) (See Plate 70.)

since it has now been identified on several ancient South American objects (Howe: unpublished data; Scott: unpublished data). Gold from ancient South America always contains silver, and work by Frantz and Schorsch (1990) has shown that such gold alloys can be coloured or corroded with the formation of mixed oxides of silver and gold, such as petrovskaite, AuAgS, or uytenbogaardtite (synthetic) Ag_3AuS_2. These mixed sulphides may form deliberate surface patinas on Egyptian gold objects: there is no evidence for deliberate use of these effects from ancient South American objects, although they may occur as surface corrosion. Figure 12.16 illustrates a spectacular gold owl pectoral from the area of La Mina, Moche culture, dated to the 1st century AD (Donnan 1995). The body of the central owl figure has been worked by repoussé, creating surface depressions in which green and blue pigment has been preserved in an alternating colour sequence. This exceptional preservation of pigment, illustrates how important even slight traces of pigment may be, if found *in situ* on gold or gilded surfaces, for without proper conservation of these residues, alteration products or pigments, the technological history and burial environment of the object cannot be fully reconstructed.

Acknowledgements

Thanks are due to Professor Warwick Bray, Department of Prehistory UCL Institute of Archaeology, for sharing important information concerning the San Agustin plaques, to Professor Christopher Donnan, Director, Fowler Museum of Cultural History, UCLA, for information on Moche cultural material, to Ellen South and Karen Sexton Josephs of the Getty Conservation Institute for secretarial support, and to Michele Cave, Department of Environment, UCL Institute of Archaeology, Ellen Howe, Department of Conservation, Metropolitan Museum of Art, Ken Klassen, Bogotá, Charles Craig, Santa Barbara, and Lesley Moorcroft.

References

Alva, W. and Donnan, C. 1993. '*Royal Tombs of Sipan*'. Los Angeles, Calif.: Fowler Museum of Cultural History.

Bergsøe, P. 1937. '*The Metallurgy and Technology of Gold and Platinum among the Pre-Columbian Indians*', F. C. Reynolds (trans.). Ingeniorvidenskabelige Skrifter No.A46. Naturvidenskabelige Samfund I Kommission hos GEC Gad. Copenhagen.

Bouchard, J-F. 1979. 'Hilos de Oro Martillado Hallados en la Costa Pacific a del Sur de Colombia'. *Boletin: Museo del Oro* **2**: 14–17.

Bray, W. 1978. *The Gold of El Dorado.* London: The Royal Academy of Art and Times Books.

Bray, W. 1985. 'Ancient American Metallurgy: Five Hundred Years of Study'. In *The Art of PreColumbian Gold: The Jan Mitchell Collection*, 98–118. New York: Metropolitan Museum of Art.

Bray, W. 1992. '1992 Sitio Conte metalwork in its pan-American context'. In *River of Gold: PreColumbian Treasures from Sitio Conte,* P. Hearne and R. J. Sharer (eds), 32–47. Philadelphia, Pa.: University Museum.

Bray, W. 1993. 'Techniques of gilding and surface-enrichment in pre-Hispanic American Metallurgy'. In *Metal Plating and Patination: Cultural, Technical and Historical Developments*, S. La Niece and P. Craddock (eds), 182–92. Oxford: Butterworth-Heinemann.

Bray, W. 1994. Personal communication.

Bushnell, G. H. S. 1951. *The Archaeology of the Santa Elena Peninsula in Southwest Ecuador.* Cambridge: Cambridge University Press.

Butler, A. R., Glidewell, C., Glidewell, S. M. *et al.* 1987. 'The solubization of metallic gold and silver: explanation of two sixth century AD Chinese protochemical recipes'. *Polyhedron*, **6**: 483–8.

Caley, E. R. 1926. 'The Leiden Papyrus X: An English translation with brief notes'. *Journal of Chemical Education*, **3**: 1149–66.

Cave, M. 1995. Personal communication.

Christensen, R. T. 1954. 'A recent excavation in Southern Coastal Ecuador'. *Bulletin of the University Archaeology Society*, **5**: 30–54.

Donnan, C. 1995. Personal communication.

Elrlij, T. 1974. 'Studies of some Vicus-style metalwork from the Stewart Collection', unpublished BSc dissertation. Department of Conservation and Materials Science, Institute of Archaeology, University College London.

Evans, C. and Meggers, B. J. 1955. 'Preliminary report on archaeological investigations in the Guayas Basin, Ecuador'. *Cuadernos de Historia y Arqueología*, **5**: 308–328.

Fabrizi, M., Ganiaris, H., Tarling, S. and Scott, D. A. 1989. 'The occurrence of sampleite, a complex copper phosphate, as the principal corrosion product on ancient Egyptian Bronzes from Memphis, Egypt'. *Studies in Conservation*, **34**: 45–51.

Frantz, J. H. and Schorsch, D. 1990. 'Egyptian Red Gold'. *Archaeomaterials*, **4**: 133–52.

Forty, A. J. and Durkin, P. 1980. 'Micromorphological study of the dissolution of silver-gold alloys in nitric acid'. *Philosophical Magazine*, **42**: 295–8.

Gettens, R. J. and Mooradian, V. G. 1937. 'Report on metallographic and microchemical investigations of gilded objects from Cocle, Appendix III: Cocle: An archaeological study of central Panama', S. K. Lothrop (ed.). *Memoirs of the Peabody Museum of Archaeology and Ethnology*, **7**. Cambridge, Mass.: Harvard University Press.

Graf, L. and Ata, H. O. K. 1973. 'Untersuchung der Spannungskorrosionsempfindlichkeit von Ein-und Vielkristallen aus Messing und Kupfer-Gold-Legierungen'. *Zeitschrift fur Metallkunde*, **64**: 366–71.

Grieder, T. 1978. *The Art and Archaeology of Pashash.* Austin, Tex. and London: University of Texas Press.

Grossman W. J. 1972. 'An ancient gold worker's tool kit'. *Archaeology*, **25**: 270–5.

Hosler, D., Lechtman, H. and Holm, O. 1990. 'Axe-monies and their relatives, 22–3'. Studies in Pre-Columbian Art and Archaeology, 30. Washington DC: Dumbarton Oaks Research Library.

Lechtman, H. N. 1973. 'The gilding of metals in pre-Columbian Peru'. In *Application of Science in the Examination of Works of Art*, W. J. Young (ed.), 38–52. Boston, Mass.: Museum of Fine Arts.
Lechtman, H. N. 1979. 'A pre-Columbian technique for electrochemical replacement plating of gold and silver on copper objects'. *Journal of Metals*, **31**: 154–60.
Lechtman, H. N., Erlij, A. and Barry, E. J. 1982. 'New perspectives on Moche metallurgy: Techniques of gilding copper at Loma Negra, Northern Peru'. *American Antiquity*, **47**: 3–30.
Lothrop, S. K. 1942. *Cocle: An archaeological study of Central Panama: Part II*. Memoirs of the Peabody Museum of Archaeology and Ethnology, 9. Cambridge, Mass.: Harvard University Press.
Merkel, J. F., Seruya, A. I., Griffiths, D. and Shimada, I. 1995. 'Metallography and microanalysis of precious metal objects from the Middle Sicán elite tombs at Batan Grande, Peru'. In *Materials Issues in Art and Archaeology IV*, J. R. Druzik, I. Freestone and P. Vandiver (eds), 105–26. Pittsburgh, Pa.: Materials Research Society.
Oddy, W. A. 1982. 'Gold in antiquity: Aspects of gilding and of assaying'. *Journal of the Royal Society of Arts*, **130**: 730–43.
Ogden, J. 1993. Personal communication.
Peckholt, T. 1899. 'Heil und Nutzpflanzen Brasiliens'. *Berichte Deutsche Pharmaceutische Gesellschaft*, **8**: 45–8.
Percy, J. 1880. *Metallurgy, Silver and Gold Part I*. London: John Murray.
Plazas de Nieto, C. 1979. 'Orfebrería Prehistoria del Altiplano Nariñense'. *Revista Colombiana de Antropología*, **XXI**: 197–244.
Rackham, H. 1968. *Pliny: Natural History*, vol. 9. London: Loeb Classical Library.
Rapson, W. S. 1982. 'Effects of biological systems on gold'. *Gold Bulletin*, **15**: 19–20.
Rivet, P. and Arsandaux, H. 1946. 'La metallurgíe en Ameríque preColombienne'. Travaux et Memoires de l'Institut d'Ethnologíe, Vol. XXXIX. Paris: Musée de l'Homme.
Rovira, S. 1994. 'Pre-Hispanic goldwork from the Museo de America, Madrid: A new set of analyses'. In *Archaeometry of Pre-Columbian Sites and Artefacts*, D. A. Scott and P. Meyers (eds), 323–52. Los Angeles, Calif.: Getty Conservation Institute.
Sahagun, B. de 1565. 'Book 9 of the General History of the Things of New Spain: Florentine Codex', A. J. O. Andersen and C. E. Dibble (trans.). Monographs of the School of American Research 14(10), 1959. Santa Fe, N. Mex.: School of American Research.
Saville, M. H. 1907–1910. *The Antiquities of Manabi, Ecuador: Contributions to South American Archaeology*. New York: George G. Heye Expedition.
Scott, D. A. 1982. 'Pre-hispanic Colombian metallurgy: studies of some gold and platinum alloys', unpublished PhD thesis. Department of Archaeological Conservation and Materials Science, Institute of Archaeology, Univeristy College London.
Scott, D. A. 1983a. 'Depletion gilding and surface treatment of gold alloys from the Narino area of ancient Colombia'. *Journal of the Historical Metallurgy Society*, **17**: 99–115.
Scott, D. A. 1983b. 'The deterioration of gold alloys and some aspects of their conservation'. *Studies in Conservation*, **28**: 194–203.
Scott, D. A. 1986a. 'Gold and silver alloy coatings over copper: an examination of some artefacts from Ecuador and Colombia'. *Archaeometry*, **28**: 33–50.
Scott, D. A. 1986b. 'Fusion gilding and foil gilding in pre-Hispanic Colombia and Ecuador'. In *Metalurgia de America Precolombina*, C. Plazas (ed.), 283–325. Bogota: Banco de la Republica.
Scott, D. A. 1991. 'Technical examination of some gold wire from PreHispanic South America'. *Studies in Conservation*, **36**: 65–75.
Scott, D. A. 1994. 'Goldwork of pre-Columbian Costa Rica and Panama: A technical study'. In *Materials Issues in Art and Archaeology IV*, P. Vandiver, J. Druzik and I. Freestone (eds). Cancun, Mexico: INAH/Getty Conservation Institute, Los Angeles.
Shedrinsky, A. S., Stone, R. E. and Baer, N. S. 1991. 'Pyrolysis gas chromatography studies on Egyptian archaeological specimens: Organic patinas on the "Three Prin-

cesses" gold vessels'. *Journal of Analytical Applied Pyrolysis*, **20**: 229–38.
Smith, C. S. 1977. 'Some constructive corrodings'. In *Corrosion and Metal Artefacts: A dialogue between conservators and scientists – corrosion*, A. Brown, P. Burnett, T. Chase *et al.* (eds), 43–155. National Bureau of Standards Special Publication no. 479. Washington DC: US Department of Commerce.
Tammann, G. 1919. 'Die chemischen und galvanischen Eigenschaften von Mischkristallreihen und ihre Atomverteilung'. *Zeitschrift fur Anorganische und Allgemeine Chemie*, **107**: 1–240.
Tammann, G. and Wilson, C. 1928. 'Die Anderung des galvanischen Potentials der Metalle usw'. *Zeitschrift für anorganische Chemie*, **173**: 156–63.
Tamalameque 1555. *Averiguaciones en Tamalameque Sobre los Manyllas que Mando Hazer de Azpeleta a los Yndios de su Encomienda de Anpihuegas. Justicia*, Legajo 610–12. fos 2520–25. Seville: Archivo General de Indias.
Ubelaker, D. H. 1981. 'The Ayalan Cemetery: a Late Integration Period Burial Site on the South Coast of Ecuador'. *Smithsonian Contributions to Anthropology* **29**. Washington DC: Institution Press.

13 The Characterisation of Gold Layers on Copper Artifacts from the Piura Valley (Peru) in the Early Intermediate Period

Silvia A. Centeno and Deborah Schorsch

Abstract

Investigations of gilded copper metalwork from the Moche cemetery at Loma Negra have shown that the gilding method used differs significantly from that observed on Moche metalwork from centres further to the south. A comparative analysis of gold layers on copper sheet artifacts from Loma Negra, and those attributed to the Vicús people, their neighbors in the Piura Valley (north coast Peru), was carried out at The Metropolitan Museum of Art using metallography, energy dispersive X-ray spectroscopy (EDS) and scanning electron microscopy (SEM). The results suggest that the local Vicús population and the Moche at Loma Negra shared some aspects of their metallurgical traditions.

Introduction

Various techniques for altering the surface appearance of metals were widely employed by metalworkers of pre-Columbian South America, particularly in the northern and central Andean regions of Peru, Ecuador and Colombia. One feature of great interest was surface colour, and many of the sophisticated metallurgical techniques developed in these regions were applied to the production of gilded or silvered surfaces on metal substrates (Bergsøe 1938; Scott 1983, 1986a,b; Lechtman 1988).

In 1969, a large group of Moche metal objects, most of which were made of gilded or silvered copper sheet, was found by *huaqueros* at Loma Negra, located in the Vicús area of the upper Piura Valley on the northern coast of Peru (Disselhoff 1972) (Figs 13.1–13.3). The discovery was of great importance not only because it increased the small number of metal objects at that time assigned to the Moche culture (*c.* AD100–800), but because it deepened our understanding of the Early Intermediate Period in Peru (*c.* 200BC–AD600) as one of intense innovation in metal technology (Jones 1975; Lechtman 1979; Lechtman *et al.* 1982; Schorsch 1998). Loma Negra is, in fact, situated among habitation and burial sites attributed to the Vicús culture (*c.* 250BC–AD650) and separated from the main sphere of Moche occupation to the south by the Sechura Desert.

Recent discoveries of Moche metalwork at the centrally located sites of Sipán (Alva 1988; Alva and Donnan 1993) and La Mina (Donnan 1990: 32; Naváez 1993) provide further support for our high estimation of Moche metallurgical achievements, as well as parallels for many of the finds at

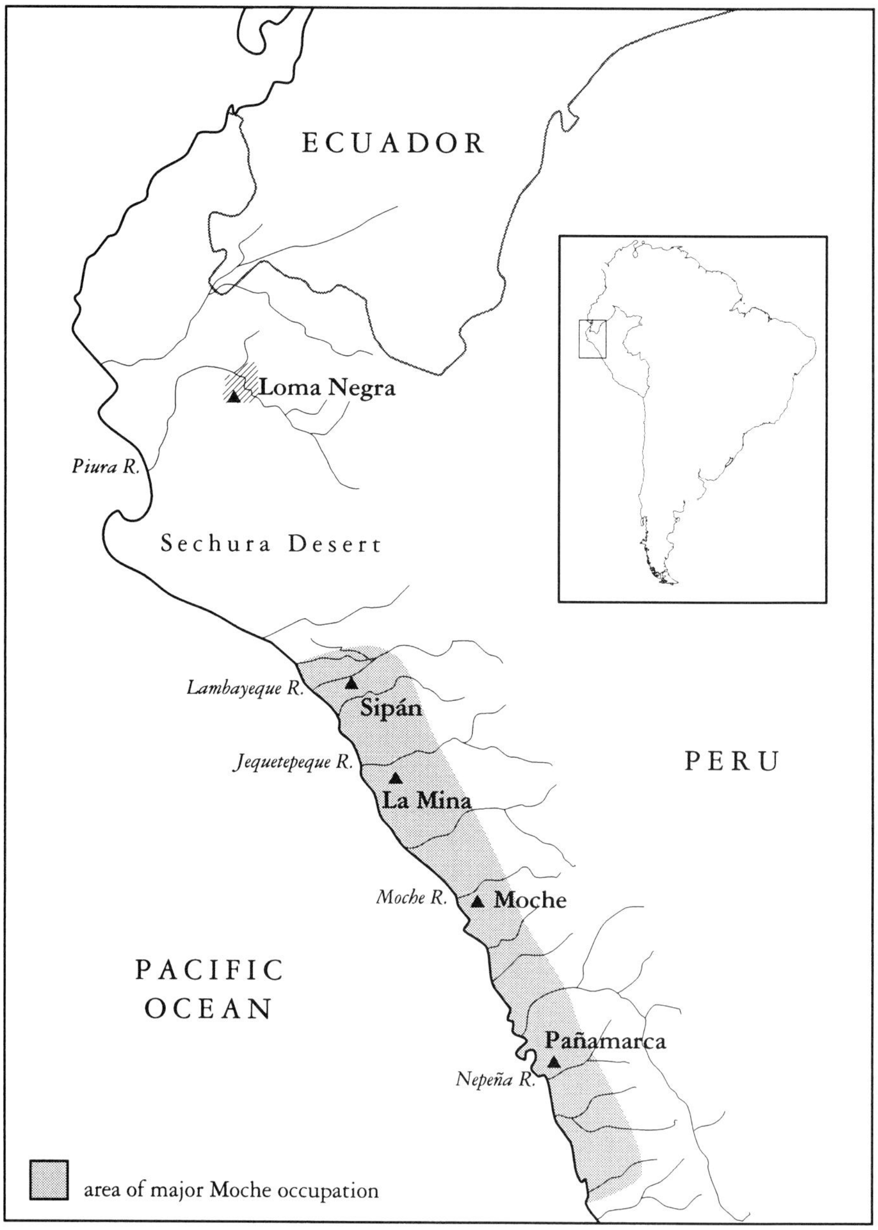

Figure 13.1 Map of the north coast of Peru, showing the major area of Moche occupation and locations of selected Moche sites (▲). The area where Vicús culture sites have been discovered so far is also indicated (courtesy of Adam Hart).

Loma Negra, and at the same time suggest that some innovations employed there may be local developments of the Piura Valley rather than characteristic of Moche metalwork as a whole.

Looting of archaeological sites in the Piura Valley in the region of Cerro Vicús had long preceded the discovery of Moche metalwork at Loma Negra (Murro Mena 1994). In the mid 1960s several archaeologists were sent to the area to investigate reports of widespread illegal excavations. The distinctive finds from these burials could not be attributed to any known Andean culture and were assigned the name Vicús by Matos Mendieta (1965–66). Guzman Ladrón de Guevara and Noriega (1964) subsequently excavated 41 tombs at Cementario Yécala, but their results, except for a preliminary report, were never published. A few of the hundreds of metal finds from Tomb 11 of their excavation were studied by Ríos and Retamozo (1993). Disselhoff (1969, 1971) carried out a series of small excavations that yielded Vicús ceramics but few metal objects. He published the first radiocarbon dates from charcoal samples, placing the Vicús culture (*c.* 250BC–AD650) within the Early Intermediate Period, and more or less contemporaneous with the Moche. In 1988, as part of the Upper

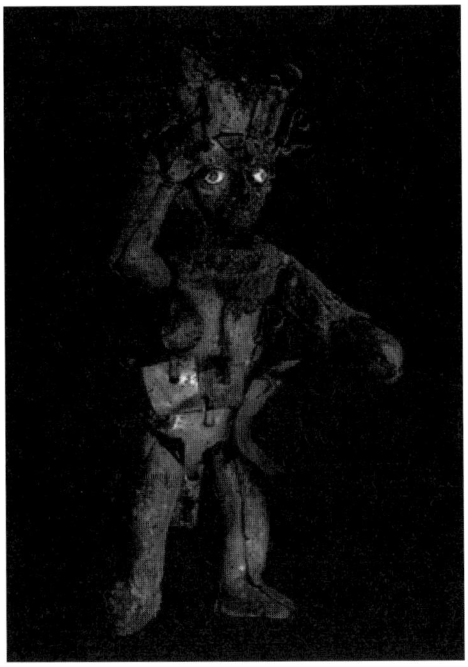

Figure 13.2 Gilded copper warrior figure, Moche, from Loma Negra. The Metropolitan Museum of Art, Gift of Jane Costello Goldberg, from the collection of Arnold I. Goldberg, 1981 (1981.459.32). (Photograph: E. G. Howe). (See Plate 71.)

Piura Archaeological Project of the Pontificia Universidad Católica del Perú co-directed by Kaulicke and Makowski (1990), a metallurgical survey of the region was undertaken by Shimada (1988). No mines or smelting sites were found in the area of Cerro Vicús, but a furnace found at Pampa Juarez was later reported by Makowski (1994).

The quality of interaction between the Moche at Loma Negra and their Vicús neighbors, and between the inhabitants of the Piura Valley and those of the Moche heartland, remains to be clarified (Bawden 1994; Kaulicke 1994; Shimada 1994). A preliminary examination of gilded and silvered surfaces on copper artifacts from the Piura Valley suggests that the Moche at Loma Negra and the local Vicús population shared some aspects of their

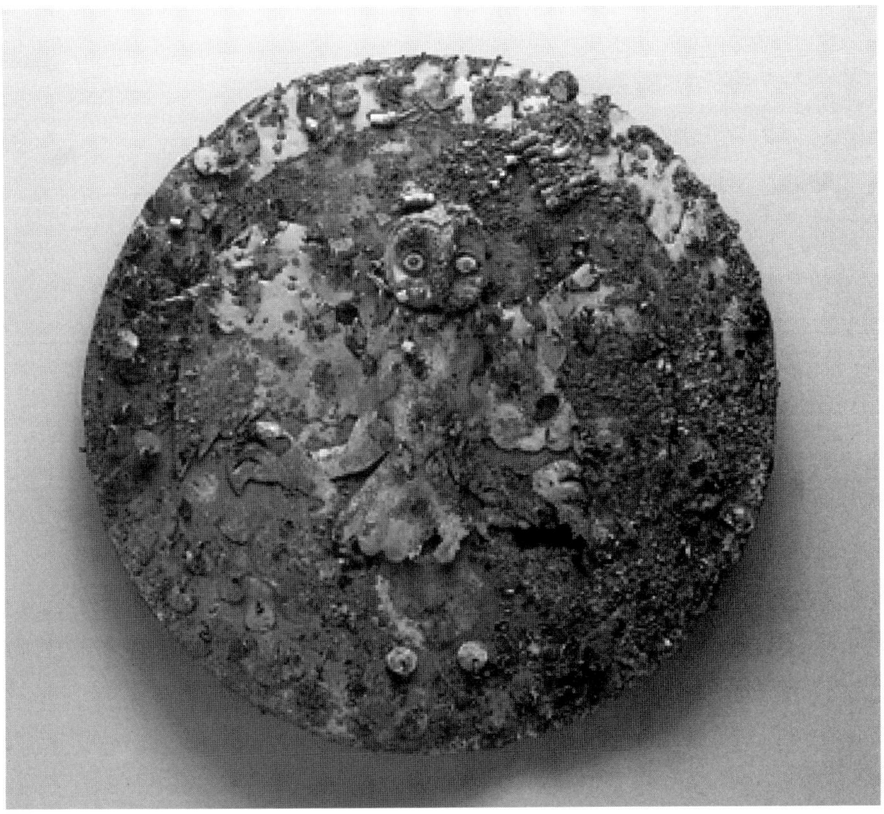

Figure 13.3 Gilded and silvered copper disk, Moche, from Loma Negra, obverse. The Metropolitan Museum of Art, Bequest of Jane Costello Goldberg, from the collection of Arnold I. Goldberg, 1986 (1987.394.56). (Photograph: B. J. Schwarz). (See Plate 72.)

respective metallurgical traditions, though their visual repertoires are quite distinct (Figs 13.2-13.7).

Whereas cast artifacts have been attributed to both cultures, most Moche and Vicús metalwork was made from hammered metal sheet of gold, silver, copper, silvered gold and gilded or silvered copper. As a rule, Vicús artifacts of gold and those that combine silver and gold, are nose ornaments; objects entirely of silver are rare. The copper sheet used for Vicús artifacts, both with and without precious metal surface layers, is usually somewhat thicker than that produced by Moche metalworkers. Moreover, unlike the elaborate three-dimensional objects constructed from multiple sheets of metal often favoured by the Moche (Fig. 13.2), most Vicús metal objects consist of a single sheet and have only shallow relief or superficial linear decoration. 'Plaques' made in a limited number of geometric and organic shapes, as well as crowns, discs and pectorals, are the most typical products of Vicús metalworkers (Figs 13.4–13.7)

In common with much ancient Peruvian metalwork, many of the Vicús copper sheet artifacts have dangles, usually circular or rectangular in

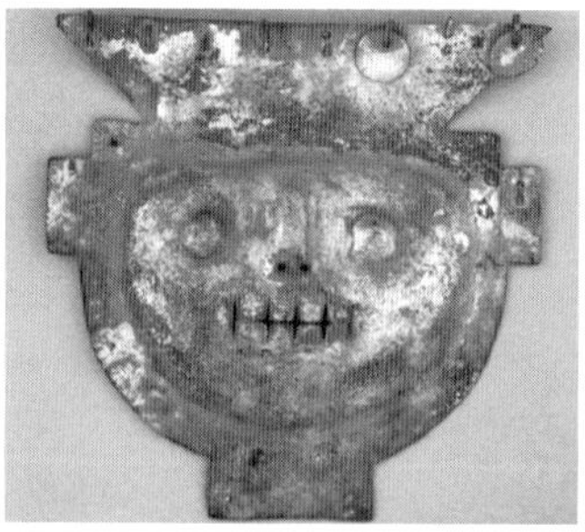

Figure 13.4 Gilded copper plaque, Vicús, obverse. The Metropolitan Museum of Art, Bequest of Jane Costello Goldberg, from the collection of Arnold I. Goldberg, 1986 (1987.394.169). (See Plate 73.)

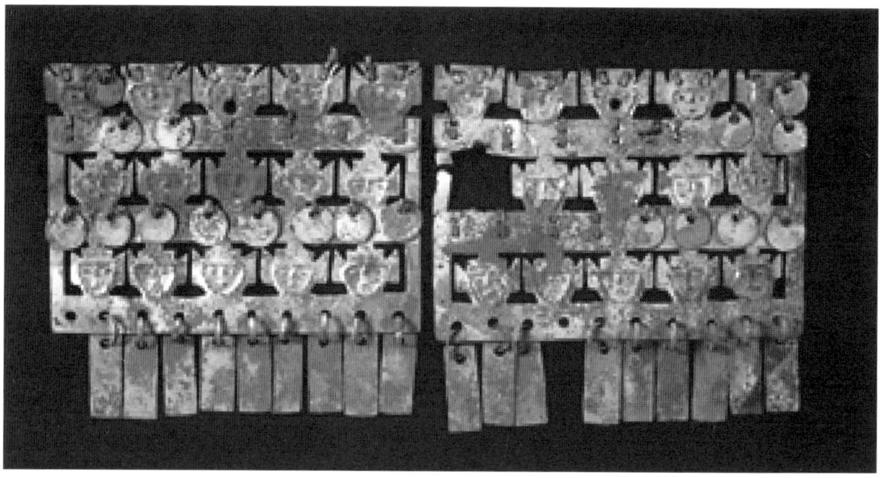

Figure 13.6 Gilded copper plaques,Vicús, obverse. The Metropolitan Museum of Art, Bequest of Jane Costello Goldberg, from the collection of Arnold I. Goldberg, 1986 (left: 1987.394.252, right: 1987.394.251). (See Plate 75.)

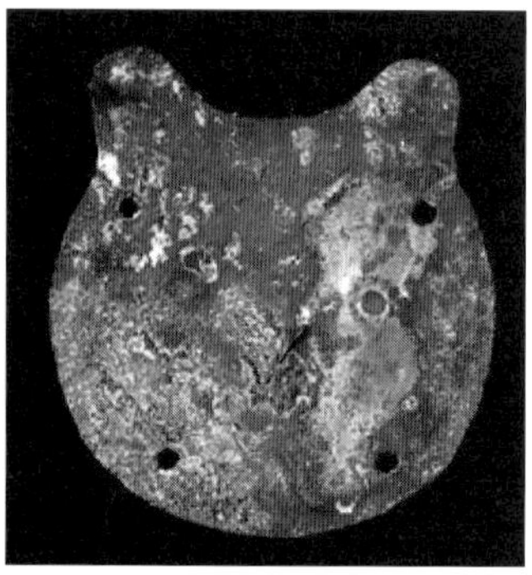

Figure 13.5 Gilded copper plaque,Vicús, obverse. The Metropolitan Museum of Art, Bequest of Jane Costello Goldberg, from the collection of Arnold I. Goldberg, 1986 (1987.394.599). (See Plate 74.)

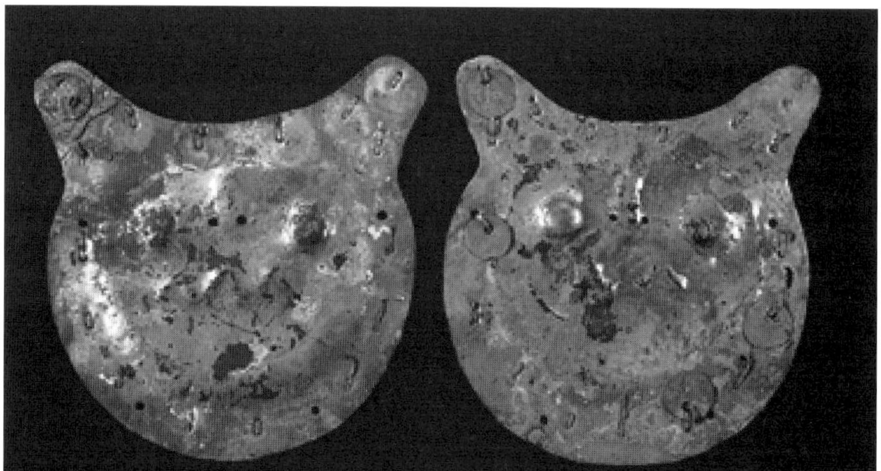

Figure 13.7 Gilded copper plaques,Vicús, obverse. The Metropolitan Museum of Art, Bequest of Jane Costello Goldberg, from the collection of Arnold I. Goldberg, 1986 (left: 1987.394.598, right: 1987.394.596). (See Plate 76.)

shape, that were attached to the substrates using rectangular-section wires, and that jingled and shimmered when the objects were moved. All of the plaques have small holes that allowed their attachment to textiles or other substrates. We often lack specific archaeological contexts for the various artifacts, and due to the general paucity of scholarship in the field of Vicús archaeology, it cannot be said how the plaques and disks might have been used or displayed.

Previous studies of gilding techniques in the Piura Valley

Studies to determine the methods of gilding and silvering used on copper-based metalwork found at Loma Negra were undertaken by Lechtman (1979) and by Lechtman *et al.* (1982). On the basis of their investigations, the authors proposed that the gold and silver layers had been applied using an electrochemical replacement plating technique, suggesting the following sequence of steps. First, plating solutions were prepared by dissolving gold and silver in aqueous solutions of corrosive minerals available in the arid coastal environment of northern Peru. The availability of the raw materials alone, however, can not explain how this sophisticated technology took root and developed. Precious metal layers of varying compositions were deposited onto the copper substrates by dipping them into the solutions, which had been neutralised by the addition of sodium carbonate. Under these conditions a redox reaction takes place, in which some copper atoms from the substrate are oxidised and gold or silver ions in solution are reduced and deposited on to the copper surface. The authors proposed annealing to secure the layers to the substrates and burnishing to increase their lustre as final steps. Modern replication samples prepared by Lechtman and her colleagues reproduce the macroscopic characteristics observed in the ancient samples, including the extreme thinness and uniformity of the precious metal layers, as well as some features of their microstructures.

Several studies of Vicús gilded copper objects have been undertaken. Ríos and Retamozo (1993) examined some of the metalwork found in Tomb 11 at Cementerio Yécala by Guzman Ladrón de Guevara and Noriega in 1964. Among the finds are a number of gilded copper artifacts, including plaques, crowns and feather-like attachments that may have functioned as plumes. Based on the examination of samples taken from two copper artifacts that appeared to have gilded surfaces, Ríos and Retamozo report that the layers average 10 μm in thickness and are composed principally of copper containing 'only traces of gold, silver and silicon'. It is not indicated how the elemental analysis was carried out, but, in any case, it is likely that most of the copper detected in the surface layers is due to the proximity of copper corrosion products from the copper substrate, and that the surfaces were fully intended to appear golden. Ríos and Retamozo suggest that a fusion gilding technique had been used.[1]

In a general survey of Vicús metalwork, Diez-Canseco (1994), mentions the examination of one sample from a Vicús site at Pampa Juárez, excavated by Makowski (Kaulicke and Makowski 1990) in 1988 and 1989. Unfortunately no metallographic sections or analytical data are presented, but the sample is described as an unalloyed copper substrate supporting an irregular gold layer averaging 8 μm in thickness. The author proposes the use of an electrochemical replacement plating technique although surface layers thought to have been applied using such methods are generally much thinner.

Lechtman *et al.* (1982: 29) and co-workers describe the preliminary examination of some Vicús gilded copper plaques with pierced openwork designs. They report the gilding layers to be 'quite thick', and otherwise lacking features observed on the Loma Negra samples, and suggest that the plaques had not been gilded using an electrochemical replacement process.

Studies of extremely thin gold and silver layers on copper artifacts from archaeological contexts have proven to be difficult, in part because the lateral resolution of most conventional electron microprobes does not allow the compositional analysis of the layers without interference from signals originating in the metal substrate. In addition, the ubiquitous presence of archaeological copper corrosion products both on the surfaces of the artifacts and within the precious metal layers, and the solid state diffusion of elemental gold, silver and copper across the interfaces, add to uncertainties in establishing the original compositions of the gold and silver layers.

Loma Negra gilding layers

Presented here are the results of visual, radiographic and metallographic examinations carried out on samples of Vicús and Loma Negra gilded copper sheet at the Sherman Fairchild Center for Objects Conservation at The Metropolitan Museum of Art. These examinations were supplemented with scanning electron microscopy (SEM) and elemental analyses using conventional as well as high resolution energy-dispersive X-ray spectroscopy (EDS).[2] The results are compared to those reported by Lechtman and co-workers (Lechtman 1979; Lechtman *et al.* 1982), and to observations and data obtained from many of the Loma Negra artifacts examined previously at the Sherman Fairchild Center (Howe *et al.* 1993; Schorsch *et al.* 1996).

Gold and silver layers on artifacts from Loma Negra examined in the course of the present study were found to be of the same order of magnitude in thickness as those studied by Lechtman *et al.*, that is 0.5–2 µm (Fig. 13.8). Each layer is quite uniform in thickness and covers the entire surface of the substrate, including the inner walls of perforations and the edges of the thin hammered sheet, as well as all surface irregularities. The layers show no evidence of having been applied using metallurgical methods

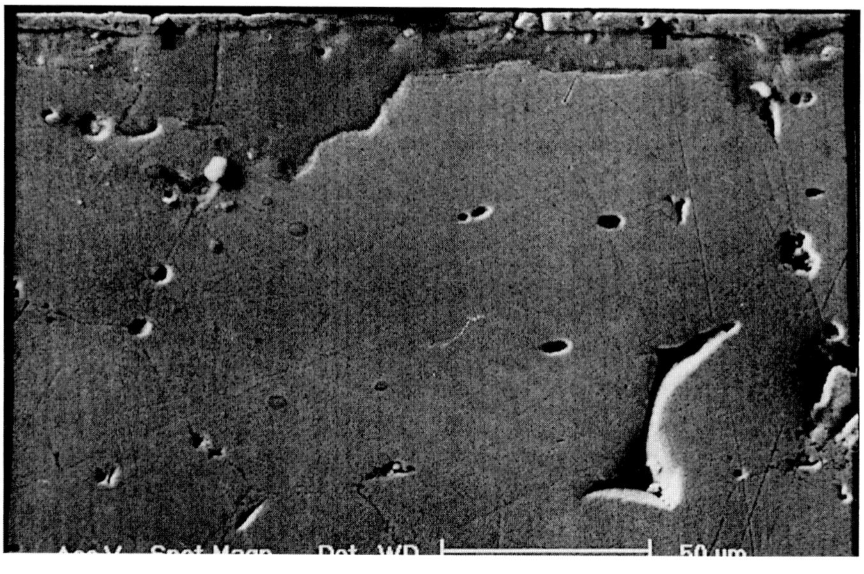

Figure 13.8 Loma Negra unassociated dangle, backscattered image of metallographic section. Original magnification: 500× (Photograph: L. Carapia).

known to have been used in the New World, such as fusion or diffusion gilding,[3] or strictly mechanical methods involving the application of leaf or foil (Scott 1986b). Furthermore, as the samples contain no mercury, amalgam gilding was ruled out.[4] Depletion gilding, a method often observed on Andean metalwork that has been reported for gilded copper artifacts from the Moche site of Sipán (Eckmann 1993),[5] was also eliminated as a possibility because copper was the only metal detected in the Loma Negra substrates that were analysed at the Sherman Fairchild Center.[6]

EDS elemental analysis using an electron microprobe with a high lateral resolution (*c.* 0.5 µm) was carried out on samples removed from two representative artifacts. The beam was focused directly on the gilding layers, avoiding excitation of adjacent copper substrates and surface corrosion. The first object is a large disk that displays an owl (MMA1987.394.56) (Fig. 13.3). The owl itself is made of gilded copper, as are the border applied to the circumference of the disk, the dangles and the dangle wires. The disk itself is silvered copper. All of the various components were joined mechanically using tabs and slots.[7] The sample was removed from a tab that holds the owl to the disk. The second sample was taken from an unassociated Loma Negra dangle.

The analytical results for the two Loma Negra samples are shown in Table 13.1. Precious metal surface layers ranging in composition from 100% silver to 98% gold with 2% silver have been observed on the other Loma Negra artifacts examined (Schorsch *et al.* 1996).[8] Based on results of plating experiments carried out at the Sherman Fairchild Center, it can be stated that the ratio of gold to silver in a given layer is not necessarily identical to the starting ratio in the solution used for its deposition.

Table 13.1 Thickness and weight percent compositions of precious metal layers on Loma Negra objects. The weight percents Au, Ag and Cu were normalized to total 100.

accession number	wt %	Ag : Au wt %	thickness (µm)
1987.394.56	Au: 85.6 Ag: 11.0 Cu: 3.4	1 : 7.8	1–1.5
unassociated dangle	Au: 63.5 Ag: 18.8 Cu: 17.7	1 : 3.4	0.8–1.2

EDS point analysis was also used to study the compositional variation across the gold/copper interfaces. The distance to which gold and silver had diffused into the substrate was determined to be 2–4 µm on the two Loma Negra artifacts examined. The presence of a diffusion zone indicates that the artifacts – which had been repeatedly hammered and heated during their fabrication – were heated for the last time during or after the gilding had been applied.

Lechtman and co-workers noted pores in the precious metal layers in sections of replication samples prepared following the electrochemical replacement plating method they proposed. They suggested that the pores might result from the oxidation of copper atoms from the substrate during deposition of the precious metal layer but had difficulty in locating analogous pores in sections prepared from the ancient specimens because the interfaces between the copper substrates and the gold surface layers are invariably heavily corroded (Lechtman 1979: 159–60; Lechtman *et al.* 1982: 25). They also suggested that the extreme thinness of the ancient gold layers would make it difficult to locate pores that might be present.

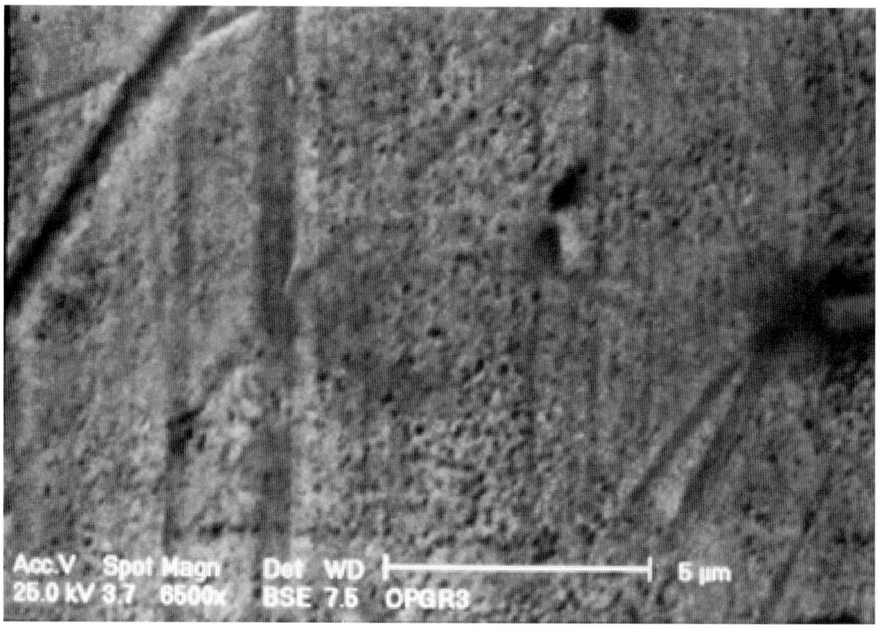

Figure 13.9 Replication sample (100% gold layer on a copper sheet substrate), prepared according to the electrochemical replacement plating technique proposed by Lechtman *et al.* (1982), backscattered image of the surface. Original magnification: 6500× (Photograph: L. Carapia).

An even distribution of pores (*c.* 0.1 μm in diameter) on the surfaces of replication samples prepared at The Metropolitan Museum of Art was observed using SEM(Fig. 13.9). Similar pores, though larger and less uniform in diameter (*c.* 0.2–0.4 μm), were seen on the surfaces of ancient samples (Fig. 13.10). The degree of porosity and the size of pores present in corrosion-free films formed by electrochemical replacement processes are affected by a large number of interdependent factors, including the initial condition of the metal substrate surface, in particular its cleanness and texture, and stirring conditions during the deposition process. Both the temperature and the duration of thermal treatments after deposition are also important variables (Visco 1974). With ancient artifacts it is very difficult to evaluate quantitatively the size and density of pores in the gilding layers at the time of application, especially because of physical alterations associated with the subsequent formation of corrosion. The three-dimensional morphology of the surface pores formed by electrochemical replacement processes differs markedly from that of channels and islands common in layers formed by depletion gilding tech-

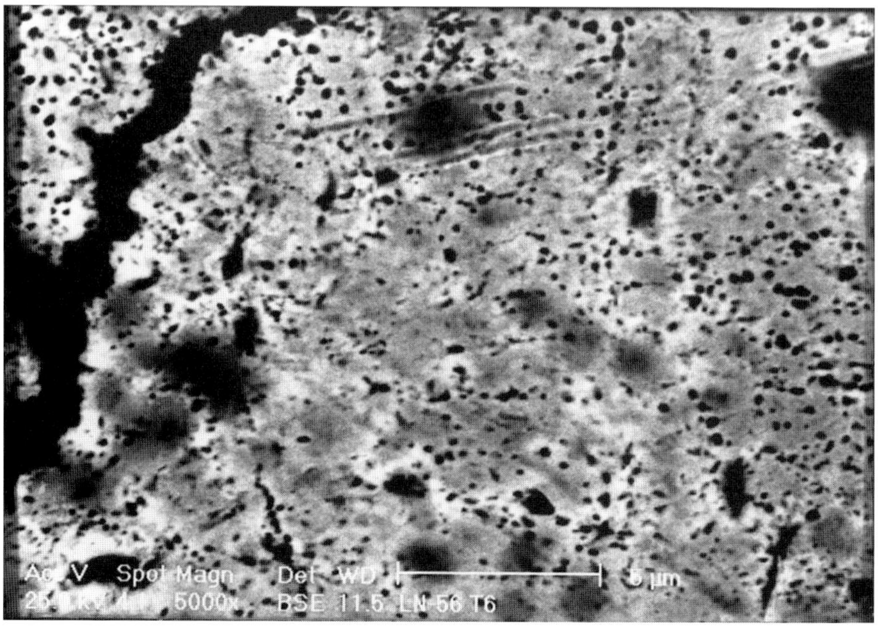

Figure 13.10 Loma Negra gilded copper sheet (MMA 1987.394.56), backscattered image of the surface of a sample. Original magnification: 5000×. (Photograph: L. Carapia).

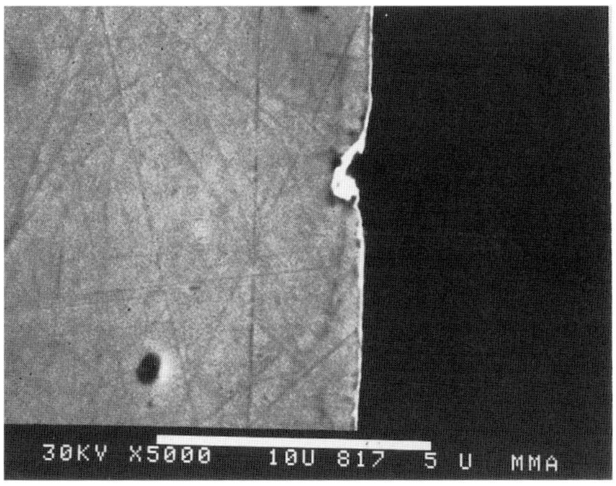

Figure 13.11 Replication sample (100% gold layer on copper sheet substrate), prepared according to the electrochemical replacement plating technique proposed by Lechtman *et al.* (1982), backscattered image of metallographic section. Original magnification: 5000× (Photograph: M.T. Wypyski).

niques, which also appear pit-like when the surface is viewed from above (Forty 1979; Forty & Rowlands 1981).

Of main concern to Lechtman and her colleagues (1982: 25) was sometimes finding ancient gold layers even thinner (*c.* 0.5 μm) than those on the replication samples they prepared. Following their technique, and varying deposition times and concentrations of the plating solutions, we were able to deposit, onto a copper substrate, uniform gold layers as thin as the thinnest layers observed on ancient specimens (*c.* 0.5 μm) (Fig. 13.11).[9]

Vicús gilding layers

The Metropolitan Museum of Art houses a small but representative collection of metal objects attributed to the Vicús culture (Figs. 13.4–13.7). Even though their provenance has not been established through scientific excavation, the artifacts have stylistic characteristics that make this proposed cultural association credible.

Preliminary metallographic studies of polished sections from several Vicús gilded copper artifacts indicate that the gilding layers are quite similar to those observed on objects from Loma Negra: they seem to be uniformly thin and cover all external and internal edges, including the inner walls of perforations, and surface irregularities entirely. As in the case of the Loma Negra materials studied, the gold and silver layers show no evidence of having been applied using any of a variety of mechanical or metallurgical methods known to have been used in the Old and New Worlds in ancient times.

More than twenty gilded or silvered copper Vicús artifacts in The Metropolitan Museum of Art and the American Museum of Natural History in New York were examined using an optical microscope. 'Wrinkles' were observed, particularly in areas of relief, on the surfaces of a number of these objects (Fig. 13.12). The wrinkles are due to a failure of the bond between the gold layer and the substrate but the exact cause of the failure is unknown. Similar wrinkles have not been observed on gilded and silvered copper artifacts from Loma Negra.

Three Vicús ornaments were selected for metallographic study and elemental analysis. The first is a rectangular plaque with pierced openwork (MMA 1987.394.252) (Fig. 13.6: left) with gilded copper dangles suspended from flat wires of gilded copper. A sample was removed from a dangle wire located in the second row from top on the proper right side of the plaque. Two other plaques (MMA 1987.394.596 & 1987.394.598) (Fig.

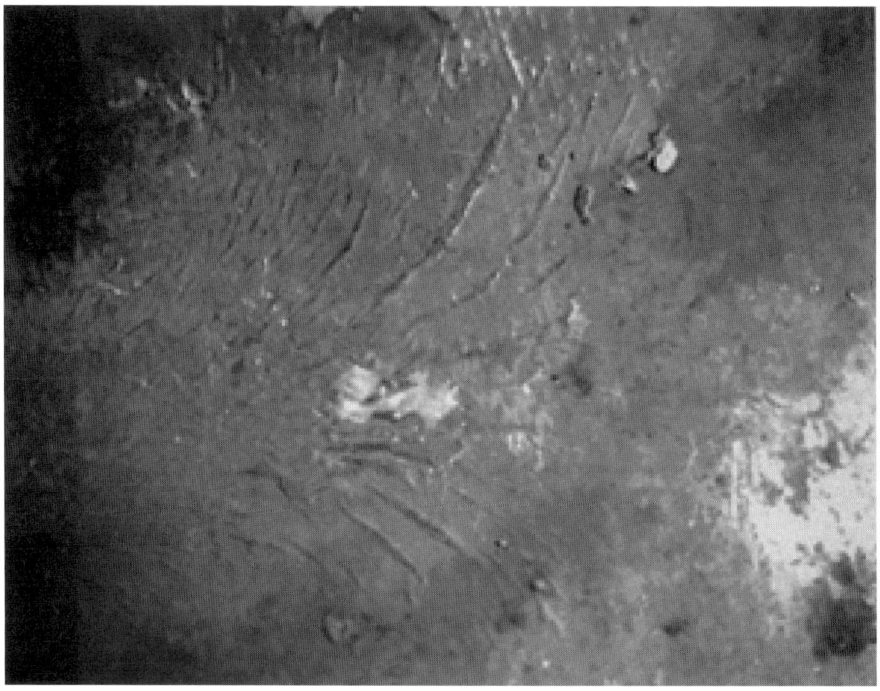

Figure 13.12 Gilded copper plaque, Vicús (MMA 1987.394.597), photomicrograph of the surface showing the presence of wrinkles. Original magnification: 35× (Photograph: E.G. Howe). (See Plate 77.)

13.7) were also sampled. Each depicts the head of a feline, with the contours of the face, as well as the eyes, nose and mouth, slightly raised from a flat sheet. Gilded copper dangles, almost all of which are now lost, were attached near the perimeter of each plaque using wires of gilded copper. Each plaque also has four holes distributed along its edges that served to mount it onto a supporting material. Two larger holes below each of the raised noses appear to represent nostrils. The samples were removed from a dangle wire on the proper right side of MMA 1987.394.596, and from a dangle on MMA 1987.394.598.

Examination of the three sections revealed precious metal layers measuring in thickness on the order of those observed on artifacts from Loma Negra, that is 0.5–2 μm (Fig. 13.13a–c and Table 13.2). In each case the interfacial region between the copper substrate and the gold layer is corroded. The bulk of the copper substrate in the sample from the second feline plaque (MMA 1987.394.598) is completely mineralised and above the gilding layer a thick crust of copper corrosion products can be observed, averaging 20 μm in thickness (Fig. 13.13c).

Elemental analyses of the more well preserved gilding layers on the geometric plaque and the first feline plaque indicate that they are primarily gold, silver and copper with minor amounts (1–2 wt %) of impurities such as silicon, iron and aluminum (Table 13.2). The results of analyses of multiple spots along the highly corroded sample from the second feline plaque

Table 13.2 Thickness and weight percent compositions of precious metal layers on Vicús objects. The weight percents of gold, silver and copper were normalised to total 100.

accession number	wt %	Ag : Au wt %	thickness (mm)
1987.394.252	Au: 43.4 Ag: 12.0 Cu: 44.6	1 : 3.6	1–1.5
1987.394.596	Au: 48.0 Ag: 12.2 Cu: 39.8	1 : 3.9	0.5–1
1987.394.598	–	1: 3.0	0.8–1.2

(a)

(b)

(c)

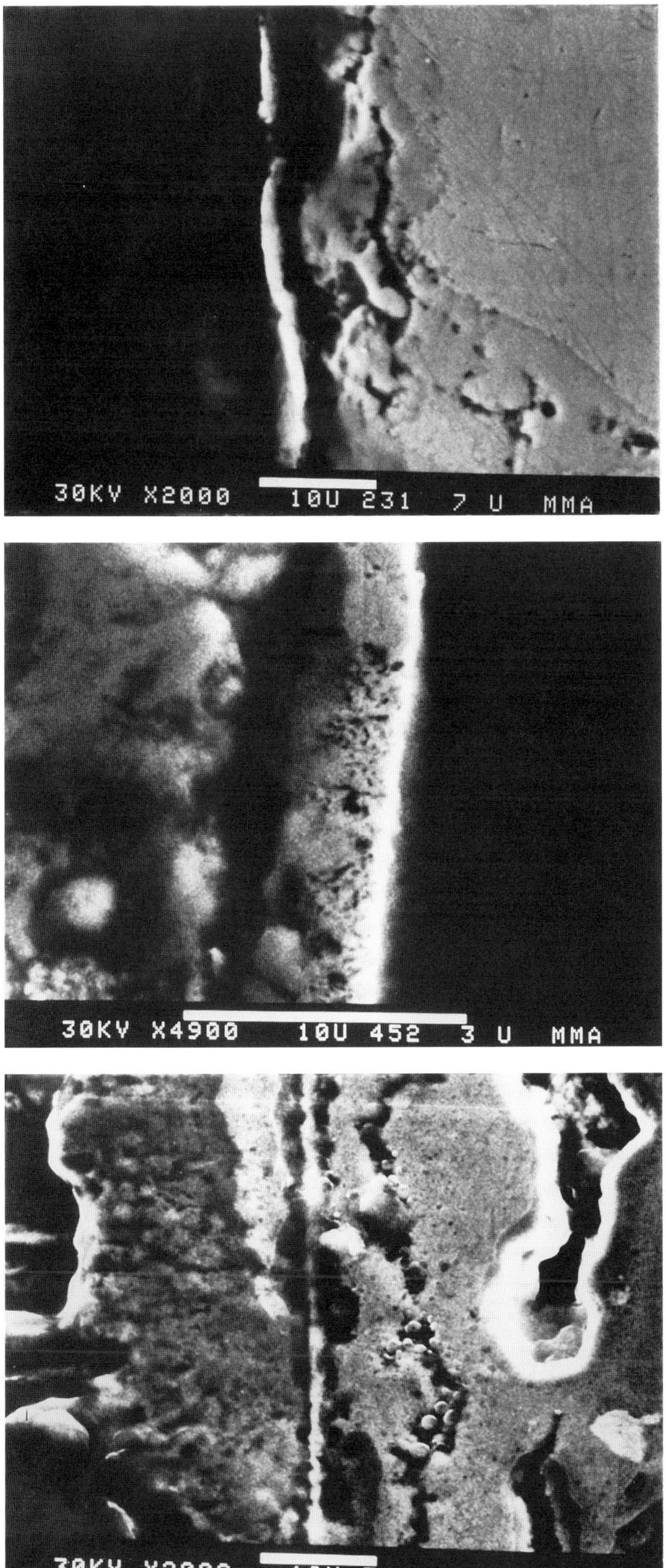

Figure 13.13 Gilded copper sheet, Vicús, backscattered images of metallographic sections: (a) MMA1987.394.252 (2000×); (b) MMA1987.394.596 (4900×); (c) MMA1987.394.598 (2000×). (Photographs: M.T. Wypyski.)

(a)

(b)

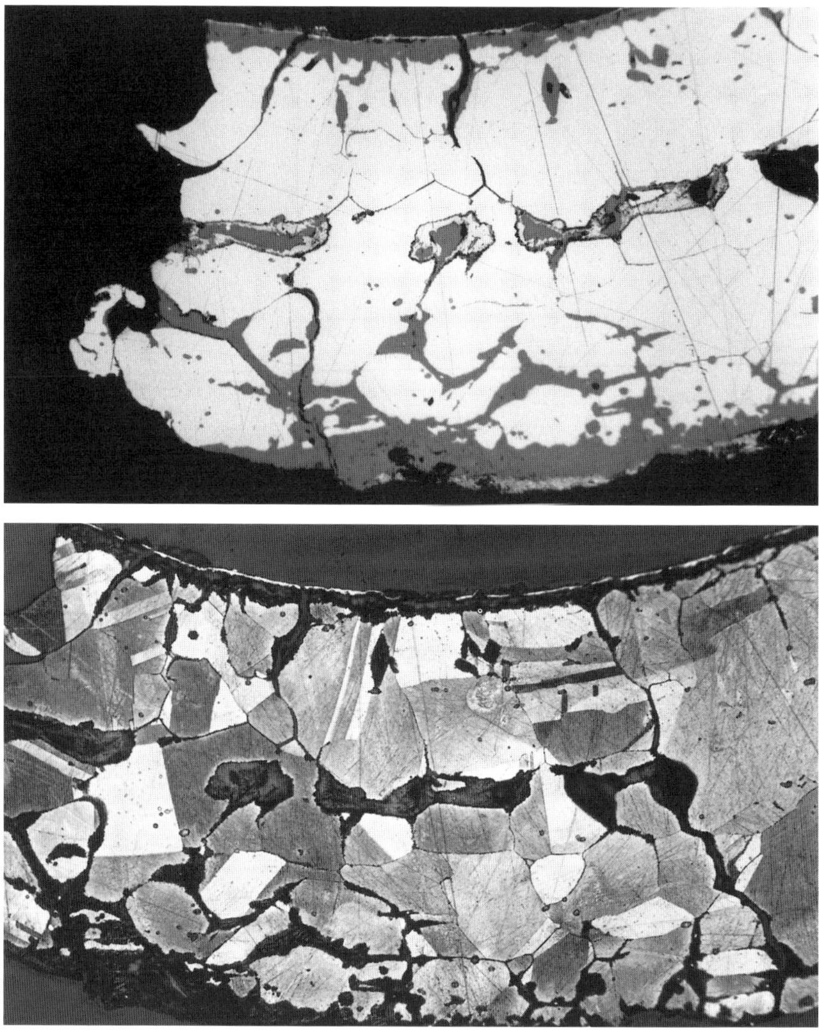

Figure 13.14 Gilded copper sheet, Vicús (MMA1987.394. 252), metallographic section observed before (a) and after etching with an alcoholic ferric chloride solution (b). Magnification: 200×.

(MMA 1987.394.598) were so variable that only the average gold to silver ratio, which was found to be fairly consistent, is reported (Table 13.2). Copper is the only metal detected in the interiors of all three substrates.

The production of metal sheet from cast ingots requires repeated hammering and reheating. The presence of recrystallised grains with abundant twinning within the copper substrate of the openwork plaque (Fig. 13.14a,b) indicates that heating was the final step in its manufacture. EDS

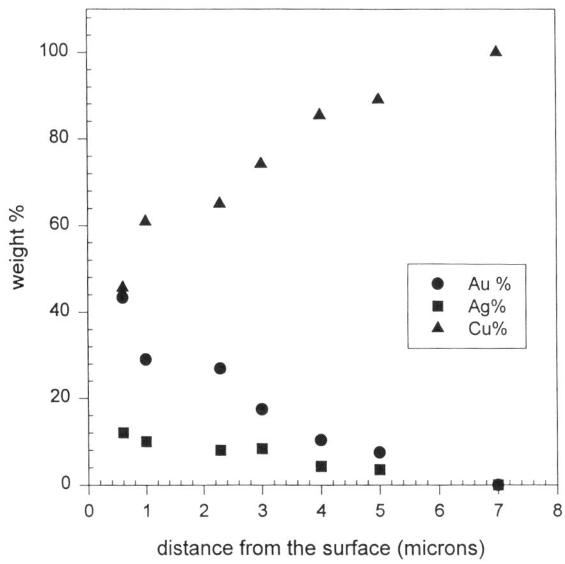

Figure 13.15 Weight percents of gold, silver and copper (normalised to total 100%), as a function of the distance from surface (MMA1987.394.252), as detected using EDS.

spot analyses were carried out on the two better preserved Vicús gilded copper sections in order to study compositional variations across the interface between the gold layer and the substrate, and the results for the openwork plaque were plotted in relation to the distance from the surface (Fig. 13.15). The diffusion of gold into both substrates to a depth of 5–6 μm suggests that, as in the case of the Loma Negra pieces, the gold was applied before the final reheating.

SEM was used to characterise and record the surface morphology of the Vicús samples before they were mounted for metallography (Fig. 13.16). In backscattered images of the gold layer on the openwork plaque (MMA 1987.394.252) a crack in the gilding and dark regions of corrosion can be

(a) (b) (c)

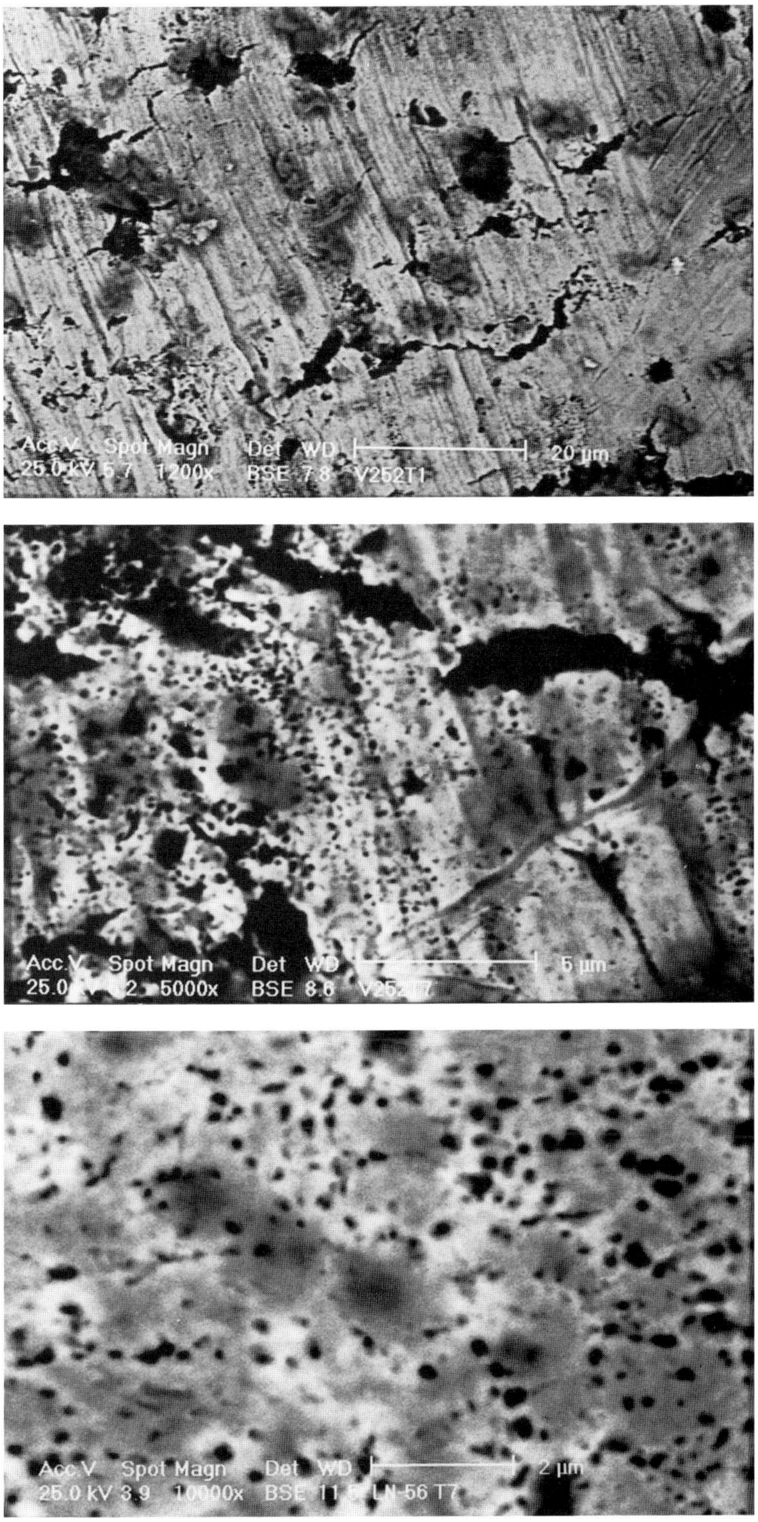

Figure 13.16 Gilded copper plaque, Vicús (MMA 1987.394.252), backscattered images of the surface: (a) 1200×; (b) 5000×; (c) 10 000×. (Photographs: L. Carapia.)

observed, as well as scratches on the surface. It was not possible to determine if these scratches were introduced during manufacture or subsequent to the removal of the plaque from its burial environment. The most important feature observed on the surface is a non-uniform distribution of pores (*c.* 0.1–0.2 μm) in diameter, similar to those observed on the Loma Negra surfaces (Fig. 13.10) and on the replication samples gilded using an electrochemical replacement technique (Fig. 13.9).

Conclusion

Metallographic examination and SEM and EDS analyses have revealed that gold layers on Vicús and Loma Negra copper substrates share important features, such as surface morphology, including the presence of pores, as well as extreme thinness, uniformity, and the presence of an interdiffusion zone at the copper-gold interface. None of the gold layers on the Vicús or Loma Negra objects studied show any evidence of having been applied as leaf or foil, or using amalgam, fusion, diffusion or depletion gilding techniques.

Two differences in the precious metal layers on artifacts associated with Loma Negra and those assigned to the Vicús culture can be noted. 'Wrinkles' were observed only on the latter objects. The wrinkles result from local failure in the bonding between the gilded layer and the copper substrate, but whether this occurred at the time of manufacture, or developed subsequently during use or burial, is difficult to establish.

A second difference is the amount of copper detected in the gold layers of the Loma Negra and the Vicús samples (Tables 13.1 and 13.2). Far more copper, up to a maximum of 44.6 per cent, was detected in the Vicús samples; in the two Loma Negra gold layers analysed under identical operating conditions using the same high-lateral resolution electron microprobe, the copper content does not exceed eighteen per cent. It is probable that a significant contribution to the EDS signals comes from copper that has migrated into the gold layer as a result of corrosion processes.[10] In addition, copper atoms diffuse through the gold layers as a consequence of thermal treatments – the resulting bond being necessary to improve the adhesion of the gold to the substrate – both along grain boundaries and through defects in the film (Guillet *et al.* 1985).

As mentioned previously, gold and silver diffused 5–6 μm and 2–4 μm into the copper substrates of the Vicús and Loma Negra samples respectively, as measured from their surfaces. It is possible that longer thermal treatments were applied to the Vicús pieces, which would explain, although only in part, the higher copper content detected in the gold layers on Vicús objects. It should be noted that the composition of the precious metal layers and the approximate distances that gold and silver diffused into the substrates have been calculated for only two samples from each group; more representative values might be established using a larger number of samples.

In spite of these two differences, the evidence obtained thus far suggests that the same method was used in the manufacture of both Moche gilded and silvered copper artifacts from Loma Negra and those made by their Vicús neighbors. The experimental reexamination of the electrochemical replacement process proposed by Lechtman and her colleagues allowed us to recreate gilding layers as thin as the ones on Moche and Vicús objects from the Piura Valley, and to relate other properties, such as porosity, to features observed in the gilded surfaces of the ancient artifacts.

The method of gilding used on the Loma Negra artifacts differs from the depletion gilding process observed on Moche copper objects from Sipán, which was used by other Central and Northern Andean cultures as well. The use of the same method by both the Moche at Loma Negra and their Vicús neighbors suggests that the electrochemical replacement plating process is part of the shared metallurgical tradition of the Piura Valley during the Early Intermediate Period.

Further analysis using high resolution transmission electron microscopy and electron diffraction is in progress in order to determine crystalline structure, morphology and particle size in gilding layers from the Piura Valley and in the replication samples. [11]

Acknowledgements

We gratefully acknowledge the generous contributions of our colleagues. Mark T. Wypyski of The Sherman Fairchild Center for Objects Conservation carried out the SEM examination of the Vicús and Loma Negra sections and M. José Yacaman, Leticia Carapia and Jesus Arenas of the Universidad Nacional Autónoma de México and the Instituto Nacional de Investigaciones Nucleares (ININ), Mexico, helped with the EDS analysis and the SEM surface analysis. Our thanks to Julie Jones of the Department of the Arts of Africa, Oceania and the Americas, The Metropolitan Museum of Art, James H. Frantz, Richard E. Stone and Ellen G. Howe of the Sherman Fairchild Center, and Lawrence Becker of the Worcester Art Museum for their help and support in the course of this research. Also thanks to Craig Morris, Judith Levinson and Sumru Aricanli of the American Museum of Natural History. This research was supported in part by the L.W. Frohlich Charitable Trust.

Notes

1. Fusion gilding describes a process of applying a molten gold alloy to metal or alloy with a higher melting point. The substrates can be dipped into a molten bath or, if only one surface or area is to be coated, the gold can be flushed on locally. The use of these techniques by pre-Columbian metalworkers at La Tolita, Ecuador and in the Nariño region of Colombia has been studied by Bergsøe (1938) and Scott (1986a,b). The thickness of gold layers applied in this manner are typically upwards of 100 µm (Scott 1986a,b).
2. Conventional EDS analysis was carried out by Mark T. Wypyski at The Sherman Fairchild Center for Objects Conservation, and analysis using a high resolution electron microprobe was carried out at Instituto Nacional de Investigaciones Nucleares (ININ), Mexico, by Leticia Carapia and Jesus Arenas.
3. Diffusion gilding involves the application of foil or leaf to a metal surface, followed by burnishing and heating. This technique has been reported for pre-Columbian metalwork from Ecuador (Scott 1986b).
4. Amalgam gilding has never been found on pre-Columbian metalwork from South America (Bray 1993: 183) although the use of cinnabar in a variety of Andean contexts , including the Vicús site at Cementerio Yécala, has been reported (Ríos & Retamozo 1993: 46).
5. Exceptionally, the backplates and shafts of a pair of Loma Negra earflares (MMA 1979.206.1245 and 1246) (Lechtman *et al* 1982: 7), as well as a pair of appliqués that were probably earflare frontal ornaments (MMA 1979.206.1234 and 1235), were gilded using a depletion gilding method (Schorsch *et al.* 1996: 153).
6. Lechtman *et al.* (1982: 12) reported finding approximately 1 per cent silver by weight in copper substrates of Loma Negra gilded and silvered objects that they

examined. Copper came into widespread use in the Central Andes during the beginning of the Early Intermediate Period (Lechtman 1980: 289), but arsenical copper did not appear until the subsequent Middle Horizon (*c.f.* Lechtman, 1975; Lechtman 1988, 1991)

7. For the construction of this object described in depth, see Schorsch (1998: 115–16).
8. Because of their inability to resist attack from the burial environment, silver-rich layers on the Loma Negra objects often survive only as isolated particles on the copper substrate. For this reason silvered copper samples were not chosen for detailed measurements of thickness or composition in this study.
9. The gold layer in the section shown in Figure 13.10 was obtained by dipping the copper sheet into the plating solution for a few seconds. Lechtman *et al.* (1982: 16) use a deposition time of five minutes in order to obtain a 1 μm thick layer. It is also possible to obtain thinner precious metal layers by lowering the concentration of gold and/or silver ions in the plating solution through the addition of water.
10. The oxidation of copper at the surface of the gilding layer has a direct effect in the extent of diffusion because it removes copper atoms, enhancing further diffusion from the substrate (Pinnel 1979).
11. This work is being undertaken in collaboration with M. José Yacaman and J. Rendón from the Universidad Nacional Autónoma de México and the Instituto Nacional de Investigaciones Nucleares (ININ), Mexico.

References

Alva, W. 1988. 'Discovering the New World's richest unlooted tomb'. *National Geographic,* **174**: 510–49.

Alva, W. and Donnan, C. B. 1993. *Royal Tombs of Sipán.* Los Angeles: Fowler Museum of Cultural History, UCLA.

Bawden, G. 1994. 'La paradoja estructural: la cultura Moche como ideología política'. In *Moche: propuestas y perspectivas*, S. Uceda and E. Mujica (eds), 389–412. Lima: Universidad Nacional de La Libertad, Trujillo.

Bergsøe, P. 1938. 'The gilding process and the metallurgy of copper and lead among pre-Columbian Indians'. *Ingeniørvidenskabelige skrifter*, **A46**: 5–58.

Bray, W. 1993. 'Techniques of gilding and surface enrichment in pre-Hispanic American metallurgy'. In *Metal Plating and Patination: Cultural, Technical and Historical Developments*, S. La Niece and P. Craddock (eds), 182–92. London: Butterworth-Heinemann.

Diez-Canseco, M. 1994. 'La sabiduría de los orfebres'. In *Vicús*, K. Makowski, C. B. Donnan, I. Amaro Bullon *et al.* (eds), 182–209. Lima: Banco de Crédito del Perú.

Disselhoff, H-D. 1969. 'Seis fechas radiocarbónicas de Vicús'. *Verhandlungen des 38. Internationalen Amerikanistenkongresses,* **1**:341–5. München: Kommissionverlag Klaus Renner.

Disselhoff, H-D. 1971. *Vicús: Eine neu entdeckte alteperuanische Kultur*. Volume in series *Monumenta Americana VII*, G. Kutscher (ed.). Berlin: Gebr. Mann Verlag.

Disselhoff, H-D. 1972. 'Metallschmuck aus der Loma Negra, Vicús (Nord-Peru)'. *Antike Welt*, **3**(2): 43–53.

Donnan, C. B. 1990. 'Masterworks of art reveal a remarkable pre-Inca world'. *National Geographic*, **177**: 16–33.

Eckmann, C. 1993. 'Gilding processes and surface treatment of gold–silver–copper alloys from the royal tombs of Sipán'. Paper presented at the colloquium Andean Royal Tombs, Works of Art in Metal, The Metropolitan Museum of Art, 10 May 1993.

Forty, A. J. 1979. 'Corrosion micromorphology of noble metal alloys and depletion gilding'. *Nature*, **282**: 597–8.

Forty, A. J. and Rowlands, G. 1981. 'A possible model for corrosion pitting and tunneling in noble metal alloys'. *Philosophical Magazine*, **A43**(1): 171–88.

Guillet, M., Guglielmacci, M. J., Benhenda, S. *et al.* 1985. 'Transport of copper through a thin protective film of gold'. *Applied Surface Science*, **24**: 1–10.
Guzman Ladrón de Guevara, C. and Casafranca Noriega, J. 1964. 'Vicús. Informe preliminar de excavaciones'. *Informaciones Arqueológicas 1*. Lima: Comisión Nacional de la Cultura.
Howe, E. G., Schorsch, D., Wypyski, M. T. *et al.* 1993. 'Technical overview of Loma Negra Metalwork'. Paper presented at the colloquium Andean Royal Tombs, Works of Art in Metal, The Metropolitan Museum of Art, 10 May 1993.
Jones, J. 1975. 'Mochica works of art in metal: a review'. In *Pre-Columbian Metallurgy of South America*, E. P. Benson (ed.), 53–104. Washington DC: Dumbarton Oaks Research Library and Collections, Trustees for Harvard University.
Kaulicke, P. 1994. 'La presencia Mochica en el Alto Piura: problemática y propuestas'. In *Moche: propuestas y perspectivas*, S. Uceda and E. Mujica (eds), 327–58. Lima: Universidad Nacional de La Libertad, Trujillo.
Kaulicke, P. and Makowski, K. 1990. *Informe del Proyecto Arqueológico del Alto Piura (1988–1989)*. Lima: Instituto Nacional de la Cultura.
Lechtman, H. N. 1975. 'Issues in Andean metallurgy'. In *Pre-Columbian Metallurgy of South America*, E. P. Benson (ed.), 1–40. Washington DC: Dumbarton Oaks Research Library and Collections, Trustees for Harvard University.
Lechtman, H. N. 1979. 'A pre-Columbian technique for electrochemical replacement plating of gold and silver on copper objects'. *Journal of Metals*, **31**: 154–60.
Lechtman, H. N. 1980. 'The central Andes: metallurgy without iron'. In *The Coming of the Age of Iron*, T. A. Wertime and J. D. Muhly (eds), 267–331. New Haven, Conn.: Yale University Press.
Lechtman, H. N. 1988. 'Traditions and styles in Central Andean metalworking'. In *The Beginning of the Use of Metals and Alloys*, R. Maddin (ed.), 344–78. Cambridge, Mass.: MIT Press.
Lechtman, H. N. 1991. 'The production of copper–arsenic alloys in the central Andes: highland ores and coastal smelters?', *Journal of Field Archaeology*, **18**: 43–76.
Lechtman, H. N., Erlij, A. and Barry, E. 1982. 'New perspectives on Moche metallurgy: techniques of gilding copper at Loma Negra, Northern Peru'. *American Antiquity*, **47**: 3–30.
Makowski, K. 1994. 'Los Señores de Loma Negra'. In *Vicús*, K. Makowski, C. B. Donnan, I. Amaro Bullon, *et al.* (eds), 83–141. Lima: Banco de Crédito del Perú.
Matos Mendieta, R. 1965–66. 'Algunas consideraciones sobre el estilo Vicús'. *Revista del Museo Nacional*, **34**: 89–130.
Murro Mena, J. A. 1994. 'Arqueólogos y huaqueros'. In *Vicús*, K. Makowski, C. B. Donnan, I. Amaro Bullon *et al.* (eds), 3–21. Lima: Banco de Crédito del Perú.
Narváez, A. 1993. 'La Mina: una tumba Moche en el Valle del Jequetepeque'. In *Moche: propuestas y perspectivas*, S. Uceda and E. Mujica (eds), 59–81. Lima: Universidad Nacional de La Libertad, Trujillo.
Pinnel, M. R. 1979. 'Diffusion-related behavior of gold in thin film systems'. *Gold Bulletin*, **12**(2): 62–71.
Ríos, M. and Retamozo, E. 1993. 'Investigaciones sobre la metalurgia Vicús'. *Gaceta Arqueológica Andina*, **7**(23): 33–66.
Schorsch, D. 1998. 'The manufacturing methods of composite gold and silver artifacts from Loma Negra'. *Metropolitan Museum of Art Journal*, **33**: 105–36.
Schorsch, D., Howe, E. G. and Wypyski, M. T. 1996. 'Silvered and gilded copper metalwork from Loma Negra: manufacture and aesthetics'. *Boletín del Museo del Oro*, **41**: 145–63.
Scott, D. A. 1983. 'Depletion gilding and surface treatment of gold alloys from the Nariño area of ancient Colombia'. *Journal of the Historical Metallurgy Society*, **17**: 99–115.
Scott, D. A. 1986a. 'Gold and silver coatings over copper: an examination of some artifacts from Ecuador and Colombia'. *Archaeometry*, **28**: 33–50.
Scott, D. A. 1986b. 'Fusion gilding and foil gilding in pre-Hispanic Ecuador and Colombia'. In *Metalurgia de América Precolombina*, C. Plazas (ed.), 307–25. Bogotá: Museo del Oro.

Shimada, I. 1988. 'A metallurgical survey in the Vicús region of the far north coast of Peru'. Paper presented at the Seventh Northeast Conference on Andean Archaeology and Ethnohistory, Amherst, Massachusetts, 6–7 November 1988.
Shimada, I. 1994. 'Los modelos de organización sociopolítica de la cultura Moche: nuevos datos y perspectiva'. In *Moche: propuestas y perspectivas*, S. Uceda and E. Mujica (eds), 359–87. Lima: Universidad Nacional de La Libertad, Trujillo.
Visco, R. 1974. 'Gold'. In *Modern Electroplating,* F. Lowenheim (ed.), 224–44. New York: Wiley Interscience.

14 Gilding Techniques of the Renaissance and After

Andrew Lins

Abstract

This essay describes developments in gilding on metals from the Renaissance to the present, particularly electrochemical gilding, mercuric nitrate priming for fire gilding, gold electroplating, and dip or immersion gilding. These developments reflect the evolution of applied chemistry in metal production, an evolution that is often poorly represented in conservation studies of historic metals. Recent experimental work is presented in the Appendix, reproducing several of the very numerous processes published during this period.

Historical background up to AD1500

The traditional means of gilding through the antique and medieval periods, West and East, have been described in preceding papers. They are:

- leaf gilding with adhesives
- leaf gilding with mechanical bonding – either cold or with heat to create diffusion at the gold/substrate interface
- leaf or foil gilding with lead and/or tin solder – close plating
- fire gilding or amalgam gilding
- plaquage or rolled gold
- electrochemical displacement reactions, using powders, pastes, or solutions
- surface enhancement/depletion gilding.

While the most ancient of these techniques – the attachment of foil or leaf to a substrate by simple mechanical means or with adhesives[1] – persisted into the Renaissance and later periods, it was largely displaced by more durable methods for applying thin layers of gold to various metal substrates. Fire gilding using gold amalgams or pastes was the predominant technique at the beginning of the Renaissance. The use of mercury as an adhesive for leaf also continued into this period. Both techniques employ heat to drive off excess mercury, but leave behind a finite and measurable trace of mercury.[2]

Concurrent with these techniques was the ongoing use of leaf or foil, which was applied directly to the substrate and burnished into a matted substrate. The mechanical join was sometimes heated to create a diffusion bond of the gold with the substrate. This technique became known in its later manifestations in the eighteenth and nineteenth centuries as gilding *au feu* or as *French gilding* in the English-speaking world. As described by Stalker and Parker (1688) and well illustrated by Diderot and D'Alembert (1751–80), leaf or foil was applied to surfaces at blue heat, and burnished into the surface; on cooling the contraction of the mechanical bonds were thought to improve adhesion.[3]

Common in the medieval period and later was the use of soft solder (tin and lead), tin or lead to enhance the adhesion of gold or silver to a metal substrate. In these soldering techniques (close-plating), heat was used – probably with a soldering iron in most cases – to join the gold foil or thick leaf to the 'tinned' surface. Sometimes silver was laid down first to provide a better surface for gilding. This means of applying gold is particularly suited to ferrous substrates, which resist direct amalgamation with mercury.[4] How common gold close-plating is during 1500 to 1850 remains to be evaluated, as guild and assaying regulations in the West restricted the decoration of many kinds of metalwork, particularly in the early part of this period.

Other techniques are certainly known at an early time. Not all of them appear to have been adopted throughout the metalworking trades or guilds. For instance, the use of plaquage or rolled gold – in which a thin sheet of gold was clamped or pressed on to a thicker sheet of silver or copper (later brass) and then fused – has medieval or earlier origins in metal wire and textile work.[5] The techniques are very similar to the manufacture of Sheffield plate dating to the mid-eighteenth century, yet plaquage is little documented in the silversmiths or goldsmiths trades of 1500–1800.[6] This example serves as a reminder that much of the specialised craft training and guild apprenticeship was very conservative of the status quo, very narrowly focused, or both.

Electrochemical gilding by paste or powder

This historical review of gilding practices after AD1500 starts with gilding by electrochemical displacement. In displacement reactions, a dissolved metal (gold) is reduced and plated on to the substrate by an electrochemical reaction, which, in equal measure, oxidises part of the substrate surface. Strong mineral acids – with additives or impurities that we would now identify as oxidants or complexing agents – are required to dissolve gold. Recipes for pastes or powders containing dissolved gold were known in the medieval period. For practical use these recipes required acids of dependable and consistent strength to be available.

The origins of nitric acid manufacture are of historical and technical significance in creating such pastes or powders, since nitric acid is the likeliest mineral acid solvent for gold alloys when combined with common salt or sal ammoniac (ammonium chloride). In the West, evidence points at least as far back as the so-called Latin Geber, who clearly described the manufacture of nitric acid in his *Investigation of Perfection*, a text now dated to thirteenth-century Spain:

> First take vitriol of Cyprus [copper sulfate], one pound; of saltpeter, two pounds; and of Jamenous allom, one fourth part; extract the water with redness of the alembic (for it is very solutive). . . This is also made much more acute if in it you shall dissolve a fourth part of sal ammoniac; because that dissolves gold, sulfur and silver.[7]

A number of medieval texts mention the manufacture, dissolution, or precipitation of gold, including Jabir (the Arabic writer of the ninth

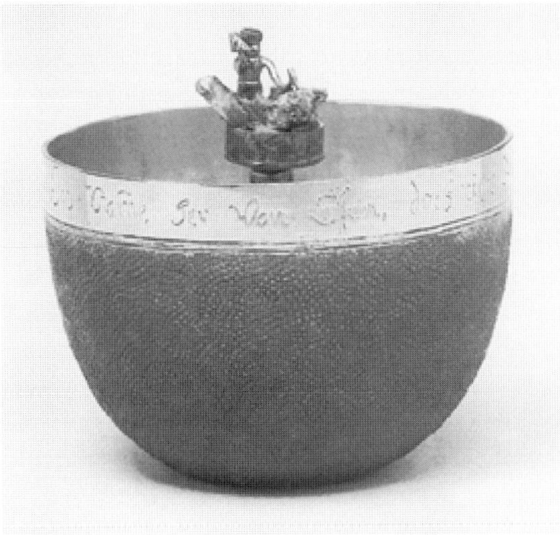

Figure 14.1 Herrengrun beaker (Turner 1974: plate 1) (See Plate 78.)

century), Theophilus, Roger Bacon, Albertus Magnus, and Girard of Cremona.[8] Interest in the dissolution of gold and its reappearance as a precipitate continued (not surprisingly) into the Renaissance and later.

Illustrating the point is a series of metal objects from Hungary, dating from the middle of the seventeenth century, variously inscribed in German or Latin: 'Iron I was, copper I became, with gold am I covered'.[9] Most of these objects are travellers' novelties in the form of small cups or tobacco holders, made of copper precipitated from copper-rich sulphate solutions on iron substrates (Fig. 14.1). The inscriptions on the metalwork bear witness to the persuasiveness and pervasiveness of the notion of permutation expressed in alchemical texts: that is, that one substance or combination of substances can be made to transmute into another state or substance (Fig. 14.2). Mystical contexts notwithstanding, alchemical texts

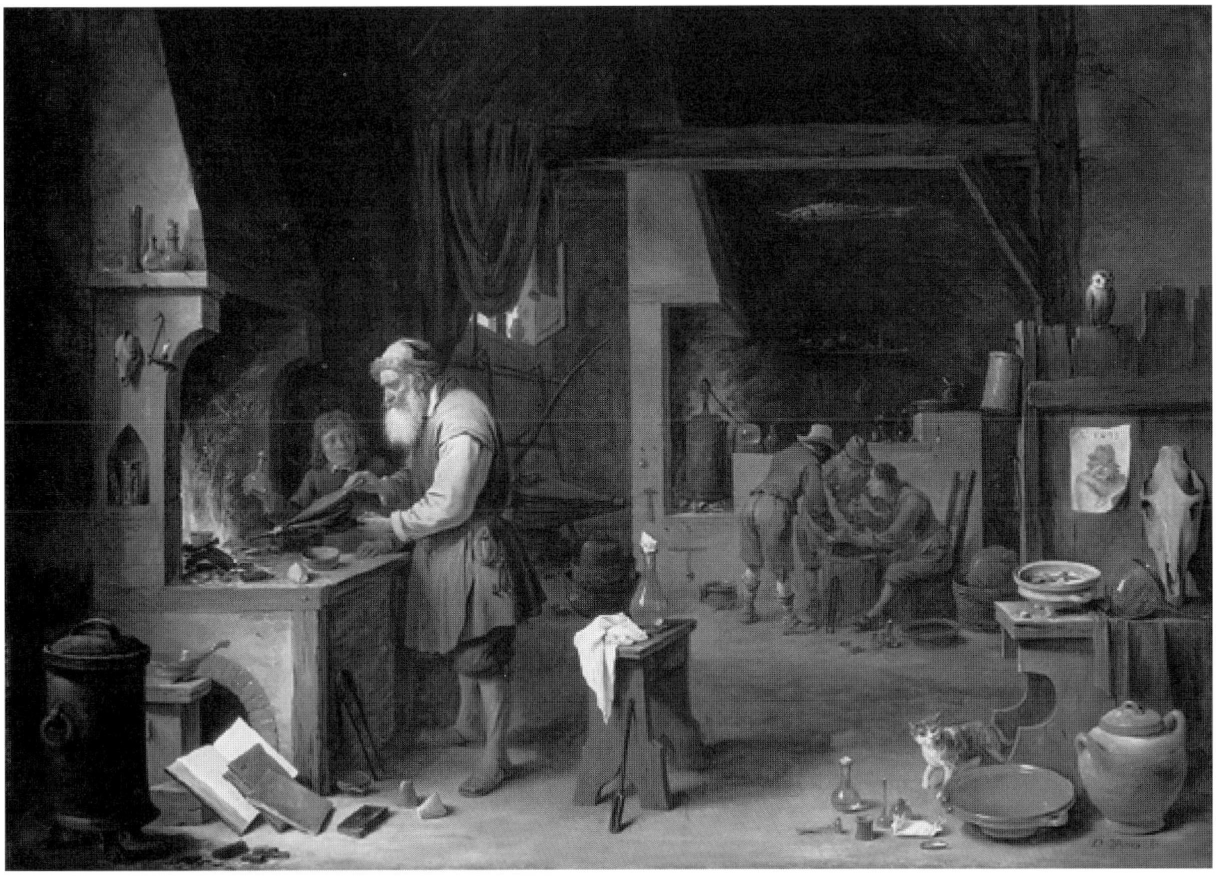

Figure 14.2 *The Alchemist*, David Teniers, 1649 (Philadelphia Museum of Art). (See Plate 79.)

of 1500–1750 formed part of the growing body of practical knowledge about how one metal could be dissolved and then precipitated onto another.[10] A number of examples are certainly also found in the non-alchemical, 'standard' texts on mining, metallurgy, and chemistry in the sixteenth and seventeenth centuries (Biringuccio 1540; Agricola 1556; Cellini 1568; Ercker 1580; Barba 1640; Glauber 1656–61), about which present-day readers may tend to attribute more widespread popularity and influence than they formerly enjoyed.

The use of gold-containing powders, pastes, and solutions are described repeatedly in encyclopedias of the arts, which appeared in ever larger numbers in the early eighteenth century and continued throughout the nineteenth century. One of many recipes – these were often copied or plagiarised from other earlier authors – is from an English patent application of 1698 by a Robert Southwell, for making a powder that could create a very thin layer of gilding, similar to pastes and powders used for silvering the faces of clocks and instruments at this time and known in the earlier medieval period: 'steeping linen rags in a solution of gold, then burning them; with a piece of cloth dipped in salt water, rubbing ashes over the silver intended to be gilt'.[11]

Ultimately such recipes – whether this one is actually workable I do not know – resulted in the development of liquid gold paints around 1800, detailed later in the writings of Brogniart at Sevres and others.[12] How frequently such pastes or powders were used in the metal trades prior to the eighteenth century remains an open question, the answer to which is complicated by the difficulties of examination with current analytical techniques on very fragile, very thin surface layers that have been altered by time.[13]

Modifications of fire gilding

Use of mercuric nitrate

One paste or solution that developed from electrochemical exploration is the mercuric nitrate primer applied before gilding in the traditional fire gilding process, clearly the predominant gilding technique in the West until the middle of the nineteenth century. The practice of priming or 'quicking' a metal before fire gilding is described in texts such as Stalker and Parker (1688) and Chambers (1728)[14] and in books about the large-scale manufacture of buttons, toys, military decorations, and ormolu wares (Matthew Boulton) that developed around Birmingham by the middle of the eighteenth century.[15] The technique was probably known much earlier in gold working centres like Paris, London, Nürnberg, Strasbourg, Pforzheim, Florence, and others.

While nitric acid or variants of it had been used for some time in the cleaning and preparation of metal surfaces for gilding, solutions of mercury in dilute nitric acid easily and quickly produced a mirror-like layer of mercury on copper and copper alloys by a simple electrochemical displacement reaction:

$$Hg^{2+} + M^0_{\text{substrate}} \rightarrow Hg^0_{\text{deposition}} + M^{2+}$$

The layer of mercury is made thick or thin, depending on the concentration of the salt, the time of exposure, and temperature of the bath or solution. Its particular effectiveness as a surface preparation was that the so-called quicked surface allowed gold amalgam to be spread very easily and rapidly, the surface being already wetted with mercury. Further advantage was gained by the savings in gold and by the increased reliability in spreading gold amalgams very thinly on to a quicked surface. An alternative in controlling the costs of gilding was to lay leaf, or a couple of layers of leaf, on such a quicked surface and then to heat gently in the fire, in order to keep the gold thickness even, reduce the waste from puddling, and eliminate the need to make an amalgam. These factors were not overlooked by metalworkers: a number of surviving account books from the eighteenth century deal as much or more with costs and economics of the processes as they do with materials and artistic techniques.[16]

Coloring and finishing

Modification of the traditional fire-gilding technique by the use of a mercury primer was certainly not the only modification of the process by applied chemistry during this period. There were, for instance, many recipes for altering the colour of fire gilding or ormolu after a piece was removed from the furnace. These included the use of acids and other chemicals to create matte or 'dead' gilding, which was popularised at the end of the eighteenth century. The processes, of which there are many in the literature, may be divided into two main groups: subtractive processes and additive processes.[17] The former are clearly related to the older ideas of *mis-en-couleur* or depletion gilding, while the latter (additive) processes are based on inorganic colourants as well as organic waxes and dyes and pigments, some of which diffuse into or bond on to the gold surface with heat.

Electroplating

Chemists and cyanides

Emerging ideas about electricity and solutions were brought together in the investigations of Volta and many others, leading to the invention of gold electroplating in the period *c.*1800–1840.

The sciences grew more empirically and academically based toward the end of the eighteenth century. A very strong interest in the charging and discharging of electricity drew experimenters in many directions. In papers published in 1800, Volta described the first usable cell or battery – constructed of zinc and silver plates – to produce electric current.[18] The reactions in aqueous solutions causing electric current flow soon became known as electrochemical reactions. Luigi Brugnatelli is currently credited with depositing the first intentional decorative electrogilding on a work of art, two silver medals, variously dated between 1801–1803, the actual physical confirmation for which is still lacking.[19] Many people – academics and amateurs – were attracted to this kind of experimentation. Among the most relevant for gold electrochemistry in this and the following years were: A. Volta, silver and zinc battery with salt electrolyte (1800); W. Nicholson and A. Carlisle, hydrogen and oxygen discharge from gold and platinum (1800); W. Cruickshank, silver and copper precipitation from solution (1800); K. W. Böckmann, silver, copper and tin precipitation (1801); J. L. W. Grüner, silver deposition from pile (1801); W. Hisinger and J.J. Berzelius, electrolytic decomposition of salt solutions (1803); and Sir

H. Davy, summary of his previous work on electrolytic decomposition of water and the effect of salts (1806–1807).

By the 1830s, such outstanding scientists as Faraday, Berzelius, and Leibig were exploring the electrochemical phenomena that resulted ultimately in the practical manufacture of gold electroplate. The most significant work was done by: M. Faraday, *Experimental Researches,* which provide the basic 'laws' of electro-deposition (1833) ; J. von Leibig, reduction of a noble metal salt by an organic reducing agent, specifically silvering by deposition from $AgNO_3$ solutions by means of sugars (1835); J. J. Berzelius, preparation of Au and Ag cyanides (1836); J. F. Daniell, battery or self-polarising cell with copper and zinc electrodes using sulphate solutions (1836); G. R. Elkington and H. Elkington, patents for dip plating gold (1836 and 1837): M. H. von Jacobi, work on galvanoplasty, *Atheneum* publication (4 May 1838); J. C. Jordan, galvanoplasty, English publication (8 June 1839); T. Spenser, galvanoplasty (paper read 13 Spetember 1839); G. R. Elkington and H. Elkington, British patent application number 8447, for gold plating using cyanides and a supply of current (25 March 1840, granted in September 1840); and C. de Ruolz, French patent application for gold plating using cyanides and a supply of current (1840).

Essential to this new process of gilding are cyanides. Alkaline cyanide solutions dissolve gold:

$$4Au + 8CN^- + O_2 + 2H_2O \rightarrow 4[Au(CN)_2]^- + 4\,OH^-$$

Gold cyanide solutions enable gold to be effectively plated on to metal surfaces made negative by connection to a battery or batteries. The inclusion of cyanide salts in gold plating solutions represents a significant improvement from many contemporary and earlier recipes, which were typically based on partial neutralisation of the reaction products of gold with strong mineral acids – such as aqua regia, usually HCl and HNO_3 in the ratio 3:1. The older gold solutions were more difficult to control. Experimenters described the gold deposits as burned, stained or uneven – the result of too rapid and uncontrolled attack on the substrate by the plating solutions. The use of cyanides redresses many of these shortcomings. While Jacobi in Russia and others had already described the use of plating baths using external current, G. R. Elkington and H. Elkington were the first to make commercial application for the use of cyanides in gold plating with an external current, in a British patent of 1840. They had close competition from the Frenchman de Ruolz who claimed, in his French patent filed just a couple of months after the Elkingtons filed theirs, that he was the originator of the process of gold electroplating using cyanides. The economic impact of the process must have seemed overwhelming at the time; the dispute over patent rights grew very heated and voluminous legal claims were filed. Enormous publicity about the various claimants spread news of the technology so that within a matter of months it was known in many parts of the world.[20]

By 1850 there were few parts of the European-visited world of metalworking that had not heard of and applied some part of the new technology. In most of the older centres of metals production, the fire gilding process was almost completely displaced, propelled by the recognition – apparent to metalworkers since the medieval period – that fire gilding was a serious health hazard.[21] Gold electroplating became the primary method of gilding from the second half of the nineteenth century and retains still this dominance.

In order for the new electroplating technology to spread, it was necessary for a sufficient and economic supply of cyanide salts to be available.

Initially this seems to have been supplied by plant extracts.[22] By the 1830s a number of ferro- and ferricyanide reactions were produced by heating tannery wastes (decomposing hooves) with carbon (charcoal) and iron (rust). The pigment Prussian blue was formed by such reactions, as were metal cyanide/ferri or ferrocyanide salts for electroplating baths. The investigations leading to the synthesis of cyanides include: Macquer's 'Memoirs on Prussian blue' (1749–52), describing work with cyanogen compounds; C. W. Scheele, 'Experiments concerning the colouring principle in Prussian blue' (1782–83), including the production of HCN from NH_3 and precipitation of $Hg(CN)_2$; Clouet, confirmation of Scheele's work (1791); Gay-Lussac, 'Recherches sur l'acide prussique' (1812–15), including the production of anhydrous HCN from $Hg(CN)_2$); Berzelius, double cyanides salts of metals (1819–20); Desfossès, KCN production from N_2 heated over potassium carbonate and charcoal (1828); and Glassford and Napier, 'On the cyanides of the metals, and their combinations with cyanide of potassium. Part 1. Cyanide of gold' (1844).[23] A number of synthesis routes were worked out in the following years, none much simpler than that of Wohler, 1850:

$$K_4Fe(CN)_6 + 3\,H_2SO_4 \rightarrow 2K_2SO_4 + FeSO_4 + 6\,HCN$$

Basic processes of electroplating

The basic mechanisms of electroplating are significantly different from previous gilding techniques. In the typical set-up for electroplating, the object to be plated is connected to the negative pole of a battery or other electrical supply and is defined as the cathode.[24] The opposing *electrode* in the plating solution is connected to the positive pole of the power supply and termed the anode. In thermodynamically well behaved solutions, each electrode undergoes reversible changes during the plating process (Fig. 14.3).

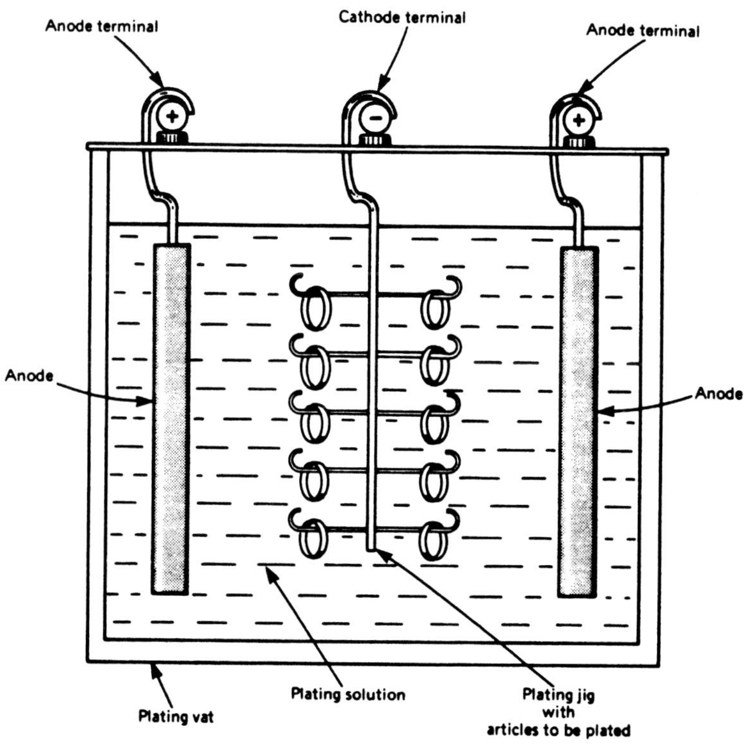

Figure 14.3 Typical electroplating bath and power supply.

The net negative charge on the cathode draws simple, positively charged metal cations to its surface. There, by electron exchange, the cation is reduced to metal, depositing or plating out on the surface of the cathode:

$$M^{z+}_{\text{solution}} + ze^{-}_{\text{cathode}} \rightarrow M^{0}_{\text{deposition}}$$

At the anode (gold in most gold plating baths), the oxidation reaction is:

$$M^{0}_{\text{anode}} \rightarrow M^{z+}_{\text{solution}} + ze^{-}_{\text{anode}}$$

Table 14.1 Standard reduction potentials (vs SHE at 298K)[25]

Reaction	Potential (V)
$Au^{+} + e^{-} \leftrightharpoons Au$	1.692
$Hg^{+2} + 2e^{-} \leftrightharpoons Hg$	0.851
$Ag^{+} + e^{-} \leftrightharpoons Ag$	0.7996
$Hg_2^{2+} + 2e^{-} \leftrightharpoons 2Hg$	0.7961
$Cu^{+} + e^{-} \leftrightharpoons Cu$	0.522
$Cu^{2+} + 2e^{-} \leftrightharpoons Cu$	0.3402
$Pb^{2+} + 2e^{-} \leftrightharpoons Pb$	–0.1262
$Sn^{2+} + 2e^{-} \leftrightharpoons Sn$	–0.1375
$Ni^{2+} + 2e^{-} \leftrightharpoons Ni$	–0.257
$Fe^{2+} + 2e^{-} \leftrightharpoons Fe$	–0.447
$Zn^{2+} + 2e^{-} \leftrightharpoons Zn$	–0.7628

In some cases, the deposited metal initially copies the crystal lattice spacings of the basis metal, a phenomenon known as epitaxial growth. A reasonable example of reversible (Nernstian) behaviour is shown by the copper sulphate bath in which electroforming or galvanoplasty is usually performed. It is this reversible sort of electrode process that is listed in the standard tables of reduction potentials (Table 14.1).

Gold cyanide solutions behave less predictably and reversibly. There is a substantial overpotential necessary for deposition to occur and the deposition process itself becomes more complicated as a result of the negatively charged cyanide complexes.[26] The mechanism of how negatively charged species (in particular $[Au(CN)_2]^{-}$) plate out on a negatively charged cathode is not susceptible to simple explanation or illustration. That gold plates out from $[Au(CN)_2]^{-}$ is indisputably clear. The mechanism of direct deposition of gold from the complex cyanide at the cathode has been exhaustively studied in the mining and refining fields. Direct deposition appears to be the generally accepted mechanism, growing out of earlier work by Blum, Hogaboom, Raub Brenner and others:

$$[Au(CN)_2]^{-} + e^{-} \rightarrow Au + 2CN^{-}$$

Gold cyanides have been used in many different formulations for gold-plating baths (solutions) from the 1840s to the present. It is fair to say that many nineteenth century texts on this subject suffer from incomplete understanding of how the baths operate, including unsure identification of double cyanide gold salts that are quite difficult to purify and to characterise accurately. These shortcomings in understanding did not in any way diminish the popularity and dominance of alkaline cyanide solutions for gold plating until the late 1930s or 1940s.

Matte, dendritic deposits are typically produced by alkaline cyanide gold-plating solutions. In addition to the problems of finishing such deposits (scratch-brushing, polishing, buffing or burnishing), alkaline cyanide gold plating suffers from two common material defects: attack on the substrate metal, and extreme softness of the gold layer. Brinnel Hardness Numbers of 40 are common for alkaline cyanide plate, compared to values of 65–85 for leaded brass/bronze, 65–70 for annealed 85Cu15Zn, 60–150 for cast and work-hardened sterling silver alloys, and 100–270 for cast and wrought 14 carat gold alloys with varying amounts of copper and silver.[27] The toxicity and environmental hazards posed by cyanide solutions are additional problems of use, with health regulations and public awareness more established now than in previous generations.

Plating developments since 1930

These drawbacks of traditional alkaline cyanide baths, coupled with the need for very high purity, pore-free surfaces, are the primary reasons that new plating solutions have been devised for the increasingly stringent requirements of electronic, aerospace, and 'high tech' manufacturers. Buffered acid baths formulated with phosphates/citrates were introduced in 1939, following the discovery that $KAu(CN)_2$ remains stable at pHs as low as 3.5. Earlier attempts had been made to formulate bright plating solutions for gold, similar to those implemented around 1930 for nickel and tin. In contrast to the primarily organic brighteners in nickel and tin plating baths, the additives to gold solutions tend to be inorganic – primarily silver based, though later nickel and cobalt salts were introduced to both harden and brighten the gold layer (Fig. 14.4). Further improvements in the purity, reduction of porosity and hardness occurred as the electronics and other industrial applications of gold electroplating expanded in the late 1950s.[28] Spot or partial plating is routinely carried out with masks and other techniques, in which thickness of a micrometer and more can be built in less than 3 seconds on high speed, high volume production lines, while very low porosity gold layers can be created with plating thicknesses of 100 nm or less (Table 14.2). It has been common electroplating practice for many decades to interpose a layer of copper, silver or nickel (more recently), instead of the mercury layer, to improve the adhesion of the gold layer on base alloys.

- Traditional hot alkaline cyanide (excess cyanide)
- Alloy gold cyanide (non-free cyanide, often with cobalt or silver)
- Acid gold pH 4.0–4.5 (buffered citrate/ phosphate baths)
- Acid/neutral fine gold pH 6.2–6.8 (citrate/phosphate based)
- Non-cyanide electrolytes (alkaline sulfites)

Figure 14.4 Gold electroplating solutions

Table 14.2 Common thickness dimensions for gold electroplating

0.010 in	=	254 μm
0.001 in	≡	1 mil = 25 μm
0.0001 in	=	2.5 μm
0.00001 in	=	0.25 μm/ 254 nm
0.000001 in	≡	1 microinch = 0.0254 μm/ 25 nm

10 millionths of an inch, or 10 microinches (0.000 01 in), is equivalent to 254 nanometers or 2540 Ångstroms – just over 1000 layers of gold atoms in the 100 plane/direction (d_{100} – 2.36 Å, JCPDS #4-784).

Dip/Immersion plating: main processes

In the absence of an external source of electric current, there are two electrochemical routes for depositing gold in solution on base metal substrates:

Method 1: $Au^+ + M^0 \rightarrow Au^0 + M^+$, where M is the metal substrate
Method 2: $Au^+ + R^0 \rightarrow Au^0 + R^+$, where R is a reducing agent added to the bath.

Many of the early solutions devised for plating very thin layers of gold are based on electrochemical displacement reactions given by Method 1. The Elkington patents of 1836–37 employ such solutions. An improved immersion solution or dip, often called a *flash* gilding solution,[29] is specified in patent 1837, #7304, 'Coating Metals' under the name of Henry Elkington.[30] Like many of the earlier immersion recipes, this gilding solution is based on partial neutralisation of the reaction products of gold with strong mineral acids – such as aqua regia, HCl and HNO_3 in the ratio 3:1; the Elkington patent employed potassium bicarbonate to neutralize the acid. In the absence of any externally applied current, the older recipes suffered from too rapid and uncontrolled an attack on the substrate, resulting in uneven deposition, corrosion of the substrate and staining. The Elkingtons' recipes were less aggressive to the substrate and represent a continuation of techniques that they had been using for some time in jewellery, frames for eyeglasses, and other metalwork. To mitigate the aggressiveness of the baths, the Elkingtons reported quicking objects as a pretreatment for dip or immersion gilding.[31] Later baths employ cyanides of gold and of mercury. Layers much thinner than electroplate are formed, typically less than 1 μm. In current immersion practice by Method 1, the layers are commonly 0.1–0.3 μm thick, and are usually protected with a lacquer.[32]

Method 2 relies on the addition of a reducing agent to the bath. In this kind of bath, now usually called an autoplating bath, the substrate metal is not the source of the electrons that reduce the gold from solution; that is, the substrate is not attacked (ideally) by the solution as deposition proceeds. The plating layer can be deposited faster and thicker than with Method 1 baths. Autoplating baths can now plate thicknesses up to 100 μm at rates of 2.0 $\mu m\ h^{-1}$, compared with electrolytic rack rates of 60 $\mu m\ h^{-1}$. These autoplating baths, however, suffer from extreme sensitivity to contamination, high alkalinity, and limited operating life.

Other technological developments in the twentieth century

Technological innovations during the second half of this century have produced entirely new means of gilding, physically and chemically different from methods of the past. Most of these innovations have focused on aerospace, electronic and 'high tech' materials, some of which are now finding their way into museums of science and technology and other collections. Metallising and vacuum deposition are the primary techniques for depositing very thin and even layers of gold on different substrate materials.[33] There are a number of variants of these techniques, including sputtering, ion plating, and plasma arc metallising, all requiring very controlled conditions of application, extreme cleanliness and, usually, a vacuum. In sputtering technology, rates of deposition are in the range of 1–10 $nm\ s^{-1}$. These techniques are capable of regularly producing gold layers 10 nm thick, that is less than 50 layers of gold atoms. Liquid gold paints, inks and coatings have been formulated in a number of different vehicles – with different substrate pre- and post-application treatments – and can be manufactured in layers as thin as 0.1 μm (100 nm).

Conclusion

No matter what century or culture, techniques have been continuously sought and improved upon for creating gold layers on less costly metal substrates. Since the Renaissance, developments in gilding technology have been linked closely to advances in chemistry and metallurgy. Modifications and manipulations – the devious included – will continue to evolve, making Spenser's caveat[34] as relevant today as it was in the sixteenth century: 'Gold all is not that doth golden seem'.

Technical appendix

Laboratory findings

Fire gilding with and without a primer/quicking

A limited set of tests with mercuric nitrate solutions have been carried out. Priming/quicking proves amazingly effective in improving the wetting and spreading of gold amalgam (varying from 3Hg:1Au to 1.5Hg:1Au) on copper and on brass (85Cu15Zn) substrates. This is so clear that anyone who had seen such a primer applied would have used it, be it in the seventeenth or nineteenth century. Inevitably this observation provokes the question of how far such a technique might have spread before it was first recorded – most likely by an outsider in the form of a gentleman scholar like Diderot; someone without practical metalworking experience. This question remains unanswered.

Studies of primer application are summarised:

1. Changing the concentration, temperature, and time of immersion of copper or brass in mercuric nitrate solutions produces different initial thicknesses of mercury on copper (A) and brass (B). These layers, correspondingly, age differentially, as the mercury amalgamates into the substrate. For thin deposits (less than a micrometre, but not yet quantified reproducibly in the lab, but typical of short immersion times, weak solutions, low temperatures), the silvery layer of mercury begins to change color within the first 15 minutes and is apparently dissipated – to the unaided eye – within an hour. The loss of mercury from the surface in one example was followed by electron microprobe (EPMA with WDS). The sample was prepared by immersion of brass sheet in 1.0 % w/w $Hg(NO_3)_2$ for 15 seconds at room temperature and immediately placed under dry ice until inserted in the instrument. The results, in the absence of any quantitative WDS standards (complicated by the kinetics of the plating process), suggest that relatively little mercury (<1 μm average thickness) was deposited, although the surface appeared even and silver-coloured to the eye (see Figs 14.5 and 14.6).

2. There is little visual difference in the result between sample C, mercury-coated brass over which leaf has been laid, and sample D, quicked brass over which leaf has been laid. Both applications were followed by heating at 500 ± 15°C for 30 minutes. In samples prepared by method C, mercury was spread thinly on to the base metal with a wire brush at room tempera-

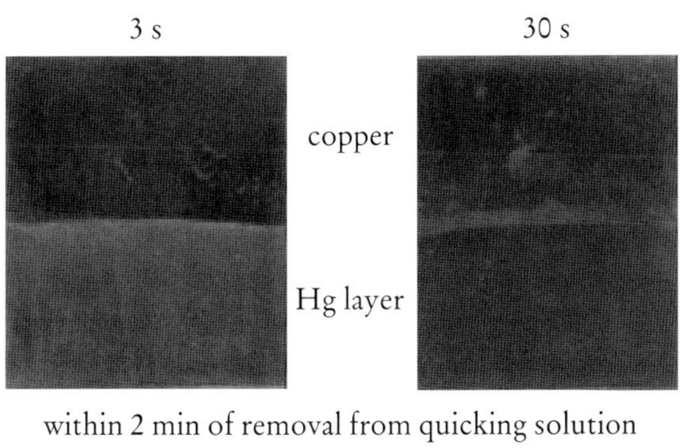

Figure 14.5 Copper samples A immersed in 1.0% w/w $Hg(NO_3)_2$ for 3 s and for 30 s at room temperature. Relatively little mercury (<1 μm average thickness) is deposited, although the surfaces appear even and silver-coloured to the eye. (See Plate 80.)

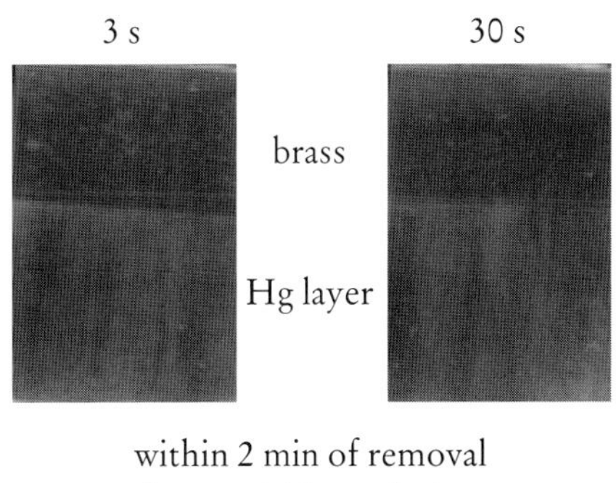

Figure 14.6 Brass samples B immersed in 1.0% w/w $Hg(NO_3)_2$ for 3 s and 30 s at room temperature and photographed within two minutes of depositing the mercury mirror. (See Plate 81.)

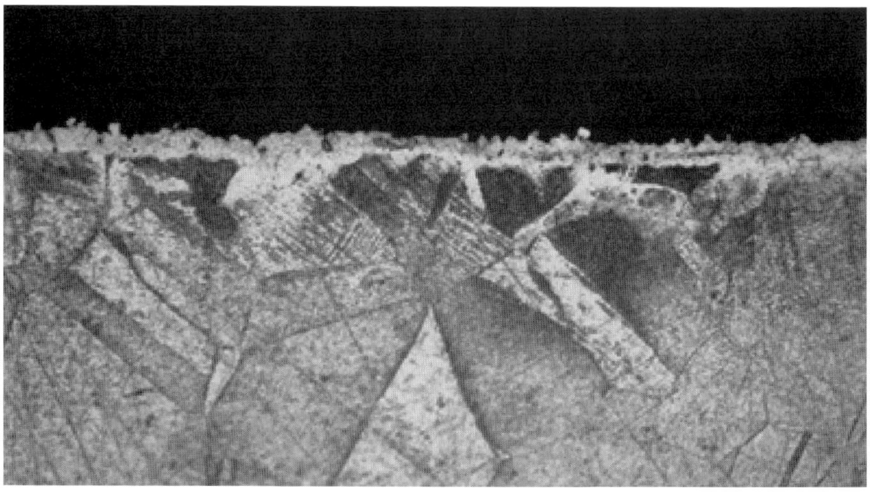

Figure 14.7 Brass sample C with mercury primer and leaf gilding, and heated at 500 ± 15°C for 30 min. As many as ten sheets of 23½ carat gold could be stuck/ amalgamated on the brass before excess leaf sloughed off. BF 500×. The average thickness of the gold layer in this sample is 12 m; the amalgam attack along some of the grain lines penetrates 25 m. (See Plate 82.)

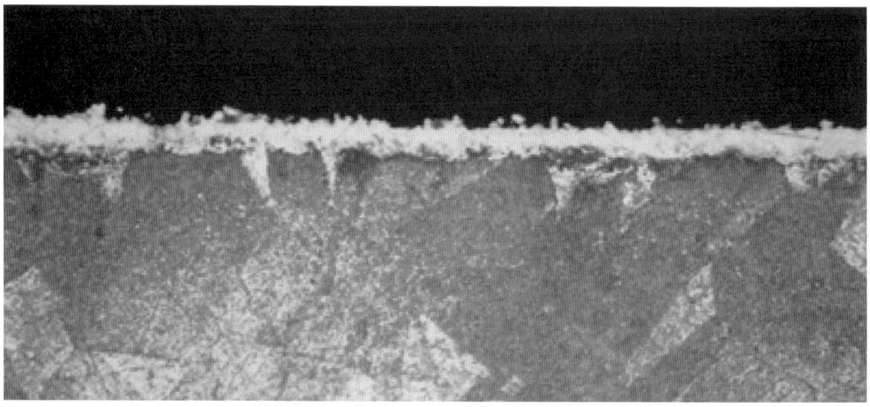

Figure 14.8 Brass sample D with 1% mercuric nitrate primer and leaf gilding, and heated at 500 ± 15°C for 30 min. The maximum capacity of the quicked surfaces was four leaves under these conditions. BF 500×. (See Plate 83.)

ture, 20–22°C. By this technique, as many as ten sheets of 23½ carat gold (Sepp Leaf Co.) could be stuck/amalgamated on the brass before excess leaf sloughed off. In technique D, the surface was quicked in a 1% mercuric nitrate solution for 30 seconds at 20–22 °C. The maximum capacity of the quicked surfaces was four leaves under these conditions. This comparison makes clear that mercury as a liquid metal cannot be applied as thinly as a 'typical' mercury priming *solution*. On heating, the thicker mercury layer from the liquid metal 'priming' tends to diffuse more into the substrate than the thinner layer from the mercury solution, creating a wider zone of amalgamation under equivalent heating conditions, evident on etching under 500× magnification (see Figs 14.7 and 14.8).

3. Amalgam (approx. 2Hg:1Au by weight) was applied directly on to brass using a steel brush; sample E. For comparison, sample F was quicked in mercury solution as described above and then the same amalgam mixture was applied as for sample E. While the mercury surface coverage produced in the priming step *per se* for sample F is incomplete,[35] the wetting action produced by the priming made the application of amalgam much easier and faster than for the amalgam alone. The two techniques 'E' and 'F' produced very similar surfaces after heating for 30 minutes at 500 ± 15°C, not readily differentiated from one another. Samples E and F, however, were differentiable from the two leafed samples (C and D), the former showing more pronounced bubbling, including spheroidal gold particles and erupted surfaces prior to burnishing. After burnishing these differences were largely lost when viewed in the binocular microscope at 50×, nor were they easily found by closer view in metallographic cross-sections at 200–500×. Long immersion times in dilute etchants and ageing of some

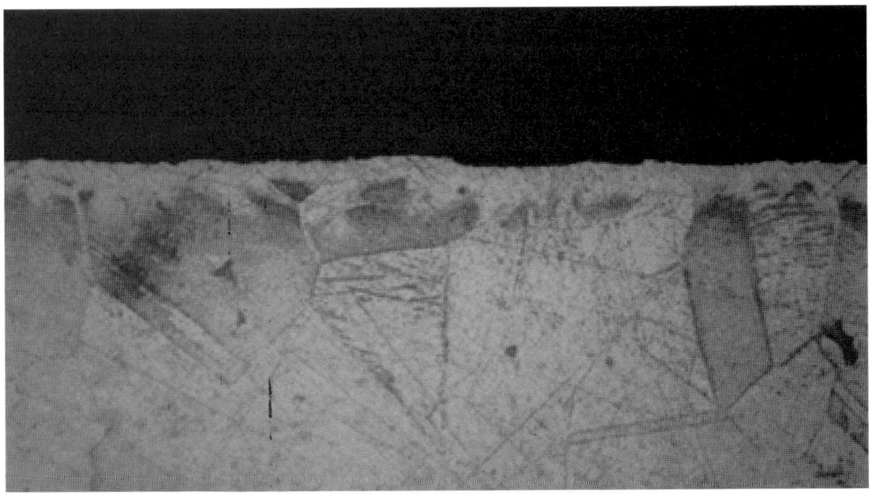

Figure 14.9 Brass sample E with amalgam gilding without primer. BF 500×. (See Plate 84.)

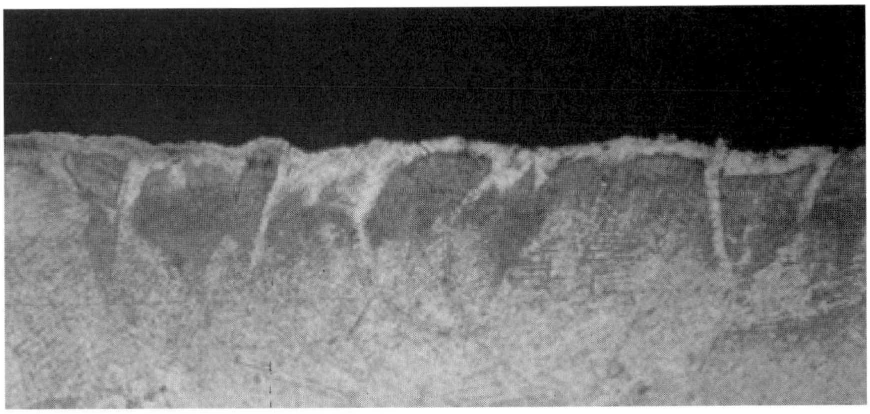

Figure 14.10 Brass sample F quicked in mercuric nitrate solution (as described for sample D) with amalgam gilding. BF 500×. (See Plate 85.)

etched sections indicate that the amount of mercury applied per unit area ($g\ cm^{-2}$) determines how diffuse and deep the penetration of the amalgam layer will be into the substrate (see Figs 14.9 and 14.10.).

From EPMA line and spot analyses, the depth of detectable penetration of mercury into the brass substrates in these experiments (samples C and D) appears less general and diffuse than has been observed for silver substrates in the literature. In brass, mercury attacks grain boundaries preferentially. This attack is promoted by heating. Etching ($FeCl_3$/HCl-based and NH_4OH/H_2O_2-based) brings out the chemical differences in the interface structures, in some cases more clearly than EPMA can resolve. The grain size within the gold layer is related to duration and temperature of heating, as etching of the *surface* illustrates; however, the relationship between surface grain size and different application techniques is not yet clear.[36]

Electroplating: 'quicking' procedures

Adding complexity to discussions and analysis of gold electroplating are two procedures described in several nineteenth century texts on gold electroplating. First is the notion that a 'new' and less toxic way to produce *ormolu* employs mercury plating baths, usually based on cyanide solutions,[37] followed by gold plating baths. As described by Charles Christofle for his manufactory,[38] the plated object is then heated at a very low heat in a remote muffle furnace to drive off the mercury. This procedure has been copied in our laboratory and the layering formed by heating is – not surprisingly – similar in appearance to that for traditional fire gilding; sample G. A comparable result is obtained when a surface is dipped in a mercuric nitrate bath, electroplated with gold, and then heated; sample H. By XRF

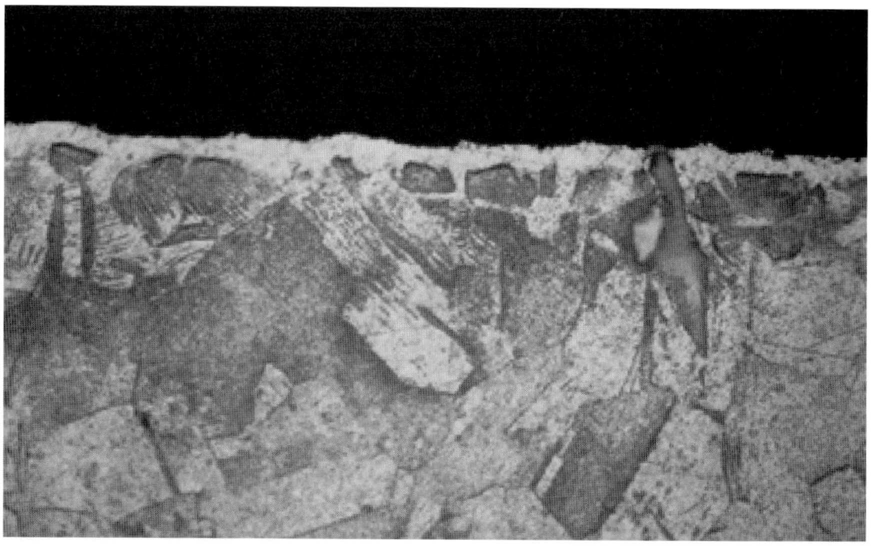

Figure 14.11 Brass sample G plated with mercury, plated with gold, and heated. BF 500×. (See Plate 86.)

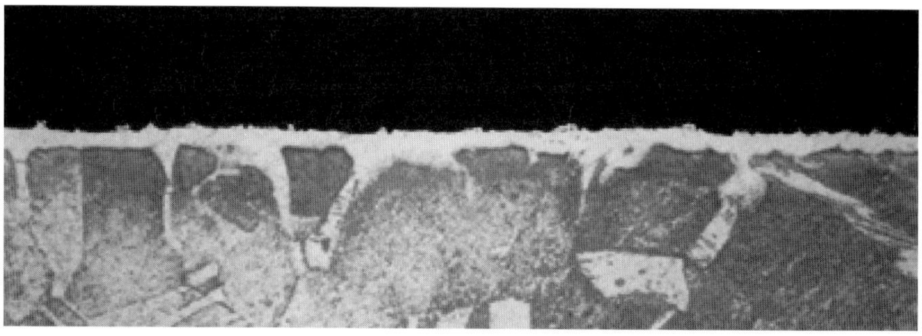

Figure 14.12 Brass sample H quicked in mercuric nitrate solution, gold plated, and heated. BF 500×. (See Plate 87.)

analysis, the gold layer would be expected to show varying mercury content depending on the thickness of the initial mercury layer, the thickness of the gold layer, and the temperature and the duration of heating to drive off the mercury (see Figs 14.11 and 14.12.).[39]

Second is the regular use of mercuric nitrate solutions, typically called 'quicking' solutions in this context, as a pretreatment for gold and silver electroplating. Dating as early as 1845,[40] mercury deposits from mercuric nitrate solutions (applied as above) were employed to improve the adhesion of the gold plate in early baths, a problem quite frequently noted in the literature of the time. These solutions remained in common use until at least the 1950s, where they are still listed as one of the options in the British Standard BS2816:1957 for 'Electroplated Coatings of Silver for Engineering Purposes'.[41]

The appearance of a brass surface that has been dipped in a quicking solution and then electroplated with gold is not easily distinguishable from gold plating alone; sample I. XRF analysis of the bulk surface may not detect the underlying mercury if the mercury layer is very thin (<1 μm) or the gold layer is very thick ($\gg 5$ μm) (see Figure 14.13).

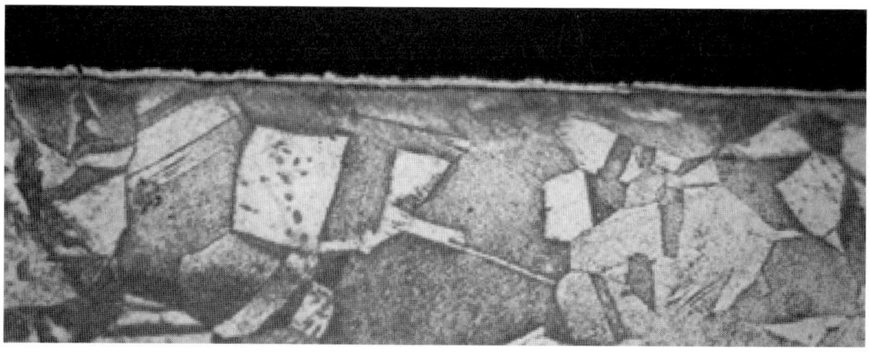

Figure 14.13 Brass sample I gold plated without a mercury primer or quick. The gold thickness ranges from 2 μm to 2.5 μm. BF 500×. (See Plate 88.)

How effective a mercury layer is in preventing poorly adherent electrochemical deposition or pitting corrosion of the substrate – corrosion that might undermine the plating – has never been firmly established.

In recent laboratory work, attempts have been made to determine whether brass substrates in cyanide electrolytes are protected by precoating with mercury. The experimental work thus far has not been able to overcome a basic kinetic property of the coating – it diminishes and amalgamates into the substrate surface faster than the time required to introduce the sample into the bath, to stabilise the equilibrium current and potential, and to make an anodic or cathodic sweep to measure corrosion reactions by typical potentiostatic methods.[42]

Immersion/dip gilding results

The practice of employing priming or quicking solutions to develop thicker, better-adhered immersion plating layers was examined in the laboratory. In several references on nineteenth-century immersion plating, practitioners state that better adhesion and more durable gilding layers could be produced by alternating mercury and gold dipping baths. Adjustment of pH, electrolyte concentration, type of complexing agent, temperature, and stirring are all important factors in the performance of such baths.

Attempts to reproduce dip gilding with nineteenth century recipes utilising quicking solutions as pretreatment have given uneven results:

- In cyanide solutions with low concentrations of electrolytes, the displacement reactions are very slow, possibly due to the formation of insoluble cyanide films on the mercury-rich surfaces.
- In near neutral, cyanide-free solutions in which gold was initially dissolved in aqua regia, spotting and burning were common defects, caused by too rapid and uneven deposition in some areas.

quicking, 5×

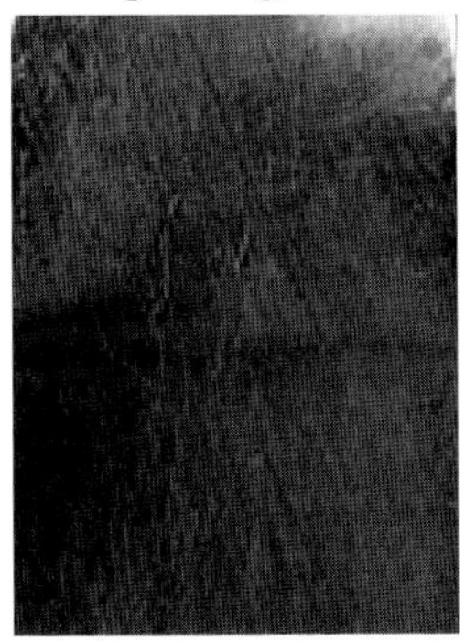

immersion plating

Figure 14.14 Brass sample J prepared by alternately dipping in mercuric nitrate and gold cyanide. (See Plate 89.)

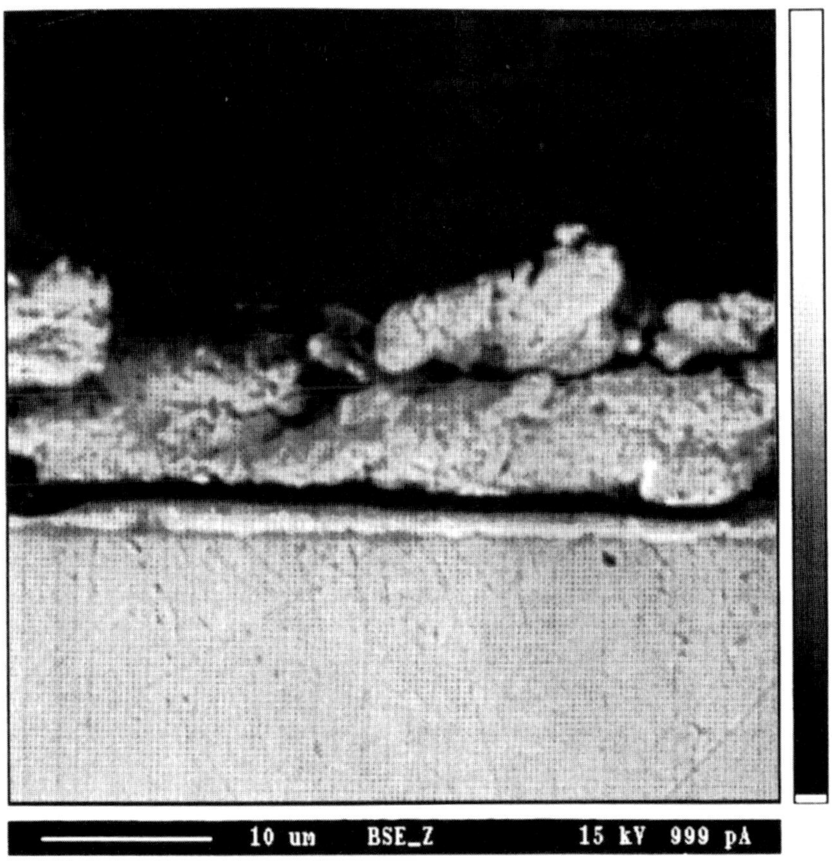

Figure 14.15 Back scattered electron image K of brass sample J alternately quicked and gilded by immersion.

One of the more successful dip plating samples is illustrated here in sample J. The layering was produced by alternately dipping (5×) brass into two baths: first 1% w/w $Hg(NO_3)_2$, followed by gold cyanide solution based on 5 g $KAu(CN)_2$ and 15 g KCN per litre of water, with the temperature between 20°C and 60°C. These baths produced a red gold layer from the dissolved copper and underlying oxide, the layering of which proved very uneven and discontinuous when examined by EPMA. In sample K, the thickest deposits were 3–5 µm thick. In analysis of the metallographic cross-section, the electron microprobe with WDS was not able to resolve the distribution of mercury within the gilding layers, although XRF analysis of the bulk surface had previously detected the presence of mercury quite easily.

From these limited studies, it is clear that the analytical detection of mercury in the examination of gilded metalwork is not a sufficient sole criterion for supposing that the object was fire-gilded, and that quantifying the relative amounts of gold and mercury and their distribution will be necessary to discriminate between some of the application methods that originated in the nineteenth century (see Figs 14.14 and 14.15).

Instrumentation

Zeiss M-5 inverted stage metallographic microscope, Cameca SX50 electron probe microanalyser equipped with wavelength dispersive spectrometers (WDS) and energy dispersive spectrometers (EDS); X-ray diffraction: Philips PW1729 X-ray generator equipped with a PW1840 diffractometer; atomic absorption: Varian SpectraAA-5; X-ray fluorescence: Kevex energy dispersive X-ray fluorescence (Ag secondary target, 40kV, 1.7 mA).

Acknowledgements

I am extremely grateful to Sally Malenka (Philadelphia Museum of Art, Philadelphia, PA) for her help with this research. I am indebted also to Ed Vicenza, Princeton Materials Group, Princeton, NJ for EPMA-WDS and -EDS and Janice Carlson, Winterthur Museum, Winterthur, DE for X-ray fluorescence analysis.

Notes

1. Unusual variants of this practice are found as late as the nineteenth century. For instance, in its large-scale production of lighting devices, the firm Cornelius and Co. (Philadelphia) employed finely ground gold in lacquer in some of its metal finishes for copper alloys, the results giving a matte and dark ochre color of gold.
2. In fire-gilding technique, an amalgam of gold and mercury is applied to the well cleaned metal surface of an object. The object is then heated, causing the mercury to sublime/evaporate and leaving behind a gold-rich surface. If the surface retains organic soils, inorganic depositions or corrosion products such as common air-formed oxides, the mercury will not wet the surface during application of the paste and the gilding will not form in those areas.

   The weight of mercury remaining in the gold layers after the 'firing' step will depend on a number of factors, particularly the time and temperature of the firing, the composition and thickness of the amalgam application, and the subsequent cleaning and finishing procedures, which may involve chemicals, heat, and sometimes more mercury (see note 19 below).

Written evidence for the use of cold amalgamation – that is mercury amalgam or paste applied without heating – as a method for joining very thick leaf or foil to metal substrates in the Renaissance and later has not yet been located.

3. Stalker and Parker (1688 [1971]: 67); Diderot and D'Alembert (1751–80: book 1, 643–4); Veitch (1909: 212). It may be that some variation of this technique, with longer, higher heating, was responsible for the production of the cheap 'gold' bullion bars which appeared in Europe in the 1730s and 1740s. These bars were made up of the then unwanted platinum from the Colombian placer deposits, combined with silver and copper, and covered with gold. Since the density of platinum is close to that of gold, these gold bricks were apparently successful enough that legislation had to be enacted against their manufacture. [M. L.] (1785: 6); McDonald (1960: 14).
4. Iron surfaces cannot be amalgamated easily, so this close-plating technique is particularly suitable for wrought iron and steel surfaces. Hansen reports that mercury is insoluble in solid iron, while the solubility of iron in mercury is less than 5×10^{-7} by weight near room temperature; Hansen (1958: 661); Lins (1974).
5. Biringuccio describes the general techniques in Book IX, Chapter 8, entitled "Concerning the practice of making gold into wire and also silver, iron, copper, and brass" and the following chapter "Concerning the method for preparing gold for spinning", Smith and Gnudi (1959: 377–81, 381–2). It is clearly described in similar terms in nineteenth-century texts, for example Tabart and Co. (1807: 56–63), where the distinction between wiresmiths and wiredrawers in the Nurnberg guilds AD1351 is even discussed, and by Bigelow (1847: 216–17), where a traditional discussion of the use of flatting rollers for converting wire into textile wrapping strip is reproduced. Several papers have documented the structures of such threads: for example Stodulski *et al.* (1985: 76–91).

   The technique of rolled gold is 'rediscovered' in the jewellery and watch trade in the mid nineteenth century (Smith 1930).

   An early description of plaquage is found in Theophilus, Book 3, chapter 77:

   > Silver, weighing twelve pennyweight, is refined and hammered out into a narrow piece the length of half the little finger. Then cemented gold, weighing one pennyweight, is hammered out to the same width and length, and these two are soldered with the gold solder mentioned above [essentially a copper oxide paste] until they stick fast to each another. Stuck together like this they should be hammered out until a very thin flat sheet is made. This work has the appearance of silver that has been gilded on one side, although a sheet of such length could not be so brilliantly gilded [probably fire-gilded] with two or three pennyweight of gold.

   The monk's remarks on the relative costs of production (one versus two- or three-pennyweight gold) should not be overlooked, nor should the need for relatively pure starting materials, in order that the metals respond appropriately under the hammer.
6. The date of Thomas Bolsover's discovery of Sheffield plate is usually given as 1742. The process consists of bonding sterling silver to a copper (alloy) substrate by means of heat. The product is rolled to an appropriate thickness (giving it also the old trade name of 'copper rolled plate' Wyllie (1913: 7), and then formed.
7. Holmyard (1957: 136). The product of this distillation is not pure nitric acid. In the recipe quoted, the addition of sal ammoniac (NH_4Cl) would precipitate silver as AgCl or form a chloride skin on the silver, while it promotes the stability of gold chlorides in solution. Agricola (1556 [1950]: p.460, note 20) preferred the term 'aqua valens' in distinction to the 'aqua fortis' of other authors. Agricola's illustrations and discussion of the production are given in Book X (pp. 439–43).

   The translators and editors of Agricola (1556 [1950]) describe four methods of separating gold from silver through AD1100–1500, the first being parting by nitric acid, and stated that among the first descriptions of saltpeter now known in the West are those given by Roger Bacon (1214–94) and by Albertus Magnus (1205–80) (Agricola 1556 [1950]: 458–60, note 21). To keep this development in perspective, it should be noted that Theophilus did not mention the use of acids in his work *Diversarum artibus schedula* and that Forbes (1970: 181) states blankly that 'Parting methods involving the use of nitric acid are not used before Agricola's time'.

The chemical composition of nitric acid was first determined by J-L. Gay-Lussac and C-L. Berthollet in 1816.

8. Among the early, non-classical writers on chemical processes concerned with metals are: al Razi - Abu Bakr Muhammad ibn Zakariyya, Avicenna, Robert of Chester (1144: 'The Book of the Composition of Alchemy'). These are well described in E.J. Holmyard's work, which gives a clear account of the impact of Arab and Islamic science on Western chemistry.
9. Steiskal-Paur (1979: 50–66). For example, 'Eisen war ih, Kupffer bin ih, Gold trage ih' (p. 56), or 'Eisen war die Ankunft mein, iezt zirt mich golt, Bin Kupffer fein' (p. 56). This process was called copper cementation in most seventeenth and eighteenth century texts. The method by which the gilding was applied on these objects from the Herrengrund region has not yet been determined by analysis, fire gilding being the most likely candidate.
10. How popular these texts were is partly indicated by the number of alchemical publications. In 1654, Pierre Borel published a catalogue of 276 pages listing 4000 titles:

    > Bibliotheca chimica sev catalogvs librorvm philosophicorvm in qvu qvatvor millia ciriter Authorum Chimicorum, vel de transmutatione Metallorum, re Mineral,& Orcinus, tam manuscriptorum, quam in lucem editorem, cum eorum editionibus, vsque ad annum 1653 continentur. Authore Petro Berellio, Parisiis, MDCLIV.

    Among the more interesting of these are Andreas Libavius, *Alchemia*, 1597 the works of Basil Valentine (especially *Triumphant Chariot of Antimony*) many of which were not published until the early seventeenth century and Thomas Norton, *The ordinall of alchimy*, 1477.
11. Veitch (1909: 211). While this recipe of Southwell's suffers from considerable vagueness as to quantities and proportions, it – or other sources from which it was plagiarised – is copied for at least the next two centuries in recipe books and handbooks describing metalworking: for example A. R. (1813: 43), where the powder produced by burning the gold solution soaked linen is called 'Goldzunder'. This recipe is very similar to, but not as detailed as one published anonymously in Nürnberg in 1708 (*Der Kunstliche Gold-und-Silber-Arbeiter von einem Sondebaren Liebhaber vieler Kunst-und-Wissenschaften*, British Museum Library no. 1043. c. 4, p. 58):

    > One takes good and perfect gold /this is dissolved in Aqua Regis to which salt has been added/and one half of the Aqua-Regis is allowed to fume away [abrauchen]/then let the glass be set in a damp cellar in cool sand/ overnight and the gold thusly develops into beautiful crystals/these are taken out and allowed to dissolve again in distilled vinegar/and thereafter the vinegar is allowed to evaporate by half/ set the glass again in the cellar in damp sand/ again overnight crystals develop these are dissolved in rain-water, half is evaporated off, again crystals develop/then one crushes these up with a knife/ put this [powder] into a hard-boiled egg from which the yolk has been removed/set this in a cool damp place/thus does it develop overnight into an oil/which you paint on a silver [piece] only with the thinnest [coating] and allow to dry/ such a thing is gilded in the highest way.

12. Brogniart (1830, 1844). The fulminates of gold were undoubtedly known for many centuries before the production of a 'safe' precipitate was given by Bergman in 1780. A not too different method is mentioned by Nicolas Lemery in his *Traite de la Chymie*, 1669; see Holmyard (1931: 128–31).
13. Part of the difficulty in analysis lies in the fact that in the older recipes, the surface metal must undergo oxidation or dissolution during the electrochemical deposition of gold. This marks the starting point for the failure of the coating itself, as the corrosion that begins with the deposition step is then accelerated by the gold/base metal couple that is created by the plating and exposed to the envi-

ronment. The gold layers that formed in earlier baths were incomplete, soft, easily damaged layers, most of them probably much less than a micrometre thick. The chances for survival of such deposits are not bright over several decades, much less centuries.

14. Stalker and Parker (1688: 65) describe the coating of a scratch brushed surface:

> make ready quick-silver, by mixing it with a very small quantity of aqua fortis [nitric acid] in a vial, which should always stand by you; three or four drops only of Aqua fortis, is sufficient I assure you for an ounce of quick-silver. With this quicken your work, that is with your finger or a fine rag rub this mixture onto your metal, till tis all over silvered or toucht with the said quick-silver.

They give another recipe for regilding (p. 66) which uses a mixture of 'quicksilver and aqua fortis' to prime a surface, though as with the quote above, it is not clear how much mercuric nitrate the mix consisted of or if the authors actually had performed any of this work themselves. It is not likely, however, that the authors intend to describe mercury that has merely been cleaned by nitric acid, since cleaned (elemental) mercury cannot be easily applied with 'a finger'.

See also Chambers (1728). Other authors commonly cited are Diderot and D'Alembert (1751–80: 'Dorure'), and Smith (1755). Wyllie (1913: 43–4), terms the fire-gilding process using a mercuric nitrate primer for Sheffield Plate 'water gilding', in which the amalgamated articles are 'laid, gilt side uppermost, on an iron pan over a coke fire, and the mercury gradually evaporates, leaving a beautiful gold surface', a heating process possibly less vigorous than usual, to reduce the possible delamination of the plate. This terminology – 'wash gilding' – is repeated in other contemporary texts, for example Tomlinson (1854: vol 1., 264). In German the term 'Quickwasser' is found; A. R. (1813).

15. Turner (1974: 14); Goodison (1974: 69–70); Chapman (1994: 232). The distinctions made by the last author between leaf and amalgam gilding are not evident in the final product.
16. Goodison (1974) refers repeatedly to the Boulton's notebook, ledgers, journals, letters and diary in his clear account of how costs and manufacturing efficiencies affected Boulton's work.
17. These recipes abound in the literature, from Stalker and Parker, Chambers, and Diderot, to recent work by Turner, Goodison and Chapman. It is, perhaps, not surprising that mercury and mercury salts were listed frequently in such colouring recipes, particularly for matting or deading the gilding and for whitening the gold.
18. Volta's 1796–97 experiments were not published until 1800 by the Royal Society London, due to the war. Many other scientists and amateurs had been working on electric phenomena and the passage of charge through water and aqueous solutions, including Priestly, Cavendish, Lavoisier, Beccaria, Bockmann, von Arnim, Gilbert, Erman, and Gruber. It may be useful to note here that attempts to find archeological evidence of electrochemical gilding before the Renaissance in the West have not been productive. One recurrent candidate, the so-called Baghdad cell found in the Baghdad Museum, has been demonstrated clearly to be a nineteenth century contraption: see Eggert (1995). Electrochemical gilding methods are found in sites dated before AD1500 in South and MesoAmerica, described in the work of Lechtman, Bray, Scott and others.
19. Baldini (1972: 494) assigns a date of 1802. Hunt (1973: 17–18) gives a date of 1803. Raub (1993: 186, 190) gives a date of 1801, based on a publication that I have not had the opportunity to verify (Brugnatelli 1801). Smee (1845) gives references known in the mid-nineteenth century for Brugnatelli's work : Van Mons 1803, *Journal de Chimie*, p. 357; Brugnatelli, 1805, *Annales de Chemie*, XXII, Pavia, p. 22 and Brugnatelli, 1807, *Bibliotheque de Campagna*, X, Milan, p. 185; and also an extract from a letter of Brugnatelli to Van Mons in *Philosophical Magazine*, XXI, 1805.

Some of Brugnatelli's experimental work from the 1790s , his *Giornale fisico-medico*, is reported in the Bulletin Acad. Bruxelles, 1835, **ii**, p. 275ff. Attempts in the 1820s and 30s (Van Mons, Becquerel and others) to duplicate Brugnatelli's process, using gold-ammonium ('ammoniuret') derivatives precipitated from solutions of aqua regia, were not successful.

20. This controversy may be too remote in time and technology for present day readers to appreciate. In the many books that were published on electroplating in the nineteenth and early twentieth century, however, this dispute about the history and origins of the patent and related processes often occupies a very substantial part of the book. Even as late as 1911, for a re-revision of Watt's 'Electro-Deposition', Arnold Philip covers over 20 pages on the controversy (1911: 77–98). In 1851 Christofle published most of the documents surrounding the controversy in a book of more than 400 pages, entitled *Histoire de la Dorure et de l' Argenture Electro-chimique*. The first experimenter to record regular use of cyanides for plating precious metals was Spenser, whose work was absorbed by the Elkingtons in their 1840 patent. Hunt (1973) describes some of the intricacies well.
21. In Russia, 60 workmen are reported to have died from mercury poisoning during the gilding of the exterior dome for St. Isaac's in St. Petersburg. By the 1850s, Jacobi, the Russian scientist, and his successors were electroplating gold on a grand scale for the interior of the dome: in one example, 'copper plates for the Church of the Redeemer, Moscow required 498 kg of gold'. This appears to be the first time that a modern form of specification and quality control testing was used. The specification called for 28.44 g m^{-2} of gold + 20% (0.000,182 in. + 0.000,036 in.) and required that two plates out of every hundred be checked for gold thickness by a strip and weigh technique. If the two plates failed to meet the specification, the entire hundred plates were rejected. The capacity of the tanks for processing of these plates was 1,500 (US) gallons, or 5,677 l Weisberg (1974: 10).
22. In the period 1801–1803, Buccholz, Bohm, Schrade, Gehlen and Vauquelin are said to have isolated HCN from various plant materials (Partington **date?**: 234, 583). The name 'acide prussique' has been attributed to Guyton de Morreau (*ibid.*: 527). This term is already employed by Lavoisier when he lists prussiates of gold, silver and mercury (1790: 288–9).
23. Haynes (1954: vol. 1, 29ff.):

> Prussian blue . . . made in England by Bramwell and Company in 1770 . . . sold for $10 a pound. About 1825 Charles Tennant, who was promoting Prussian blue as a dye to the Scottish calico printers, cut down this price by lowering the cost of the basic material, prussiate of potash, from $1.25 an ounce to $1.25 a pound. He accomplished this by improving the methods of treating nitrogenous waste (blood , hooves, tannery scrap, etc.), and taking a leaf from his book, Carter and Scattergood brought the first yellow prussiate of potash into this country about 1834. Twelve years later they made the first red potassium prussiate. The chief use of these cyanide chemicals was in the color field, a market soon to be much curtailed by coal-tar dyes.
>
> After 1887, however, a big new employment appeared for cyanides, the solvent extraction of gold from its ores, discovered by J.S. MacArthur and the Forrest brothers. Few chemical processes have had such wide economic effects . . .
>
> Since the 1860s the old waste nitrogen products had been largely replaced as a source material for prussiates by spent oxides from municipal gasworks. This resulted in products of better purity, but the new demand inspired an even more noteworthy technical triumph. The Cassel Company in England adopted the Beilby process involving the interaction of ammonia, charcoal, potassium carbonate, supplanting it soon by Castner's method of producing sodium cyanide from ammonia, charcoal, and metallic sodium in Castner's own electrolytic process . . .

The two principal prussiates of commerce in the nineteenth century, besides prussian blue, are: yellow prussiate of potash – potassium ferrocyanide – $K_4Fe(CN)_6$, and red prussiate of potash – potassium ferricyanide – $K_3Fe\ (CN)_6$. There are numerous other salts and hydrated forms. L. Gmelin discovered the red prussiate of potash which he named – 'rothes cyaneisensaurenkalium', in the

Journal of Chemistry, 1822, **XXXIV**: 325. For additional discussion, see Williams (1978).

24. The development of better, more powerful, more longer lasting batteries was a concern of many of the writers on electroplating during 1836–1880. There were a number of early attempts to find alternative sources of power. After the development of the electric dynamo in the 1860s, production plating shops increased output dramatically. Among the first to install such electrical supplies in Europe were: Christofle et Cie, Paris, 1872, machine of Theophil Gramme; Grohe in Berlin, 1875, machine by Siemens and Halske, (Krämer *et al.* 1959: 33); and H. Möhring in Frankfurt am Main, machine of Siemens design (Binder 1884: 39).
25. *Handbook of Physics and Chemistry*, (1989–1990), 70th edn, Boca Raton, Fla.: CRC Press, D-151-158. Only two listings of species typically found in plating baths is now given in the tables of 'reversible' electrode processes:

    $AgCN + e \leftrightharpoons Ag + CN^-$ −0.017 V

    $Ag_4[Fe(CN)_6] \leftrightharpoons 4Ag + [Fe(CN)_6]^{4-}$ 0.1478 V

    $AuCl_4^- \leftrightharpoons Au + 4\,Cl^-$ 1.002 V

    Cotton and Wilkinson (1966: 1039) give 'standard potentials' for the reactions below, which are similar to those that can be calculated from the appropriate equilibrium constants, but they state curiously that 'standard potentials are of limited utility, particularly since these depend on the nature of the anion':

    $Au(CN)_2^- \leftrightharpoons Au + 2\,CN^-$ −0.6 V

    $Ag(CN)_2^- \leftrightharpoons Ag + 2\,CN^-$ −0.31 V

    It is to be noted that the effect of the cyanide complexes are to substantially depress the potential at which reduction of the metal occurs.
26. Gold cyanide solutions typically show good throwing power (that is efficient plating coverage and weight of metal deposited for current consumed), but there are problems with the ageing and decomposition of the electrolyte and with the passivation of the anode. (Kirk *et al.* 1980).
27. These values were assembled from a number of sources and are intended only to provide a comparative notion of typical values that have been observed, including: Reid and Goldie (1974); and Nobel (1982). Characteristics of the plating layer that affect durability include: thickness, hardness, porosity, composition, underlayers, galvanic coupling and the presence of a protective coating.
28. For a review of these recent developments, see Vivian (1970). There are many non-cyanide electrolytes for gold electroplating prior to this period, in fact as far back as the 1840s, but these did not develop into production baths to the same extent as these newer baths.
29. At this time the E. R. Elkington and Company was largely involved in the manufacture of glasses and in their company advertisement for their Newhouse Street, Birmingham, referred to itself as 'water gilders' (Binder 1884: 114).
30. The specification for the patent #7304, 1837, was published by the Great Seal Patent Office, 1856.
31. Patent # 7304, p. 6, lines 14–16: 'What I claim as the second part of my Invention is, the using of a solution of mercury and a solution of gold in place of the amalgam of gold and mercury heretofore employed'. H. Elkington describes the preparation of a quicking solution (p. 5, line 9–p.6, line 8), but states that there are many ways to prepare the solutions. Smee (1845: 325ff).
32. Siman (1993: 15). The most common reducing agents in modern baths are $NaBH_4$ and DMAB (dimethylaminoborane).
33. For an entertaining introduction to this subject, see Green (1973: 9ff.). General descriptions of the technology are provided in *The Metals Handbook* (1982: 361–426), as well as in later editions.
34. *Fairie Queen*, 1590, book 11, canto 8, stanza 14.
35. See Chapter 15, this volume.
36. Some of the difficulties and limitations of etchants for this dscrimination are discussed by Weiner (1962) and Weiner and Fuhr (1964).
37. A great many variations on these basic methods have been omitted, for the sake

of brevity, in this general review of gilding. One variant that appears frequently is a mixed gold–mercury or gold–silver solution in nitric acid that deposits on the substrate, initially forming a very thin black deposit that is rendered bright and golden by heating: Savary (1760: vol. II, 126–30). This text is repeated by Jamieson (1829: 390). In fact, Elkington's of 1837 contains a passage about heating the metal and plating to obtain a desirable appearance.

38. *La Galvanoplastie dans la manufacture de MM Christofle et Cie* (1866), Paris: Veuve Boulhard-Huzard. *Note à Consulter sur l'Établishment de MM. Christofle & Cie* (1889) Paris: Chaix, section on 'Dorure Électrochimique au mercure'. Christofle (1851).
39. The quantity of mercury remaining in and under the gold plating layer is subject to many variables. The porosity of the over layer of gold is different in electroplated samples from amalgam applications. Mercury may not be easily and completely evaporated/sublimated from under a thick layer of electroplated gold, while the coalescence of Au_xHg_y phases ($0< x,y < 2$) during the heating of an amalgam present a very different, continuously varying, and much more open structure from which mercury can vaporise. Metallographic evidence – from a limited number of samples – suggests that for longer heating times and high temperatures, the effect of a mercury pre-treatment under gold electroplating is to promote the penetration of mercury along grain boundaries in brass alloys without producing the general diffusion zones evident at the substrate interface in amalgam gilding. Grain boundary attack is just one feature of the different structures that could be further elucidated. Other avenues of exploration are: selective etching of the cross-sections; and etching to bring out the grain size and boundaries of the surface of the plating.
40. Smee borrows from M. Bequerel's *Eléments d'Electro-chimie* in his *Nouveau Manuel Complet de Galvanoplastie, ou Éléments d' Électro-Métallurgie* (1845: 299):

    > *MM. Elkington et Becquerel ont eu l' heureuse idée de combiner, en qulelque sorte, l'ancien procédé de dorure au mercure avec le nouveau, et de utiliser l'extrême affinité qui existe entre l' or et le mercure, en déposant une couche infiniment mince de ce dernier métal sur les objets que l 'on veut dorer par voie galvanique. Après avoir parfaitement décapé les objets, on les plonge dans une solution étendue de proto-nitrate de mercure; on les lave ensuite à grande eau, puis on les frotte avec une peau très douce, pour bien étendre la couche mercurielle. On recommence l'opération jusqu' à ce que la pièce soit entièrement amalgamée et prèsente un aspect mat ou poli, suivant l'effet qu' on veut obtenir et d' une couleur blanche. Il faut se rappeler que la couche de mercure doit être excessivement mince, et juste suffisante pour se combiner avec le métal à recouvrir. Un exces de mercure produirait un effet contraire a celui qu' on veut obtenir, car il se formerait alors un amalgame blanc d' or et de mercure, qui nuirait à la fois à l' adherence et à la couleur de la dorure.*

    Smee's 1843 edition in English did not contain the above passage. The passage, however, is copied almost verbatim by Villon (1887: 124). The earliest record of this practice has yet to be confirmed in the archives of the Elkington or Christofle firms. The Elkington patent of 1840 does not list the use of a quicking solution as part of the procedure.
41. Note: Anonymous (1957). See also Bottomley (1949), in which the author recommends a very dilute quicking solution, 15.5 g HgO per gallon of solution with excess cyanide, and Evans (1948: 708).
42. Quicking solutions – as these solutions of soluble mercury salts became known – have not been recommended for substrates such as zinc, aluminum or cold worked copper, for the last of which it represents a method for testing stress corrosion cracking or excessive residual stress from mechanical forming operations. Some of the positive effects may be compared to that of the transitory dropping mercury electrode in polarography, where mercury as a cathodic surface is not readily attacked in many aggressive solutions due to its high overpotential for hydrogen discharge. Mercury is expected to be relatively stable in cyanide solutions when coupled to copper or zinc. See Chapter 15.

References and selected reading

Agricola, G. 1556. *De Re Metallica*, H. C. Hoover and L. H. Hoover (trans.). New York: Dover, 1950 (re-publication of 1912 translation).

Anonymous 1957. 'Silver plating for engineering purposes'. *Electroplating and Metal Finishing*, **11**: 283–4.

A. R. 1813. *Kunst und Wunderbuch, oder verbogene Geheimnisse welche ein sterbender Vater seinen Kindern ubergeben*. Nürnberg: C. H. Zeh'schen.

Baldini, U. 1972. 'Luigi Valentino Brugnatelli'. In *Dizionzario Biografico degli Italiani*. Rome: Istituto della Enciclopedia Italiana.

Barba, Alvaro Alonso 1640. *Arte de los metales, en que se enselia el verdadero beneficio de los de oro, y plata por acogue*. Madrid.

Bergbuchlein 1949. A. G. Sico (trans.). New York.

Berzelius, J. J. 1836. Lehrbuch der Chemie, ubersetzt v. F. Wohler, 4, verbesserte. Orignal Auflage Bd. 4. Dresden und Leipzig: 623–99.

Bigelow, J. 1847. *The Useful Arts, considered in connexion with the Application of Science*. New York: Harper and Bros.

Binder, F. 1884. *Handbuch der Galvanoplastik*, 5th edn. Weimar: B. F. Voigt.

Biringuccio, V. 1540. *The Pirotechnia of Vannoccio Biringuccio*, C. S. Smith and M. T. Gnudi (trans.). Cambridge, Mass.: MIT Press, 1959. Based on the 1540 text; the first of three published by Curtio Navo or di Navò (1540, 1550, 1558/9).

Bottomley, J. E. 1949. 'The plating shop "quick"'. *Electroplating*, December: 119 20.

Brenner, A. 1963. *Electrodeposition of Alloys, Practices and Principles*, volume 1. New York: Academic Press.

Brogniart, A. 1830. *Essai sur les Arts Ceramiques*, Paris.

Brogniart, A. 1844. *Traité des Arts Ceramiques ou des Poteries*, 2 vols, Paris.

Brugnatelli, L. 1801. 'Chemische Bemerkungen über die electrische Säule'. *Annalen der Physik*, **6**: 7105–6.

Cellini, Benvenuto 1568. *Du trattati, vno intorno alle otto principali arti del I'oreficeria. L'altro in materia dell' arre della scultura*, C. R. Ashbee (trans.). London, 1898 (from the Italian edition edited by Milanesi, 1857).

Cennini, Cennino 1437. *Il libro dell' arte*, D. V. Thompson, Jr. (trans.), 2 vols. New Haven, Conn., 1932–33.

Chambers, E. 1728. *Cyclopaedia: or, an universal Dictionary of Arts and Science*. London (D. Midwinter *et al.* 1738, 2nd ed.; Innys 1751–52, 7th ed.).

Chapman. M. 1994. 'Techniques of mercury gilding in the eighteenth century'. In *Ancient and Historic Metals*. Singapore: the Getty Conservation Institute.

Christofle, C. 1851. *Histoire de la Dorure et de l"Argenture électrochimiques*. Paris: Duverger.

Cotton, F. A. and Wilkinson, G. 1966. *Advanced Inorganic Chemistry*. New York: Interscience.

Diderot, D. and D'Alembert 1751–80. *Encyclopédie (ou dictionnaire) raisonné des sciences des arts et des métiers par une société des gens de lettres*. Neuchatel, Paris.

Eggert, G. 1995. 'On the origin of a gilding method of the Baghdad silversmiths'. *Gold Bulletin*, **28**(1): 12–16.

Ercker, Lazarus 1580. Treatise on ores and assaying, A. G. Sisco and C. S. Smith (trans.). Chicago, 1951. (First English translation by John Pettus, London, 1683.)

Evans, U. L. 1948. *Metallic Corrosion Passitivity and Protection*. London: Longman, Green and Co.

Forbes, R. J. 1970. *Studies in Ancient Technology*, volume 8. Leiden: E. J. Brill.

Glauber, Johann Rudolph 1656–61. *Des Teutschlandts Wohlfart.* Amsterdam. (Also Latin edition, *De Prosperitate Germaniae*, Amsterdam, 1656–61. Included in collected English edition of Glauber's works, C. Packe (trans.), London, 1689.)

Goodison, N. 1974. *Ormolu, the Work of Matthew Boulton*. London: Phaidon.

Green, T. 1973. 'The noble metal'. *Illustrated London News*, December, Christmas edition: 9ff.

Hansen, M. 1958. *The Constitution of Binary Alloys*, 2nd ed. New York: McGraw-Hill.

Haynes, W. 1954. *American Chemical Industry*, New York: D. van Nostrand.

Holmyard, E. J. 1931. *Makers of Chemistry*. Oxford: Oxford University Press.

Holmyard, E. J. 1957. *Alchemy*. Harmondsworth: Penguin Books.

Hunt, L. B. 1973. 'The early history of gold plating: a tangled tale of disputed priorities'. *Gold Bulletin*, **6**: 16–67.

Jamieson, A. 1829. *A Dictionary of Mechanical Science, Arts, Manufactures, and Miscellaneous Knowledge*. London.

Kirk, D. W., Foulkes, F. R. and Graydon, W. F. 1980. 'Gold passivation in aqueous Alkaline cyanide'. *Journal of the Electrochemical Society*, **127**(9): 1962–9.

Krämer, O. P., Weiner, R. and Fett, M. 1959. *Die Geschichte der Galvanotechnik.* Saulgau: Eugen G. Leuze Verlag.

Lavoisier 1790. *Elements of Chemistry in a New Systematic Order Containing all the Modern Discoveries*. London: Wm. Creech. (Republished by Dover, New York, 1965, R. Kerr (trans.))

Lins, A. 1974. 'A History of Metal Coatings on Metal: A Report to Sir John Cass College'. London: unpublished.

Lins, P. A. and Oddy, W. A. 1975. 'The origins of mercury gilding', *Journal of Archeological Science*, **1975**: 2.

McDonald, D. 1960. *A History of Platinum*. London.

The Metals Handbook 1982. Volume 5: *Surface Cleaning, Finishing and Coating*. Metals Park, Ohio: American Society for Metals.

[M. L.] 1785. *Mémoire sur la Platine ou Or Blanc*. Paris: Académie Royale des Sciences.

Nobel, F. I. 1982. 'Gold plating'. In the *Metals Handbook*, 9th edn, vol. 5, 281–4. Metals Park, Ohio: American Society for Metals.

Norton, T. 1477. *The ordinall of alchimy*. Facsimile edition, E. J. H. Holmyard (ed.). London, 1928. (First printed in Elias Ashmole's collection, *Theatrum Chemicum Britannicum*, London, 1652.)

Paracelsus Hohenheim, Philippus Aureolus Theophrastus Paracelsus Bombastus von 1537. *De natura rerum IX*. Strasburg: Bücher. 1585.

Partington (1961, 1962) *History of Chemistry*, vols 2 and 3. London: MacMillan.

Pavlova, O. I. 1963. *Electrodeposition of Metals: A Historical Survey*, S. A. Pogodin (ed.). Moscow. (English translation, U.S. Department of Commerce. Israel Programme of Scientific Translations, Jerusalem 1968.)

Percy, John 1880. *Metallurgy: the art of extracting metals from their ores . . . : Silver and gold*, part I. London.

Perez De Vargas, Bernardo 1569. *De re metalica, en el qval se tratan mvchos y diversos secretos del conocimiento de toda suerte de minerales*. Madrid.

Philips, A. 1911. *The Electro-plating and Electro-refining of Metals, being a new edition of Alexander Watt's 'Electro-deposition'*, 2nd ed. New York: van Nostrand.

Porta, G. B. 1560. *Magia naturalis oder Hab, Kunst, und Wunder Buch/nach dem vermehrten in XX Büchern bestehenden lateinischen Exemplar ins Hochteutsche jbersetzt etc.* (*Magiae naturalis sive de miraculis rerum naturalium libri quatuor*. Antwerp 1560). 2. Aufl., Njrnberg 1715.

Probierbuchlein 1949. A. G. Sico and C. S. Smith (trans.). New York.

Raub, C. 1993. 'The history of electroplating'. In *Metal Plating and Patination*, S. La Niece and P. Craddock (eds), 284–90. Oxford: Butterworth-Heinemann.

Raub, E. and Mueller, K. 1967. *Fundamentals of Metal Deposition*. Amsterdam: Elsevier.

Reid, F. H. and Goldie, W. 1974. *Gold Plating Technology*. Ayr: Electrochemical Publications Ltd.

Savary, P. L. 1760. *Dictionnaire Universel de Commerce, d'Histoire Naturelle, & des Arts et Metiers*. Copenhagen.

Siman, F. 1993. 'Deposition of Gold without External Current Source'. *Gold Bulletin*, **26**(1): p.15.

Smee, M. 1845. *Nouveau Manuel Complet de Galvanoplastie, ou Éléments d' Électro-Métallurgie*. Paris: Manuels-Roret. (Republished in 1977, Paris: Chez L. Laget.)

Smith, C. S. and Gnudi, M. T. 1959. *The Pirotechnia of Vannoccio Biringuccio*. Cambridge, Mass.: MIT Press. (Based on the 1540 text, the first of three published by Curtio Navo or di Navò; 1540, 1550, 1558/9.)

Smith, E. A. 1930. 'Rolled gold: its origin and development'. *Journal of the Institute of Metals*, **XLIV**(2): 175ff.

Smith, G. 1755. *The Laboratory: Or School of the Arts*, 4th ed. London: Hodges.

Stalker, J. and Parker, G. 1688. *A Treatise of Japanning and Varnishing*. Oxford. (Republished by A. Tiranti, London, 1971.)
Steiskal-Paur, R. 1979. 'Herrengrunder Kupfergegenstände'. In *Barockes Kupfer aus Herrengrund und Ornamentale Vorlageblätter*, G. Egger (ed.). Vienna: Österreichisches Museum für angewandte Kunst.
Stodulski, L. P., Mailand, H. F., Nauman, D. *et al.* 1985. 'Atomic emission spectrographic and scanning electron microscopic-energy-dispersive X-ray studies of European, Middle Eastern, and oriental metallic threads'. In *Application of Science in Examination of Works of Art*, P. A. England and L. van Zelst (eds). Boston: Museum of Fine Arts Boston.
Tabart and Co. 1807. *The Book of Trades, or the Library of the Useful Arts, part III*, London: C. Squire.
Theophilus n.d. Diversarum artium schedula. An essay upon various arts. (Latin text and English translation with notes by Robert Hendrie, London, 1847.)
Tomlinson, C. 1854. *Cyclopaedia of Useful Arts and Manufactures*. London.
Turner, E. 1974. *An Introduction to Brass*. London: Her Majesty's Stationery Office.
Valentinus, Basilius 1694. *Chymische Schriften*. Hamburg (reprinted 1740).
Veitch, H. N. 1909. *Sheffield Plate: Its History, Manufacture and Art*. London.
Villon, A-M. (ed.) 1887. *Nouveau Manuel Complet de Dorure. De MM. Maigne et O. Mathey*. Paris: Chez L. Laget, 1980.
Vivian, P. G. L. 1970. 'Precious metal plating'. *Electroplating and Metal Finishing*, November: 21ff.
Weiner, R. 1962. 'Metallographic studies of rolled gold and on electroplated gold'. *Metall*, **16**(1): 11–13.
Weiner, R. and Fuhr, E. 1964. 'Metallographic investigations of rolled-gold coatings'. *Metall*, **18**(6): 619–20.
Weisberg, A. M. 1974. *Gold Plating Technology*, F. H. Reid and W. Goldie (eds). Ayr: Electrochemical Publications Ltd.
Williams, G. 1978. 'The cyanides of gold, the history of their key role in electroplating'. *Gold Bulletin*, **11**: 56–9.
Wyllie, B. 1913. *Sheffield Plate*, 2nd ed. London: B. T. Batsford.

15 The Use of Mercury Salts in Gold Electroplating

Andrew Lins and Sally Malenka

Abstract

The use of mercury salts in nineteenth and twentieth century electroplating literature has been reviewed and is summarized in this chapter. Experiments were conducted using sample coupons prepared in mercuric nitrate solution. These coupons were evaluated visually and analysed by metallography, X-ray diffraction (XRD), and electron probe microanalysis (EPMA) to determine if mercury applied to a metal substrate by quicking was detectable. Analyses by atomic absorption and potentiodynamic polarisation measurements were carried out to investigate if a layer of mercury affects the corrosion of the substrate during plating and thereby the adhesion of the plating. A museum object, an electrotype of the Milton shield, was examined by X-ray fluorescence (XRF) and mercury was detected.

Introduction

The use of mercury salts in gold electroplating is recorded in the electroplating literature throughout the nineteenth and into the twentieth century. Solutions containing mercury salts, most frequently nitrates, were used in a process called 'quicking', 'bluing' or 'amalgamating'.[1] Typical stages of the electroplating process included the cleaning of the object to be plated by scratch brushing or dipping into compound acids, quicking, and then immersion in the electroplating bath. When the object was 'quicked' in a mercury salt solution, a layer of mercury deposited electrochemically onto the metal. Quicking is most frequently described as a pretreatment before silver or gold electroplating of copper and copper alloys, sometimes including German silver.

Review of nineteenth and twentieth century electroplating literature

Although written references do not always reflect actual workshop practices, the frequent mention of quicking with mercury salts in many practical guidebooks and articles indicates that their use was common.[2] Before the development of electroplating, mercury salts were used in fire gilding to improve the flow and even application of the mercury-gold amalgam

(Stalker 1971 [1688]: 65; Chambers 1752; Spon 1878: 194; Watt 1887: 203). The technology from fire gilding transferred into the innovations and industrialisation of gold and silver electroplating: 'electroplaters have borrowed from the gilders by fire an operation which will be found important in practice . . . [plunging into a solution of nitrate of binoxide of mercury]' (Roseleur 1872: 32–3). E. R. Elkington & Co. held the patent to use mercury and gold salts for immersion plating (Elkington 1856 [1837]: 5–6), three years before their patent of 1840 for electroplating by applying an external current to plating baths containing cyanides.[3]

The chemistry of the quicking process and the necessity of using mercury salts were not experimentally established, nor agreed upon by many authors. Some authors omit quicking entirely from discussion; others state that it is not always necessary (Roseleur 1872: 20; Watt 1887: 175; Langbein 1902: 111, 351; Weston 1915: 81; Graham 1945: 97). The primary reasons for quicking were to improve adhesion and durability of the plating, including resistance to atmospheric attack. Adhesion and durability of electroplated gold compared to fire gilding were a prevailing concern in the electroplating industry (Smee 1845: 249; Becquerel 1864: 478–9; Roseleur 1872: 119). The problem of poor adhesion was related to the difficulties encountered in controlling all aspects of the plating operation: supplying a constant and well regulated power source, maintaining the electroplating bath, and controlling the rate and quality of deposition.

> The want of adhesion between the silver and the metal beneath arose partly from the employment of too many cells in the battery, but chiefly from want of perfect cleanliness of the receiving surfaces. It was eventually overcome by cleaning the surfaces with very great care, dipping the cleaned article in a very weak solution of mercury immediately before placing it in the depositing liquid, and lessening the number of battery cells (Gore 1884: 22–3).

(See also Gore 1854: 49; Watt 1889: 237; Smith 1891: 28–9; Langbein 1902: 92, 111–12.)

As Gore and others clearly understood, adhesion was related directly to the cleanliness of the surface to be plated. An even and reflective film of mercury after quicking would be possible only if the surface were clean. Whatever the contributions of mercury to the subsequent metal deposition, the ability to assess and assure the cleanliness of the surface was an essential factor for good adhesion in all plating processes.

> We may lay down as a principle that the solution of nitrate of binoxide of mercury is the touchstone of the cleansing operations. With a perfect cleaning, the piece will, after passing through a quite strong mercurial solution, be perfectly white and bright. On the contrary, there will be a damascene appearance, or various shades of color, if the cleansing has not been properly done (Roseleur 1872: 33–4).

Quicking was often mentioned particularly for delicate items that were plated by immersion gilding, and for items such as watch parts that were to appear grained and matte (Roseleur 1872: 130; Spon 1878: 226; Wahl 1883: 174; Watt 1887: 185, 210; Weston 1915: 167). Quicking was also recommended for areas of objects that were difficult to plate, particularly over soldering where the gold would not deposit without the mercury pretreatment (Roseleur 1872: 124; Gore 1884: 22; Weston 1915: 182). Similarly, mercury salt solutions were found to be useful in the replating of old objects: 'the main object is simply to coat with mercury, spots that will not plate without it and the less mercury employed to amalgamate the surface the better for the work' (Weston 1915: 167–8).

The colour of gold plating, which tended to be more red than fire gilding, could be modified by the use of mercury salts (Roseleur 1872: 132). As sometimes described, the procedure corresponded remarkably to fire gilding.

> When the deposit made in cold [electro-gilding] baths is unsatisfatory in appearance, although the quantity is sufficient, the proper shade may be imparted by [the following:] The gilt article is steeped in a solution of nitrate of binoxide of mercury, until it has become white. It is heated afterwards to volatilize the mercury, and scratch-brushed (Spon 1878: 194).

(See also Gore 1884: 138; Langbein 1902: 357; Wahl 1883: 169; Watt 1887: 175; 1911: 183)

Twentieth-century sources continue to refer to the use of mercury salts, although generally not in the form of the recipes and recommendations of the nineteenth century as the industry developed (Coggeshall 1920: 186, 245; Evans 1948: 708; Bottomley 1949: 97; Schwahn 1960: 90). The use of mercury salts is cited more often for silver plating than for gold plating.[4]

Experimental study

In order to determine if a mercury layer from quicking would be detectable and to gain some understanding of the process of quicking and its effects on the electroplating of gold, several experiments and analyses were carried out, using copper and brass coupons prepared to approximate actual workshop practices.

Procedures

Quicked brass and copper coupons were prepared by abrasive polishing, bright dipping in a solution of sulfuric and nitric acids, rinsing thoroughly to minimise contamination of the quicking and plating baths, quicking in a mercuric nitrate solution at room temperature, rinsing, and gold plating in a heated bath.[5] Unquicked samples, that is, samples prepared omitting the quicking step, were also made for comparison.

For these experiments, time of immersion (3 s; 30 s; 300 s) and concentration of the quicking solution (0.1%; 1.0%; 10.0% $Hg(NO_3)_2$ wt/vol) were varied in order to observe the changes in the mercury deposition and to understand better the detection limits. While the lowest percentage concentration most closely approximates the period recipes, one goal of the experiments was to assess the detection limits of the mercury by using higher concentrations and longer immersion times.[6] After quicking, the coupons were rinsed thoroughly with deionised water and submerged in the plating bath already connected to the DC power supply (that is 'live'). The process of moving from the bright dip to the quicking solution to the plating bath is invariably described in the literature as rapid, although no fixed amount of time is stated. The gold plating solution was prepared as 2.3% potassium cyanide, 0.5% gold (I) cyanide and 500 ml water, weight to volume.[7] As is generally recommended for alkaline gold cyanide baths, the bath was heated (50°C or 80°C) to improve the conductivity.

The quicked surfaces were examined by visual observation, in plan and cross-section with a metallographic microscope and by EPMA (wavelength dispersive spectroscopy, WDS, and energy dispersive spectroscopy, EDS) to study the physical characteristics of the mercury and gold layers and the penetration or amalgamation of the mercury with the copper and brass substrates. Atomic absorption analysis was undertaken to measure copper and zinc dissolution in quicking solutions. Polarisation measurements were carried out to determine E_{corr} and i_{corr} of the various prepared substrates in cyanide baths.[8]

Results

Visual observations

In the quicking experiments, mercury formed thin mirror-like layers within seconds in the solution; if the surface were imperfectly cleaned, a discontinous and discoloured film was readily seen. If left in the air after withdrawal from the quick, mercury was no longer visible to the unaided eye within hours, depending on the mercury concentration, time of immersion and surface preparation. In concentrated solutions of mercuric nitrate, in prolonged immersions or in immersion at elevated temperature, the mercury layer became so thick that droplets formed. Wiping and rubbing the surface with a laboratory tissue removed the thicker layers of mercury; mercury on copper could not be rubbed away as readily as mercury on brass.

In initial plating tests no visually observable difference was noted between the plating of the quicked and unquicked samples. All samples, whether quicked or unquicked, emerged from the gold plating solution with a matte gold surface over the entire coupon.

Metallography

The thinness of the mercury layers (less than 2 μm) made examination of cross-sections through the metallographic microscope up to 500× difficult.[9] Some of the physical characteristics of the gold layers could be studied (see Chapter 14, this volume). A careful selection of etchants and the ageing (oxidation) of the sample allows chemical differentiation between the layers when other microscopic techniques (such as EPMA) are not available.

EPMA-WDS and -EDS

Backscattered images of the quicked surfaces in plan showed that the mercury layer is not continuous, and in thicker deposits begins to form droplets (Fig. 15.1).[10] By EPMA-WDS and -EDS of cross-sections, mercury could be detected both in the brass substrates and in the gold layers. Although line scans show mercury and gold overlapping over a distance of five micrometers (Fig. 15.2), spot scans show that mercury is more concentrated at the interface with the brass substrate. For samples that were quicked at the lowest concentrations and shortest immersion times, the mercury line was much weaker, but still detectable. X-ray area scanning (elemental mapping) similarly indicated the presence of mercury in the gold and brass layers (Fig. 15.3). Mercury appears to be more clearly associated with the gold layer and less clearly amalgamated with the brass substrate. For all these analyses it should be noted that establishing the location of elements for thin layers is not completely reliable by EPMA: the focusing limits and excitation volume of the electron beam do not permit

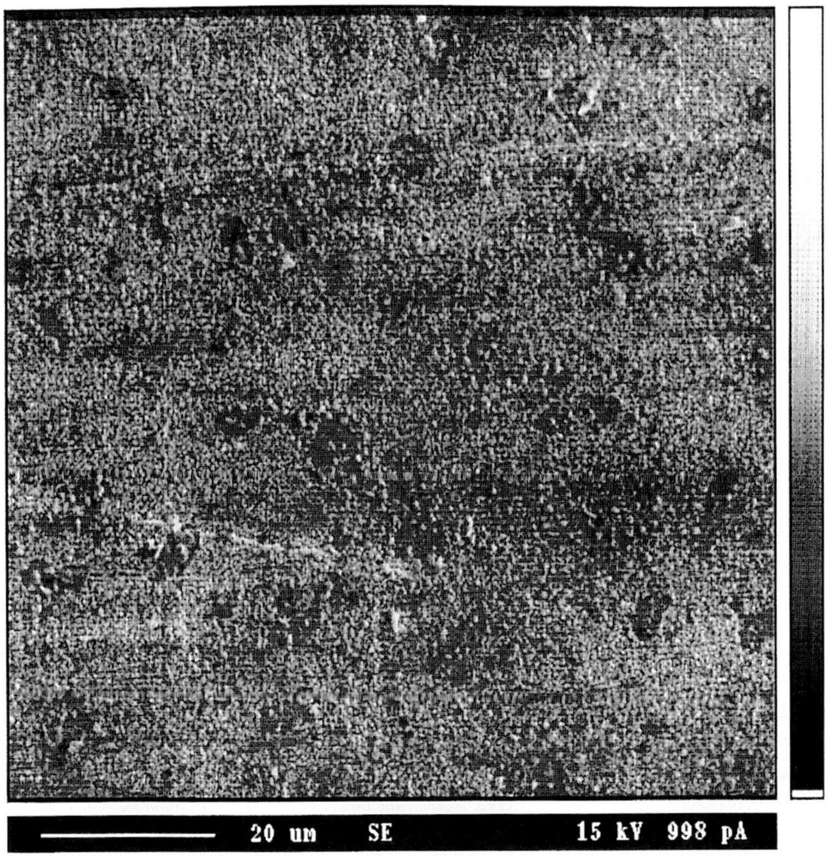

Figure 15.1 Backscattered electron image of a brass coupon prepared by a 15" immersion in 1.0% mercuric nitrate (w/v). Mercury, having a high atomic number (Z) appears white; in contrast, the substrate, composed of elements (Cu and Zn) having lower atomic numbers, appears dark. The mercury is discontinuous across the surface and forms droplets.

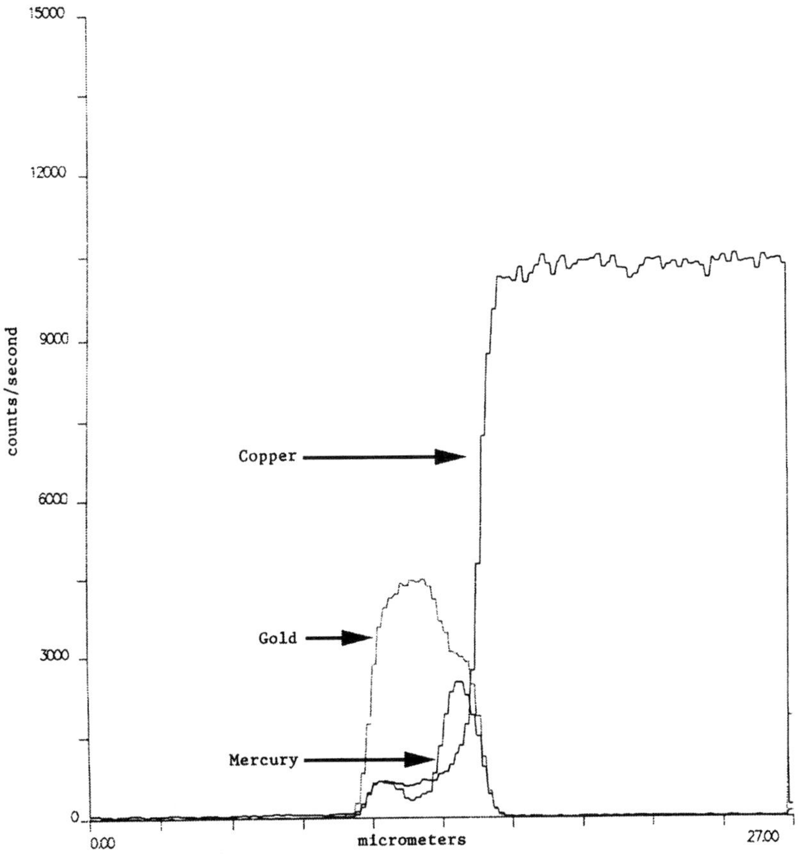

Figure 15.2 EPMA-WDS line scan of a brass coupon prepared by 15" immersion in 1.0% mercuric nitrate (w/v) and gold-plated for one hour. The horizontal axis is marked in 2.7 μm increments; the vertical axis is counts per second. Focused spot analysis in the same section indicated that the mercury is not uniformly amalgamated with the gold, but is in higher concentrations at the interface with the brass substrate.

resolution better than one micrometer in most current instruments for gold, mercury, copper and silver analyses. In EDS analysis (as for X-ray fluorescence) the separation of gold and mercury signals may be difficult, as the energy bands for mercury are close to those of gold and may not

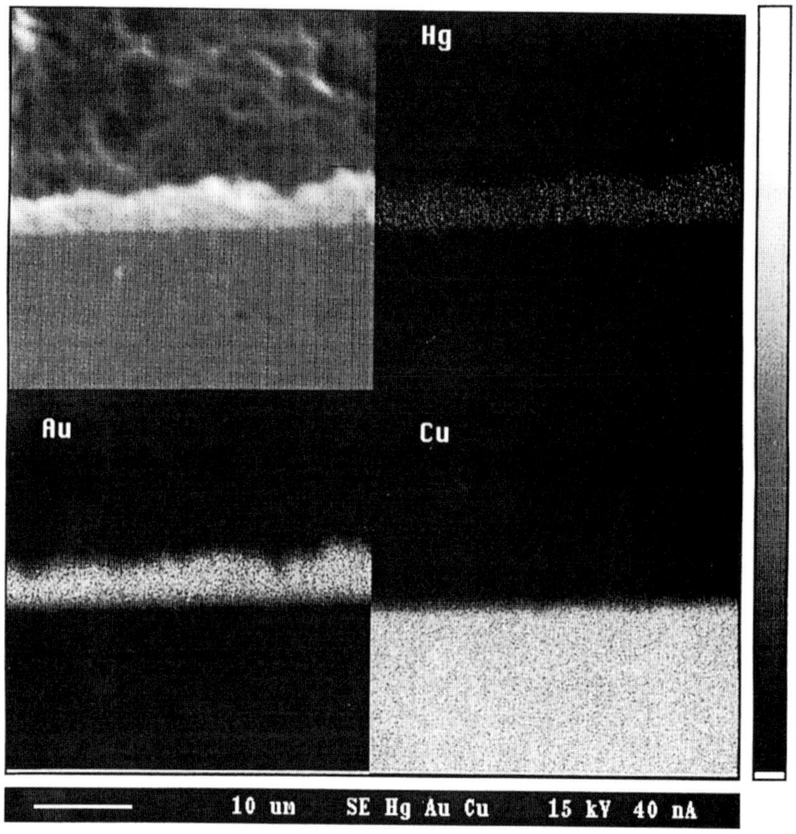

Figure 15.3 Back scattered image and X-ray area scanning of a brass coupon prepared by 15" immersion in 0.1% mercuric nitrate (w/v) and gold plated for one hour.

always be well resolved; in WDS, the separation of mercury and gold is more reliable (Heinrich and Newbury 1986).

X-ray diffraction
X-ray diffraction of the coupons on the diffractometer in plan (not in section) did not prove to be a useful technique. Only gold and brass could be detected reliably, though more careful mechanical isolation of the interface, or anodic stripping of the gold or the substrate, may yield better results.

Atomic absorption
Atomic absorption analysis was carried out to determine the dissolution of copper and zinc by the attack of mercuric nitrate solutions on the brass substrate (Fig. 15.4).[11] The concentration of copper and zinc in solution increased in proportion to the increases in the concentration of mercuric nitrate, or the increase in time. The rate of zinc dissolution dropped off slightly at higher concentrations and longer immersion times.

Potentiodynamic polarisation measurements
Polarisation measurement, one of several applicable electroanalytical techniques, was selected to assess if mercury forms a protective layer for gold electroplating. Corrosion potential (E_{corr}) and current (i_{corr}) were measured for quicked and unquicked brass immersed in cyanide solutions exposed to air.[12] The relationship between electrode potential and current density is recorded graphically.[13]

In this experiment, as an initial approach to a complex problem, the number of variables were limited: quicked and unquicked brass coupons were the test electrodes, a saturated calomel electrode was the reference electrode, and potassium cyanide (0.5M KCN) was the electrolyte.[14]

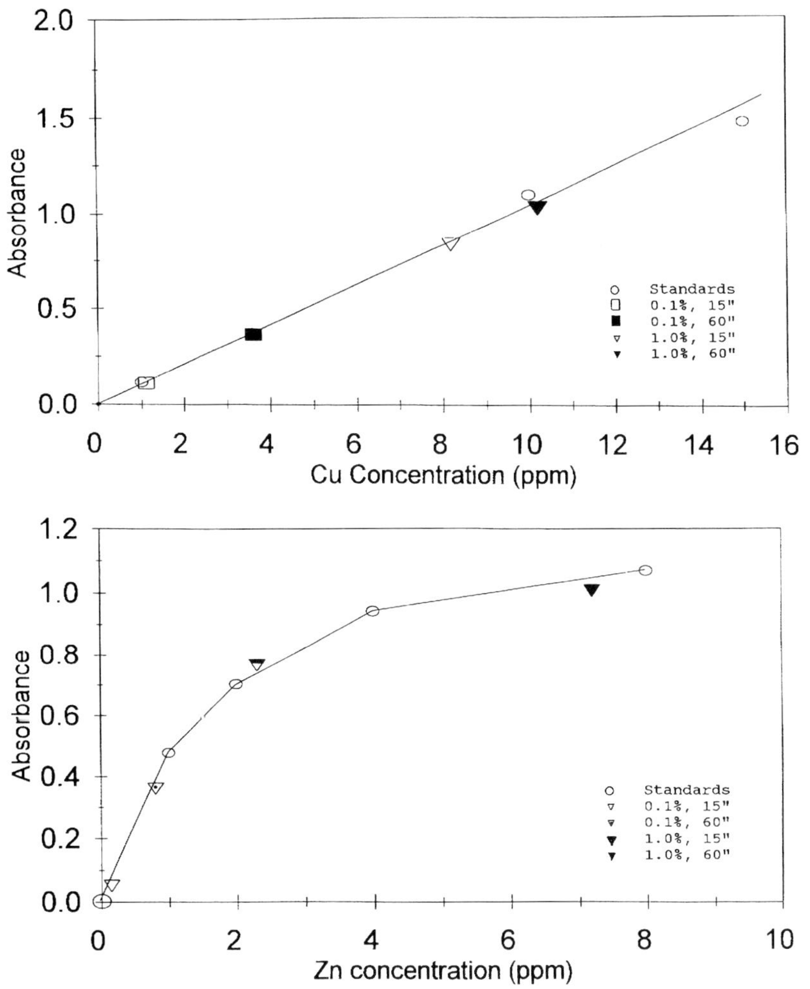

Figure 15.4 Copper and zinc dissolution during immersion of brass coupons in mercuric nitrate baths, as measured by atomic absorption. Copper and zinc dissolution increased with concentration of and immersion time in mercuric nitrate solution. The dissolution of copper and zinc indicated that the mercuric nitrate did not form a continuous film on the substrate.

Potassium cyanide was used to mimic the action of cyanide solutions containing gold. In a typical plating bath, gold and other ions in solution affect the conductivity, potential and performance of the solution in different ways.[15] Plating solutions normally operate with an excess of cyanide.

The curves from the experimental measurements, following stabilisation for 600", are shown in Figure 15.5, and E_{corr} and i_{corr} values are found in Tables 15.1 and 15.2. Polarisation curves for quicked samples in 0.5M

Table 15.1 Calculated corrosion potential (E_{corr}) and corrosion current (i_{corr}) values in 0.5M KCN: Samples varied by concentration of mercuric nitrate†

Sample	Stabilisation time (s)	E_{corr} (V)	i_{corr} (A cm^{-2})
unquicked brass blank	600"	–1.2083 V	6.180×10^{-5}
brass w/0.1% quick (1st run)	600"	–828.4 mV	2.699×10^{-7}
brass w/1.0% quick (1st run)	600"	–869.5 mV	2.243×10^{-5}

† calculated from experimental data and the Tafel equation: $\eta = \beta \log(i/i_0)$

Table 15.2 Calculated corrosion potential (E_{corr}) and corrosion current (i_{corr}) values in 0.5M KCN: Sample measured during ageing of quicked layer†

Sample	Stabilisation time (s)	E_{corr} (V)	i_{corr} (A cm^{-2})
brass w/0.1% quick (1st run)	600"	–828.4 mV	2.699×10^{-7}
brass w/0.1% quick (2nd run)	600"	–885.4 mV	2.049×10^{-4}
brass w/0.1% quick (3rd run)	600"	–957.0 mV	5.876×10^{-5}

† calculated from experimental data and the Tafel equation: $\eta = \beta \log(i/i_0)$

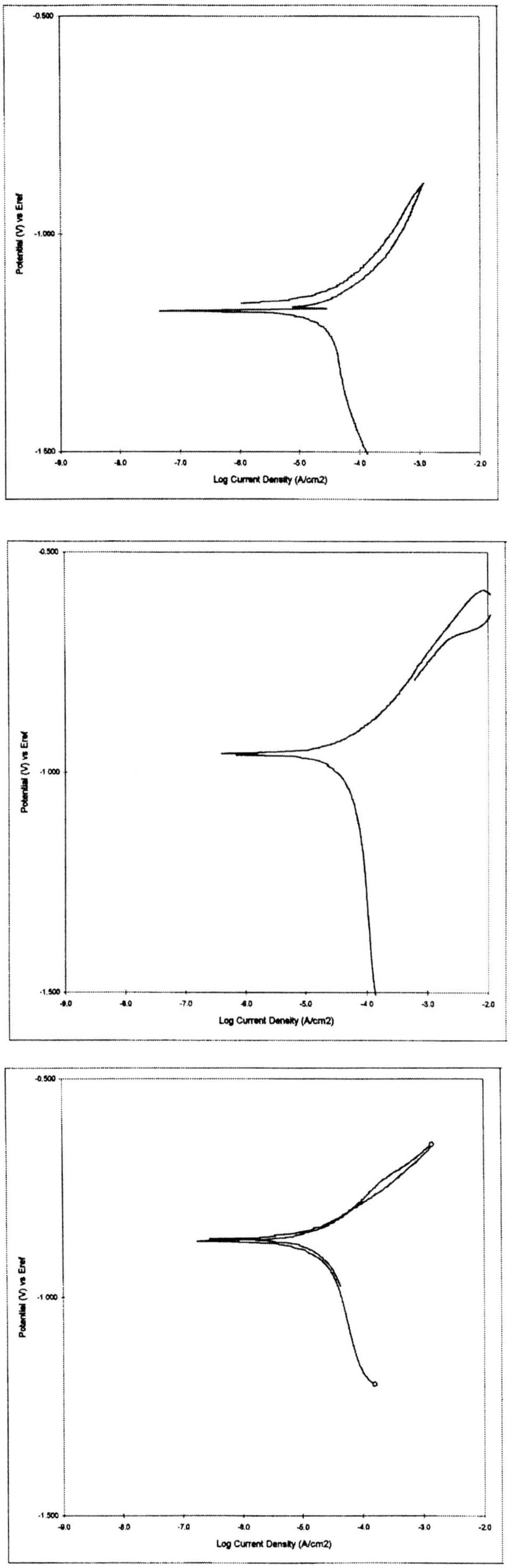

Figure 15.5 Potentiodynamic polarisation curves for brass and quicked brass coupons (0.1% mercuric nitrate and 1.0% mercuric nitrate). The quicked coupons show a higher potential (E_{corr}) than the unquicked brass coupon. Note that the E_{corr} measured is a mixed potential summing the contributions of three components (Hg, Cu and Zn) in the case of the quicked coupons.

KCN, showed that mercury raises the potential in the positive, more noble direction compared to the brass blank. The potential of the quicked brass, however, becomes more negative over time. The experimental corrosion rate values (i_{corr}) vary significantly and require further study to interpret the results.

Discussion

The detection of mercury in cross sections, even at low concentrations of mercuric nitrate and short immersion times in the quicking solution, was possible by X-ray area scanning and EPMA-WDS. Analysis of surfaces in plan for techniques like EPMA-EDS and XRF is dependent on the thickness of the gold plating (see Table 15.3).

The disappearance to the unaided eye of the mercury film is primarily due to amalgamation of mercury with the substrate.[16] Mercury penetrates and amalgamates with brass to a shallow depth, less than 2 μm in these experiments, depending on time, temperature and composition of the substrate. The emphasis in the literature on moving rapidly from the quick to the gold plating bath must relate to maintaining the sample free from oxidation, but may also suggest that the benefit of mercury, if any, is related to its presence on the surface before it amalgamates to a great degree with the substrate.[17]

Thick layers of mercury, illustrated by quicked samples prepared at high concentrations of mercuric nitrate and prolonged immersion times, presented several problems for electroplating gold.[18] Most significantly, although gold could be plated readily on these samples, it would not withstand scratch brushing, polishing or burnishing, the common finishing techniques for the matte deposition that results from alkaline gold cyanide baths. Mercury would also affect the colour of the gold deposit; if too thick, the mercury layer would have made the gold white.

If deposited as a continuous layer, mercury would be expected to protect the substrate from attack by the solution. In conjunction with the evidence from the backscattered electron images, atomic absorption analysis indicated that the mercury film was not continuous and that the substrate continued to be accessible to attack.[19]

Complex cyanide solutions lower the apparent potential of copper, brass, zinc and silver and the metal substrate corrodes. This initial corro-

Table 15.3 The X-ray range, z†, the depth of penetration of electrons that will excite X-ray photons, as calculated from a current formula given by the American Society of Metals (ASM) and compared to the values from Cameca software.

Metal‡	V_0 = 15kV z (μm)	V_0 = 20kV z (μm)	V_0 = 40kV z (μm)	V_0 = 80kV z (μm)
ASM: Au over Au′ $L\alpha_1$	0.08	0.18	0.72	2.43
ASM: Au over Hg $L\alpha_1$	0.09	0.17	0.72	2.42
Cameca: Au over Hg $L\alpha_1$	0.09	0.28	1.40	5.00
ASM: Au over Hg $M\alpha_1$	0.15	0.25	0.79	2.50
Cameca: Au over Hg $M\alpha_1$	0.30	0.50	1.62	5.21

† z is the distance from the specimen surface to the level at which the electrons cease to excite X-ray photons. $z = 0.035(V_0^{1.65} - V_q^{1.65})\{\rho^{-1}\}$ μm, where V_0 is the operating voltage of the instrument in KeV, V_q is the excitation voltage for the emitted X-ray line, and ρ is the density of the material in g cm^{-3}.
Values from Bearden (1964): for gold $L\alpha_1$, E = 9.713 KeV, λ = 1.276Å; for mercury $L\alpha_1$, E = 9.989 KeV, λ = 1.241Å; $M\alpha_1$, E = 2.196 KeV, λ = 5.645Å.
‡ The metals are listed as the gold layer over a substrate metal, gold or mercury. The data refers to the depth of gold which will stop an electron that is capable of exciting a gold or a mercury atom to emit an X-ray photon.

sion would have caused poor plating adhesion in cyanide baths, but it could be partially overcome by introducing the object to be plated into the bath 'live', that is, already connected to the current, and by reducing the concentration of gold cyanide in solution, that is, using a 'strike' bath. These modifications would have pushed the potential of the substrate into the immunity regions where the metal is thermodynamically favored, but battery performance, especially prior to 1870, did not allow good voltage regulation. The coupons that were quicked raised the mixed potential of the substrate in the 0.5M KCN solution.[20] The voltage applied in the plating bath would put such quicked surfaces into the immunity region more easily than an unquicked substrate. It is also clear from the experimental data that the potential of the quicked brass becomes more negative over time, as the mercury layer on the surface diminishes by amalgmation into the substrate surface, bringing more copper and zinc to the surface.

Examination of an electrotype of the Milton Shield

Evidence for quicking may be found by the detection of mercury for some objects. An electrotype object, the Milton Shield, in the Philadelphia Museum of Art (1984-133-1) was examined for evidence of gilding and quicking (Fig. 15.6). The original shield, designed and crafted by Leonard Morel-Ladeuil while he was working for Elkington and Company, is made of three silver repoussé panels set into a steel repoussé and gold inlaid frame. When shown at the Paris Exposition of 1867, it was considered a masterpiece of composition and craftsmanship. The South Kensington Museum, now the Victoria and Albert Museum, purchased the original shield in 1868, and it remains part of the permanent collection there.

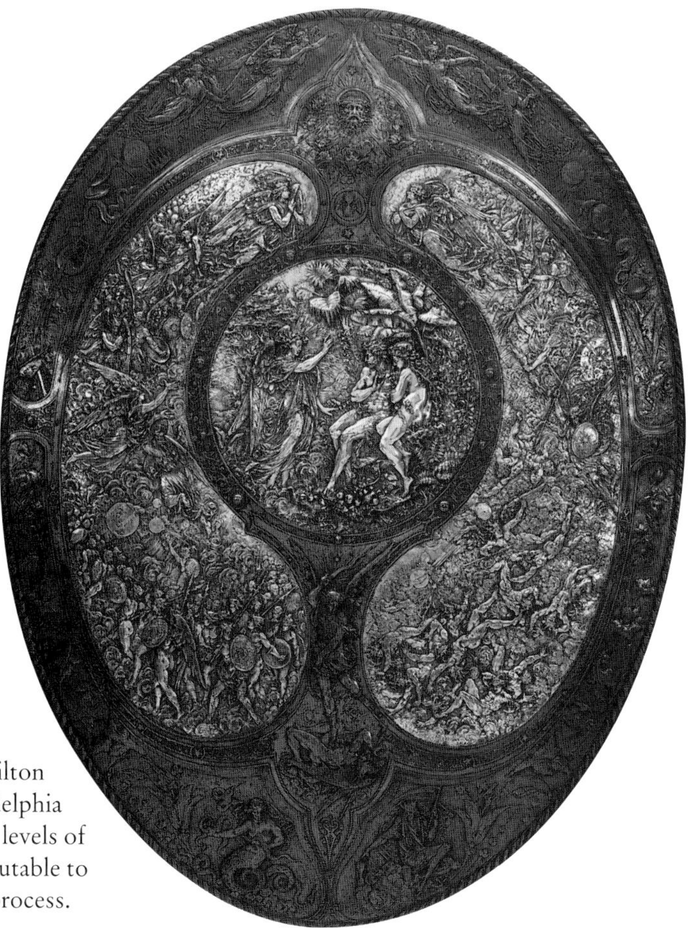

Figure 15.6 An electrotype of the Milton shield (accession 1984-133-1) (Philadelphia Museum of Art, Joe Mikuliak). Trace levels of mercury were detected by XRF, attributable to a quicking step in the electroplating process.

The South Kensington Museum established an agreement with Elkington and Co. to produce electrotype copies of original objects, which were then sold to other institutions and the public. In a catalogue of reproductions, the shield is described as 'silvered and oxydised with bands gilt where damascened [inlaid] in the original' (Victoria and Albert Museum 1873: 34). The electrotype in the PMA collection is a copper shell assembled in four sections, each section reinforced with solder from the back and silver plated on the front. The object has been overpolished and there is little visual evidence of gold.

X-ray fluorescence analysis (XRF) was carried out on design areas that would have been gold and/or silver plated. The results indicated trace levels of mercury.[21] There was no visual evidence, such as mercury spread, drip marks or corrosion differentiation, common to many mercury gilded silver pieces, to support a conclusion of fire gilding. It is a reasonable conclusion for an object of this type that the source of mercury is from a quicking step either in the silver or gold electroplating process.

Conclusion

The detection of mercury is not evidence only for fire gilding: mercury may also be detected for electroplated surfaces. Electroplating is a complex process, and the many efforts, including quicking, to control the quality of the resulting plate are described extensively in nineteenth- and twentieth-century sources. These shop practices have not been adequately considered in previous discussions of gilding technology. Techniques such as X-ray area scanning and EPMA-WDS of cross-sections are useful to detect low levels of mercury in museum objects when sampling is possible. Similarly, EPMA-EDS and XRF may be useful if the gold (or silver) layer is not too thick. X-ray diffraction was not useful in detecting mercury even for samples prepared with high concentrations of mercuric nitrate. Similarly, metallography up to 500× was not useful to examine cross-sections with layers of mercury that were typically less than 2 μm.

From the experimental work presented here, the presence of a thin film of mercury on the substrate as a test for cleanliness would have undoubtedly improved the adhesion of the gold plate. The mercury layer from quicking was found not to be continuous: the substrate, even if quicked, would not have been protected by a continuous barrier layer of mercury from contact with and possible dissolution by corrosive plating solutions. However, polarisation measurements indicate that mercury appears to raise the potential of the surface, making possible a reduction in substrate corrosion at the cathode; this would have been a significant factor in nineteenth-century practice where battery performance did not allow good voltage regulation. The reduction of substrate corrosion in the cyanide bath would have improved the adhesion of the deposited metal. Further experimental work is needed to elucidate the complex chemical reactions that may underlie the long history of use of mercury salts in electroplating.

Acknowledgements

Janice Carlson, Winterthur Museum, Winterthur, DE for X-ray fluorescence analysis. Rick Corbett, Corrosion Testing Lab, Newark, DE for potentiodynamic polarization testing. Rick Sobel, Community College of Philadelphia, for atomic absorption analysis. Ed Vicenza, Princeton Materials Group, Princeton, NJ for EPMA-WDS and -EDS.

Notes

1. Many nineteenth-century recipes start with elemental mercury from which the cyanide, nitrate, or occasionally chloride salts are made. The chemical names of the period include: 'nitrate of binoxide of mercury', 'nitrate of protoxide of mercury', 'binoxide of mercury', 'bichloride of mercury' and others. For the experimental work in this study, mercuric nitrate, $Hg(NO_3)_2 \cdot H_2O$, (Fisher ACS grade M-168) was used for the quicking, probably approximating the solutions made from 'nitrate of binoxide of mercury'; it is the most soluble of the mercury salts in water.
2. It is sometimes difficult to identify primary references. Much information is repeated in different publications without citations.
3. A list of patents given by Gore (1884: 371) includes one dated 1 June 1842, H. B. Leeson, *Electro-depositing processes . . . 'quicking' their surfaces. Claims 430 different salts*. Elkington (1856 [1837]: 5–6) describes his patent:

   > I shall now describe a method of gilding certain metals by means of solution of gold and a solution of mercury. Dissolve gold in nitro-muriatic acid, as before directed, and evaporate the solution to about one fourth or until it assumes a deep red color and crystallized on becoming cold. The solution may then be diluted with sufficient water to render it convenient to measure off any quantity of the gold contained in the solution. Prepare also a solution of mercury as follows: Dissolve any quantity of mercury in nitric acid, and to from two to four pennyweights of mercury in solution add a gallon of water and three pounds of carbonate of potash, and boil them until the potash is dissolved. The mercury at first will be precipitated by the potash, but when boiled together will afterwards re-dissolve. Then having the articles to be gilded made perfectly clean, take any convenient quantity of them, and having measured off as much of the gold in solution as is considered sufficient, put it into a clean earthen vessel capable of holding sufficient water (I usually employ from two to six grains of gold to one pint of water); then dip the articles into the boiling solution of mercury and potash, when they will immediately be coated with mercury (a very slight coating will suffice), wash them in clean water and pass them into the solution of gold diluted, as before described, keeping the liquid in motion and occasionally stirring about the articles. The gold will be seen to alloy itself with the slight coating of mercury, and the articles will become of a greenish bronze appearance; when this appearance is the same over the whole surface of the goods, they may be passed again into the solution of mercury and potash, and the process repeated, as before, until they are sufficiently gilt. They must then be put into the solution of mercury and potash (which may have more mercury in this instance), and when uniformly white over the whole surface, they must be heated over a clear fire until all the mercury is expelled (a process called heightening, and well understood in the ordinary method of gilding). In gilding by these means, the operator must guard against having the solution of gold too strong, and the coating of mercury should be proportioned to the strength of the solution of gold, all which he will best understand by a little practice, as the quantities employed in both cases may be varied considerably with the same result, or nearly so. I would remark, that I prefer to gild by this process only such articles are required of an inferior quality, and for such goods this process has many advantages over the old method of gilding by amalgam of gold and mercury, the gold being more uniformly spread over the whole surface of the articles, and the operation is performed without injury to the workman employed . . . What I claim as the second part of Invention is, the using of a solution of mercury and a solution of gold in place of the amalgam of gold and mercury heretofore employed.

4. Silver cyanides are more aggressive than gold cyanides for many substrates. A typical silver cyanide plating bath will dissolve gold if it is placed in the bath without an applied cathodic voltage.
5. The brass and copper coupons were prepared as follows: polish with 180 grit to

600 grit silicon carbide paper; buff on a rouge wheel; wipe clean with mineral spirits; clean with a dental steamer; dry with acetone. When not used immediately, samples were wrapped in acid free tissue and stored in chambers desiccated with silica gel.

6. Examples of recipes for quicking:

- 2½ gal H_2O; ⅓ oz nitrate of binoxide of mercury; ⅔ oz sulphuric acid (Spon 1878: 205)
- 10 l H_2O; 100 g binoxide of mercury; and a quantity of sulphuric acid sufficient for dissolving the peroxide of mercury (Fontaine 1885: 147)
- 1 lb nitric acid; add three times its volume of distilled water; 1 oz. mercury; dilute with 1 gal water (Trevert 1891)
- 1 oz quicksilver in nitric acid 1:3 water; add 1 g distilled water with 2 oz sulphuric acid (Watt 1889: 236)
- 1 oz mercury in nitric acid, 1:3 water; add cyanide of potassium until no more precipitate. Filter. Dissolve in solution of cyanide until dissolved. Add water to one gallon (Watt 1889: 236)

7. Other additives found in period recipes were eliminated for this experiment in order to limit the number of variables, but it should be noted that these additives affect the conductivity and throwing power of the solution, as well as buffer the metal ion concentration. Also, many period recipes begin with elemental gold or gold chloride that is then converted to gold cyanide; the carry-over of impurities would have also affected the conductivity of the solution, its performance, and its useful life.

   Examples of recipes for gold plating:

- Becquerel's solution: 1 part terchloride of gold; 10 parts potassium ferrocyanide in 100 parts water; filter; add 100 parts saturated solution of potassium ferrocyanide; dilute mixture with one to two times volume of water (Gore 1854: 74)
- 31 g and 25 centigrams of oxide of gold; 5 hectograms of potassium cyanide; 4 l of water; boil for half an hour; or 10 parts ferrocyanide of potassium; 1 part terchloride of gold in 100 parts water; boil for two to three hours until precipitate; filter; add three times the volume in water (Gore 1854: 75)
- 21 oz crystallized phosphate of soda; 3½ oz bisulphite of soda; ⅓ oz pure potassium cyanide; ⅓ oz gold chloride; 2 ⅓ gal distilled water (Spon 1878: 194)
- From Roseleur: 2.11 oz sodium phosphate; 0.35 oz sodium sulphide; 30.86 grains potassium cyanide; 15.43 grains gold chloride; 1 quart water (Langbein 1902)

8. Equipment used: Cameca SX50 electron probe microanalyser equipped with wavelength dispersive spectrometers (WDS) and energy dispersive spectrometers (EDS); X-ray diffraction: Philips PW1729 X-ray generator equipped with a PW1840 diffractometer; atomic absorption: Varian SpectraAA-5; X-ray fluorescence: Kevex model 8050 750 Energy dispersive X-ray fluorescence analyzer (Ag secondary target, 40 kV, 1.7 mA).
9. The mercury layer measured by metallographic cross-section, after the time for sample preparation and embedding, approaches an equilibrium condition. The section would not indicate the initial film thickness immediately after quicking. Aged samples show variation in appearance and penetration of mercury from the results observed within days of manufacture.
10. In EPMA instruments the sample is run in vacuo and the beam heats the sample. As part of the experimental work, a 30 kV beam was focused on a spot and mercury signals were measured; the beam was left on the spot for more than one minute and the mercury signal was recounted. A minor diminution in the mercury signal was observed.
11. The preparation of samples for atomic absorption analysis was as follows:

- 1" × 2" copper and brass coupons were polished as describe above, excepting the rouge polishing
- a coupon in 100 ml deionized water was used as a blank for each test
- samples of each metal were submerged for intervals of 15" and 60" in 100 ml mercuric nitrate solution prepared in concentrations of 0.1% and 1.0% (w/v)
- copper (1–15 ppm) and zinc (1–8 ppm) standards were run in order to plot absorbance vs. concentration (ppm)

12. Absorbed oxygen and carbon dioxide were not eliminated from the solution

prior to taking measurements, approximating typical electroplating conditions. Many electrochemical studies eliminate absorbed gases by bubbling nitrogen, helium, or hydrogen through the solution before measurements begin.

13. Polarisation of the potential, ϕ, is measured between a test electrode and reference electrode connected to a potentiometer. If the potential is displaced continuously from its equilibrium value in the anodic and cathodic directions, the measurement is called potentiodynamic. In cyclic potentiodynamic polarisation, the potential of the quicked brass is ramped continuously positive (anodic polarisation) and negative (cathodic polarisation) to observe the corrosion current changes, determining the corrosion potential (E_{corr}) and the corrosion current (i_{corr}) (Wranglén 1985: 25–7).
14. The initial amalgamation of the mercury layer into the substrate is faster than the time required to introduce the sample into the bath, to stabilise the equilibrium current and potential, and to make an anodic or cathodic sweep to measure corrosion reactions by potentiostatic methods.
15. A study of the mercury layers in gold cyanide solutions is complicated by the fact that gold begins to plate out immediately on the substrate – in the absence of applied voltage – by simple electrochemical displacement.
16. The mercury–copper, mercury–zinc, and mercury–copper–zinc phases are not stable amalgams over time. In cross-sections prepared for this work in 1995, and re-examined in 1997, it could be seen that the mercury preferentially attacked the grain boundaries of the brass substrate layers with areas of diffusion into the grains. Other chapters address the time/temperature dependence of amalgamation.
17. Mercury will not easily wet or bond to surface oxides or other corrosion products.
18. Roseleur advised thicker layers of mercury for objects that were to be more heavily plated, such as eating utensils, and less thick for objects which required only thin deposits of gold, such as jewelry (Roseleur 1872: 32–4). Roseleur's advice is not consistent with the comments of other authors; they note that a thick layer of mercury would not allow the gold to withstand scratch brushing or burnishing. Trevert (1891) emphasised the importance of the thinnest coating possible, as a thick coating of mercury will cause metal to strip after the plating is done.
19. A logical follow-up to this experiment would be to measure absorbance for copper, zinc and mercury after immersion in a gold cyanide plating bath for quicked samples, and for copper and zinc for unquicked samples to detect substrate corrosion during the plating process.
20. Hydrogen or oxygen overvoltage, the polarisation of the electrode necessary to cause hydrogen evolution or oxygen reduction, may also play an important role. In systems of two or more metals, where the more noble metal shows low hydrogen overvoltage for the given solution, the more noble metal will act as an efficient cathode and will accelerate the corrosion of the less noble metal (and will increase the rate of hydrogen evolution). Mercury has a high hydrogen overvoltage; it will not serve easily as a cathode surface for hydrogen evolution in corrosion reactions and thus tends to reduce the corrosion current (i_{corr}). These measurements are reported as follows where η is the activation polarisation (overpotential) and i_0 is the exchange current density (Uhlig and Revie 1985: 43–4):

Metal	Temperature	Solution	i_0 (A m^{-2})	η (V)
Au	20°C	0.1N NaOH solution	5×10^{-9}	0.47 V
Au	20°C	1N HCl solution	5×10^{-3}	0.15 V
Ag	20°C	0.1N HCl solution	5×10^{-3}	0.30 V
Cu	20°C	0.1N HCl solution	2×10^{-3}	0.44 V
Cu	20°C	0.15N NaOH solution	1×10^{-2}	0.36 V
Zn	20°C	1N H_2SO_4 solution	1.6×10^{-7}	0.94 V
Hg	20°C	0.1N HCl solution	7×10^{-9}	1.10 V
Hg	20°C	0.1N H_2SO_4 solution	2×10^{-9}	1.16 V
Hg	20°C	0.1N NaOH solution	3×10^{-11}	1.15 V

21. The Winterthur Museum Analytical Laboratory, no. 2691, 1/22/92.

References

Bearden, J. A. 1964. *Report NYO 10586: X-ray Wavelengths*. Oak Ridge, Tennessee: US Atomic Energy Commission.

Becquerel, M. 1864. *Elements d'Electro-Chimie*, 2nd edn. Paris: Librairie de Firmin Didot Freres, Fils et Cie.

Bottomley, J. E. 1949.' The plating shop "quick"'. *Electroplating*, December, 1949: 119–20.

Chambers, E. 1752. *Cyclopaedia: or, an univeral Dictionary of Arts and Science.* London: Innys. (First edition 1728.)

Coggeshall, K. M. 1920. *The Modern Electroplater.* New York: The Norman W. Henley Publishing Co.

Elkington, H. 1856. *Coating Metals, No. 7304, 1837.* London: Great Seal Patent Office.

Evans, U. R. 1948. *Metallic Corrosion Passivity and Protection*. New York: Longmans, Green & Co.

Fontaine, H. 1885. *Electrolysis: a Practical Treatise.* London: E. & F. N. Spon.

Gore, G. 1854. *Theory and Practice of Electro-Deposition.* London: Charles Griffin and Company.

Gore, G. 1884. *The Art of Electro-Metallurgy*, 2nd edn. New York: D. Appleton and Co.

Graham, A. K. 1945. *Plating and Finishing Guidebook.* New York: Metal Industry Publishing Co.

Heinrich, K. F. J. and Newbury, D. E. 1986. 'Electron probe X-ray microanalysis'. In *Metals Handbook: Materials Characterization,* 9the edn, vol. 10, Ruth E. Whan (ed.), 516–35. Metals Park, Ohio: American Society for Metals.

Langbein, G. 1902. *A Complete Treatise on the Electro-Deposition of Metal,* 4th edn, with additions by William T. Brannt. Philadelphia: Henry Carey Baird & Co.

Nobel, F. I. 1982. **(not cited)** 'Gold plating'. In *Metals Handbook: Surface Cleaning, Finishing, and Coating,* vol. 5, 9th edn, William G.Wood (ed.), 281–4. Metals Park, Ohio: American Society for Metals.

Roseleur, A. 1872. *Galvanoplastic Manipulations: A Practical Guide for the Gold and Silver Electroplater and the Galvanoplastic Operator,* A. A. Fesquet (trans.). Philadelphia: Henry Carey Baird.

Schwahn, C. 1960. *Workshop Methods for Gold- and Silversmiths.* New York: Chemical Publishing Co., Inc.

Smee, A. 1845. *Nouveau Manuel Complet de Galvanoplastie, ou éléments d' Éléctro-Metallurgie,* E. de Valicourt (trans.). Paris: Chez Leonce Laget, 1977.

Smith, S. A. 1891. *Source and Nature of Electricity and Its Application to the Electroplating Process.* Providence, Rhode Island: Livermore & Knight.

Spon, E. 1878. *Workshop Receipts, for the use of Manufacturers, Mechanics, and Scientific Amateurs.* London: E. & F. N. Spon.

Stalker, J. and Parker, G. 1971 [1688]. *A Treatise of Japanning and Varnishing.* London: A. Tiranti.

Trevert, E. 1891. *A Practical Hand-Book of Electro-plating.* Lynn, Mass.: Bubier Publishing Company.

Uhlig, H. H. and Revie, R. W. 1985. *Corrosion and Corrosion Control: An Introduction to Corrosion Science and Engineering.* New York: John Wiley & Sons.

Victoria and Albert Museum 1873. *Illustrated Catalogue of Electrotype Reproductions of Works of Art from Originals in the South Kensington Museum.* London: Eyre and Spottiswood for HMSO.

Wahl, W. H. 1883. *Galvanoplastic Manipulations: A Practical Guide for the Gold and Silver Electroplater and the Galvanoplastic Operator.* Philadelphia: Henry Carey Baird & Co.

Watt, A. 1887. *Electro-Deposition: A Practical Treatise,* 2nd edn. London: Crosby Lockwood and Co.

Watt, A. 1889. *Electro-Metallurgy: Practically Treated,* 9th edn. London: Crosby Lockwood and Son.

Watt, A. 1911. *The Electro-Plating and Electro-Refining of Metals,* 2nd edn, revised by Arnold Philip. London: Crosby Lockwood and Son.

Weston, J. H. 1915. *The Electro-Platers Hand-Book.* Chicago: Frederick J. Drake & Co.
Wranglén, G. 1985. *An Introduction to Corrosion and Protection of Metals.* New York: Chapman and Hall.

16 The Fabrication of Gilt Bronze Mounts for French Eighteenth-century Furniture

Brian Considine and Michel Jamet

Abstract

This chapter details the steps in the production of a French eighteenth-century furniture mount by the sand casting and amalgam mercury gilding processes. There is a brief introduction to the subject of the production of gilt bronze mounts in eighteenth-century Paris, covering the chemistry of the bond between the copper alloy substrate and the gold and mercury amalgam as well as the social and commercial context for the production of these mounts. Fifty-two photographs illustrate the steps involved in the casting, chasing, gilding and patination of a gilt bronze mount from a model by Dubois. The photographs are explained with captions that give the details of each step, the materials involved, the tools used and the procedure for each step.

Introduction

Although the origins of mercury gilding remain unknown, Lins and Oddy (Lins and Oddy 1975) have demonstrated that the Chinese were practising it as early as the third century BC, that in Persia it was used from the third century AD and that the Romans were using it in the later Roman Empire. Jett (1993) even traces mercury gilding in China to the Warring States Period in China (480–422BC). Its use in the medieval period is documented by Theophilus in his twelfth century *Treatise on Divers Arts* (Chapman 1994). In the seventeenth and eighteenth centuries, mercury gilding on copper alloys is described first by Stalker and Parker in their *Treatise of Japanning and Varnishing* published in London in 1688 (Chapman 1994) and later by Chambers in his *Cyclopedia*, published in London in 1728 and then in Paris in Diderot and d'Alembert's *Encyclopédie ou Dictionnaire Raisonné des Sciences, des Arts et des Métiers,* published in 1751–7 (Fig. 16.1).

The actual bond between the gold and the copper alloy is achieved by creating an interdiffusion bond during the heating process. This interdiffusion between the gold and the copper alloy, however depends on the formation of mercury–copper intermetallic compounds at the interface where the mercury is preferentially retained, rather than diffusing more extensively as it would into silver. The formation of the mercury–copper intermetallic compounds means that the mercury cannot be completely driven off during the heating. The mercury that remains becomes bound up with the copper, which is stable at the boiling point of mercury. In cross-section, the

Figure 16.1 *Metal Gilder's Workshop* from Denis Diderot, *Pictorial Encyclopedia of Trades and Industries* (Diderot 1763).

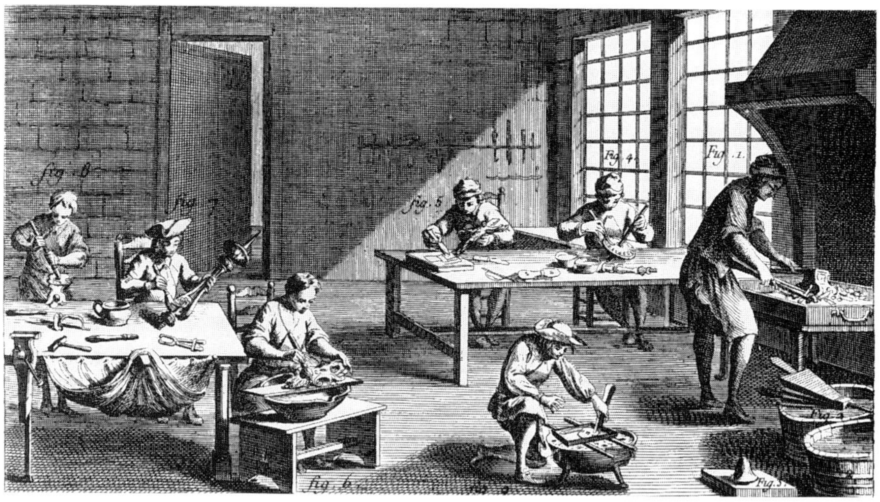

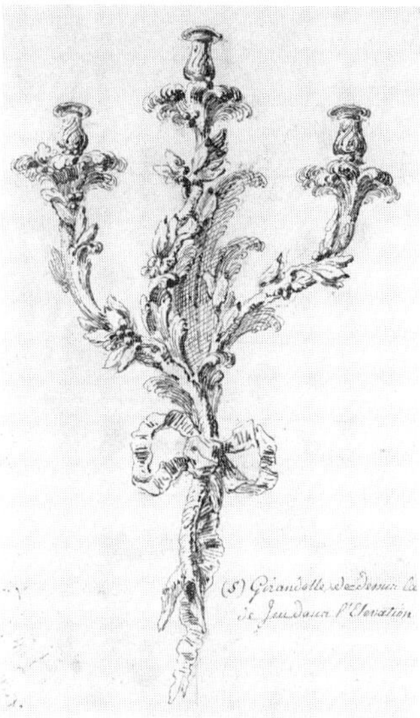

Figure 16.3 Drawing for a Wall Light, *c.* 1760–80, attributed to Contant d'Ivry. (The J. Paul Getty Museum, Los Angeles. Photo: Jack Ross.)

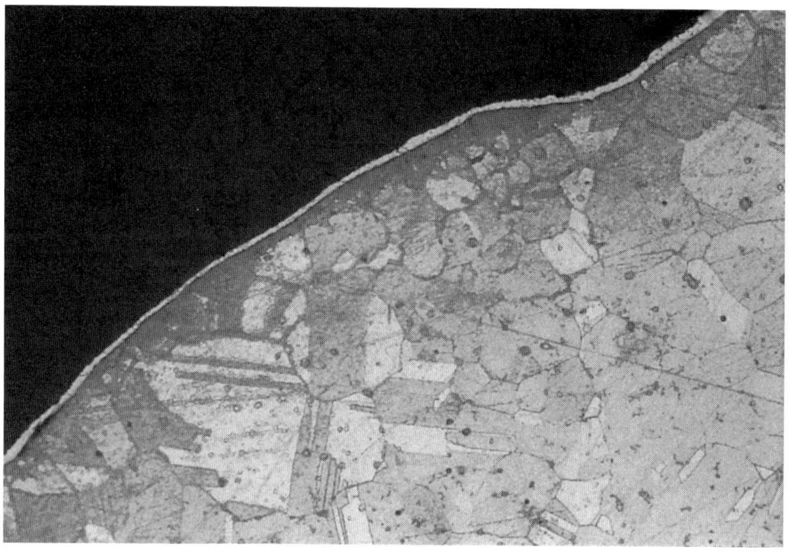

Figure 16.2 Cross-section of mercury gilding on bronze, taken from the mount illustrated in this article. (Photo: David Scott.) (See Plate 90.)

gold is usually elevated immediately at the surface of the amalgam, with a slowly increasing concentration of mercury through the lower levels of the coating and slightly into the copper. The copper shows a very low concentration at the surface, gradually increasing until the copper substrate is reached (Scott 1998) (Fig. 16.2).

The art of the mercury gilder reached its fullest expression during the eighteenth century in France, where it was used to luxurious effect decorating the sumptuous furniture found in the Royal *châteaux* and palaces as well as the homes of the aristocracy and the *haute bourgeoisie*. Louis XIV had established Paris as the international centre of craftsmanship for the luxury furnishing trades, such as furniture making, tapestry weaving, porcelain manufacturing and gilt bronze making. Not only were objects such as furniture, stone vases and porcelain embellished with gilt bronzes, but articles such as firedogs, clocks and wall lights, chandeliers and candelabra were made almost entirely of gilt bronze. The commissions began with drawings by the leading designers, such as André Charles Boulle, Gilles-Marie Oppenordt, Juste Aurèle Meissonier, Contant d'Ivry (Fig. 16.3) and later Jacques and Philippe Caffiéri to name a few. From there, a model would be carved by a sculptor in wax, wood or terracotta (Fig. 16.4). These were sometimes presentation models, but could also serve as the models for the casting. Some sculptors of great renown, such as Houdon, Pajou, Boizot and Clodion, supplied models to *fondeurs* (Verlet 1987). The mod-

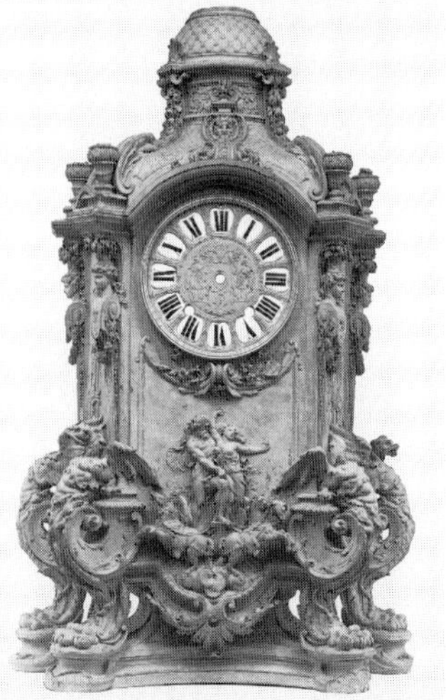

Figure 16.4 Terracotta model for mantle clock, *c.* 1700. (The J. Paul Getty Museum, Los Angeles. Photo: Jack Ross.)

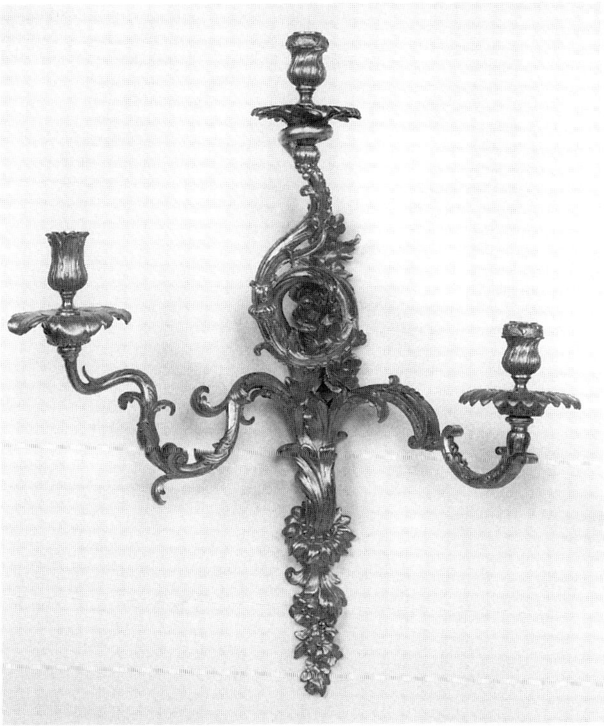

Figure 16.5 Wall light, *c.* 1720, attributed to Boulle. (The J. Paul Getty Museum, Los Angeles. Photo: Jack Ross.)

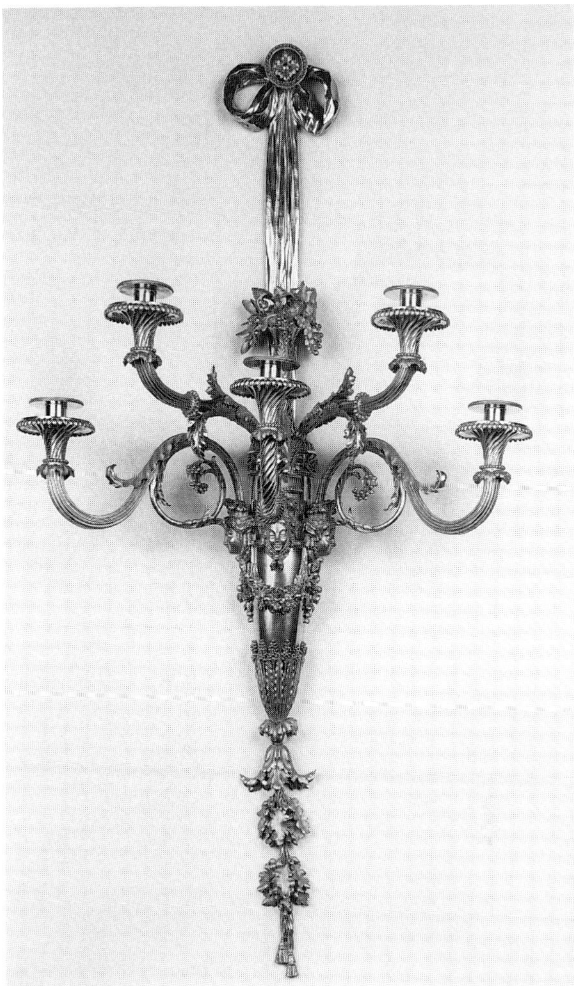

Figure 16.6 Wall light, *c.*1787, attributed to Pierre-Philippe Thomire. (The J. Paul Getty Museum, Los Angeles. Photo: Jack Ross.)

ellers and the casters were part of the same *corporation* as the *fondeurs-ciseleurs*, but there were ongoing struggles between them and the *ciseleurs-doreurs,* about which *corporation* had the authority to do the chasing. The chasing was the step that took the bronze from the crude cast to the sharp, refined stage with definition and texture, ready for gilding (Figs 16.30–16.40). As in all of the French decorative arts, there was a clear evolution of style during the eighteenth century from the bold, gestural chasing of Boulle (Fig. 16.5) to the highly refined and detailed work of Gouthière and Thomire (Fig. 16.6).

The mercury gilding was done by one of two methods. The first method was to coat the base metal with a layer of mercury and then lay the gold leaf onto the mercury, before evaporating the mercury in the fire. This technique is known as leaf gilding and produced the overlays of the leaf that are now sometimes visible as the gold wears thin. The second technique, illustrated in this article, is known as amalgam gilding or *or moulu* (ground gold) where the mercury and the gold are melted together to form the amalgam, which is then brushed onto the copper alloy. In both techniques, the objects being gilded were then returned to the fire for the bonding of the gold to the base metal. Documents reveal that the number of reapplications of the amalgam was sometimes even specified in the order. As is the case for wood gilding, different colours of gold were also used either exclusively or with the common yellow gold to achieve variations of colour. Mounts were also silvered, particularly during the Rococo period. In a step known as the *mis en couleur d'or*, the colour of the gilded mounts was adjusted by treating them with chemicals, such as ammonium chloride, potassium nitrate, cupric acetate and zinc sulphate, that matted some areas and added tonal differences, perhaps more subtle than those achieved by the use of different coloured gold (Chapman 1994). This step of tonal adjustment, sometimes necessary so that all of the mounts on a single piece of furniture, or a set of furniture were a perfect match, should not be confused with the use of acids to brighten copper alloy mounts done as an inexpensive alternative to gilding, but known by the same name. These different processes were the domains of two different professional corporations in eighteenth century France. The *fondeurs-ciseleurs* were allowed to brighten their mounts instead of gilding them, but only the *ciseleurs-doreurs* were allowed to tone them once gilded (Verlet 1987).

The study of French gilt bronzes is made even more interesting by a law recorded by Parliament on 5 March 1745, which required that objects weighing more than two ounces containing copper be stamped by a crowned C, indicating that a tax had been paid. The purpose of the edict was to raise money, but it was revoked on 4 February 1749(de Bellaigue 1974). During that time, however, all articles '*de cuivre pur, de fonte, de bronze & autres des cuivre mélangé, fondu, moulé, battu . . .*' had to be stamped before they were sold, gilded or turned. Since the edict applied to both new and old objects, the appearance of the stamp on an object can indicate either that it was made, or that it was in the stock of a dealer during the tenure of the edict. Several models of stamps are known. The most common is the three pointed crown in one of three sizes, but others were also used (Fig. 16.7). Counterfeiting the stamps is as old as the law requiring them. During the enforcement of the law, a counterfeiter dodged the tax and since its revocation, it has been done to add an indication of date to counterfeit mounts. There are also a great many mounts that were not stamped simply because the maker was intent on dodging the tax. Therefore, neither the presence nor the absence of a stamp on a *bronze d'ameublement* of mid-eighteenth century French origin is a guarantee of its age, but another interesting clue to its identity.

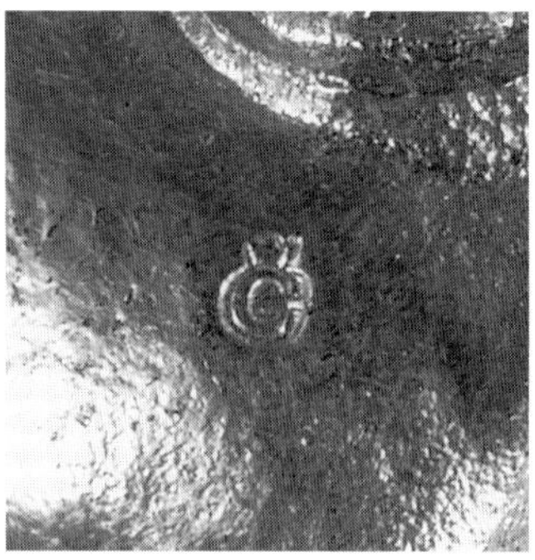

Figure 16.7 Detail of a crowned C stamp from one of a pair of Coffres on stands, *c.*1684–89, attributed to André-Charles Boulle. (The J. Paul Getty Museum,Los Angeles. Photo: Jack Ross.)

Louis XIV was fully aware of the prestige and political power which could be engendered through the production of works of art. To that end, he endeavoured to make Paris the artistic center of Europe and, through his Minster Colbert, he set up Royal Manufactories, such as the Gobelins for the production of luxury furnishings for the royal châteaux and for diplomatic gifts. Throughout the eighteenth century, these craftsmen produced works of such refinement and skill that all of the Courts of Europe took inspiration from Paris as the undisputed centre of artistic accomplishment. Many of these manufactories, like the Gobelins and the Sèvres Porcelain Manufactory are still functioning. Similarly, private craftsmen, like those who produced these mounts, continue to exercise their skills at an unsurpassed level, ensuring the continuation of Louis XIV's vision.

The illustrations that follow document the casting, chasing and gilding of a copy of a gilt bronze mount. The project was commissioned by the J. Paul Getty Museum and organized by the Chaser René Schaeffer under the direction of Michel Jamet in Paris. Three copies of the mount were produced (Figures 16.56–16.59): one was left unchased and ungilded; one was chased, but not gilded; and the third was chased and gilded. The entire process was photographed to produce a record of each step and the three mounts were then displayed in a temporary exhibition at the J. Paul Getty Museum.

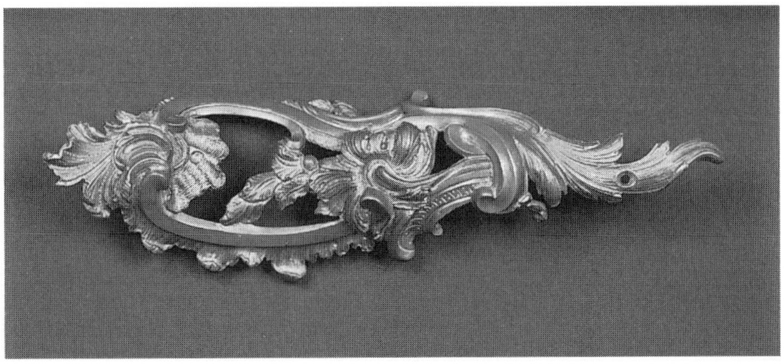

Figure 16.8 The mount to be reproduced, a *chute* by Dubois in the Rococo style. This foundry uses an alloy of 86.6% copper, 6.6% tin, 3.3% lead and 3.3% zinc. (Photo: Abby Hykin.) (See Plate 91.)

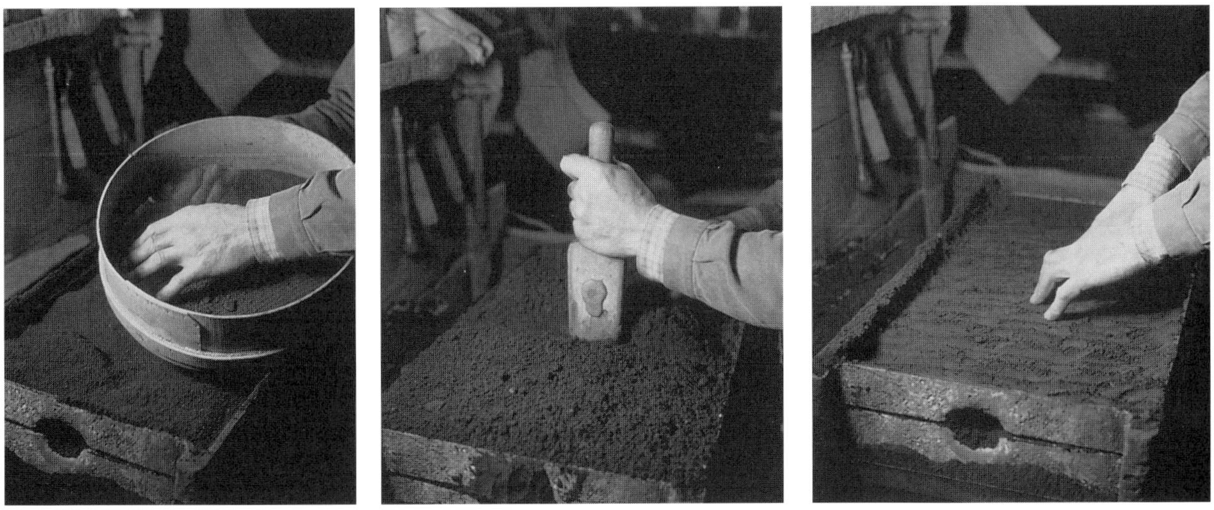

Figure 16.9 To cast the bronze mount, the founder fills the steel chassis with sifted sand. The pour holes for the molten bronze are visible on the near edge of the chassis. The sand is called Fountenay aux Roses. It is very fine and a yellow colour before it has been used and is composed of silica and aluminum. (All photos in Figs 16.9–16.54 are by Francis Razon, Paris, except where noted.) **Figure 16.10** The humid sand is compacted and will become very hard. **Figure 16.11** The sand is evenly spread out to have the same consistency everywhere. The bronzes to be recast will be placed on this bed of sand and then the counter caul will be placed on top. (See Plates 92–4.)

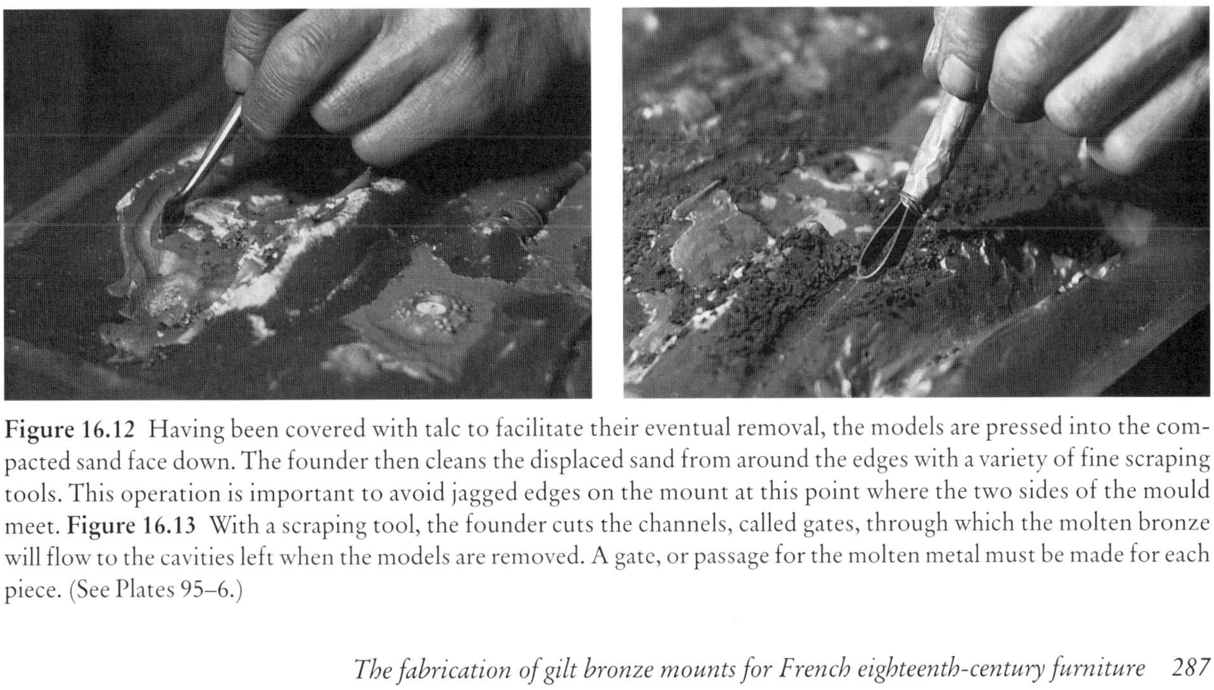

Figure 16.12 Having been covered with talc to facilitate their eventual removal, the models are pressed into the compacted sand face down. The founder then cleans the displaced sand from around the edges with a variety of fine scraping tools. This operation is important to avoid jagged edges on the mount at this point where the two sides of the mould meet. **Figure 16.13** With a scraping tool, the founder cuts the channels, called gates, through which the molten bronze will flow to the cavities left when the models are removed. A gate, or passage for the molten metal must be made for each piece. (See Plates 95–6.)

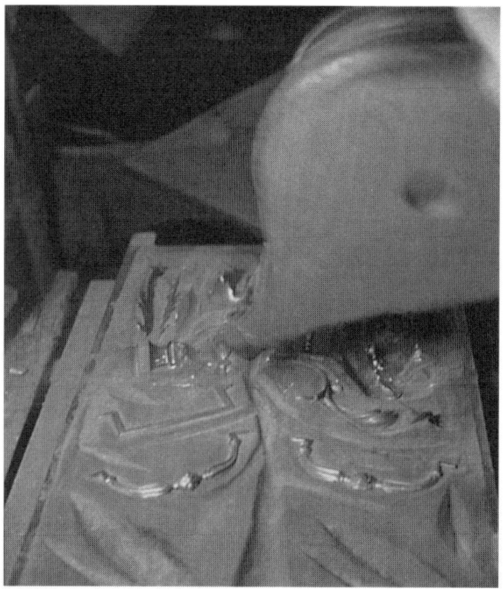
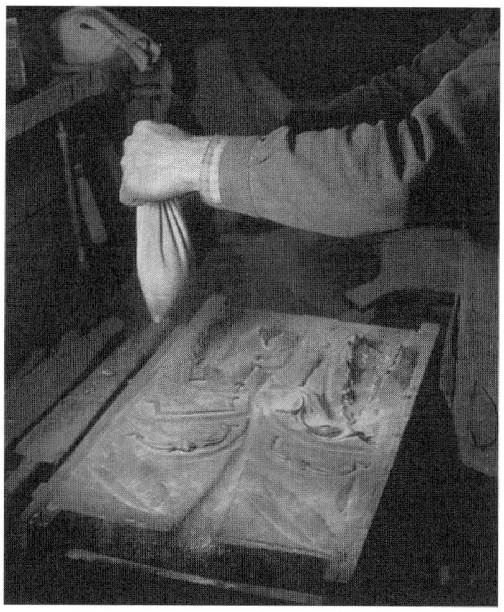

Figure 16.14 The final cleaning of the moulds is done with a bellows to avoid interfering with the surface of the sand. You can see all of the gates for each piece. **Figure 16.15** After the gates have been cleaned up, the mould is coated with wheat starch sifted through a cloth bag. This interface prevents the sand on the two sides of the mould from sticking together. (See Plates 97–8.)

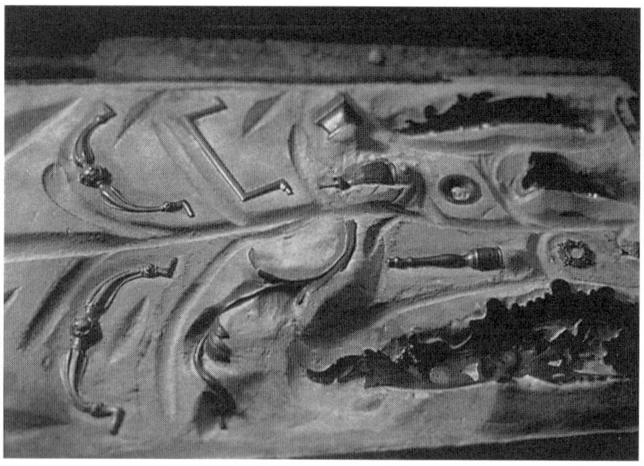
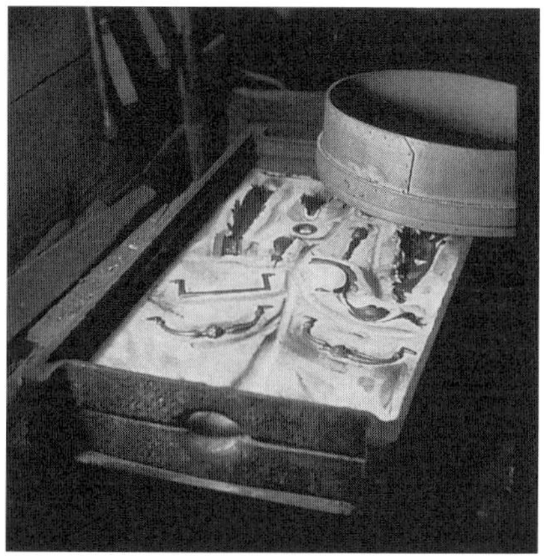

Figure 16.16 The first half of the mould is now ready. You can see both the gates for the flow of the molten metal and the channels cut to allow the gasses to escape. These are called the vents. The mould is now ready for the application of the counter mould, which is a faster operation since you now have only to stack another chassis on the first one and fill it with compressed sand. **Figure 16.17** The second steel chassis is stacked on top of the first and held in place by wooden rods. As with the first chassis, the sand is sifted into the chassis to form the counter-mould. (See Plates 99–100.)

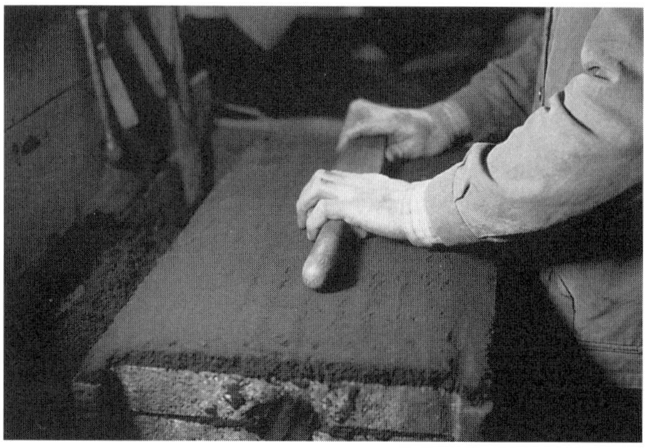
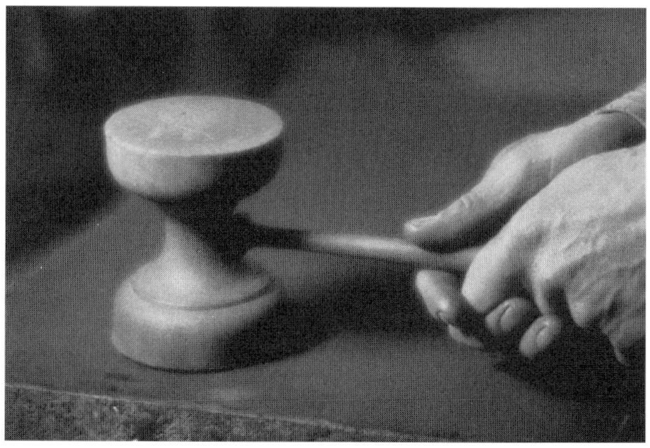

Figure 16.18 A wooden bat is used to pound the sand down evenly so that it marries the exact form of the back side of the bronzes underneath. Once the sand is compacted, it is scraped off level with the edges of the chassis. **Figure 16.19** With a boxwood mallet, the sand is tapped to release the counter mold from the mould. This is critical to separating the two halves of the mould. (See Plates 101–2.)

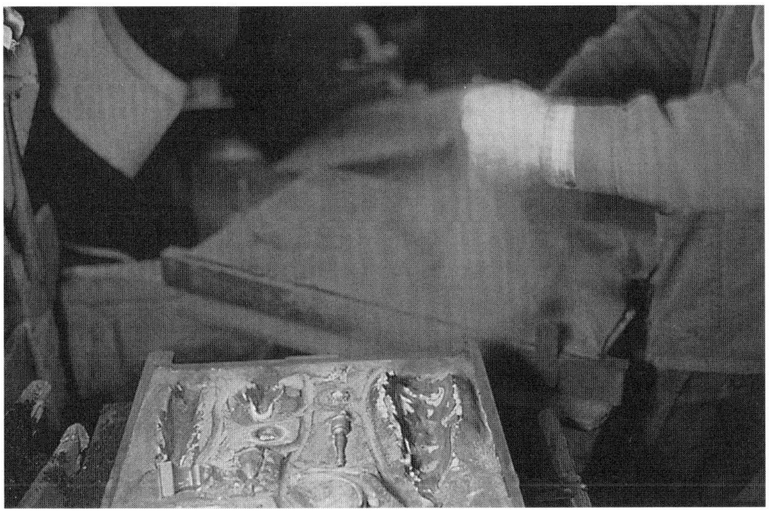

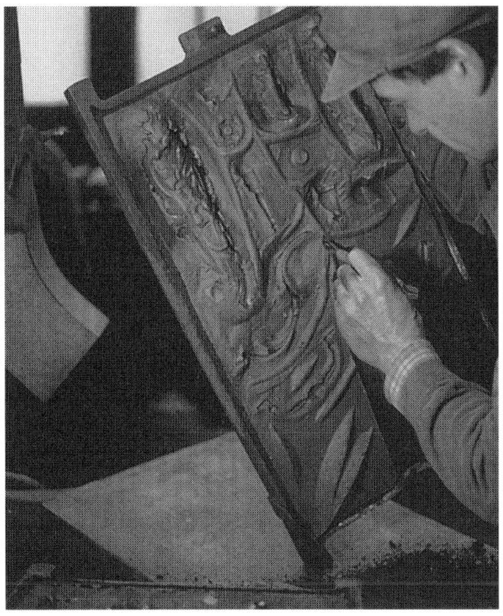

Figure 16.20 The two moulds are separated very delicately. **Figure 16.21** The gates and vents that were moulded convex in the counter mould are now scraped out in the counter molds since they need to be concave on both halves of the mould. (See Plates 103–4.)

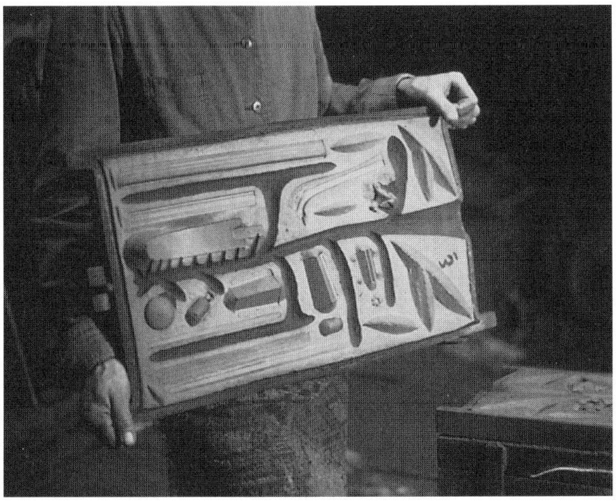

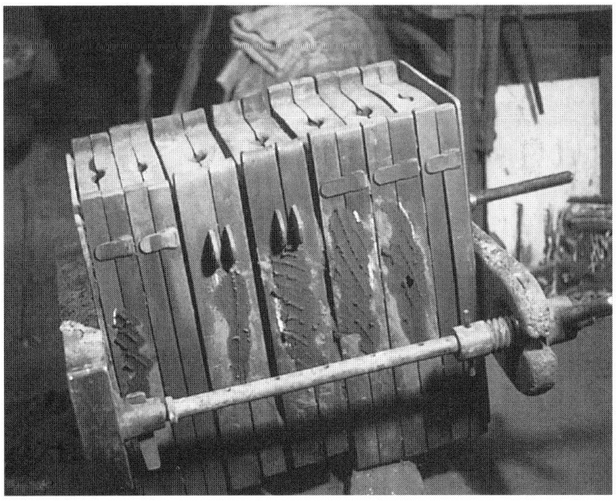

Figure 16.22 The moulds, showing the gates and the vents, are now ready for the pour. **Figure 16.23** The moulds are then stacked together and held with screw clamps to prevent deformations during the pour. The stack of moulds are put in a hot room for twelve hours to dry the sand and then stood on end to receive the poured molten metal. (See Plates 105–6.)

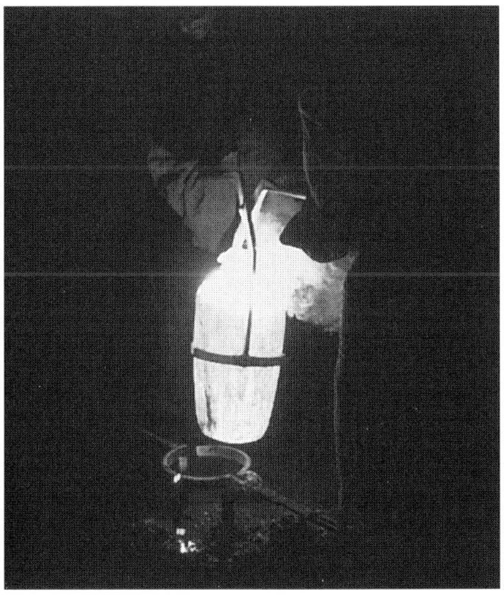

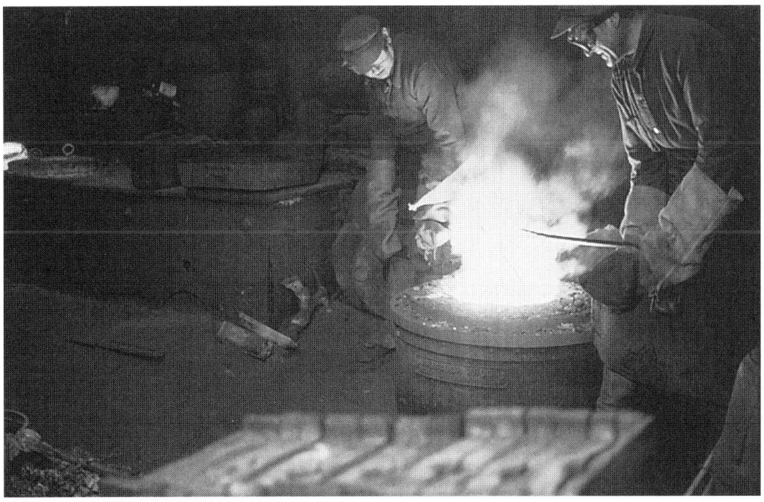

Figure 16.24 The alloy is melted in a crucible, which is put in a furnace. Scraps from previous pours are also added to the new alloy. **Figure 16.25** The two founders take the crucible out of the oven with the stack of molds in the foreground. The ambiance of the foundry has not changed since the eighteenth century. (See Plates 107–8.)

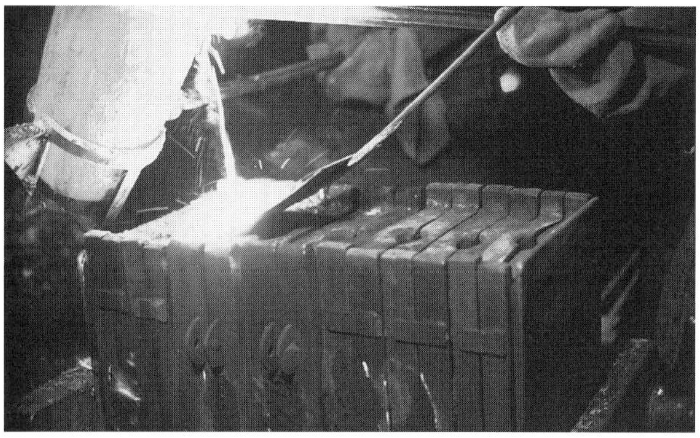

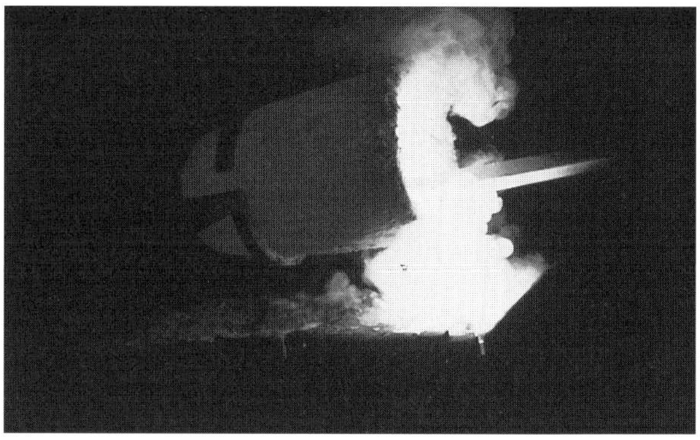

Figure 16.26 While two founders manipulate the lantern, a third holds a small spade to guide the molten metal, which must be prevented from splashing into the next mould to be poured. If this happens, the alloy will cool and clog the gates prematurely, thereby blocking the flow of the molten bronze. You can also see the flames of the escaping gasses burning on the sides. **Figure 16.27** The pour is completed. It has to progress very quickly or the metal will not flow completely into the depths of the last mould. The colour of the crucible shows that the metal has cooled considerably. (See Plates 109–10.)

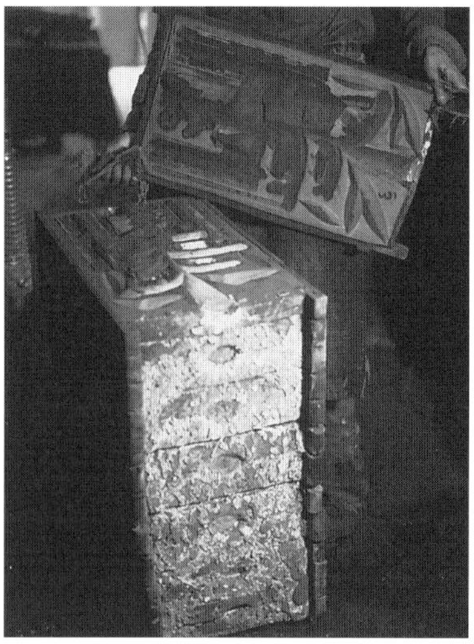

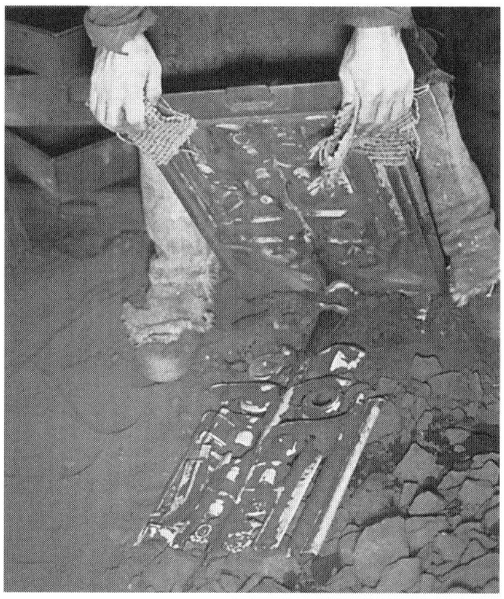

Figure 16.28 The moulds are broken apart immediately after the pour. You can see that the metal is still red hot. **Figure 16.29** The bronzes fall on the floor, but are still attached by the alloy filling the gates and vents. These will be cut to separate the bronzes. The sand moulds are only used once since they are destroyed when they are opened after the pour. The sand is then passed through a screening machine before it is reused for other moulds. (See Plates 111–12.)

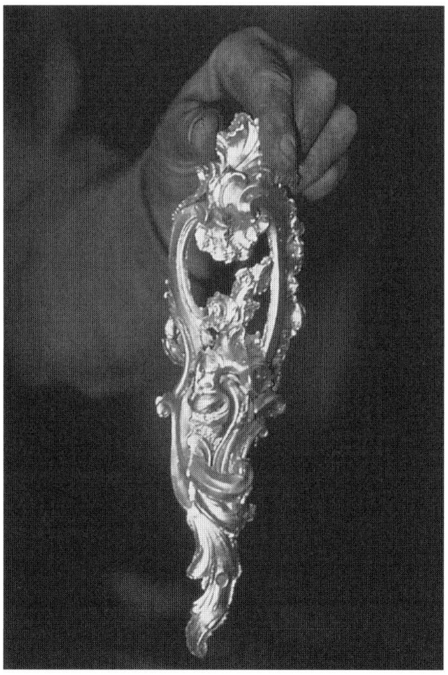

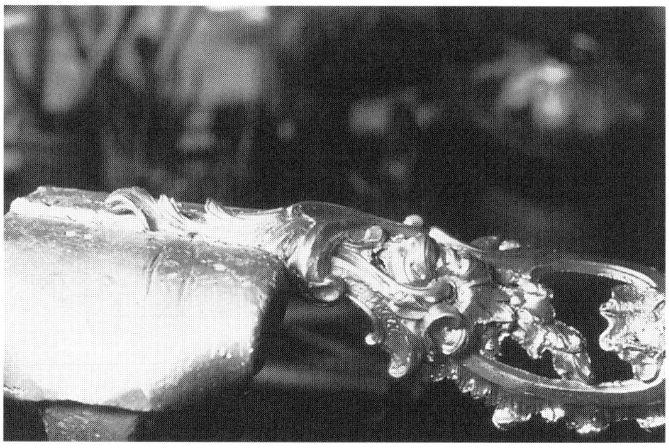

Figure 16.30 The bronzes are then brushed on a wheel to remove any remaining sand and the fire-skin. They are then ready to be sent to the chaser. **Figure 16.31** The chaser mounts each bronze in a vice on which the jaws have been lined with lead to avoid damaging the bronze. He will sharpen all of the edges and embellish the bronze with details that cannot be achieved in the casting. (See Plates 113–14.)

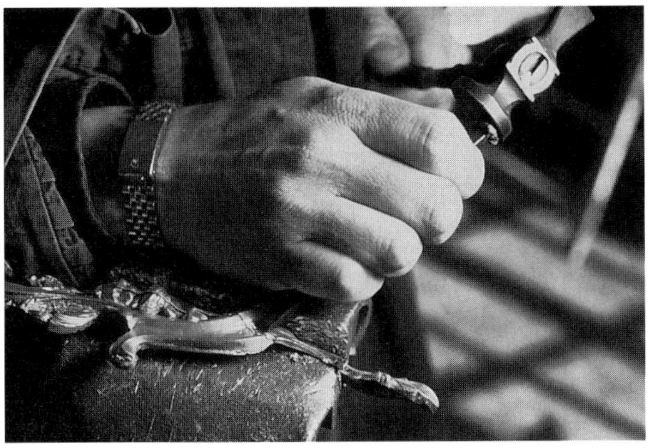
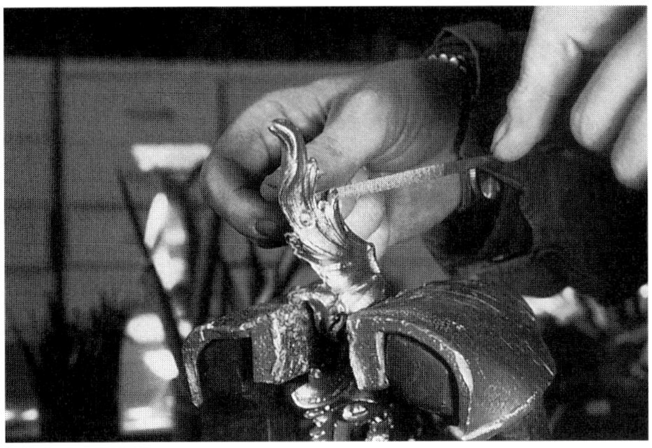

Figure 16.32 The burr along the edges, called the stitches, are removed with a burin and a hammer. **Figure 16.33** The edges are cleaned and sharpened with a file. Files with a wide variety of profiles are necessary to get into all of the tight curves of the bronzes. (See Plates 115–16.)

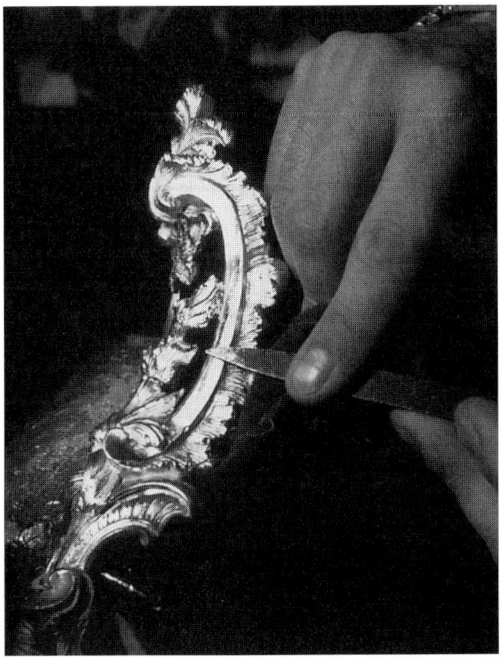
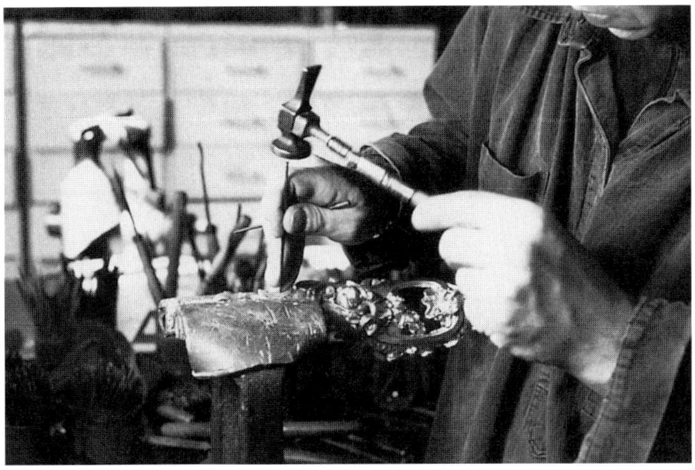

Figure 16.34 The chaser smoothes the flats of the mount with a riffler and then a scraper. The riffler is passed perpendicular to the direction of the strap to smooth and polish the surface that will be burnished after the gilding. **Figure 16.35** The veins of the bronze are sharpened with a burin. (See Plates 117–18.)

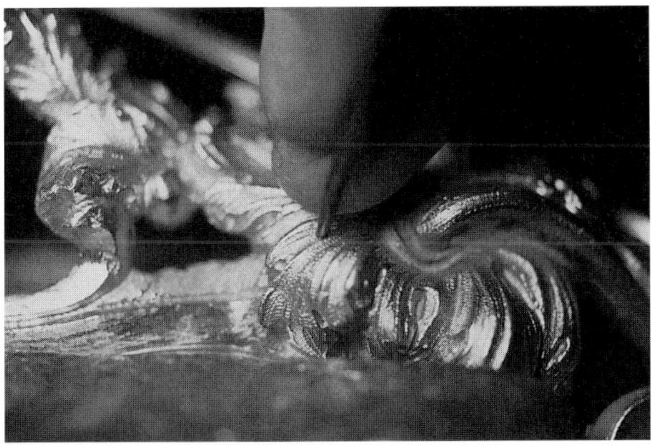
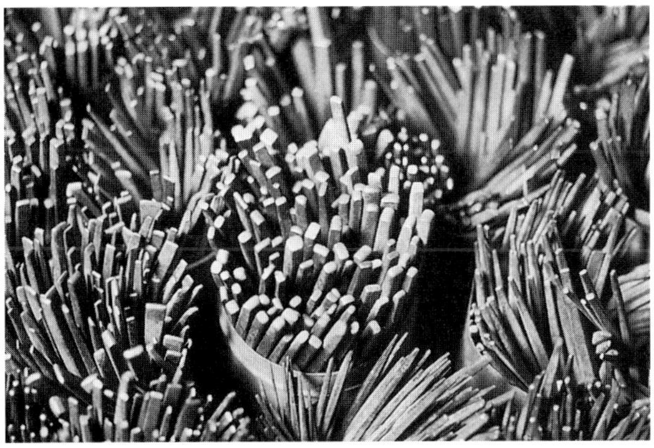

Figure 16.36 Punches are used to give texture. They do not remove metal, but displace it to give detailed form and volume. **Figure 16.37** A good chaser will have over 3,000 punches. (See Plates 119–20.)

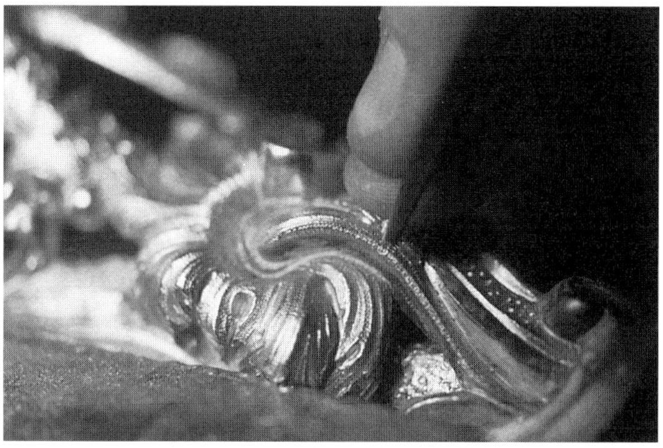

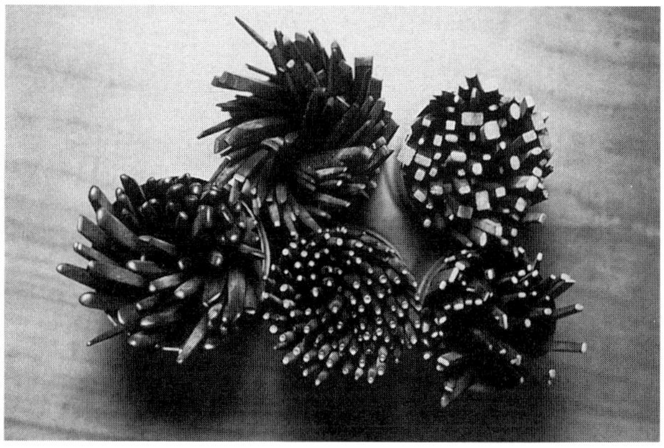

Figure 16.38 Every pattern or texture is produced by a separate punch, which must be the appropriate style and size. **Figure 16.39** These are the five families of chasing tools: top left, the mats; top right, the burins; bottom left, the tracers; bottom centre, the pearlers, and *bouterolles*. (See Plates 121–2.)

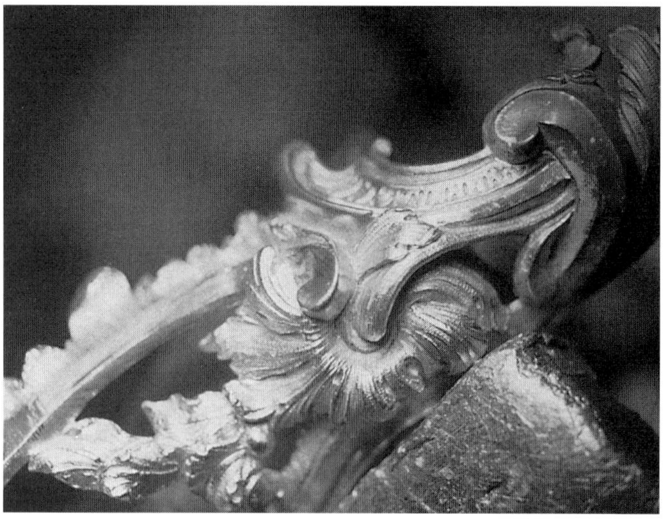

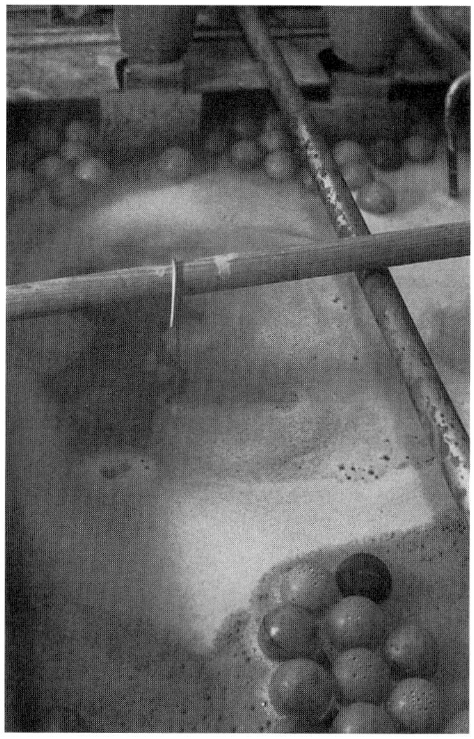

Figure 16.40 Every surface, both textured and smooth are produced by the chasing tools, which takes many hours. The bronze is then ready to go to the gilder. **Figure 16.41** The first step of the gilding is to pickle or clean the bronze in nitric acid or by electrolysis. (See Plates 123–4.)

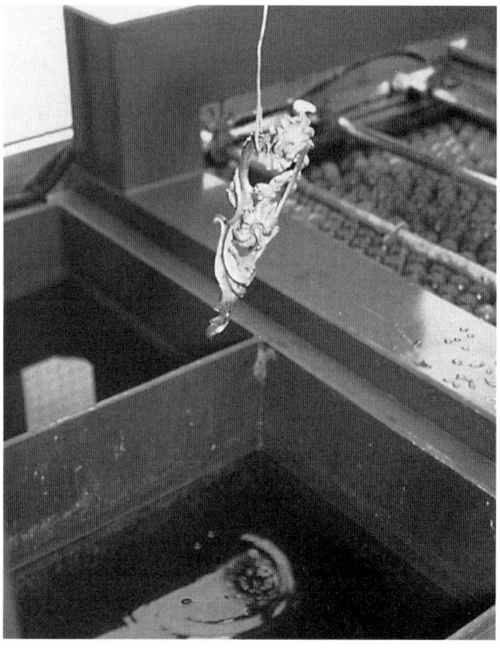

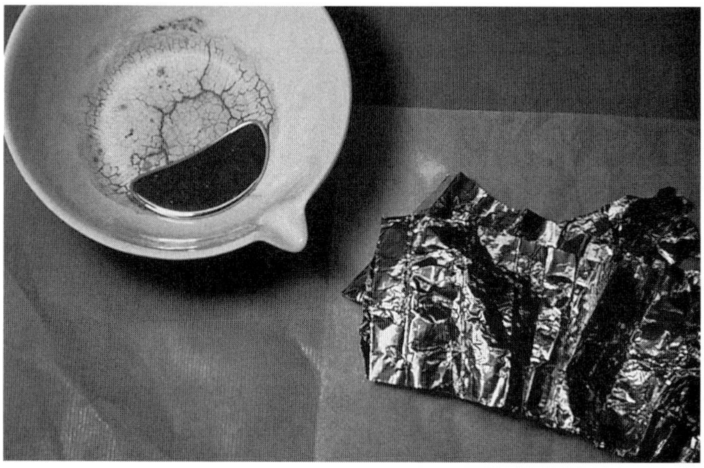

Figure 16.42 It is then rinsed, several times in distilled water, of any impurities that could interfere with the adhesion of the gilding. **Figure 16.43** The amalgam is a mixture of eight parts of mercury to one part gold leaf, which is thicker than the leaf used for wood gilding. (See Plates 125–6.)

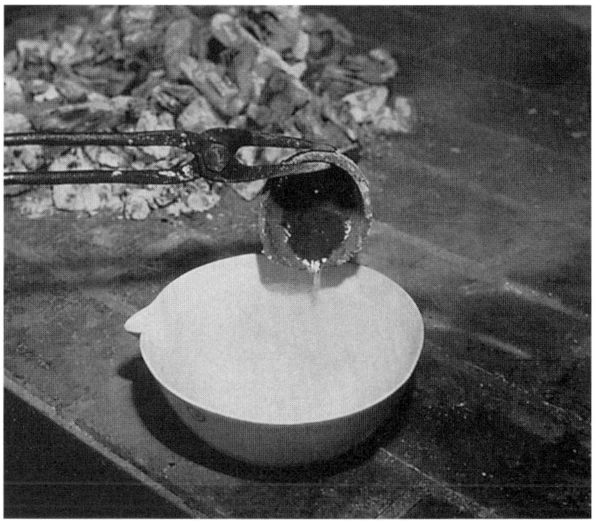

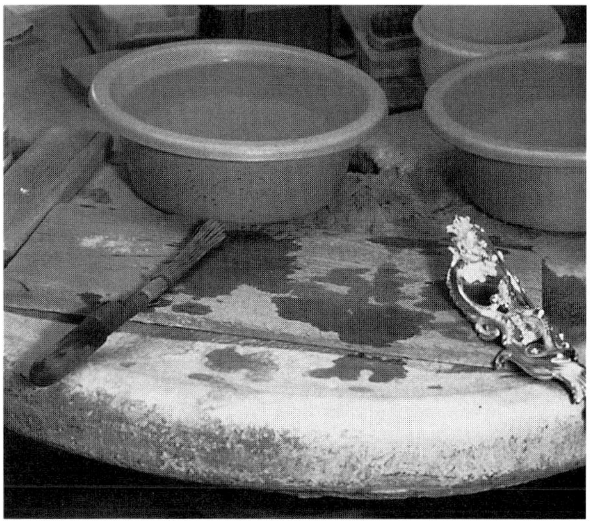

Figure 16.44 The mercury and gold are melted together in a crucible on a coal fire to make the amalgam, which is then rinsed several times in distilled water to remove any impurities. **Figure 16.45** This shows the tongs and the brush that will be used to apply the amalgam. The brush, called a bat, has brass strands. (See Plates 127–8.)

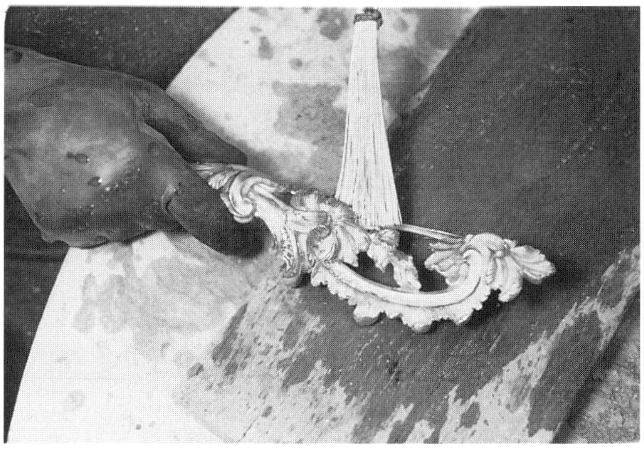

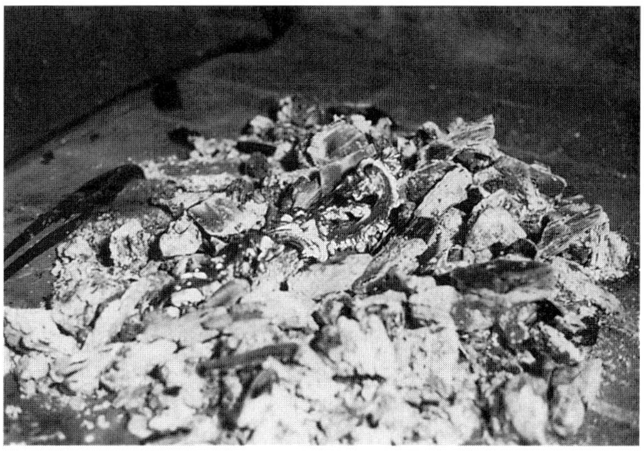

Figure 16.46 The amalgam is applied to the bronze with a bat. A strong tapping motion is used to drive the amalgam into the interstices of the bronze. **Figure 16.47** The bronze is then placed into the coal fire to evaporate the mercury. Under the influence of the heat, the metal dilates and the gold is deposited into the matrix of the metal. The bronze is recoated with the amalgam a number of times, depending on the budget and the details of form. (See Plates 129–30.)

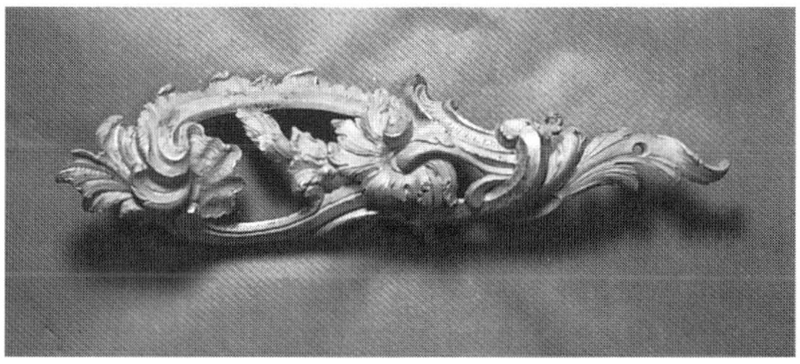

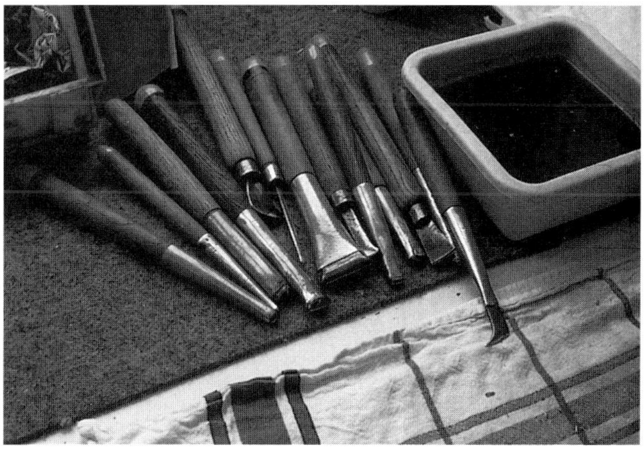

Figure 16.48 The gilded bronze is a dull color after the gilding and requires further finishing. It is brushed under running water to give it some life and shine and then dried in sawdust. **Figure 16.49** The highlights are then burnished. Here a steel burnisher is being used; they can also be made of hematite stone. (See Plates 131–2.)

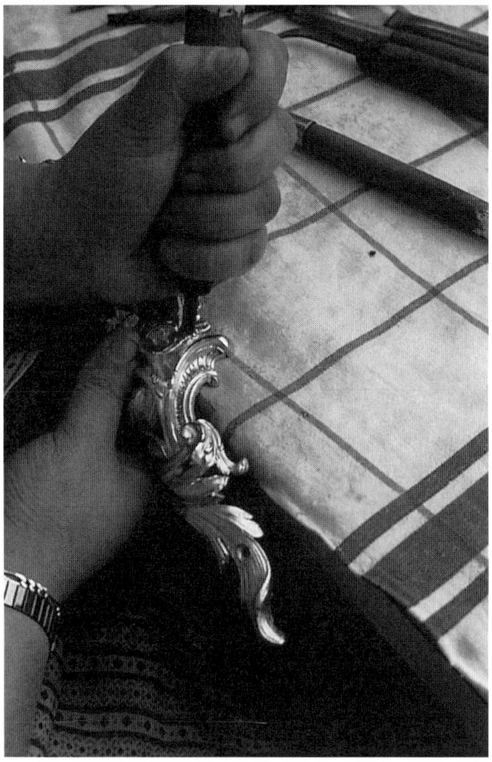

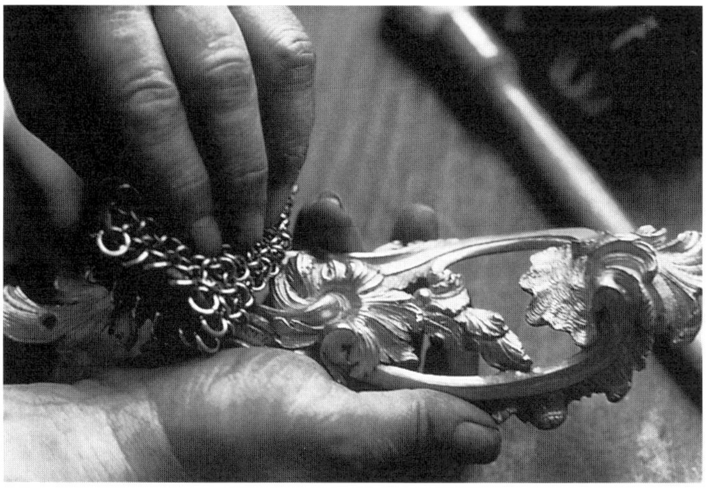

Figure 16.50 Polishing with the burnisher makes the flat surfaces brilliantly reflective. This is a delicate operation because you must not slip and polish a matte area, but it requires a great deal of pressure. **Figure 16.51** Chain mail is used to soften the edges and age the bronze. (See Plates 133–4.)

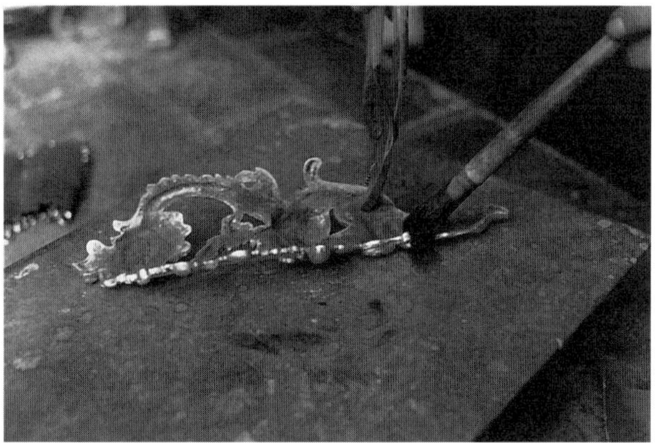

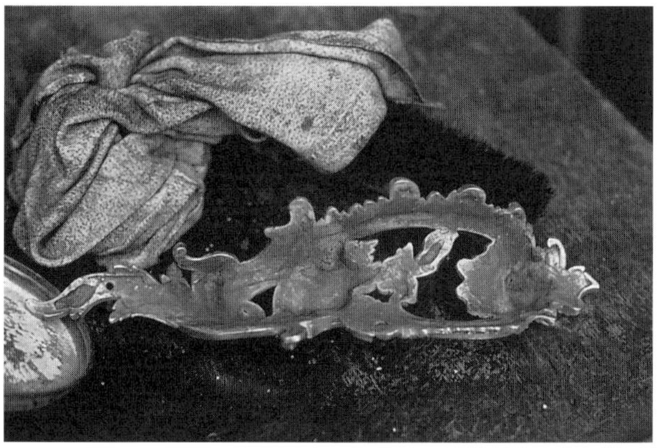

Figure 16.52 The backs of bronzes are patinated with acids to artificially age them. **Figure 16.53** The spills of mercury can be seen on the back of the bronze. (See Plates 135–6.)

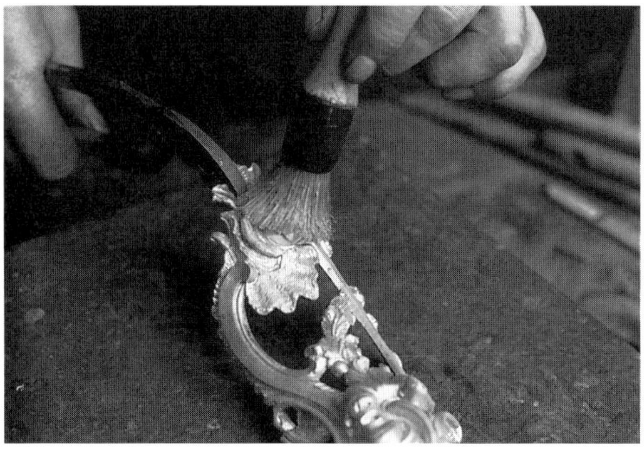

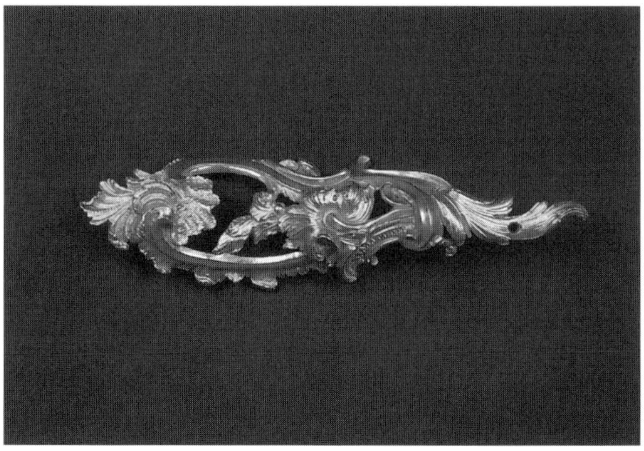

Figure 16.54 The bronzes are patinated or 'mis en couleur d'or' with chemicals as varied as ammonium chloride, potassium nitrate and cupric acetate to achieve the desired tone and unify the colour among a set. **Figure 16.55** The finished bronze. (Photo: Abby Hykin.) (See Plates 137–8.)

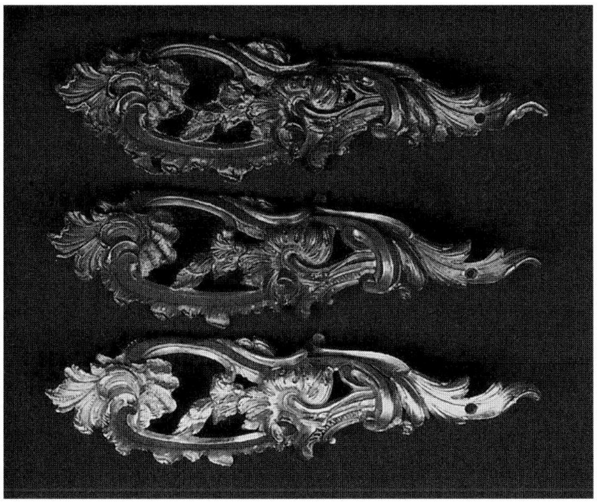

Figure 16.56 The three bronzes reproduced, from top to bottom, the cast before chasing, the cast and chased mount, and the cast, chased and gilded mount. (Photo: Abby Hykin.) **Figure 16.57** Detail of the cast mount without chasing or gilding. (Photo: Abby Hykin.) (See Plates 139–40.)

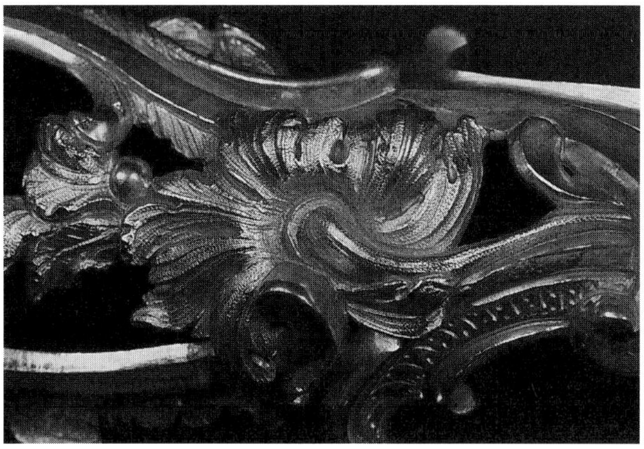

Figure 16.58 Detail of the cast and chased mount. (Photo: Abby Hykin.) **Figure 16.59** Detail of the cast, chased and gilded mount. (Photo: Abby Hykin.) (See Plates 141–2.)

References

Chapman, M. 1994. 'Techniques of mercury gilding in the eighteenth century'. In *Ancient and Historic Metals*, D. Scott, J. Podany and B. Considine (eds), 229–39. Los Angeles: The Getty Conservation Institute.

de Bellaigue, G. 1974. *The James A. De Rothschild Collection at Waddesdon Manor, Furniture, Clocks and Gilt Bronzes*, 31. Fribourg, Switzerland: Office du Livre.

Diderot, D. 1763. *Pictorial Encyclopedia of Trades and Industries*, C. Coulstrom (ed.). Gillespie, New York: Dover Publication, 1959.

Jett, P. 1993. 'A Study of the gilding of Chinese Buddhist bronzes'. In *Metal Plating and Patination: Cultural, Technical and Historical Developments*, S. La Niece and P. Craddock (eds), 193–200. Oxford: Butterworth Heinemann.

Lins, P. A. and Oddy, W. A. 1975. 'The origins of mercury gilding'. *The Journal of Archaelolgical Science*, **2**: 365–73.

Scott, D. February, 1998. personal communication.

Verlet, P. 1987 *Les Bronzes Dorés du XVIIIe Siècle*. Picard: Grands Manuels Picard.

17 The Many Golds of Fabergé: Recognition and Guidelines for Care

Carol Aiken

Abstract

One of the most important elements in the Fabergé repertoire was gold. Golds of many shades and hues were used for construction but also to finish the surfaces of other metals. As a result, in Fabergé objects it can be difficult to distinguish between parts made of gold and those that merely look like gold. Platings and finishing treatments successfully contribute to carefully constructed illusions. The technical differences among the golden surfaces must be regarded as significant by anyone responsible for the maintenance or protection of a Fabergé object. Some of the golden surfaces are as fragile as they are beautiful. Through the recognition of what they are, informed decisions can be made to protect the golden surfaces of Fabergé from being diminished, damaged or lost.

The House of Fabergé earned renown in pre-Revolutionary Russia under the direction of Carl Peter Fabergé. From 1870 to 1918 Fabergé presided over a workforce having a formidable technical proficiency. It skilfully re-interpreted established traditions to create extraordinary objects, sometimes in unusual materials. Cork, blued steel and sandstone were combined, for example, with multicoloured golds and faceted stones. Fabergé was proud of the artistry that set his business apart from other purveyors of luxury goods. In 1914 he stated:

> If you compare my things with those of such firms as Tiffany, Boucheron and Cartier . . . you will find that the value of theirs is greater than mine . . . expensive things interest me little if the value is merely in so many diamonds or pearls. (Snowman 1983: 11)

By placing an emphasis on design and high standards of execution, Fabergé had granted his workmen an unusual freedom. They were able to choose materials for decorative qualities or inherent properties, such as strength, instead of solely on the basis of intrinsic value. The design and production of cigarette boxes, opera glasses, desk sets, frames, jewellery, and silverware for the table – all found among the diverse products created in large numbers by the company – provided a broad scope for Fabergé workmen to develop, then refine, unique effects enhanced by their approach to the use of materials.

Coloured golds

Kenneth Snowman described the planning process when important new objects were made. Fabergé called 'a round-table conference of all the various specialists who would be required to contribute in some way, at one stage or another, to the final achievement of the job in hand'. The use of gold was an essential part of the discussion.

> The goldsmith could decide with the enameler exactly how and when his individual work should be done. Almost certainly the gilder would have to consult both these craftsmen – the goldsmith in order to ascertain the nature and purity of the metal to be used, the enameler to help him decide on the best colour.
>
> (Snowman, 1952: 50)

Snowman's comments make a distinction between the goldsmith, who made the alloys and formed the objects, and the gilder, who applied finishes to the surfaces. The raw materials for both practices required preparation by the goldsmith. Through the alloying process, the goldsmith determined the hardness, working qualities and colour of each mixture that he created.

The elemental composition of an alloy determines its properties and colour, but the *fineness* or the actual quantity of metal that is present is the basis for its identification. In pre-Revolutionary Russia, gold standards were expressed in *zolotniks*. Fine or pure gold had the designation of 96 zolotniks, based on the total number of parts by weight that determined the gold standard, in contrast to 24 parts by weight used as the standard in the USA, and expressed in carats. Hence, 96 and 24 are both designations of gold in its purest state. According to the Russian system, 40 parts by weight of metal or metals were added to complete a 56 gold alloy. The specific metal or metals that were added determined the final color and working properties of a 56 gold. Golds of standards 56 and 72 were commonly used by Fabergé.

Russian standards for silver also were based on 96 parts by weight, with the most frequently used proportions being 84, 88 or, sometimes, 91. Parts of objects made in gold, silver and their alloys were stamped as each was made, before assembly. Stamps found on the separate pieces of a complex Fabergé object can help to distinguish the sections that are made in different metals.

Examples of representative compositions of coloured golds are listed in many jewellers handbooks. As a general rule, alloying with copper results in a red gold and alloying with silver in a green gold or white gold. Alloying with platinum can also result in a white gold. Golds with distinct hues of white, red, yellow and green are found in many works by Fabergé. Shades of purples, greys, blues and oranges can be identified as well. One of the more ubiquitous elements in the Fabergé products is the use of a technique derived from eighteenth-century gold boxes and other luxury items, known as gold *à quatre couleurs* (four-colour gold). Many of the decorative motifs that embellish Fabergé surfaces are based on the use of three, four and five colours of gold. A striking example of the use of multicoloured golds is the sumptuous swags on the Imperial Fire-Screen Frame in the Forbes Magazine Collection, New York (Fig. 17.1). This interpretation of an eighteenth century style contrasts dramatically with a gothic-inspired Double Marriage Cup in the collection of A La Vieille Russie, New York (Fig. 17.2). The cups are made in four colours of gold and decorated with graved lines. The decoration was applied to the surface after the cups were

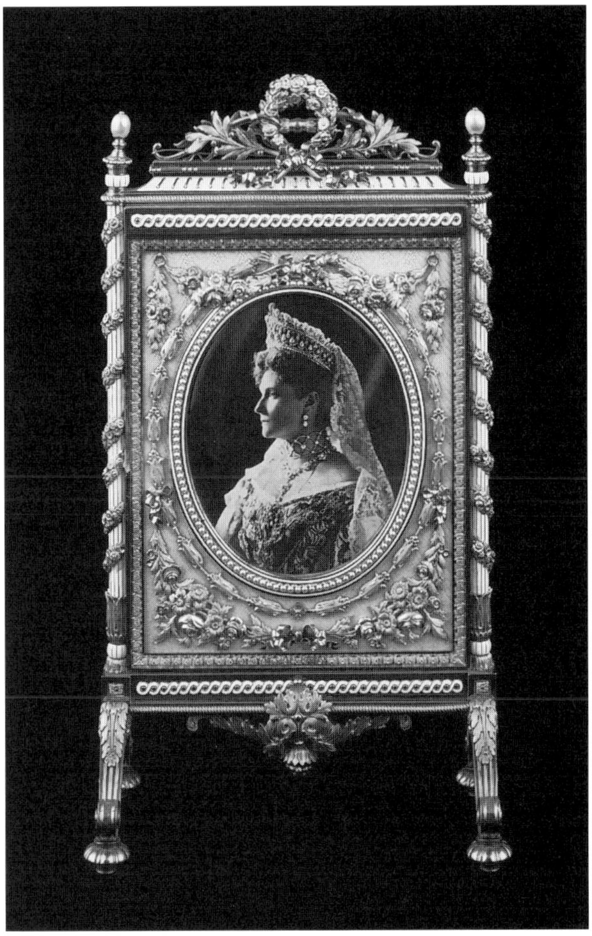

Figure 17.1 Imperial Fire-Screen Frame. Double-sided frame with photographs of Czar Nicholas II and his wife Czarina Alexandra Feodorovna. 7⅞" tall. (Photograph used by permission of The Forbes Magazine Collection, New York. All rights reserved. Photo by H. Peter Curran.) (See Plate 143.)

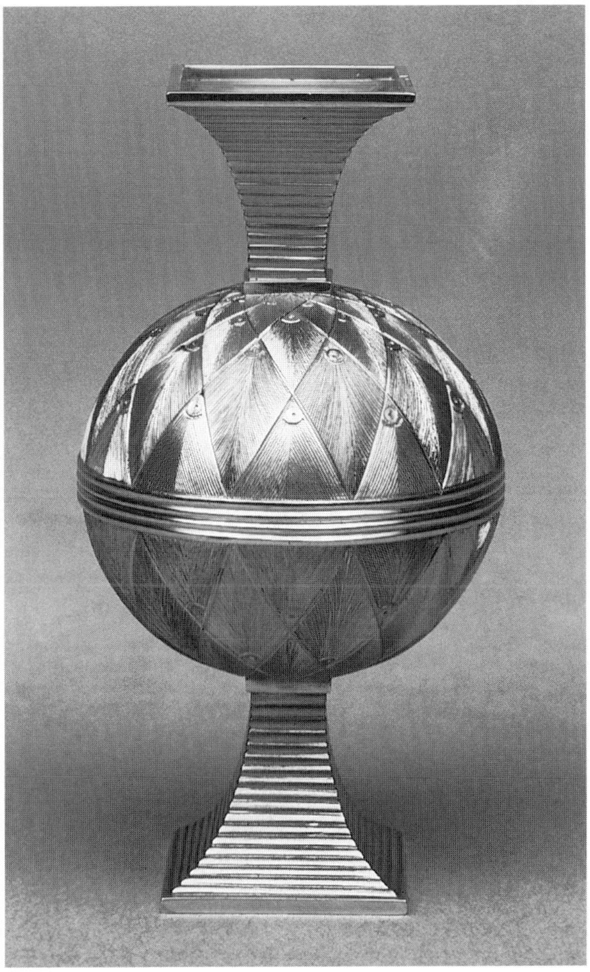

Figure 17.2 Double Marriage Cup. 3⅝" tall. (Photograph used by permission of A La Vieille Russie, Workmaster Michael Perchi, St. Petersburg, before 1896.) (See Plate 144.)

constructed. The swags of the fire screen, however, are made of individual cast elements that were chased or polished before they were joined together.

Structural elements, including pedestals, bases, supporting legs and frames, are rarely made entirely of gold alloys. In larger objects the need for qualities such as strength or toughness, as well as the considerations of economy, dictated that different materials be used. Ormolu (gilded bronze) is one of the robust materials chosen for structural elements. Strength combined with beauty made ormolu a natural choice for the floral stems as well as the legs of the Imperial Lilies of the Valley Easter Egg in the Forbes Magazine Collection, New York (Fig. 17.3). The large green leaves clustered around the base of the enameled shell are made in ormolu as well but, unusually, the gilded surface is covered with a translucent layer of green enamel. This is a good example of a non-traditional use of the material: the golden surface of the ormolu contributes to the richness of the leaf color, but is not recognised in this context for its own qualities. Vermeil (silver gilt), having a more lustrous surface than ormolu, forms the cap of each lily-of-the-valley blossom. The blossoms, each composed of a single pearl, are mounted along the stems in graduated sizes. Vermeil also is used for the frames of the miniature portrait surprises which emerge from the egg. The lustrous silvery-toned surface of vermeil discolours over time, as does the rich yellow-gold surface of ormolu.

Alloys of gold and platings of gold discolour when corrosion forms in response to the surrounding environment. The character of every addition to a gold may affect how an alloy eventually changes. Solders that are colour-matched to the metals that they join often discolour at different rates than surrounding metal surfaces, so that solder-rich interfaces become visible, although they were never intended to be seen. Gold alloyed with

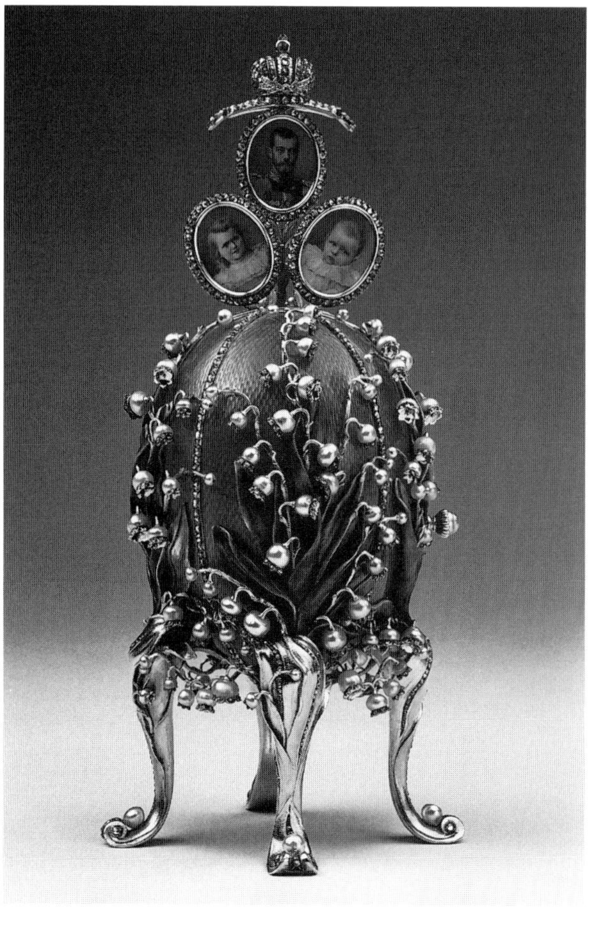

Figure 17.3 Imperial Lilies of the Valley Easter Egg. 7⅞" tall. (Photograph used by permission of The Forbes Magazine Collection, New York. All rights reserved. Photo by Joseph Coscia, Jr.) (See Plate 145.)

silver develops a continuous black layer similar in appearance to the tarnish that forms on a piece of silver. When gold is plated over silver, corrosion products from the substrate layers migrate through microporous surface layers to form depositions on top of the plated surfaces. References to black gold that are occasionally found in literature about Fabergé appear to have mistaken some of these undesirable changes for an intentional effect (see Habsburg 1986: 154, #165).

Fabergé intentionally used black-coloured metallic surfaces that were not based on gold alloys to achieve dramatic decorative effects. Blued or blackened steel features prominently in the Romanov Tercentenary Egg of 1913 in the Armory Museum, Moscow. Within the Tercentenary Egg is a blued-steel globe divided into two halves. One depicts the map of the Russian Empire in 1613, and the other is the map in 1913. The steel represents the sea, while land masses are formed in coloured golds. The Steel Military Egg of 1916, also in the Armory Museum, consists of a shell supported on the points of four miniature artillery shells ringed with yellow gold bands. Snowman noted that Fabergé's son Eugene said the Military Egg originally had a blackened steel shell. By 1952, however, the surface was a brightly polished steel (Snowman 1952: 109).

Snowman did not offer reasons for the change in the surface of the Military Egg. A conservator might assume that cleaning had been responsible: perhaps someone did not recognise that the dark finish was intentional. If the egg was cleaned for this reason, the bright results could have been viewed as improvement. In contrast, if the egg was cleaned with care, but no understanding of the special methods required for plated or patinated surfaces, the change might have been unexpected and viewed as unfortunate. The professional conservator is expected to recognise and understand characteristics of different surfaces, and to know the most appropriate methods, materials and techniques for the care and treatment of metal surfaces. All caretakers of Fabergé objects should seek to develop an informed understanding about materials and methods, in accordance with their own levels of responsibility. Only through the active participation of all interested parties can high standards of care be ensured for these objects on a continuing basis.

Change and response: guidelines for care

Changes that take place in metal surfaces as the result of exposure to the environment were, perhaps, tacitly acknowledged and addressed by the original makers of Fabergé objects. The most complex Fabergé objects usually can be dissembled with relative ease, because originally they were joined with mechanical fasteners such as metal tabs, slender pins and threaded shanks. The apparently seamless facades completely conceal this manner of construction. Fabergé's competitors, in contrast, demonstrate a preference for semi-permanent methods of assembly, including jeweller's stitches and soft solders.

For many years Carl Peter Fabergé was personally responsible for the maintenance of the Imperial treasury. From this experience, he must have been aware of the advantages implicit in the dissassembly of an object before attempting cleaning or repair. This knowledge as well as his desire for artistic and decorative effects, may have contributed to the manner of construction that characterises Fabergé objects (Aiken 1993). Whether a deliberate feature or not, the fact that Fabergé objects can be dissassembled has helped to ensure that each material in a complex construction can be cared for as it requires.

Few Fabergé objects in public or private collections have completely escaped past attempts at cleaning. Because the tarnishing of silver and silver gilt and the oxidation of copper alloys with gilded surfaces are common problems, residues of cleaning products are frequently found in the fine details of metal mounts, or under or around them. Large pieces of silver and silver gilt sometimes have deteriorated surface coatings. Objects may also show the effects of no efforts at past maintenance. They exhibit accumulations of dust and dirt that leave dull surfaces and hazy haloes around perimeters and boundaries of mounts. Whether conditions result from poor maintenance or the complete lack of maintenance, treatment can often improve the appearance of a Fabergé object.

Cleaning procedures

Recommendations for specific methods and materials for care are beyond the scope of this article. Choices for immediate and long-term care must be individually made for each object, on a case-by-case basis. Among the considerations that contribute to choices are existing problems or damages, the risk of potential problems, the treatments that are available, and priorities for the long term, within a specific context. References cited in following discussions provide information to assist in making these types of decisions. The references also include information on products and methods of care, with details about relative advantages and disadvantages.

Fabergé objects should not be cleaned without a clearly defined reason for treatment. Repeated cleaning, or the use of improper methods and materials in a single cleaning, can dramatically thin and remove the plated surfaces favoured by Fabergé, as can be seen in the detail of an icon oklad (Fig. 17.4). Some cleaning methods selectively deplete metals from surfaces of alloys. Comparisons of large numbers of similar objects within single collections or loan exhibitions will usually demonstrate characteristic patterns of surface damages.

Damages to surfaces are sometimes recognised through patterns of colour relationships. It is rare, for example, to find original silver mounts on objects decorated with deep red enamel. When the combination is noted, a careful examination of the metal surfaces may disclose traces of gold plating in the most protected recesses. Locally diminished or thinned surfaces are identified through a comparison of exposed surfaces to protected areas. Because routine maintenance can pose dangers to vulnerable surfaces, basic guidelines for cleaning and maintaining the surfaces of Fabergé objects need to be observed.

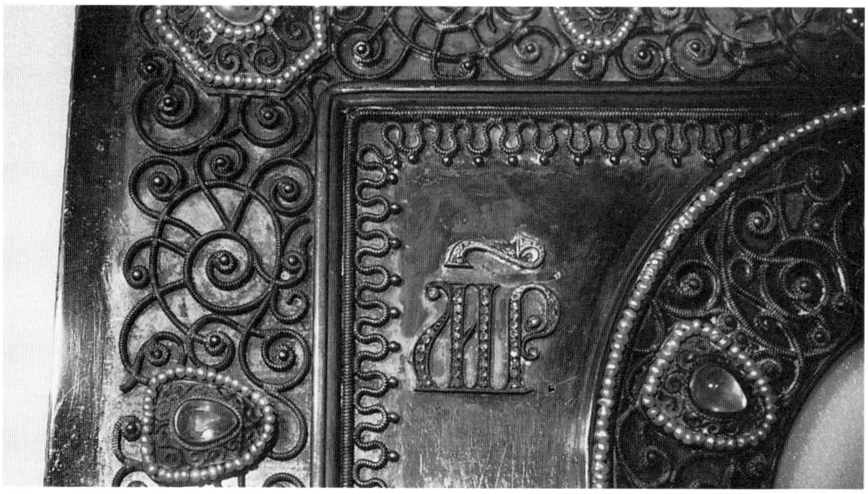

Figure 17.4 Detail of damaged gilding, before cleaning, on oklad of Fabergé Icon of Our Lady of Iberia. 11⅞ × 10⅛". (Virginia Museum of Fine Arts, Richmond, VA. Bequest of Lillian Pratt Thomas.) (See Plate 146.)

Each material in a single object made by Fabergé may require the use of a different cleaning method or material. Unless it is possible to improve a condition through strictly superficial and localised attention, it may be prudent, if not essential, to disassemble a complex Fabergé object before it is cleaned. This permits the appropriate methods to be used for each material present. Unless objects are taken apart, cleaning solutions redeposit dirt in areas that are difficult, if not impossible, to reach. Solutions run underneath mounts to leave residues. Liquid cleaners that travel under cover crystals may damage or in some instances destroy miniature paintings and photographs. (Miniatures painted in watercolour on ivory are regular features of Fabergé objects. The watercolour medium readily dissolves in aqueous solutions.) Some liquid cleaning solutions for metals cause immediate damage to materials such as lapis lazuli and pearls.

Objects that cannot be disassembled for treatment, because of their design or because the problems of maintenance are too minor to justify this approach, are treated locally with semi-dry or dry methods that leave no residues. A light polishing using a 100% cotton jeweller's cloth, such as the Hagerty 100% Cotton Jewellers Cloth, removes dull films. These cloths are available from jewellery shops and good housewares departments. Proprietary waddings, impregnated with cleaning solutions, are useful for routine cleaning, as are proprietary flannel cloths impregnated with mild cleaning agents and tarnish inhibitors (Selwyn and Costain 1991: 7–14). Only light pressure should be applied during the use of any cloth, to avoid the risk of localised burnishing or undesirable polishing.

In general, abrasive cleaners and polishes need to be avoided, whether metal elements are cleaned *in situ* or after their removal from an object. Used correctly, and without long periods of contact, non-abrasive chemical solutions and dips are satisfactory for cleaning the delicate plated surfaces in Fabergé objects. It is recommended that ammonia-containing products be avoided because of potential harm to copper alloys. In spite of concerns in some of the literature about metal cleaning procedures, and opinions held by conservators who may not clean decorative plated metals on a routine basis, the use of dips has not been shown significantly to alter surfaces. Selwyn and Costain offer a comparative study of different products, and a discussion of alterations caused by dips (Selwyn 1990: 17–18; Selwyn and Costain 1991: 10–11). Precipitated chalk, a favoured and often recommended polishing agent, can be used when a gentle abrasive is required for superficial cleaning. But chalk must be used with more care than the literature is apt to specify: it will cut through thin platings and may burnish some surfaces to higher levels of polish than they were intended to have.

Metal components removed from an object may require de-greasing with a solvent before they can be cleaned with aqueous solutions, or coatings may have to be removed from other pieces. After cleaning with aqueous solutions has been completed, all pieces require a thorough rinsing with de-ionised water then polishing with a soft cloth until dry. Ornate or highly detailed pieces sometimes benefit from de-watering in an alcohol solution, after rinsing and before drying. Blown heated air is useful to assist with drying when not all surfaces are accessible to a cloth.

Guidelines for care

Repairs on plated metals are executed more successfully if damaged parts can be removed from composite structures. The use of solder to repair a Fabergé object is generally not a treatment option. A high risk of damage is associated with every type of heating that is required for soldering. Discoloration of original surfaces and the melting of delicate elements are among the potential problems. Carefully chosen adhesives usually effect acceptable repairs, although some situations may also require inconspicuous external supports to provide extra support for a damage. In one example of this situation, the axle of a silver motor car desk ornament showed signs of several previous solder repairs. After the axle was successfully repaired using adhesive, the car was placed on a mount of acrylic sheet (Plexiglas), constructed to support the weight of the car. The removal of soft solders – sometimes found in old repairs to Fabergé objects – can be extremely difficult to accomplish, and should not be attempted without prior experience, in order to avoid causing further damage.

Gilded surfaces that are already diminished or thinned may benefit from the protection offered by clear surface coatings (Heller 1983). But the application of coatings to objects with intact platings is not generally recommended, since platings in good condition generally offer sufficient protection when they are otherwise well cared for. Coatings can be difficult to effectively apply when gems or other materials are present in combination with metal surfaces. The eventual need to remove coatings as they age or are damaged must be considered when initial decisions are made concerning whether or not to use surface coating. Selwyn offers a good discussion of surface coatings (Selwyn 1990: 14–15). Alternatives to surface coatings include measures that can be practised to enhance long-term protection (Oddy 1975; Lins 1989).

A growing awareness of the need for long-term preservation has helped to establish routine practices in museums. Fabergé objects are handled with gloves. Cotton or nitrile gloves (not latex rubber) are recommended, and are available from personal and laboratory safety suppliers. If Fabergé objects are handled without gloves, untreated jewellers' cloths of 100% cotton can be used to remove minor smudges and spots before objects are placed on exhibition or in storage. Long-term-shine cloths are recommended to remove light discolouration and provide additional superficial protection (Selwyn and Costain 1991: 13).

Fabergé objects in storage may be protected individually by wrapping in good-quality neutral pH tissue paper that contains no chemicals, including corrosion inhibitors or buffers, and no sulphur (Gilberg and Cook 1987: 14–17; Selwyn 1990: 15–16, 19–20). Tarnish-proof tissues and fabrics developed for silver storage are recommended for use. Additives in these products react with sulphur-containing gases to form metal sulphides, which prevent gases from reaching metal surfaces (Selwyn 1990: 15). Silver-protecting fabrics can be purchased as pre-made bags of various sizes, or by the yard, to be customised for wrapping materials and drawer linings.

Tarnishing rates for silver and silver alloys depend on levels of relative humidity. The rate increases under humid conditions. Monitoring and controlling humidity levels can contribute to slowing reactions between metals and their environment. Storage areas should be selected to avoid the presence of tarnishing gases (Selwyn 1990: 13). Exhibition cases, old and new, are a primary source of exposure to the materials that cause rapid discolouration in alloys and plated objects with gold and silver contents (Oddy 1975). Cases and storage containers need to be chosen with care. Plywood, wood pulp boards and unsealed woods, rubber-based adhesives and sealants, and fabrics containing fire retardants, sizings, bleaches and some dyes are all potential sources of corrosive elements. These materials may contain sulphur and organic salts including formates, acetates and chlorides. Simple tests should be carried out to determine whether materials that will be associated with Fabergé objects are potentially corrosive to metals (Blackshaw and Daniels 1979: 16–19; Daniels and Ward 1982: 58–60; Lee 1996). Silica gel and activated charcoal products can be added to cases for the active removal of both water vapor and sulphur-containing gases (Selwyn 1990: 15–16, 20). For these materials to provide effective protection, however, the products must be used correctly and according to developed standards.

Conclusion

Curators, conservators and collection managers share in the responsibilities for preserving Fabergé objects. Each of these individuals should be able to recognize that different techniques were used by Fabergé, then use this knowledge as a basis for making informed decisions concerning care. Each needs to understand the policies that govern maintenance as well as exhibition and storage practices. Caretakers must be vigilant about protecting Fabergé objects to ensure that their golden surfaces survive for the enjoyment of future generations.

References

Aiken, C. 1993. 'Imperial Easter eggs: a technical study'. In *Fabergé: Imperial Jeweller*, G. von Habsburg and M. Lopato (eds), 76–84. Washington, DC: Fabergé Arts Foundation.

Blackshaw, S. M. and Daniels, V. D. 1979. 'The testing of materials for use in storage and display in museums'. *The Conservator*, **3**: 16–19.

Daniels, V. D. and Ward, S. 1982. 'A rapid test for the detection of substances which will tarnish silver'. *Studies in Conservation*, **27**(2): 58–60.

Gilberg, M. and Cook, C. 1987. 'Anti-tarnish tissues for wrapping and packaging silver'. *Journal of the International Institute of Conservation – Canadian Group*, **12**(4): 14–17.

Habsburg, G. von 1986. *Fabergé*. Geneva: Feldman Editors.

Heller, B. B. 1983. 'The coating of metal objects at Winterthur'. *Preprints to the 11th Annual (Baltimore) Meeting of the American Institute of Conservation*, **11**: 57–64.

Lee, L. R. 1996. 'Selection of materials for the storage or display of museum objects'. *British Occasional Paper V.111.*

Lins, A. 1989. 'The inhibition of silver tarnishing'. In *Preprints from the 13th International Symposium on the Conservation and Restoration of Cultural Property – Current Problems in the Conservation of Metal Antiquities*, 119–31. Tokyo: Tokyo National Research Institute of Cultural Properties.

Oddy, W. A. 1975. 'The corrosion of metals on display'. *Conservation in Archaeology and the Applied Arts, Proceedings of the IIC Stockholm Congress*, 235–7. London: International Institute for Conservation.

Selwyn, L. S. 1990. 'Historical silver: storage, display and tarnish removal'. *Journal of the International Institute of Conservation – Canadian Group*, **15**: 12–22.

Selwyn, L. S and Costain, C. G. 1991. 'Evaluation of silver-cleaning products'. *Journal of the International Institute of Conservation – Canadian Group*, **16**: 3–15.

Snowman, K. A. 1952. *The Art of Carl Fabergé.* Boston, Mass.: Boston Book and Art Shop.

Snowman, K. A. 1983. *Carl Fabergé Goldsmith to the Imperial Court of Russia.* New York: Greenwich House.

18 All that Glitters is not Gold: Other Surfaces that Appear to be Gilded

Jonathan Thornton

Abstract

This paper explores the history, technology and rationale behind imitation gilding from antiquity to the present, including mention of surfaces that may be mistaken for gilding, but were not meant to appear gilded. Topics include: organic coatings coloured gold by dyes (both natural and early synthetic), coloured resins and bile, applied over white-coloured metals and metal leafs, or over brass alloys and true gilding (in order to alter the colour); superficially 'brassed' copper; gold-coloured copper alloys such as Pinchbeck; brass leaf and brass-leaf powders ('bronze powders'); the tin sulphide called 'mosaic gold' and other metallic sulphides; gold colour as a result of optical interference phenomena produced by thin metal oxide films on metals (by corrosion, controlled heating, or 'anodising') and on mica flakes ('lustre pigments'); plastics gilded with an aluminum layer deposited by 'vacuum metallising' or 'sputtering' subsequently top-coated with a coloured resin layer. Dates of first introduction are given where possible.

'Gold, really gold?'
'Look a little and perhaps you'll make out.'

Henry James, *The Golden Bowl*

Introduction

According to our modern understanding of the values of materials, the art of gilding would seem to be motivated by the obvious desire to make things that are not gold, appear to be. Gold is a rare metal, and one that has been used by numerous cultures to mark the culturally and religiously significant, to set off and decorate elite individuals, and to hoard and transfer wealth in a concentrated form. In husbanding this rare resource, humankind has turned its ingenuity to devising a wide variety of methods by which an object composed of another metal can be covered with a thin layer of gold. Equally ingenious have been the ways in which an object can be made to look like gold without using any gold at all. In all of these techniques the theme of transubstantiation is ever present and may have offered spiritual and intellectual rewards equal to or greater than the monetary. Medieval alchemists laboured to turn base metals into gold in a quest that had as much to do with spiritual perfection as with getting rich quick. Restorers and art forgers are also familiar with the profound satisfaction of making one material look like another, or making an object seem to be something that it is not – also for what is often limited financial gain.

While the motivation for imitating gold surfaces is complex and interesting, the following account will focus on the history and techniques of imitating gold surfaces.

Gold-coloured varnishes

Some of the earliest documented attempts to imitate gold are by means of gold-coloured organic coatings over white-coloured metals. One early text, the *Mappae Clavicula,* (Smith and Hawthorne 1974) a ninth century compilation that includes some material that can be traced to Greek and Roman sources, contains many recipes for gold-coloured coatings, including one for an 'easy way of gilding'. The ingredients are 'gold coloured sandarac, orpiment, gum tragacanth, the inner part of an egg (yolk), and goat gall' (Smith and Hawthorne 1974: 36). These are all yellow-coloured materials, but how they were used and on what surfaces is not stated. Recipes for gold-coloured varnishes often form a part of subsequent craftsmen's formula books and become more explicit and easy to decipher as time goes on. They can be characterised as organic coatings based on drying oils or alcohol-soluble resins, colored with a variety of red and yellow organic dyes or resins. These materials were part of an ancient trade in drugs and dyestuffs, and both the resins and the colourants were available from apothecaries.

The colourants used in historic recipes for gold-coloured coatings included *gall* from the gallbladders of domestic animals, the natural plant-dyes annatto (*Bixa orellana L.*), saffron (*Crocus sativus L.*), turmeric (*Curcuma domestica Val.*), alkanet (*Anchusa officianalis L.*), the dyewoods fustic (*Chlorophora tinctoria L.*), brazilwood (various species of the genus *Caesalpinia*), red sandalwood (*Pterocarpus santalinus L.f.*), eagle wood (*Aquilaria agallocha Roxb.*), camwood (*Baphia nitida*), and the coloured resins gamboge (*Garcinia hanburyi*), dragon's blood (various species of the genera *Draco* and *Daemonorops*), allöes (various species of the genus *Allöe*), and accroides (various species of the genus *Xanthorrhoea*). The topic of organic dyes used in decorative coatings has been treated in greater detail elswhere (Thornton 1994). Shellac, a natural resin secreted by scale insects in India and Southeast Asia, was used as both a primary coating resin as well as a colourant. Shellac resin varies in colour due to various factors including host plants, geographic origin and time of harvest, and until the 1930s it was standard practice in India to improve the bright golden colour by adding orpiment (Angelo Brothers Ltd. 1956: 38), a practice that continued but declined after that date.

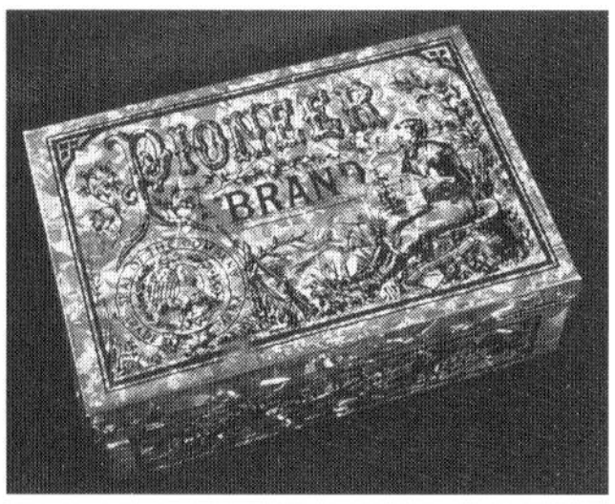

Figure 18.1 Tobacco tin made of sheet iron coated with crystallised and etched tin (moiré metallique), laquered with a dyed organic coating. (See Plate 147.)

After the advent of synthetic dyes beginning in 1857, they were used increasingly in commercially produced objects as replacements for the generally more fugitive natural dyes and colourants. Thin layer chromatography (TLC) analysis of a colourant used in the coating of a decorative nineteenth-century tobacco tin (Fig. 18.1) indicated the presence of an early synthetic dye called *Martius Yellow.*[1] Not surprisingly however, the organic colourants continued to be specified in craftsmen's manuals into the twentieth century, perhaps because they were often unquestioning compilations of earlier formulas, in a recopying tradition that can stretch back to the earliest medieval texts. Many of these recipes must have been as hard to decipher and as questionably useful as they are today, although most of the raw materials continued to be widely available from apothecaries until recently.

White-metal leafs used in imitation gilding

Various white-coloured metals were used as underlayers for gold-coloured varnishes. The fifteenth-century *Strasburg Manuscript,* among other formula compendia, gives directions for making a 'golden colour for painting over silver, tin or lead so that it looks like gold' (Borradaile and Borradaile 1966: 67).[2] Both tin and silver could be beaten into thin leaves that were then varnished to turn them gold. The *Mappae Clavicula* describes this, as well as the twelfth-century treatise by the monk called Theophilus, but these are only early examples of hundreds of such recipes and instructions that were subsequently published. In the early fifteenth century, Cennino Cennini recommended the use of tin leaf in painting because it was not prone to darkening as was silver (Thompson 1933), but he was probably referring to passages that were meant to appear silver rather than gold, and were therefore left uncoated. Silver had greater ductility than tin and could be beaten into thinner leaf. When laid over gesso and bole in a manner identical to the technique of water gilding, it could be burnished to a high shine. Burnished silver ('silvered') furniture and frames enjoyed a brief vogue in Europe during the early eighteenth century, but silver leaf is far more ubiquitous as a component of 'silver gilding' a technique that offered craftsmen a cheaper and entirely convincing alternative to gold leaf. One name for silver-coloured leaf coated with gold varnish in the seventeenth century was 'Golden Mecca', the coating being composed of shellac, gamboge and dragon's blood (Devoe 1971: 97), a recipe that is commonly given in later gilding manuals. The term *mecca* is a specific Italian word for this variety of imitation gilding, and is thought to derive from the Italian for 'cosmetic' (Reynolds 1962). It is also used colloquially by gilders in England. Other white-coloured metal leafs subsequently became available: aluminum was beaten into leaf by the early twentieth century, and was non-tarnishing but thick and not burnishable; platinum and palladium were available in leaf form from the late nineteenth century, but were more expensive than gold and an unlikely choice for a technique that was meant to be cheaper. A shortcoming of all organically coloured coatings is that they tend to fade. this generally results in the artifacts that were silver-gilded becoming less gold and more silver or brown than was originally intended. Because most organic yellow dyes are more fugitive than reds, the coatings tend to shift towards a red hue as they fade.

Imitation gold tinware

Objects made of tinned-iron plate called 'tinware' were also often gilded with varnishes. Craftsmen in the cities of Birmingham and Pontypool, where the tinware trades flourished, devised many decorative effects, including convincing gold-coloured coatings. The tobacco box previously mentioned was decorated with a beautiful crystallised and etched tin surface called '*moiré metallique*' (Smith 1960: 63–5), which was invented in 1814. The shellac based, synthetically dyed coating that covered the tobacco box is most convincingly gold on the interior where it had not been faded by light.

Gold-lacquered brass

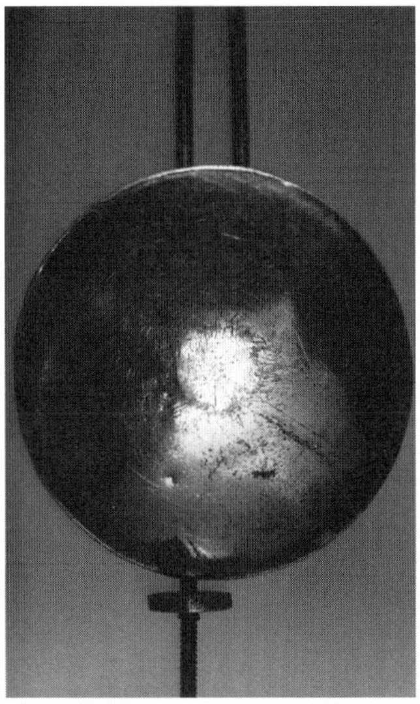

Figure 18.2 Clock pendulum bob of brass, covered with a gold-coloured lacquer. (See Plate 148.)

Gold varnishes or lacquers were also used extensively on brass to overrule the somewhat greenish cast and to make the metal more nearly resemble gold. This was certainly a common practice during the eighteenth and nineteenth centuries on virtually all objects made of polished brass, such as scientific instruments, furniture hardware and decorative objects (Fig. 18.2). An influential writer on art technique, Robert Dossie, said in 1758, 'the lacquering brass work to imitate gilding ... is a considerable object of this kind of art, and has been improved to the greatest degree of perfection', and that 'The principal body or matter of all good lacquers used at present is seed-lac' (Dossie 1758). The colourants he specifies are annatto, saffron and turmeric, as well as the naturally reddish coloured seed-lac. This and other similar recipes are so commonly specified for coating brass that conservators should question the appropriateness of any modern water-clear coating as a replacement for a missing lacquer layer. Acrylic resins and cellulose-nitrate based lacquers coloured to match gold with the *Orasol* dyes manufactured by Ciba-Geigy have been used successfully by modern conservators as substitutes.

Coloured coatings over gold

Even gold was coated to further manipulate the colour. Gold-colored lacquers have been discovered on stamped brass daguerreotype and tintype 'keepers' that had already been gilded, probably by a cold electrochemical deposition process.[3] Picture frame gilders of the nineteenth and twentieth centuries commonly applied a coloured coating called 'ormolu' to their gilded surfaces, especially over matte (unburnished) water-gilt passages. This coating, named after the mercury amalgam 'fire-gilded' brasses called ormolu (after the French *or moulu*, 'ground gold'), was commonly composed of seed-lac, gamboge, and dragon's blood dissolved in alcohol then mixed with dilute animal-glue size. The coating was used to deepen the colour of the gold as well as to even out inconsistencies and lap-marks in the gilding (Scott-Mitchell 1905).

Brass as a replacement for gold

Brass, with or without lacquer, has long been valued due to its resemblance to gold. The Romans were the first to make brass intentionally in the West. Pliny said that a material called '*aurichalcum*' (gold copper) was good for making sesterces (coins) (Hoover and Hoover 1950: 408–10), and analysis of some Roman sesterces has shown the major alloying metal to be zinc (Tylecote 1976: 59), forming an alloy that corresponds to our modern understanding of what is meant by 'brass'. The term *aurichalcum* was used for brass throughout the medieval period, and Theophilus clearly described making brass in the early twelfth century (Hawthorne and Smith 1963). The method involved heating copper with zinc carbonates and silicates called 'calamine' so that the copper absorbed the zinc. The resulting alloy melted at a lower temperature than the copper, and the molten brass was then cast. This process, called 'cementation', yielded an alpha brass of variable zinc content (but never above about 28%), that was gold-coloured, malleable, and ductile (Scott 1991: 19). This process remained the primary method of making brass until William Champion developed a process for smelting metallic zinc in commercial quantities in 1738 in

England (Tylecote 1976). During the late medieval period the German city of Nuremberg became a center for the production of brass objects because the absense of guilds there favored innovation and capitalisation (Turner 1982). The English city of Birmingham developed extensive metalworking industries for similar reasons. It lacked guilds because it was unincorporated. Because of legislated religious discrimination in incorporated towns (Hamilton 1967), it also attracted the dissenters who were such influential industrial innovators and capitalists in eighteenth-century England.

Solid copper could also be superficially 'brassed' by exposing it to zinc fumes. Andrew Ure in his early nineteenth-century technical dictionary described how this was done to brass rods, which were then drawn into wire. He states that 'The spurious gold wire of Lyons is made from such rods' (Ure 1866: vol. 1, 239). Since Lyons was a centre of silk and brocade weaving, this spurious gold may have been intended for use in textiles.

Ure also mentions that copper can be 'brassed' by boiling it in a solution of muriatic acid, 'wine stone'(presumably potassium tartrate) and zinc amalgam (Ure 1866: vol. 1, 239). A similar recipe in the fifteenth-century Bolognese Manuscript containing the ingredients tartar (potassium tartrate) *atramentum* (zinc), quicksilver and salt, appears to be a version of this method of 'gilding' (Merryfield 1849: vol. 1, 464).

Gold-coloured brass alloys

When metallic zinc became available in Europe, it became easier to accurately control the alloy content, and gold-coloured brasses became popular for small luxury items (called 'toys' in the eighteenth century) such as snuff boxes, watches, chatelaines, buttons and sword fittings. There were various names for these alloys, including Prince Rupert's Metal, Prince's Metal, Mannheim Gold, Bath Metal and Similor (Tomlinson 1862: 171). Perhaps the most famous of these was the imitation gold alloy called Pinchbeck invented in 1725 by self-described 'toy man' and watchmaker Christopher Pinchbeck. His trade card listed his address 'at Pinchbeck's Head' in Fleet Street, and claimed that he was the 'only maker of the true and genuine metal' (Heal 1968: 43). The name 'Pinchbeck' came to be synonymous with any cheap imitation, and even for falsity in character. The reproducible production of Pinchbeck and other gold-coloured brass alloys would have probably required the mixing of accurate quantities of copper and zinc. While 1725 is a few years before the Champion process, metallic zinc, known by various terms including tutenag, tuteneque, spiauter, and spelter was being imported into Europe during the sixteenth and seventeenth centuries from India and China, where it had been intentionally smelted for centuries. Metallic zinc had also been produced in Europe in small quantities as the occasional curious by-product of the smelting of other metals (Hoover and Hoover 1950).

Modern imitation-gold alloys used in costume jewellery are usually described as manganese bronze in supplier's catalogues.

Brass leaf

Gold-coloured brass could also be beaten into leaf. Such imitation leaf was called, in English, Dutch metal (a corruption of *Deutch* or German), schlag metal, composition leaf or simply metal leaf. As the names indicate, Germany was the centre of production. It is still used extensively in the picture frame industry as a cheaper replacement for gold, and as will be described, was the starting material for 'bronze powders'. Brass cannot be beaten as thin as gold, and since the resulting sheet can be handled with bare fingers without damage, it is technically a foil. This also means that it does not lay as flat on the size as gold leaf, and the tiny wrinkles in the surface are the simplest means of visual identification. Brass leaf will also tarnish over time and become a progressively less successful imitation of gold as it darkens.

Bronze powders

Most people think of 'bronze powders' as being a recent imitation of gold in the form of 'radiator paint', but these powders also have a considerable antiquity. The term 'bronze powder' probably has more to do with a primary use for these powders – imitating bronze sculpture by 'bronzing' plaster statuary and other objects – than the alloy composition. The term 'bronze' has traditionally applied to copper alloys used in sculpture regardless of actual composition. Bronze powders are generally copper–zinc alloys and would be called brass by modern nomenclature, but this restricted definition is relatively recent.

Theophilus described a mill for reducing gold leaf into powder in the twelfth century (Hawthorne and Smith 1963), and this was probably already an ancient technique. His apparatus was a simple device that operated by a strap, winding and unwinding on a shaft that turned a bronze pestle in a bronze mortar. Theophilus stated that this worked equally well for other metals. Metal powders are only reflective if they are in the form of flat, shiny flakes, and this was accomplished by first beating the metal into leaf and then grinding it into powder with the aid of an organic medium such as honey or gum to prevent the particles from sticking to each other. The milling of bronze powders continued to be a German speciality, particularly in Nuremberg with its well established brass industry.

The labour-intensive leaf grinding process made bronze powders too expensive for anything but restricted use on high quality objects, such as japanned furniture. Stalker and Parker in their 1688 treatise on japanning described a wide variety of German bronze powders, and when speaking 'of Brass-dust, which is commonly amongst the Artists called Gold-dust' say that 'This cannot be made in England fit for use though it hath often been attempted' (Stalker and Parker 1688: 5–6).

By the eighteenth century there were some English makers successfully using the hand process including a 'Mr. Warren of Birmingham' (DeVoe 1981: 4), but it was the prolifically inventive Henry Bessemer of Bessemer Converter fame that changed everything. When Andrew Ure stated that 'This Nuremberg manufacture has been successfully introduced here by Mr. Bessemer' (Ure 1866), he was mistaken, in that Bessemer's mechanised process was entirely unlike the hand process of Nuremberg. Since Bessemer carried on the manufacture of bronze powders for 40 years under the most extreme secrecy, Ure can be forgiven some confusion.

When Bessemer went to a shop to buy 'gold powder' for his sister, he was struck by the 'marvellous price' of what he guessed was 'probably, only a better sort of brass'. He thought, rightly, that the manufacture of brass powders 'offers a splendid opportunity for any mechanic who can devise a machine capable of producing it simply, by power' (Bessemer 1905). After much experimentation and examination of the German product under the microscope (surely one of the first instances of the use of microscopy in research and development), he was able to begin manufacturing and marketing his bronze powders in about 1840 (Stephen and Lee

1965) at prices that greatly undercut his competitors. He soon made a fortune that enabled him to be a completely independent capitalist for his subsequent ventures, including the revolutionary process for manufacturing steel in large quantities. The successful bronze powder manufacturing venture 'Changed the whole current of my life, and rendered possible the still greater change which the iron and steel industry of the world has undergone, and with it the fortunes of hundreds of people'.

Most of Bessemer's account of this manufacture is taken up with descriptions of his elaborate measures to deter industrial spies. He designed the machinery so that it could be erected and operated by no more than three people (the brothers of his wife who could be counted on for self-interest if not complete loyalty), and he farmed out the actual parts to widely dispersed engineering shops so that they would have no idea what they were making. The factory was built without windows, and Bessemer posed as a disgruntled worker to entertain the inevitable German spies with apocryphal tales of how the powders were made. His technical account is interesting if too lengthy to relate here, and includes descriptions of the arcane alloys he used to produce bright colours, including an alloy of copper and silver that made a beautiful peacock purple when oxidised. His finest grade of pure copper powder was supplied to the firm of Elkington in Birmingham for coating electro-type moulds in their production of 'works of art by the electro-deposition of metals' (Bessemer 1905: 70). Bessemer credited himself with thinking up the idea of 'gold paint', and he devised a paint medium (calcium succinate) that he felt would preserve the gold colour for as long as possible (Bessemer 1905: 75). It can be safely stated that before Bessemer, the idea of a lavish and extensive use of 'gold powders' would have been economically untenable.

Mosaic gold

Another 'gold' powder with a long history is mosaic gold, also called *porporino, aurum mosaicum*, and *aurum musivum*. It is actually tin sulphide – SnS_2 – made artificially in Europe as early as the ninth century and first clearly described in the Italian text *De arte illuminandi* of the fourteenth century. It was made in China by at least AD300, when it is described in a text by the alchemist Ko Hung (Butler *et al.* 1983). By the fifteenth century, numerous recipes were given for mosaic gold manufacture in European texts such as the Bolognese Manuscript, which includes six methods of making '*purpurino*' and similar gold colours (Merryfield

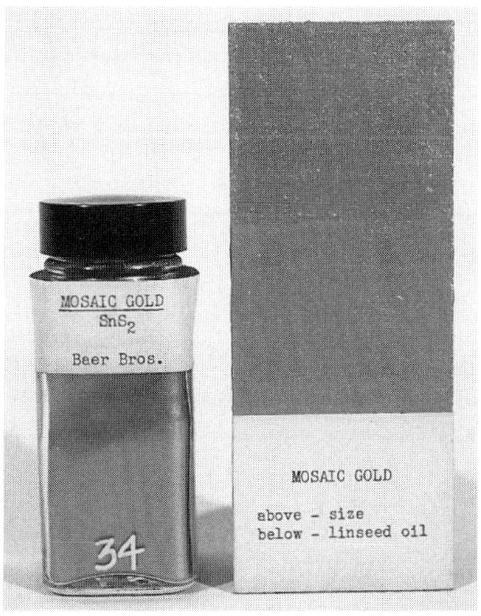

Figure 18.3 Mosaic gold and sample tile from the Edward Forbes Pigment Collection of historic pigments and media. (Straus Center for Conservation, Harvard University Art Museums, Photograph Eugene Farrell.) (See Plate 149.)

1849: vol. 2, 458–61). European methods generally involved heating tin, *sal ammoniac* (ammonium chloride), *alum* (aluminum sulphate) and mercury in a sealed vessel. The mercury was a reaction intermediate and was not a part of the final compound. Chinese alchemists apparently achieved the same result without mercury, by heating tin, alum and ammonium chloride (Butler *et al.* 1983: 132).

Mosaic gold resembles bronze powder, and was an item of commerce from the fifteenth century, when it could be purchased from apothecaries (Borradaile and Borradaile 1966: 27),[4] throughout the nineteenth century (Tomlinson 1862: 253), and as late as 1941 when E. W. Forbes bought a sample from Baer Brothers for the Forbes Collection of artist's materials (Fig. 18.3).[5]

Metal sulphides

The iron sulphide, *pyrite*, is a chemical analogue of mosaic gold, consisting of FeS_2. Some metal sulphides have a metallic lustre because their electron configurations are similar to those of metals. Intentional sulphide layers were apparently used to imitate gilding on copper alloys as early as the Roman Period. In at least one case the copper–iron sulphide *chalcopyrite* ($CuFeS_2$) has been identified on an object of that date (Oddy and Meeks 1981), apparently added as an intentional pseudo-gilding. Although iron pyrite could be mistaken for gold, and was called 'fool's gold', it seems to have never been synthesised as a pigment.

Interference colours

Thin, partially reflective films can give rise to 'interference colours' including 'gold' along with the other colours of the spectrum. The colour is produced by the interference of light waves reflecting off of the surface of a thin layer as well as off the surface underneath the layer. The emergent reflected wavelengths that are in phase (corresponding wave peaks and troughs) reinforce and are perceived as colour, while those that are not in phase cancel out. Oil on a puddle is probably the most familiar example of this phenomenon, and other examples are ubiquitous in nature. Interference colours have been created intentionally for a variety of practical and aesthetic reasons.

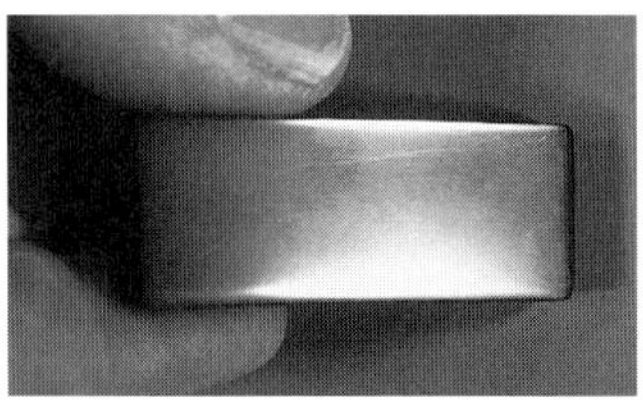

Figure 18.4 Steel coupon with 'straw' colour produced by a thin film of magnetite. (See Plate 150.)

Blacksmiths traditionally tempered steel tools 'by the colours', relying on the interference colours produced by a thin film of magnetite (ferric and ferrous oxide). This oxide layer formed as the steel was heated, and served as a guide to the temperature of the steel. The colour most resembling gold, usually called 'straw' (Fig. 18.4), was used to indicate the correct tempering heat for fine edge tools (Holtzappfel 1843: vol. 1, 245–6).

A thin film of silver-sulphide tarnish can be easily mistaken for gilding. All conservators who have worked with silver collections have observed this 'gold' layer, which is readily removed by an acidic thiourea 'silver dip', thus revealing it's true nature. This may even have been done as an intentional gold imitation as suggested by a recipe in the *Mappae Clavicula*: 'For silver to appear golden coloured: Minium, liquid alum, Cimolian earth; equal amounts. Pour these into sea water, and after thorough heating dip the silver in it' (Smith and Hawthorne 1974: 39).

Interference colours can be given to some metals by *anodising*. The metal is made the anode in an electrolytic cell, and a direct current of sufficient voltage it applied until the required oxide layer and resultant interference colour has been produced. The brightly coloured aluminum tumblers produced in the 1950s got their colours from a combination of interference colour and organic dyes that were used to colour the resultant porous aluminum oxide layer. This phenomenon was first noticed in 1857, but used commercially only since the 1920s (Child 1993). The recently popular anodised titanium and niobium jewellery shows brilliant colours, including a convincing gold, by interference colour alone (Fig. 18.5).

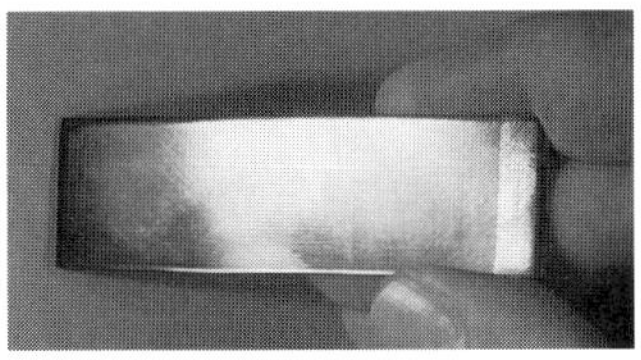

Figure 18.5 Anodised titanium metal showing gold colour. (See Plate 151.)

Metallic-lustre mica pigments

Mica-based metallic-lustre pigments that resemble gold and other metals were first patented in 1963 (Greenstein 1988), and have become commercially important as pigments in imitation-metal plastics and paints, including 'metallic' acrylic emulsion paints. They consisted initially of thin particles of mica, which had been coated with thin films of titanium, subsequently oxidised to titanium dioxide(TiO_2). Although initially intended to produce pearlescent effects, it was found that colour could also be imparted by thickness-dependent interference-colour films of titanium dioxide, and by the addition of iron oxides as tints for the titanium dioxide. The best gold colour was obtained by a mixture of TiO_2 and ferric oxide (Fe_2O_3). Bronze-coloured pigments made of ferric-oxide-coated mica became commercially available in the early 1970s (Greenstein 1988). While these pigments have the disadvantage of poor covering or 'hiding' power due to the transparency of the mica, they have the advantage over bronze powders of chemical (and therefore visual) stability. In practical terms, the best imitation for gold is achieved by an underpainting that matches a non-metallic gold colour as closely as possible, followed by mica lustre-pigments dusted on to a tacky overcoating so that the mica platelets orient parallel to the surface (Fig. 18.6). This method gives the greatest degree of specular reflectance.

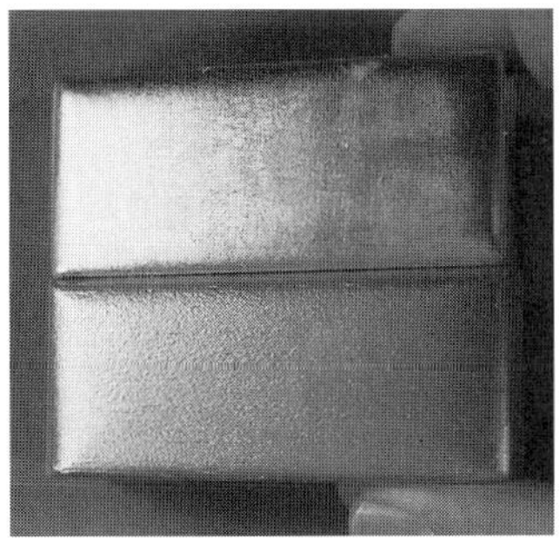

Figure 18.6 Two 'gilded' sample tiles. The top one is gold leaf on oil size. The bottom one is mica-based lustre pigments over acylic emulsion underpainting. (See Plate 152.)

Vapour-deposition plating

In addition to the older electroplating methods, since World War II metal plating has been done by depositing vaporised metal on to objects contained in a vacuum chamber. In the process, called 'vacuum metallising', the metal vapour is disseminated into a vacuum by simply evaporating the metal on a resistance filament (Mazzuca 1980–81). Another common technique, that of 'sputtering', dislodges metal atoms from a target of the plating metal by the impact of ionised atoms accelerated magnetically (Storms 1980–81). Sputtering is carried out in an inert gas (commonly argon) atmosphere. In both cases, the metal vapour deposits onto the

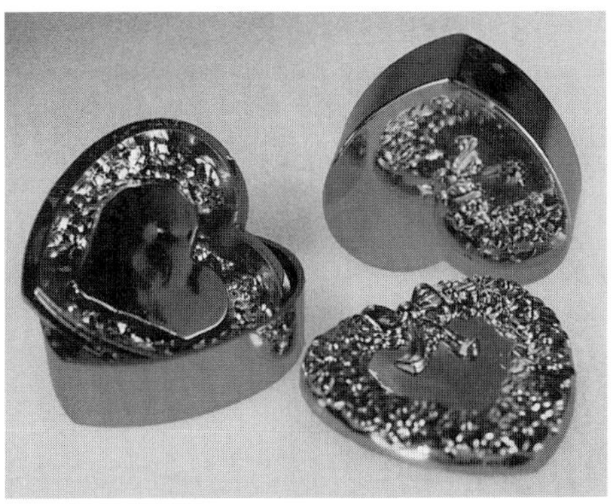

Figure 18.7 ABS plastic boxes with vapour-deposited aluminium, 'gilded' with a dyed organic coating. (See Plate 153.)

objects being plated, forming a layer as shiny as the initial surface. Metallic-plastic toys, automobile parts and other decorative 'silver' and 'gold' objects are normally plated by vapour deposition. Hard plastics such as ABS (acrylonitrile-butadiene-styrene) are preferred. Producing an imitation-gold surface involves three steps; applying a shiny lacquer coat to the plastic object, depositing the metal (most commonly aluminium), and then 'gilding' the aluminium, with a gold-coloured lacquer topcoat (Horn and Hebble 1966) (Fig. 18.7).

Conclusion

The diverse surfaces and techniques described in this paper are joined by one thread: humanity's deep-seated urge to imitate one material – in this case gold – with other materials, and to delight in applying intellect, ingenuity and even 'magic' in doing so. In Henry James' *The Golden Bowl* (James 1904), the object of the title, only appearing to be made of gold, symbolises falsity and misperception. The heroine picks up the bowl in an antique shop and says wonderingly to the dealer, "Gold, really gold?" The antique dealer hesitates, then replies, "Look a little and perhaps you'll make out."

Notes

1. Analysis performed at Buffalo State College by Scott Nolley under the supervision of Conservation Scientist Judith Bischoff.
2. This is an oil-resin varnish coloured yellow with 'Greek pitch' or two other unidentified materials.
3. Treatments carried out at the Art Conservation Department of Buffalo State College.
4. 'If you want to make a golden writing, procure *aurum musitum* from the apothecary . . .'.
5. Forbes Collection no. 34, case 3, shelf 3 (sample provided to the author courtesy of Henry Lie, Director, The Strauss Center for Conservation, The Harvard University Art Museums).

References

Angelo Brothers Ltd. 1956. *Shellac*. Calcutta: Angelo Brothers Ltd.

Bessemer, H. 1905. *Sir Henry Bessemer FRS*. London: Offices of *Engineering*.

Borradaile, V. and Borradaile, R. (trans.) 1966. *The Strassburg Manuscript*. New York: Transatlantic Arts.

Butler, A. R., Glidewell, C., Needham, J. and Pritchard, S. 1983. 'Mosaic gold in Europe and China'. *Chemistry in Britain*, February: 132–4.

Child, R. 1993. 'Modern electroplating and electrofinishing techniques'. In *Metal Plating and Patination,* S. La Niece and P. Craddock (eds), 291–300. London: Butterworth Heinemann.

DeVoe, S. P. 1981. *The Art of The Tinsmith.* Exton, Penn.: Schiffer Publishing Ltd.

Devoe, S. S. 1971. *English Papier Mache of the Georgian and Victorian Periods.* Middletown, Conn.:*Wesleyan University Press.97*

Dossie, R. 1758. *The Handmaid to the Arts*. London: J. Nourse Bookseller.

Greenstein, H. 1988. 'Nacreous (pearlescent) pigments and interference pigments'. In *The Pigment Handbook*, 2nd edn, P.A. Lewis (ed.). New York: John Wiley and Sons.

Hamilton, H. 1967. *The English Brass and Copper Indusries to 1800*, 2nd edn. New

York: Augustus M. Kelley Bookseller.
Hawthorne, J. G. and Smith, C. S. 1963. *On Divers Arts:The Treatise of Theophilus*. Chicago, Ill.: University of Chicago Press.
Heal, A. 1968. *London Tradesmen's Cards of the XVIII Century*. New York: Dover Publications.
Holtzapffel, C. 1843. *Turning and Mechanical Manipulation*, vol.1. Mendham, N.J.: Astragal Press, 1994.
Hoover, H. C. and Hoover, L. H. 1950. (An historical note on zinc) *Georgius Agricola:De Re Metallica*. New York: Dover Publications.
Horn, F. J. and Hebble, H. B. 1966. 'Decorative Coatings'. In *Vapor Deposition*, C. F. Powell, J. H. Oxley and J. M. Blocher (eds), 579–92. New York: John Wiley and Sons.
James, H. 1904. *The Golden Bowl*. New York: New American Library, 1972.
Mazzuca, 1980–81. 'Vacuum-metallizing'. In *Modern Plastics Encyclopedia*, J. Agranoff (ed.), 466. New York: McGraw Hill.
Merryfield 1849. *Original Treatises Dating From theXIIth to theXVIIth Centuries On the Arts of Painting*. London: John Murray.
Oddy, W. A. and Meeks, N. D. 1981. 'Pseudo-gilding: an example from the Roman Period'. *MASCA Journal*, **1**(7): 211–13.
Reynolds, B. (ed.) 1962. *The Cambridge Italian Dictionary*. Cambridge: Cambridge University Press.
Scott-Mitchell, F. 1905. *Practical Gilding Bronzing and Lacquering.* London: The Trade Papers Publishing Company.
Scott, D. A. 1991. *Metallography and Microstructure of Ancient and Historic Metals.* Marina del Rey, Calif.: The Getty Conservation Intitute and the J. Paul Getty Museum in association with Archetype Books.
Smith, C. S. 1960. *A History of Metallography.* Chicago: University of Chicago Press.
Smith, C. S. and Hawthorne, J. G. 1974. *Mappae Clavicula: A Little Key to the World of Medieval Techniques*. Philadelphia, Penn.: American Philosophical Society.
Stalker, J. and Parker, G. 1688. *A Treatise of Japanning and Varnishing*. Chicago, Ill.: Quadrangle Books, 1960.
Stephen, L. and Lee, S. 1903. *The Concise Dictionary of National Biography to 1900.* Oxford: Oxford University Press.
Storms, C. D. 1980–81. 'Sputtering'. In *Modern Plastics Encyclopedia,* J. Agranoff (ed.), 465. New York: McGraw-Hill.
Thompson, D. V. 1933. *The Craftsman's Handbook 'Ill Libro dell'Arte' Cennino d'Andrea Cennini*, New Haven, Conn.: Yale University Press.
Thornton, J. 1994. 'The use of organic dyes in wood polychromy'. Paper presented at the *Painted Wood* symposium, Williamsburg, Va.: AIC Wooden Artifacts Specialty Group.
Tomlinson, C. (ed.) 1862. *Cyclopedia of the Useful Arts.* London: James S. Virtue.
Turner, E. 1982. *An Introduction to Brass*. London: Victoria and Albert Museum.
Tylecote, R. F. 1976. *A History of Metallurgy*. London: Metals Society.
Ure, A. 1866. *A Dictionary of Arts Manufactures and Mines*. New York: D. Appleton and Company.

19 The Treatment of False Gilding: A Case Study

Deborah Long

Abstract

We think of coatings on metal primarily as protective materials, but in the eighteenth and nineteenth centuries organic coatings were used to provide both a protective coating and a decorative surface imitating that of gold. Because decorative arts examples of objects bearing these coatings are rare, there is a limited amount of information available regarding treatment options. This paper will discuss the analysis and treatment of a pair of argand lamps that have been coated with shellac to imitate gilding.

Historical overview

From earliest times metal objects have been gilded to increase their beauty and desirability. In the eighteenth century, the predominant method of gilding decorative arts objects was the technique of fire gilding, also known as mercury amalgam gilding or ormolu. Gold was applied to the surface of the object in a paste consisting of finely divided gold dust, dissolved in mercury to form an amalgam. When gentle heat was applied to evaporate the volatile mercury component, a well bonded gold coating remained on the surface of the object. The process could be repeated until the desired gold thickness was achieved. Fire gilding produced a very durable gold coating when compared to other methods of gold application (Goodison 1974: 69–70).

The process of mercury gilding, in spite of the excellent results obtainable, was not without problems. This type of gilding provided a beautiful surface but added considerably to the cost of the finished object. Production of amalgam-gilded objects required a great expenditure of time and expertise. In large workshops, specialists were often employed to carry out the many different stages of the preparation and gilding processes. Additionally, the mercury released during the process is highly toxic. Literature from the period often includes notes about the dangers associated with exposure to mercury vapours. Consequently, mercury gilded objects were generally produced for a small, elite, and wealthy market.

Production of objects in imitation of gilding

Because the process of mercury gilding was so time consuming, expensive and dangerous, manufacturers sought other options (Cellini 1967: 95, 104). The most successful alternative involved the application of shellac-based lacquers to copper alloy surfaces in imitation of gilding. It was far less dangerous, faster and less expensive, requiring only readily available materials.

In the middle of the eighteenth century, manufacturers in England were at the forefront in the development of products for the growing middle-class market. These shrewd businessmen, while producing objects for the upper class, also catered to the middle class, who, as in the present, desired objects indicative of wealth. Mid-eighteenth-century trade catalogues included wording implying that the look of gilding could be attained with lacquered brass. In many cases brass items could be purchased in three levels of finish from the same manufacturer: a plain polished surface, a lacquered surface, and a fire-gilded surface (Hooper 1858). Near the end of the eighteenth century, mercury gilding was outlawed in England. This made the use of alternatives to mercury gilding essential for English craftsmen. Thus, the application of shellac-based coatings became the primary means of imitating the look of gilded surfaces. While not as durable as gilding, the visual effect produced by these organic coatings could fool all but the most discerning eye (Fig. 19.1). The development of safe, commercially viable electroplating, beginning in the 1840s, quickly supplanted fire gilding (Raub 1993: 284–90), eventually reducing the necessity associated with the production of imitation gilding.

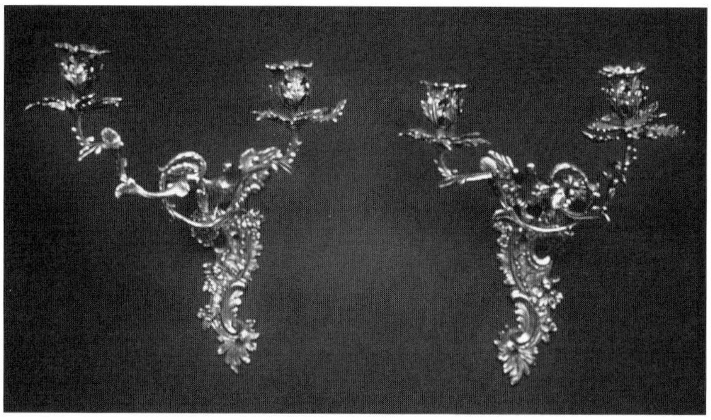

Figure 19.1 Pair of lacquered brass wall sconces. (Courtesy, Winterthur Museum 57.126.26 & 27.) (See Plate 154.)

Some examples of objects typically coated with shellac during this time include decorative arts objects such as lighting devices; furniture hardware, such as hinges, drawer pulls, and escutcheons; and small personal items, such as buttons.

Lacquer composition

Lacquers were made of several basic components including shellac, colourants, and solvent. The primary constituent was shellac, in the form of sticklac, seedlac or flake shellac, in colours ranging from red/brown to yellow. Available in Europe by the seventeenth century, shellac was a component in varnishes, adhesives and lacquers. This natural resin, secreted by Tacchardia Lacca or Laccifer Lacca, a South Asian scale insect, was collected from the branches of trees inhabited by the insects and processed to remove contaminants. A complex natural polyester, shellac contains 30–40% aleuritic acid (9,10,16-trihydroxyhexadecanoic acid) combined with sesquiterpene acids (Horie 1987: 149–50).

Resin-based colourants were sometimes added to achieve particular effects. Dragon's blood, a dark red resin from the stem of the *Dracaena* (*Agavaceae*) palm, and gamboge, a yellow gum/resin mixture from several species of trees of the Garcinia genus were the primary colourants found in early lacquer recipes. Less frequently, other colourants were mentioned (Mussey 1987: xxii). These solids were dissolved in 'rectified spirits of wine' (ethyl alcohol). Shellac-based coatings were sometimes called 'changing varnishes' because of the colour change associated with their use (Field 1883: 64–5).

Finishing and lacquering process

The initial surface preparation for objects that were going to be either amalgam gilded or shellac coated was identical. Metal objects were cast, or occasionally wrought, in a copper–zinc alloy. Typically, alloys were 75–80 per cent copper and 20–25 per cent zinc, with a small percentage of other metals added either inadvertently as impurities or intentionally to alter the working properties of the alloy (Goodison 1974). Once formed, the parts were annealed, if necessary, and pickled in a bath of 15–25 per cent nitric and/or sulphuric acid containing some copper in solution. The parts were rinsed and scrubbed to remove any remaining discolouration, then dipped briefly into a strong bath of fresh nitric acid. Finally, the parts were rinsed in several changes of water. The bite of the fresh acid produced a 'dead matte' surface (Brown 1912).

Following surface cleaning, the paths taken by gilded objects and lacquered objects diverged. Objects to be gilded proceeded to the gilding hearth prior to being burnished, while objects to be lacquered went directly to the burnisher. Burnishing produced a reflective metal surface without removing material. This highly specialised skill had to be performed without touching the work with bare hands to avoid contaminating the pristine metal surfaces. Pressure was applied to the surface of the metal with polished steel or hardstone tools, producing a bright reflective appearance. Flat smooth surfaces were generally burnished overall, while highly relieved surfaces benefited from selective burnishing (Brown 1912). The contrast between matte areas and brightly burnished areas greatly increased the visual appeal of objects. Indeed, Matthew Boulton, one of the most important English metals manufacturers of the eighteenth century considered it 'almost impossible' to make wares beautiful without selective burnishing (Goodison 1974).

Finally the objects to be lacquered reached the last stage of production –the application of the coating. There were two methods of application: hot and cold. Hot lacquering, although more time consuming, was preferred for finer wares, producing a more even and durable coating. The cold lacquering technique was used on smaller, mass-produced items that were coated in groups or sets (Mitchell 1905).[1]

When coated hot, the object was heated on a 'lacquer stove' or hot plate to the point where there was an audible hiss when the wet brush touched the surface of the hot metal (Brown 1912). The heat caused acetal cross-links to form in the shellac almost immediately. Several coats could be applied, reheating between each one, until the desired colour was achieved. The operation was one of some delicacy, requiring just enough heat to set the lacquer but not enough to blister it.

When coated cold, the object was brushed or dipped while at room temperature. It was then placed in an oven or on a hot plate to set the lacquer (Brown 1912). Whether coated hot or cold, newly lacquered surfaces were fragile and could not be handled until they had completely dried and cooled.

Deterioration of lacquered surfaces

Deterioration of the shellac coating on metal objects can be caused by a number of factors. Aging of the shellac results in further cross-linking of the resin, leading to shrinkage, cracking, embrittlement and flaking. Coating failure allows corrosion of the underlying copper alloy, with the development of associated corrosion overburden that can obscure the surrounding surfaces. As the corrosion products develop, intact coating surrounding the corroding areas can become disrupted and begin to flake. Handling causes chipping, thinning and, eventually, loss of coating on high points and in areas of heavy use. These loss areas suffer from an increase in the rate of corrosion of the exposed metal. Often, shellac coated objects are not identified as such and thus have been polished over the years as if they had uncoated copper alloy surfaces. Polishing destroys the coating and creates an uneven and unsightly appearance. The rarity of eighteenth-century shellac coatings in collections today is predominantly due to destruction by indiscriminate polishing. Furniture hardware suffers particularly from this problem, subject as it is to use and frequent polishing.

Case study

In 1991, I had the opportunity to examine and treat two objects from Winterthur's collection that still retain most of their original shellac coating. They are a pair of double armed Argand lamps made of brass, iron and glass (Fig. 19.2). Made in Boston, Massachusetts, by Henry N. Hooper, between 1839 and 1848, the lamps were retailed by William Harris and Henry B. Stanwood, also of Boston.

By the time these lamps were made, American manufacturers were starting to produce goods equal in quality and price to those of English manufacturers. Before this time English goods dominated the American

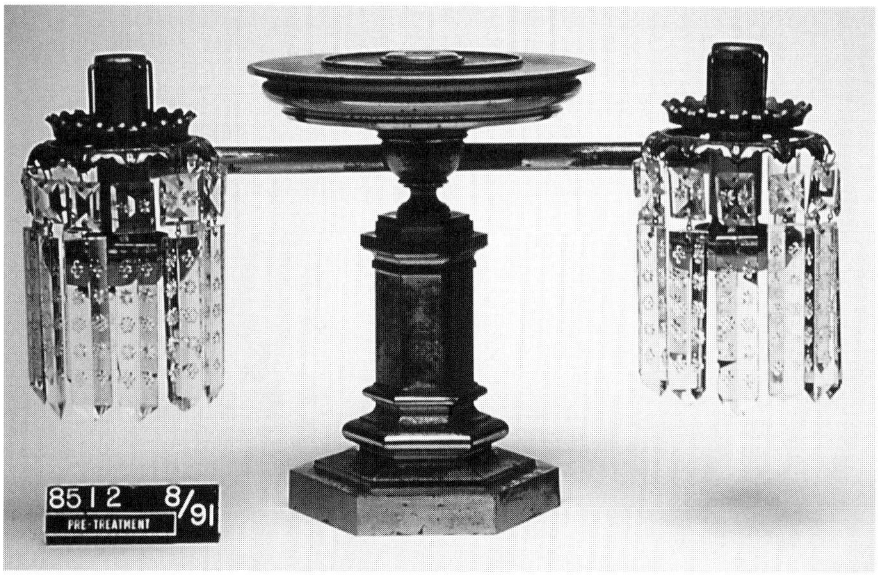

Figure 19.2 One of a pair of argand lamps by Henry Hooper, before treatment. (Courtesy, Winterthur Museum 85.1.2.) (See Plate 155.)

market because English manufacturers, with their vast distribution networks and mass-marketing skills, could undercut the prices of their American competitors even in America. While nearly identical to their English counterparts in methods of manufacture and finish, these lamps none the less act as excellent documents of the American producers' growing ability to compete in the American marketplace. Additionally, the retention of their original surfaces gave these lamps an important place in Winterthur's collection, as there are very few examples remaining in existence (Fennimore 1995).

Condition

When the lamps arrived in the Winterthur Museum Objects Laboratory, they were structurally sound but incomplete. The central decorative covers, glass shades, chimneys and a number of the original snowflake style prisms were missing. Some of the prisms had been replaced with inappropriate uncut prisms. The remaining intact prisms suffered small losses and the pins holding them to the lamps were corroded and distorted. The losses to the surface coating from use had resulted in the development of a thin, discontinuous layer of copper corrosion products in the loss areas, lifting adjacent intact coating from the surface of the metal. Surfaces, particularly around the oil fillers and burner rings were covered with oil residues, soot, dust and grime. Corrosion was somewhat accelerated in these areas, and the coating was visibly disrupted and lifting. The rear third of the shaft of one of the lamps had been painted with what appeared to be discoloured gold 'radiator' paint, in an attempt to disguise the loss of surface coating in an area of heavy wear.

Analysis and testing

The bright orange fluorescence observed under ultraviolet light examination indicated the presence of an organic coating (Fig. 19.3). A tiny sample of the coating was analysed using Fourier Transform infrared microanalysis (FTIR,) which confirmed that the composition of the coating was a natural resin, with the best spectral reference match being that of garnet lac (Carlson 1991). Energy dispersive X-ray fluorescence analysis of six

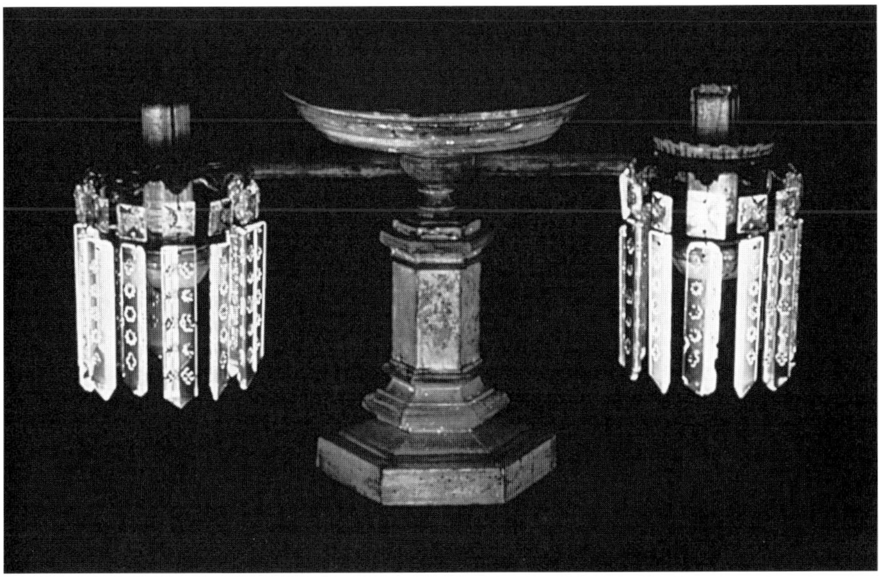

Figure 19.3 Argand lamp illuminated with ultra-violet light. (Courtesy, Winterthur Museum 85.1.2.) (See Plate 156.)

parts of the lamps showed a varied but similar elemental composition averaging approximately 78 per cent copper, 20 per cent zinc, 1½ per cent lead, with traces of tin, antimony and silver as impurities (Carlson 1986). Solubility tests indicated that the coating was still slightly soluble in ethanol but insoluble in the other solvents tested, including petroleum benzine, xylene, acetone, and water. The bronze-powder paint was not soluble in any of the solvents tested. Oily accretions, adhered dirt and copper stearate corrosion products formed in areas where lamp oil had puddled on the surface, were all soluble in petroleum benzine.

Standard methods for removing copper corrosion products include two basic techniques for treatment consideration. Mechanical techniques, such as the use of hand tools like scalpels, picks and brushes, could be tested for spot cleaning. Chemical techniques, such as the use of solvents, reagents or chelating agents could be tested for use over larger areas. Since there are no standard methods for the removal of corrosion overburden from shellac-coated surfaces and since all but the most gentle mechanical methods were obviously inappropriate, my remaining option was to test chemical corrosion removal methods.

While an effective, efficient cleaning material was required, it was equally important to cause no harm to the shellac coating and to minimise the risk to the copper alloy substrate. At this point, I hypothesised that a chelating agent would prove to be the most promising cleaning agent. However, after testing several chelating agent solutions under the microscope, I found them to be inappropriate. The excessive working time caused the original coating to soften. Additionally, the most effective working pH for chelation was unacceptably high and would have caused blanching or loss of the shellac. Basic solvents were ruled out for the same reason; they would immediately disrupt the naturally acidic shellac surface. It was interesting to note during my research that basic solutions, such as sodium hydroxide, had been recommended in several of the period treatises for the removal of damaged or worn coatings (Field 1883).

While considering possible acids, mineral acids were judged to be too aggressive, due to the strength of their action on the surface of the exposed copper alloy. A weak solution of formic acid was the only one that seemed a reasonable candidate for testing. Although it is a relatively weak organic acid, it is an effective reducing agent for copper corrosion products. Surface cleaning tests indicated that a five per cent solution would work quickly without disrupting the shellac surface except in areas where the corrosion had lifted the coating from the surface. The pH of the solution was compatible with the low pH of the aged shellac. Additionally, the volatility of formic acid eased concerns about clearing it from the surface.

After testing the solution on a small area adjacent to the site where the first sample had been removed, a second tiny sample of the coating was removed and given to Senior Scientist, Janice Carlson, for further FTIR microanalysis. She prepared a combined spectral plot so that the effects of cleaning could be visually compared. Her results indicated that the before and after treatment spectra were very similar, although the after treatment spectrum was clearer and sharper, presumably due to the successful removal of the corrosion (Fig. 19.4). The results were very encouraging.

Because of their experience working with shellac coatings, paintings conservator, Richard Wolbers and Director of Conservation, Gregory Landrey were consulted about the proposed treatment. I felt that formic acid would be effective but still hesitated to expose an original organic coating to an acid treatment. However, both conservators concurred with my reasoning and with the test results. The results of the solvent tests and the cleaning tests suggested that the primary treatment options included

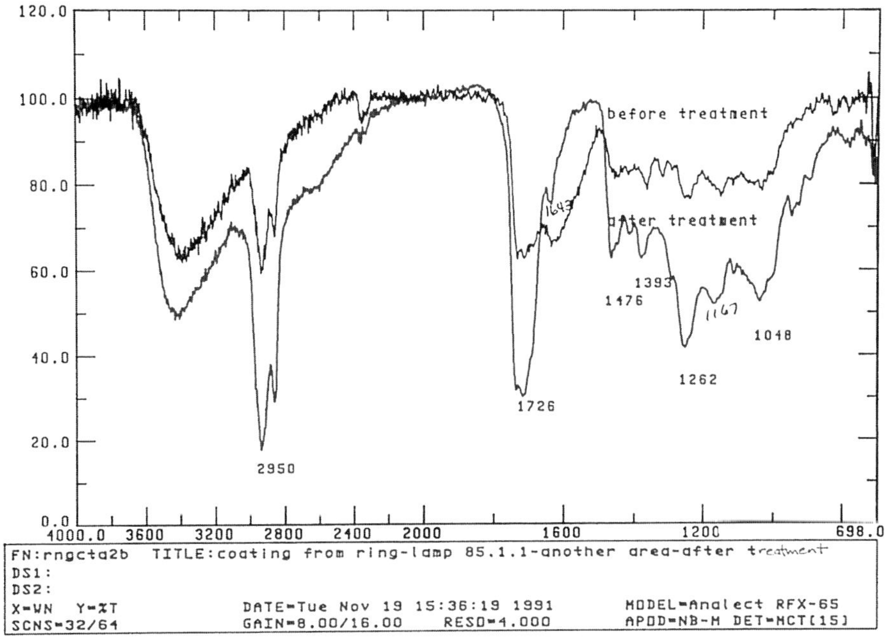

Figure 19.4 Combined FTIR spectra of shellac coating, before and after treatment. (Courtesy, Winterthur Museum Analytical Laboratory.)

using petroleum benzine to remove oily residues and formic acid to reduce corrosion.

The lamps were disassembled into their 28 component parts. Loose dust and particulates were removed with a low-power vacuum and a soft brush. Oily residues and accretions were removed with petroleum benzine. Surface corrosion was reduced by wetting the surface with a five per cent solution of formic acid applied with a sable brush. This was allowed to sit on the surface for a few seconds and then rinsed thoroughly with distilled water. Surfaces were immediately blotted dry following rinsing. The treatment proceeded very smoothly and the results were quite successful.

Surface protection

Once cleaned, the surfaces required a protective coating to minimise future anodic corrosion in the areas of shellac loss. Reversibility, of course, formed the basis of the coating selection. To minimise concerns about interactions between coatings in the future and because these objects were to be placed in permanent cases in controlled environmental conditions, a microcrystalline wax coating would provide sufficient protection. It would remain easily reversible and flexible, would help provide a moisture barrier, and would fill some of the small gaps caused by the loss of the original coating. Because there was such an improvement in appearance, visual compensation of coating loss was judged by the curator to be unnecessary.

Once the new wax coating was in place, the lamps were reassembled. New, matching glass components were marked as replacements and fitted to areas of loss to allow viewers to examine the lamps without the distraction of missing elements (Fig. 19.5). Once the lamps were completed, they were installed together in a new permanent exhibition gallery.

Future treatment implications

Remnants of shellac-based coatings are easily detected with the use of long wave ultraviolet light illumination. It is interesting to note that as

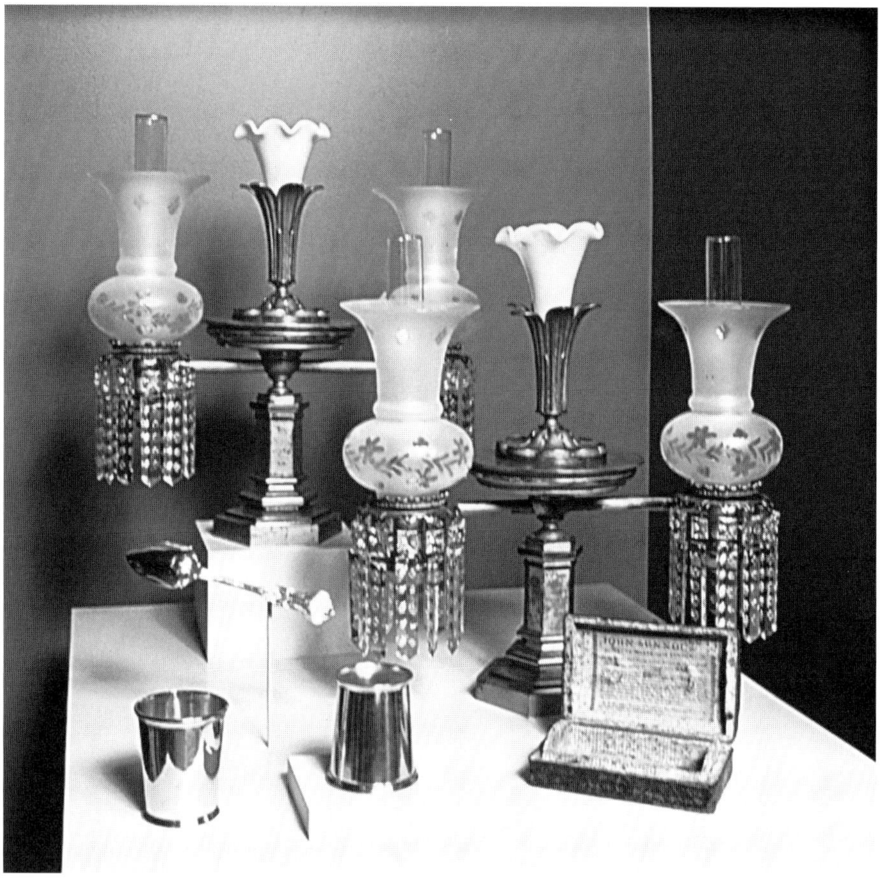

Figure 19.5 Pair of argand lamps after treatment. (Courtesy, Winterthur Museum 85.1.1 & 2.) (See Plate 157.)

these coatings age, they fluoresce under increasingly longer wavelengths. During the course of this treatment, an elaborate shellac coated copper alloy gasolier, also owned by Winterthur Museum, was examined and photographed under ultraviolet light. The entire object fluoresced when viewed under 3660 Å wavelength illumination. However, when viewed at a shorter wavelength of 2400 Å, the coating on the gasolier was found to have been recently 'touched up' by one of its previous owners because only the recent additions fluoresced at that wavelength. With detection this straightforward, there is no reason for the destruction of these rare original organic coatings.

Like any conservation project, this treatment research provided an acceptable answer to a specific set of problems. I was pleased to be able to corroborate my theoretical decision-making and practical test results with analytical data. Close collaboration with the curator enriched my knowledge of these special objects and their place in history. It was very rewarding to see beautiful objects with their cleverly created, imitation gold surfaces emerge at the end of the treatment process, and to feel that they will survive as a link between the present and the past.

Notes

1. Mitchell (1905: 39) refers to lacquer comprising garnet or stick lac dissolved in spirits of wine colored with dragons' blood and called 'ormolu'.

References

Brown, W. N. 1912. *The Principles and Practice of Dipping, Burnishing, Lacquering and Bronzing Brass Ware*, 2nd edn. London: Scott, Greenwood & Sons.

Carlson, J. 1986. *Winterthur Analytical Report #1953*. Winterthur, Del.: Winterthur Museum.
Carlson, J. 1991. *Winterthur Analytical Report #2931*. Winterthur, Del.: Winterthur Museum.
Cellini, Benvenuto. *Treatises on Goldsmithing and Sculpture*, C. R. Ashbee (trans.) New York: Dover Publications [1967].
Fennimore, D. 1995. Personal communication.
Field, George 1883. *The Painter, Gilder and Varnisher's Companion*, 16th edn. Philadelphia, Penn.: Henry Carey Baird.
Goodison, N. 1974. *Ormolu: The Work of Matthew Boulton*. London: Phaidon Press.
Hooper, Henry N. 1858. *Catalogue of Lighting*, Boston, Mass: Winterthur Library (microfilm M-2018).
Horie, C. V. 1987. *Materials for Conservation: Organic consolidants, adhesives, and coatings*. London: Butterworth.
Mitchell, F. S. 1905. *Practical Gilding, Bronzing and Lacquering*. London: The Trade Papers Publishing Company.
Mussey, R. D. Jr. (ed.) 1987. *The First American Furniture Finisher's Manual* (a reprint of *The Cabinet-Makers' Guide of 1827*). New York: Dover Publications.
Raub, C. 1993. 'The history of electroplating'. In *Metal Plating and Patination: cultural, technical, and historical developments*, S. LaNiece and P. Craddock (eds), 284–90. Oxford: Butterworth-Heinemann.

20 Richard Lippold's *The Sun* – Investigation and Treatment Considerations

Jack Flotte and Jean Dommermuth

Abstract

A gold-plated wire sculpture by Richard Lippold, *Variations Within a Sphere #10: The Sun*, has developed disfiguring green pustules on the two miles of wire that make up the piece. The authors identified the corrosion products as 'bronze disease', and conducted an investigation to determine the cause of the corrosion. The wire consists of a silicon bronze core and four layers of plating: silver, brass, nickel and, finally, gold. Three treatment options were discussed with the artist: re-making the sculpture; re-plating it; and chemically stripping the corrosion products. Tests were made on several stripping agents, including alkaline Rochelle salts, EDTA and sodium carbonate. The alkaline Rochelle salts in conjunction with ultrasonic vibration was found to remove the corrosion products, but not affect the brightness of the gold layer. Further treatment with a corrosion inhibitor such as benzotriazole may be necessary to prevent the recurrence of corrosion.

Introduction

In 1953, the Metropolitan Museum of Art commissioned the American sculptor Richard Lippold (b. 1915) to create a work of art he entitled *Variations within a Sphere #10: The Sun*. (Figure 20.1) It consists of about two

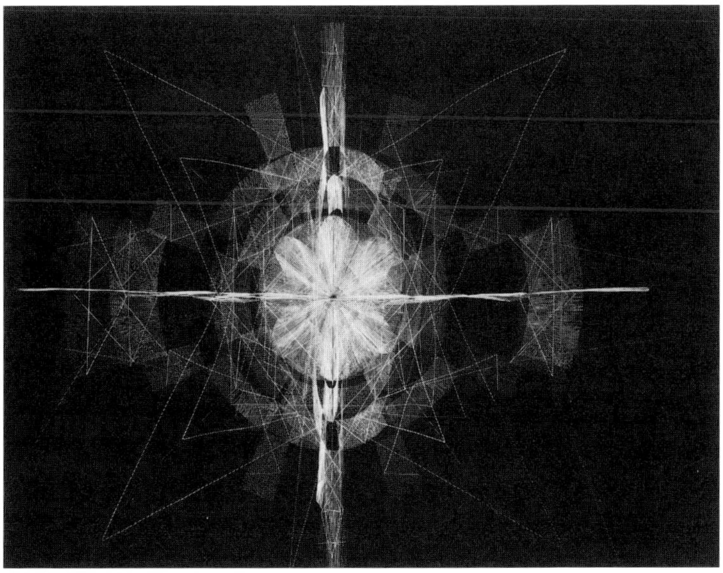

Figure 20.1 *Variations Within a Sphere, #10: The Sun* (MMA accession no. 56.106). Richard Lippold, 1956. (All rights reserved, The Metropolitan Museum of Art.)

miles of 380 μm diameter gold-plated wire, which has been welded into fans of wires that form an ethereal sphere of gold. In its assembled state, this sculpture is about 22 feet across. At the time of its creation, the sculpture was one of the most famous works of contemporary art in the country; its installation in the Metropolitan's North Wing was covered prominently in *Time, Life* and *Newsweek.* Yet within thirty years, *The Sun* had to be withdrawn from display and has languished in storage ever since. The purpose of this paper is to investigate the structure of the piece and consider treatments that will allow it to be once again on public view.

Richard Lippold is one of the generation of abstract artists who came of age after World War II. He was among the first sculptors to make use of industrial materials and methods to create works of art on a vast scale. At the time that the Metropolitan commissioned *The Sun*, Lippold was entering the prime of his career. The sculpture is a continuation of his 'Variations Within a Sphere' series. Constructed of several different types of metal wire, the variations in this series continually became larger and more grandiose. *Variation #6*, made in 1948, is only eight inches by eight inches in size; *Variation #7: Full Moon*, completed in 1950 and owned by the Museum of Modern Art, is ten feet by three feet. *Variation #10: The Sun*, is 22 feet across, 11 feet high, and five and a half feet deep (Kramer, 1956).

The construction of this sculpture has been extraordinarily well documented. Drawings, models and templates for the sculpture still exist, and contemporary accounts of its creation abound, including a story in *Art News* magazine, complete with in-process photographs of the artist at work (Campbell 1956). However, in this case, a surfeit of documentation does not necessarily ensure accuracy in the description of materials and construction.

Lippold worked on the sculpture for three years. One year was spent in the planning stage while actual construction took two. He worked out his ideas through an extensive series of drawings as well as small models in paper and clear plastic. The sculpture was to be constructed from many individual units, all but a few of which are flat fan-shaped pieces. Quarter-circle units were joined to make the disks of the central sphere; other fan-shaped units make up the corona. Full-sized drawings were used to create grooved templates of cardboard and plywood which held the wire in place during construction (Campbell 1956).

An estimated two miles of wire were needed for the construction. The wire was purchased by the artist from I. Stern and Company of Mount Vernon, New York, which manufactured decorative metal objects such as military epaulettes (Lippold 1993). The artist purchased this wire on rolls; it had to be unrolled and straightened before he could work with it. During this process, the diameter was reduced from approximately 405 μm to 380 μm . He then cut the wires to the desired length and positioned them with the aid of the templates.

Two or three structural wires run obliquely to the fans of wire. Where the two wires cross, Lippold joined them with an electric spot welder manufactured by the Sterling Utilities Company. The sculpture required approximately fourteen thousand such welds. The artist additionally used solder in some places; the 36 quarter-circle units that make up the central sphere were joined with a soldering pencil (Campbell 1956). In examining the piece it appears that two different kinds of solder have been used: a hard solder (probably a copper–silver alloy) and a soft (lead–tin) solder. In addition there is a grayish material, possibly an epoxy compound, adhering certain broken ends of the sculpture. Due to the lack of accurate records, it is unclear how much of the solder and adhesive is the result of repairs during *The Sun's* many re-installations.

Investigation

Lippold once said that 'If I should happen to fail in my work it is because I have forced a concept upon materials or because I have allowed materials and methods to impose their limitations upon me' (Campbell 1956). Unfortunately, the materials that he used, although they have not exactly failed, have manifested their limitations. The main problem with the condition of *The Sun* is the corrosion that both disfigures it and threatens its structural integrity.

Ironically Lippold had specifically chosen gold wire not only because its colour suggested the sun, but because it would not tarnish. However, small green pustules now cover the surface of every wire, creating a dull, rough and dark appearance. This completely alters the original, intended look of the sculpture. One author has written, 'light must bounce back from the intricate patterns of Lippold's wires, for otherwise the eye would pierce the work like a ray passing head-on through a prism' (Campbell 1956).

Problems with the condition of the piece were first noticed in 1981 while the sculpture was on long-term loan to the Air and Space Museum in Washington, DC. A small length of wire came loose, and it was noticed that the metal was actively corroding. When the piece was de-installed in 1984, examination revealed that the corrosion was pervasive. The examiners were in agreement that the corrosion products were copper chlorides (Kneeland 1984);[1] the present authors confirmed the presence of nantokite and paratacamite by X-ray diffraction analysis.

In order to discover the cause of the corrosion, we embedded two-centimetre lengths of the wire in epoxy resin and cut them into cross-, taper- and longitudinal-sections. Although Campbell, in his 1956 article, states that the piece is made from bronze plated with 22 carat gold, examination under the metallographic microscope revealed that the wire consisted of five distinct layers of metal. Elemental analysis with the scanning electron microscope equipped with energy dispersive spectrometry, or EDS, conducted by Mark Wypyski, Assistant Research Scientist at the Metropolitan, showed the make-up of these layers (Fig. 20.2). The 350 μm core of the wire is silicon bronze (94.1% copper and 3.6% silicon, with small amounts of manganese, aluminium and zinc). Over this is a 0.5 μm layer of silver, then a 5 μm layer of brass (93% copper, 6% zinc), an 8 μm layer of nickel, and finally a layer of gold (purity unreported) 1.5 μm thick (Wypyski 1993). The purpose of all these layers on a wire of such small diameter is unclear. It has been suggested that the silver may have been used as a lubricant in drawing the wire, or as a strike bath to prevent the deterioration of the core in the electroplating solution. Another suggestion is that nickel-plated wire might be too hard or brittle to draw thinly and evenly, and therefore a brass wire would have been drawn, and the

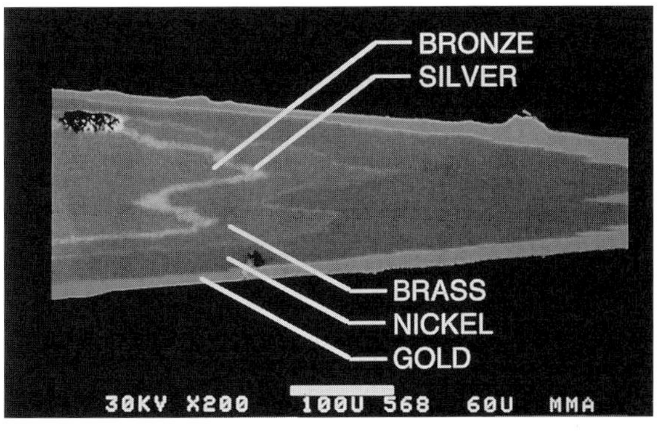

Figure 20.2 SEM photograph with added colour: taper section showing layers of plating. (Photo: Mark Wypyski.) (See Plate 158.)

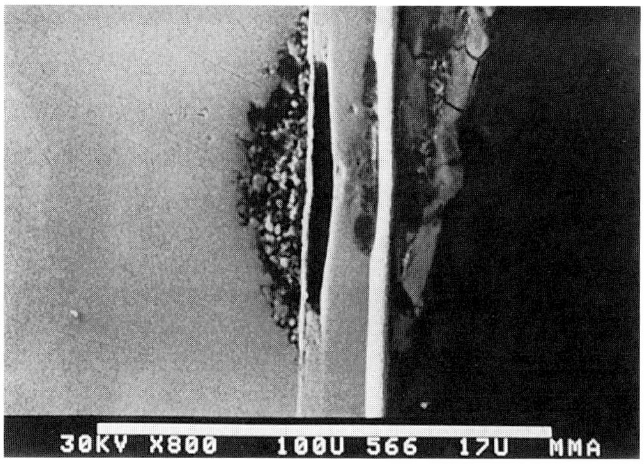

Figure 20.3 SEM photograph: longitudinal section through a pit. Note nearly intact gold and silver layers. (Photo: Mark Wypyski)

nickel and gold were plated on later (Lins 1995). However, none of these hypotheses account for all the layers. The artist has said that due to the inadequacy of the Metropolitan's commission, he was forced to use cheap materials. It is plausible that the wire provided to him by I. Stern and Company may have been intended to be sold as brass and was given a quick plating of nickel and gold.

Metallographic analysis also revealed the structure of the corrosion. Typically, the green pustules consist of mounds of loose material surmounting pits that penetrate through all four layers of plating and deep into the silicon bronze core. The copper alloys have been attacked preferentially, the nickel to a lesser degree, and the noble metals least of all; in many cases only small pinpricks have formed in the layers of gold and silver over much larger gaps in the underlying layers (Fig. 20.3).

Although it is outside the scope of this paper to explore in detail the electroplating process, certain aspects of the process bear examination (see Raub (1993) for an overview of historical aspects of electroplating, and Child (1993) for technical notes). Since its adaptation for commercial use in the mid-nineteenth century, the electrodeposition of metals on substrate surfaces has remained basically unchanged:

> Metal salts dissolved in an aqueous solution are reduced by a circuit current of electricity that is passed between two electrodes immersed in the solution. The negative electrode (cathode) has the metal plated onto it while the positive electrode (anode) dissolves into the solution (the electrolyte) replenishing the metal ions in solution.
>
> (Child 1993)

The variable factors in the practical application of this method are the composition of the metal salts, their concentration, current density, pH and temperature (Blum and Hogaboom 1949). Numerous formulae are available according to the thickness, density and composition of the desired plate (Rodgers 1959). In most cases, however, the plating baths for brass, silver and gold layers like those on *The Sun*, tend to be cyanide based, and the nickel bath is typically a 'Watts bath', which contains nickel sulphate, nickel chloride and boric acid (Burns and Schuh 1939).

The specific source of the chlorides that are causing the corrosion in *The Sun* is unclear. EDS analysis did not reveal a coherent layer of chlorides within the structure, and, in fact, most of the chlorides seem to be concentrated within the corrosion spots themselves. Chlorides may have contaminated the wire at any of several stages of the manufacturing process. In the 1950s, plating processes were, in general, much less reliable than now; some flaws could generally be tolerated for the machine parts and decorative objects being coated. Most of the advances in plating in the intervening years have been due to the needs of the aerospace and electronic industries.

One of the preparatory steps for plating is pickling; this may have been done with hydrochloric acid, one possible source of the chlorides (Child 1993). The chlorides may also have been co-deposited with metal during one of the plating processes; contamination of plating baths would not have been uncommon. The most likely source is the nickel-plating step. The balance of salts, temperature, pH and current density is essential for proper plating to occur, and if constant vigilance in replenishing the baths was not maintained, contamination of the deposited layers could have occurred (Foulke and Crane 1963). Additionally, the drawing of the plated metal would have required some sort of lubricant, which may have contained or been contaminated with chlorides.

It is also possible that the chlorides are derived from sources external to the wire. A sprayable flux may have been used during the construction process, which could have introduced chlorides to the metal.[1] It has also been noted that the sculpture is very susceptible to the collection of dust, and it may be that in the course of cleaning, chloride-containing agents may have been applied to the surface. The layered structure of the metals may have added to the problem, in that pit corrosion generated by the galvanic action of the metals may have allowed contaminants to breach the protective layers of gold and nickel and penetrate to the copper containing layers. Porosity of the plated layers may also have contributed by allowing oxygen and water, essential to the paratacamite reaction, to reach the copper alloys. If a substrate metal exhibits an irregular surface, such as may have occurred if Lippold's wire was stored for a length of time before being re-plated with nickel and gold, the incidence of porosity is greatly increased (Dini 1993).

Treatment options

After the nature of the problem was determined, we consulted with the artist to try to find a treatment plan that would be acceptable to him, to the curatorial department and to the objects conservation department. All agreed that in its present condition, *The Sun* is unexhibitable. Lippold's main concern was to see it once again installed, preferably in the museum that had commissioned it. Towards this end, several treatments have been proposed to make the surface smooth and reflective again. Local treatment was considered impractical due to the vast number of spots. We were thus forced to devise more unusual alternatives.

The first is the most radical. At one time, the idea of the artist completely remaking the piece from more stable wire was considered. He still has the original templates and could thus make an exact replica. Although he seemed to have no philosophical objection to this, Lippold was not eager to spend so much time on the project and did not seem to want to delegate this task to an assistant.

The second proposal was to replate the entire piece with a layer of nickel, followed by a layer of gold. While this would obscure the original surface, this treatment would not only be acceptable to the artist, but at the time we spoke was the option that he favoured. One fan-shaped unit was replated as a test in 1987, probably by David Epner of Epner Technology Incorporated.[1] When compared to an untreated piece, it has a dramatically improved appearance. However, this treatment would not address the problem of the active corrosion; the 'bronze disease' might well continue to eat into the copper alloys underneath the new surface. The replated wire, when examined with a binocular microscope, appears rather lumpy, with many small areas of iridescence or roughness where the new layer of gold covers the old corrosion. In longitudinal section it can be seen that a layer of nickel and gold cover the unexcavated pits. In addition, the extra strata of metal appear to be quite brittle, and shear away from the substrate wire when bent.

A third approach is to strip off the corrosion products in order to enhance the appearance of the original gilded surface. In the past, this kind of treatment was carried out on archaeological metal objects in order to remove thick corrosion layers (Organ 1961; Plenderleith 1962). This practice is generally now discouraged because the burial patina of these objects is seen as being part of the objects' history (Cronyn 1990), and because of the danger of etching the underlying metals (Merk 1978). *The Sun*, however, is a very different case, where the corrosion is simply disfiguring and destructive.

A series of experiments was conducted to determine the best treatment of *The Sun*, both in terms of its aesthetics and the underlying cause of the corrosion. Several stripping or inhibiting agents were tested: alkaline Rochelle salts (sodium potassium tartrate in an aqueous solution of 5 per cent sodium hydroxide), sulphuric acid, sodium carbonate, ethylene diamine tetraacetic acid (EDTA), benzotriazole (BTA). Samples of the corroded wire were immersed in a container of each solution for specific periods of time, rinsed in de-ionised water, and then examined under the stereomicroscope. In our initial series of tests, the most successful treatment was the application of alkaline Rochelle salts, which removed the corrosion from the surface within half an hour, leaving the gold layer unchanged. The other reagents took longer to remove the corrosion products, or else, in the case of the acid, etched the gold surface (see Table 20.1 for reaction times and concentrations of reagents.)

An additional element was added by testing corroded wire in a bath of alkaline Rochelle salts subjected to bursts of ultrasonic vibration. This treatment was even more successful, taking only minutes to remove the corrosion products. Because of the stratified structure of the pits, very little of the brightness of the gold was lost, since the diameters of the gaps in the gold under the mounds of corrosion products were so small.

Examination of a section of the treated wire under the metallographic microscope showed that the superficial corrosion had indeed been stripped off, leaving the substrate metals essentially intact. Our next step was to test one of the smaller sections of the sculpture using alkaline

Table 20.1 Effect of stripping and inhibiting agents on corroded wire from Richard Lippold's *Variations Within a Sphere #10: The Sun.*

Reagent	Time required for reaction	Effect on corrosion products	Visible effect on wire
Alkaline Rochelle salts (5 g NaOH, 15 g sodium potassium tartrate in 100 ml H_2O)	30 minutes	Completely removed	None
Sulfuric Acid (5% in H_2O)	Within 20 seconds	Mostly removed	Gold surface etched
Sodium Carbonate (5% w/v $NaCO_3$ in H_2O)	24 hours	Partially removed	None
Ethylene diamine tetraacetic acid (5% w/v disodium salt of EDTA in H_2O)	3 hours	Mostly removed	None
Benzotriazole (1% w/v BTA in H_2O)	24 hours	Very little removed	None
Alkaline Rochelle salts with ultrasound	3 minutes	Completely removed	None

Rochelle salts in the large ultrasonic bath in the Costume Institute at the Metropolitan Museum. This unit, manufactured by the Crest Ultrasonic Corporation in 1985, has an interior dimension of six feet by four feet and can be filled to a depth of between about four to eight inches and thus can accommodate most of the large pieces of Lippold's sculpture. This treatment was again quite successful in removing the surface corrosion. At this point, the option of using Rochelle salts seemed very attractive because of the savings of time and money, and because from our point of view, preserving the original surface was preferable to either replating or remaking the sculpture. When he was shown samples of the treated wire, the artist agreed that the process was successful in returning the wire to the bright surface that he desired. However, before committing ourselves to treating the large central sections of the piece we submitted our treated samples to further tests. To determine whether the corrosion would recur under conditions of elevated temperature and humidity, a segment of treated wire was placed in a sealed jar with an open container of water and subjected to temperatures of 60–66°C for a period of two months.

Unfortunately, at the end of that period, green pustules had again appeared on the wire, and small samples tested positive for chloride using the standard silver nitrate test. Although we had hoped, due to the size and awkwardness of the pieces, that a single-step treatment would be feasible, further experiments to remove the underlying cuprous chloride or to inhibit further corrosion may need to be undertaken. The option of coating the wire is also a consideration. However, we do feel that the use of both Rochelle salts and the ultrasonic bath were successful, as far as cleaning the surface corrosion from the wire. It should be noted that in addition to the problems of 'bronze disease' addressed in this chapter, *The Sun* also requires other kinds of treatment for wire that has become twisted and deformed through rough handling and poor storage conditions. A great deal of straightening, re-positioning, and rewelding needs to be done before *The Sun* can be returned to its original coruscating condition.

Conclusions

In conclusion, this project illustrates some of the problems that arise when dealing with art of the industrial age, when new materials and processes have been used whose properties the conservator as well as the artist may not entirely comprehend. Added to this factor is the effect of enlarged scale, which magnifies the problems inherent in the materials. In cases such as this one, it may be necessary for conservators to reconsider their methods, and adapt old treatments to new situations.

Although the use of stripping agents such as alkaline Rochelle salts has been out of favour recently because of the danger of etching the substrate metal, and because of the importance of preserving the patina as a record of the history of the object, the results of our experiment suggest that such agents may be of significant utility to conservators faced with a gold- or silver-plated object with underlying corrosion. Obviously, precautions must be taken to ensure that the gilded surface is not endangered or that the aesthetics of the object are not altered. Similar caveats are also in order for the use of ultrasound baths. Clearly, if the object is of a fragile or friable nature, it should not be subjected to the shock of ultrasonic treatment. However, when a piece is structurally strong, ultrasound can serve to speed treatments that would otherwise take many hours to complete.

Acknowledgements

The authors would like to thank James Frantz for his inspiration and direction of this project, as well as Peter Champe, Cindy Connelly, Hiroko Kariya, and Megan Thumm for participating in early discussions of the work, Chris Paulocik of the Costume Institute for the use of the ultrasonic unit, and finally Richard Lippold for sharing his time and expertise with us.

Note

1. MMA files. Department files. New York: Sherman Fairchild Center for Objects Conservation, Metropolitan Museum of Art.

References

Blum, W. and Hogaboom, G. B. 1949. *Principles of Electroplating and Electroforming*. New York: McGraw-Hill.

Burns, R. M. and Schuh, A. E. 1939. *Protective Coatings for Metals*. New York: Reinhold Publishing Corporation.

Campbell, L. 1956. Lippold Makes a Construction. *ARTnews* **55** (October): 30–34, 63–4.

Child, R. 1993. 'Modern electroplating and electrofinishing techniques'. In *Metal Plating and Patination: Cultural, Technical and Historical Developments*, S. La Niece and P. Craddock (eds), 291–300. London: Butterworth-Heinemann.

Cronyn, J. M. 1990. *The Elements of Archaeological Conservation*. London: Routledge.

Dini, J. W. 1993. *Electrodeposition: the materials science of coatings and substrates*. Park Ridge, N.J.: Noyes Publications.

Foulke, D. G. and Crane, Jr., F. E. (eds) 1963. *Electroplaters' Process Control Handbook*. New York: Reinhold Publishing Corporation.

Kneeland, Y. 1984. Unpublished memo. Department files. New York: Sherman Fairchild Centre for Objects Conservation, Metropolitan Museum of Art.

Kramer, H. 1956. 'Month in review'. *Arts* (October): 50–51.

Lins, P. A. 1995. Personal communication.

Lippold, R. 1993. *Conversation with Richard Lippold*. New York: Metropolitan Museum of Art (23 November 1993).

Merk, L. E. 1978. 'A study of reagents used in the stripping of bronzes'. *Studies in Conservation* **23**(1): 15–22.

Organ, R. M. 1961. 'The examination and treatment of bronze antiquities'. In *Recent Advances in Conservation, Contribution to the IIC Rome conference, 1961*, G. Thomson (ed.). London: Butterworth.

Plenderleith, H. J. 1962. *The Conservation of Antiquities and Works of Art*. London: Oxford University Press.

Raub, C. 1993. 'The history of electroplating'. In *Metal Plating and Patination: Cultural, Technical and Historical Developments*, S. La Niece and P. Craddock (eds), 284–90. London: Butterworth-Heinemann.

Rodgers, T. M. 1959. *Handbook of Practical Electroplating*. New York: Macmillan.

Wypyski, M. 1993. *SEM/EDS Elemental Analysis – Summary of Results*. Unpublished typescript. New York: Sherman Fairchild Center for Objects Conservation, Metropolitan Museum of Art.

21 Preservation of a Gilded Monumental Sculpture: Research and Treatment of Daniel Chester French's *Quadriga*

Linda Merk-Gould

Abstract

The research undertaken for the *Quadriga*'s conservation treatment, which used both historical documents and analytical methods, is presented in this chapter. The objective of the research was to understand the artist's intent and to determine the appropriate materials to use in the conservation treatment, i.e. ones consistent with the artist's intended appearance. A broader description of the entire conservation project, including the structural repair challenges, is not part of this chapter.

Gilding at the turn of the century

During the city beautification period of 1890 to 1930, thousands of sculptures were placed in US city parks, around and on government buildings. Daniel Chester French is one of the two most recognised American architectural sculptors of this time period, the other being Augustus Saint Gaudens. During this period, fewer than twenty of these monuments were gilded, including French's gilded, copper *Quadriga*. In developing the conservation plan for this sculpture, questions arose as to the artist's intent and the materials originally used.

Saint Gaudens has been credited with being the first sculptor in 1891 to gild an outdoor sculpture – *Diana* for Madison Square Garden in New York City. In *Reminiscences* (Saint Gaudens 1913) he comments, 'I am sick of seeing statues look like stove pipes' and he 'longed for the unusual combination of a gilded bronze and a cream-colored base'. However, gilding did not appear to be a permanent solution to this problem. As noted in the *Monumental News* of July 1908, after only five years of weathering, Saint Gaudens' 1903 *Sherman* sculpture in New York was 'showing great brown patches through the gilt on the flanks and shoulders of the horse'.

Although Saint Gaudens may have installed the earliest gilded statues in the USA, French appears to have been more prolific. There is no indication of such a short duration for the gilding on French's statues; however, the use of picture frame gilders without the artist's direct supervision (French 1908) in undertaking the gilding leads one to examine the materials used very closely and evaluate the effect achieved in contrast to the artist's desired intent.

Whether it was the *Alma Mater* in 1903 or the *Quadriga* in 1907, the common public reaction was a dislike for the gold surface's brilliance (French 1907) . Controversy regarding gilded statues dates back at least to the Roman Period, when Emperor Nero ordered the gilding removed from the bronze statue of Alexander and declared that it looked better without the gold – even though the sculpture was then marred with the grooves cut into it to hold the gold foil in place . To understand the concern regarding the 'overly' bright appearance, comparative photographs of viewing the same gilded monument in full sunlight (Fig. 21.1) and on an overcast day (Fig. 21.2) shows the difficulty of 'seeing' beyond a glaring surface – even when much of the gold leaf had worn off after fourteen years. The controversy that surrounded the 1989 regilding (no toning was done) of Saint Gaudens' *Sherman Monument* in New York – once again it was accused of being too bright! – illustrated the need for thorough research prior to undertaking a similar conservation treatment. Incidentally, during the research on the *Quadriga*, correspondence was located in which French was advised to tone the *Quadriga* darker than that on the *Sherman Monument* (Gilbert 1906).

Background on the fabrication of the Quadriga

The *Quadriga* was created by French and Edward Potter and fabricated at the John Williams Foundry in New York between 1903 and 1906. In his large commissions, French often collaborated with Potter. French undertook the figurative sculptures while Potter undertook the equestrian components. The John Williams Foundry, where the sculptures were made, was one of the quality US foundries of that time period. The gilding by Bazille and Partridge Inc. of St. Paul, Minnesota was undertaken in the winter of 1907. The question as to why the gilding process occurred during freezing winter months will be addressed later in this chapter.

The *Quadriga* is a sculptural group comprising a man standing on a chariot, which is pulled by four horses. Two women stand between the horses' heads. The horses are allegorical representations of the elements (earth, wind, fire and water) and the women represent agriculture and industry. The overall area occupied by this sculptural group is 21 by 13 feet and 25 feet in height. The figures' heights average between 10 feet and 12 feet. The copper has an average wall thickness of 0.059 inches (17 gauge),

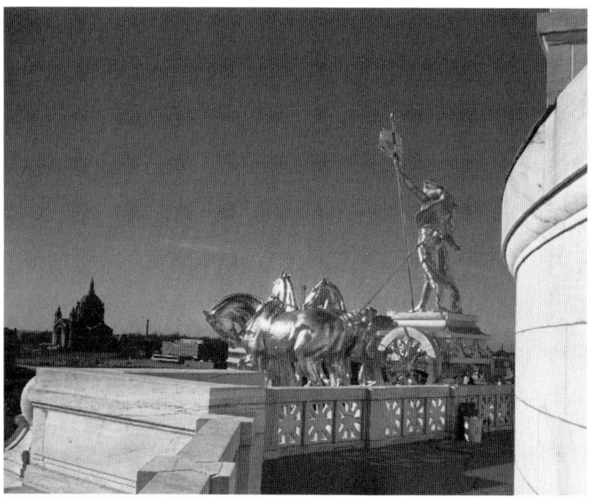

Figure 21.1 View of the *Quadriga* on a sunny day in 1993. (See Plate 159.)

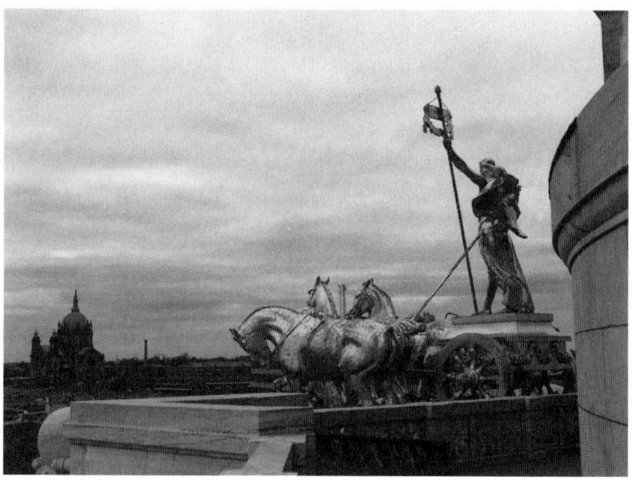

Figure 21.2 View of the *Quadriga* on an overcast day in the same week in 1993. (See Plate 160.)

which is about twice as thick as the 0.032 inch (22 gauge) copper often used on roofs.

This statuary group is one of approximately 11 gilded monuments/ sculptures by French. The D.C. French Archives at Chesterwood provided a list of seven (Styka 1994) that still exist; however, not all are presently gilded – of note, the *Alma Mater* at Columbia University. Research in other archives has increased that 'known' list to eleven. Others that may have existed have either been destroyed or are presently in private collections unknown to the archives. Nine were outdoor monumental-scale sculptures and two are likely to have been smaller statuettes, presumably for indoor display. No indoor monumental-size gilded sculptures, which could have been a useful reference for the artist's desired surface appearance, are known to exist.

The quality and importance of this statuary group within French's *oeuvre* cannot be undervalued. In a 1903 letter to Cass Gilbert, the architect for the Minnesota State Capitol building, French refers to the *Quadriga* as 'virtually a copy of the Chicago one' (French 1903). The reference to 'the Chicago one' concerns one of the monuments that French was commissioned to make for the 1893 World's Columbian Exposition held in Chicago. The composition of the 1893 Chicago sculpture is known only through an engraving as it was destroyed in a fire in January 1894. For the Minnesota *Quadriga*, the two mounted attendants on the Columbian Exposition *Quadriga* were eliminated and the central figure changed (Museum of American Art, no date)

Historic references for gilding the Quadriga

Within this framework, period records from 1903 to 1907, including French and Gilbert's letters, located in New York, Washington and St. Paul, Minnesota, were reviewed to document French's intent for the *Quadriga*'s gilded appearance. The actual work performed was documented using correspondence between the gilders, Bazille & Partridge, Capitol Board of Commissioners, the client, and the architect's St. Paul office. By combining information from these sources, an overall description of the gilding materials has been compiled. Correspondence indicates that the gilding was requested by Gilbert to be 'dull gold, and toned to a color perhaps similar to the *Sherman Statue* on the Plaza, New York or to your *Alma Mater* at Columbia' (Gilbert 1906).

From research in the Daniel Chester French Papers at the Library of Congress, it is known that the *Alma Mater* had a yellow primer used under the gilding, as its visibility was referenced in 1928 by French as an indication that it needed regilding (French 1928). However, no reference to the size used could be located nor could a period reference to the primer used on the *Quadriga* be found.

It is clear from the correspondence on all of French's gilded sculptures that he preferred a matte gold appearance. In a letter to Gilbert, French discussed the *Quadriga* and described the desired effect of the gilding: 'The glitter is removed by treating the surface after the gilding is finished. I found wax dissolved in turpentine gave the right effect' (French 1906b).

Documents from both Gilbert and Bazille & Partridge, Inc., who carried out the original 1907 gilding, indicate that it was intended that the recessed parts of the *Quadriga* be shaded (Gilbert 1906, Bazille 1906). A letter from French to Gilbert (French 1906a) discusses the gilding and that 'this effect can be easily produced by giving the gold a coat of wax and turpentine with a little oil color'. The 28 December 1906 price quote by Bazille & Partridge indicates that this work was to be done.

The records indicate that French did not include gilding in the base contract of his commissions and did not personally supervise the application of the gilding. In fact, French did not conduct a personal on-site review of the *Quadriga* before it was determined finished by the State of Minnesota Board of State Capitol Commissioners, the client and Mr. Carsley, the architect's representative.

Even with the gilding shaded in the recesses and a matte coating applied, Gilbert wrote to French that 'The color of the gilding is fine but still too bright' (Gilbert 1907). Others shared this opinion, including Channing Seabury, Vice President of the Board of State Capitol Commissioners (Seabury 1907). It is hard to know if this reaction was due to the sculpture's sharp contrast to the coal-soot darkened buildings of the early 1900s or a common reaction to such a dramatic visual appearance. Whichever, it indicates that the American public did not consider a brilliant surface aesthetically desirable at the turn of the century.

A sculptor's desire to have a gilded sculpture 'readable' in brilliant sunlight may date back to the *San Marco Horses* in Venice, which are covered with a network of fine scratches. Vittori and Mestitz state that the artist:

> must have understood that by varying the gold reflectivity according to the movement of forms he could succeed both in remarkably decreasing the glare effect and recreating the original perspective of his work of art. (Vittori and Mestitz 1975)

The maintenance history research on the *Quadriga* gives a good indication that if any such lines existed on its surface, they are likely to have been 'ground off' during the 1949 gilding campaign.

Maintenance history contribution to the gilding research and treatment decisions

With the available historic information on the *Quadriga*'s gilding detailing only some of the processes, there was a need for analytical work to determine original primer and toning colour, if possible. The maintenance/repair history files at the Minnesota State Capitol Administrative offices were important in directing the locations from which to take the samples.

Records in the Curator's office and a photograph, dated 5 August 1949, from the St. Paul Pioneer Press Archives indicate that the *Quadriga* was regilded in 1949. The photographs show outlines of 'gold-leaf squares' that remain while the workers are 'cleaning off' the surface with electric grinders. The use of grinders to take the surface to bright metal directed our analytical sampling sites to tight drapery contours into which such grinders could not reach.

Another gilding campaign occurred in 1979. The planned work involved erecting scaffolding, cleaning the sculpture, sanding to remove scratches as needed, resoldering open seams, applying primer to all areas (visible in 1993 as yellow coloured; probably zinc chromate), applying size, 23 carat gold leaf, and one top coat of exterior grade varnish (Brooks 1979).

During the 1993–94 inspections, the surface revealed extensive areas with double layers of primer-gilding, indicating that the actual work plan may have evolved into gilding on top of the existing gold leaf. The 1979 work omitted gilding of the less visible and less accessible surfaces, like the top of the chariot and the underside of the chariot pull bar. These observations also influenced the sampling site and interpretation of the cross-section information.

Quadriga *sample location and identification*

The locations from which samples for analysis were removed were the bottom edges of both ladies' drapery – close to the copper pedestal roof and difficult to reach.

The microscopic cross-section view of the sample (Fig. 21.3) from the proper right lady included copper corrosion likely to be associated with the 1906–7 winter exposure before it was gilded (Welsh 1994). With the *Quadriga*'s installation having been completed in December 1906, and snow appearing in the photographs of the new installation, it is likely that the coppery jewel of a monument was corroding green. Bazille & Partridge undertook the gilding between mid-February 1907 and early April 1907 (Carsley 1907a,b,c).

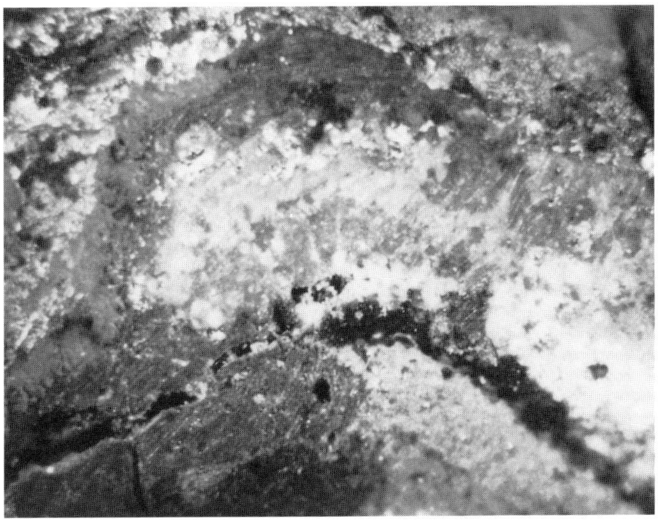

Figure 21.3 Cross-section from the *Quadriga*, removed from the proper right lady's drapery, bottom rear edge. (Source, Frank S. Welsh.) (See Plate 161.)

In the cross-section the three gilding campaigns dated from the records as 1907, 1949 and 1979 are visible as follows.

The 1907 gilding layers appear to have been:

1. reddish-orange thick lead oil paint layer
2. thin light gray oil paint layer
3. thin yellow oil size
4. gold leaf
5. evidence of surface glaze, now amber in color

The 1949 gilding layers appear to have been:

1. medium gray colored primer
2. yellow size
3. gold leaf

The 1979 gilding layers appear to have been:

1. yellow ground, presumably zinc chromate
2. yellow size
3. gold leaf
4. surface varnish (?)

With historic documents indicating 30–40 years between the gilding campaigns, the number of layers in the cross-section information confirms that this was likely. The use of a medium gray coloured primer in

1949 can be associated with an attempt to replicate the original top primer colour, which had darkened over time.

To enable future researchers to have access to original gilding samples, the man's left ear and behind his right hand were not cleaned to the copper substrate during this current project, but rather had the primer applied over the existing surface.

Client's decision for 1995 gilding materials

The artist's intent for the *Quadriga* to have been coated to reduce the gold's brilliance and his desire for toning to achieve sculptural dimensionality on an otherwise sunlit, glaring surface were a primary factor in selecting a surface treatment for the sculptural group.

With the late-twentieth-century recognition of lead hazards, there was a conscious decision not to reapply the highly preservative, but highly toxic red lead primer. The philosophy of the client, Project Architect Richard Cottle, Curator Carolyn Kompelien and Conservator Robert Herskovitz for the State of Minnesota, which formed the basis for the treatment decisions for this gilded outdoor monument, included:

- to preserve as much of the original material as possible, which was especially important in the structural work
- to replicate the original materials using the least hazardous products with proven longevity
- to remove all existing primer and gold leaf to a cleaned copper substrate in order to obtain good adhesion
- not to use experimental materials or modify products in ways not approved by the manufacturers
- to select a primer colour that would not be visually disturbing if legislative funding for regilding were delayed.

This philosophy was determined to be appropriate for the *Quadriga* and was developed based on an awareness of other recently gilded outdoor monuments. Two such gilding projects that were available for close surface inspection and discussion with the client and/or gilders were French's 1918 *Republic* in Chicago and Saint-Gauden's 1903 *Sherman Monument* in New York.

French's 25-foot high *Republic* in the Chicago Parks District was regilded in 1992 by Gold Leaf Studios. Active corrosion occurred on *Republic* after the cleaning and before the gilding was applied, which necessitated multiple cleaning processes. Using a series of cleaning methods from walnut shells to a mixture of walnut shell and limestone blasting media, to pressurised water to glass bead peening – the surface was cleaned to obtain bright metal (Dajnowski and Miller 1995). This was followed by an application of Incralac as a primer, gilding with Italian 23.5 carat leaf with no surface matting or toning (Adair 1995). Pitting through the gold leaf occurred within the first year of treatment and was visible from ground level in 1995 using a 75–300 mm telephoto camera lens.

The *Sherman Monument* was regilded in 1989 by LMC Corporation using brass brushes, pressurised water and walnut shells to remove corrosion, without cleaning to bright metal. The surface was coated using Devran 201 epoxy primer, widely used in France, and then gilded with German 23.5 carat gold leaf with a gelatin top layer (Wiart 1989). Delamination of the primer from the bronze substrate occurred on the *Sherman Monument* within the first year after its completion. Even with annual touch-up gilding, delamination of the gold leaf continues to occur both in new areas and previously treated areas (Rabinowitz 1995).

Both projects experienced problems with corrosion and/or gold leaf delamination. The sources of the problems are likely to be multi-faceted and research may be able to clarify the issues – to the benefit of future gilding projects. Awareness of the problems encountered in these projects, strengthened the plan to use a traditional primer in this conservation project.

For the *Quadriga* gilding, the client was influenced by the desire to avoid possible thermal-induced delamination of either Incralac or epoxy primers given the sculpture's location in Minnesota with its severe winters. Neither Incralac nor zinc chromate, pigmented to replicate the 1907 red lead and gray layers, were considered due to the manufacturer's warranty being voided should their products be used in such a manner.

Potential impact of materials used in outdoor gilding

Although high carat gold leaf is recognised for its resistance to discoloration or tarnishing, the beating process which produces the gold leaf typically used on outdoor sculptures has been mechanically altered to the extent that it is no longer a solid film. As noted by Lins in the gilded wood symposium publication, 'after attachment with adhesive to a substrate, leaf is further torn and suffused with size; substrates and adhesives respond differently than leaf to changes in relative humidity and temperature' (Lins 1991). Similar stress occurs in the outdoor environment, compounded by the thermal expansion as the different layers react to the sun's heat.

The resultant porosity in the gold layer has a great impact on its corrosion resistance in the outdoor environment. The longevity of the gilding can be influenced by the selection of the other materials in the gilding process. Although nontraditional materials have been used successfully on indoor objects (Thornton 1991), their use in the outdoor environment should be closely tested and evaluated before application. When problems occur, the contributing factor(s) in such a complex system can often be difficult to identify precisely. Surface preparation/cleaning, primers, size and gold-leaf quality, as well as the desirability of a protective coating, all need to be considered.

To ensure good adhesion of the new gilding, the decision was made to remove the multiple gold-leaf layers to the copper substrate. The choice of a proper primer was considered carefully. With the *Quadriga* being a site-specific sculpture – and that site being outdoors on the roof of the Minnesota State Capitol building – the properties of epoxy, synthetic resin and more traditional zinc chromate and red lead as primers were reviewed.

Given the severity of Minnesota winters and summer sun, the compatibility of the gilding layers had to be considered. The temperature variation of the copper sculptures can range from –30°F in the winter to 140°F in the summer. The various layers' coefficients of thermal expansion was a factor to consider to achieve the closest similarity as possible to reduce delamination or cracking in the layers. Marabelli's work on the gilded, Roman Period *Marcus Aurelius Monument* documented with acoustic-emission measurements the stresses caused by the sun's heat unevenly warming sections of the monument (Marabelli 1994).

A review of information in Table 21.1 shows a significant difference in thermal expansion between typical metal substrates found in gilded, outdoor sculptures and epoxy or methacrylate coatings used for isolating the copper substrate from the gold leaf. Difficulties encountered with epoxy resins when used as a coating include 'The more chemical and solvent

Table 21.1 Comparison of thermal properties of layers in a gilded sculpture* at 68°F.

Material	Thermal conductivity (cal cm^{-1}) (°C cm^{-2} s^{-1})	Thermal expansion (in/°F)	Average modulus of elasticity (psi)
Bronze (95% Cu, 5% Sn)	0.2	10×10^{-6}	16×10^{6}
Copper (99.9)	0.95	9×10^{-6}	16×10^{6}
Epoxy	0.0004	40×10^{-6}	0.5×10^{6}
Methacrylate	0.0005	50×10^{-6}	0.5×10^{6}

*adapted from Van Vlack (1967).

resistant [ones] are also less flexible, more brittle and generally have poorer adhesion . . . and that poor intercoat adhesion may result' (Tator 1983: 376), as noted in the Corrosion and Corrosion Protection Handbook.

The importance of a solid barrier layer between the cathodic gold, however porous, and the more anodic copper is essential to avoid galvanic corrosion (Van Vlack 1967: 338) where breaks in the gold layer will inevitably occur, given the nature of gold leaf. The inclusion of zinc chromate (or other colourant) in a primer allows for both a differently coloured layer to facilitate quality control for a continuous layer, and a change in the electrochemical balance of the layers. Such a barrier layer separates the dissimilar metals, thereby eliminating the metal substrate from which the sculpture is fabricated from becoming a sacrificial anode. Of course, with an unmaintained monument, any primer will eventually weather off.

In the *Corrosion and Corrosion Protection Handbook*, Tator lists characteristics for conventional inhibitive pigments such as zinc chromate to include 'neutralizes film acids; anodic passivator; renders water noncorrosive'(Tator 1983: 368). With the current state and federal regulation governing exposure to lead, red lead primer, which is part of this group of materials, was not considered for use on the *Quadriga* project.

Whatever the primer selected, application to a well cleaned, corrosion-free surface is critical to good coating performance. As for the need for a protective coating on top of a gold surface, Marcel Pourbaix's comments emphasise that with the 'combination of a polluted atmosphere and a porous gold coating, the structure will be destroyed due to the high electropotential'. A coating to seal pits and breaks in the gold layer, may be even more important when the substrate is a pitted bronze monument as he was discussing during the mid-1970s at the National Bureau of Standards conference reported by Ogburn and Passaglia (1977: 178).

Gilding – sample panels and treatment

To aid in selecting materials that would be appropriate to simulate the original surface qualities of the gilding, three sample panels were prepared for the client's review. On the panels in Figure 21.4, primer, gold, matte wax and toning colours were applied using a different (from the 1907 original) primer and different colours for the final toning layers. Two panels were prepared using US Environmental Protection Agency (EPA) approved materials, which are less toxic than the lead-based primers in use in 1907. A third, red-coloured primer was applied to a panel to simulate the appearance of the materials applied in 1907, as revealed by the analysis work. The panels were prepared by Evergreen Painting Studios of New York, who would undertake the gilding work in the author's conservation facility.

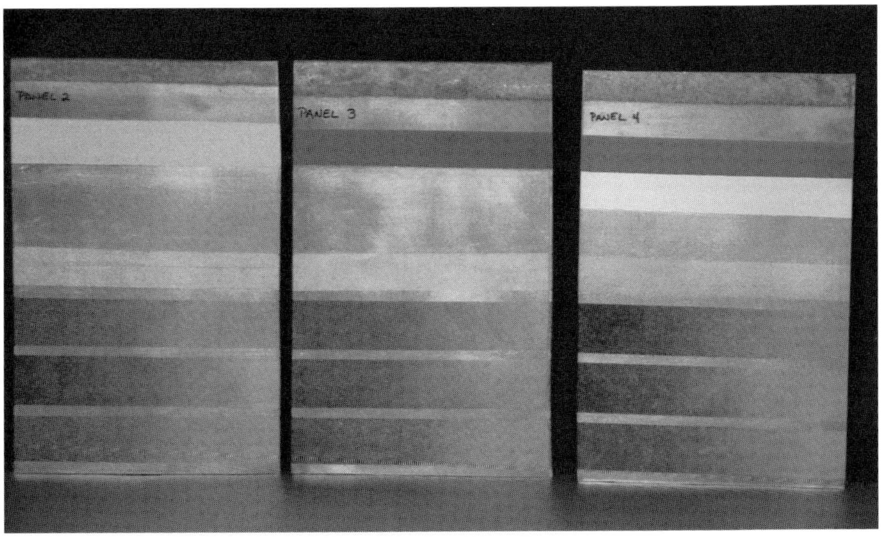

Figure 21.4 Gilding sample test panels prepared for the client: 1. corroded surface; 2. cleaned surface with residual pitting; 3. primer; 4. gold leaf; 5. wax as matting agent; 6. three colours as toning choices graded from dark to light. (See Plate 162.)

Consideration was given to preparing samples representative of both the smooth and more irregular contours. To simulate the broad flat areas, flat panels were cut from copper that had been treated with a commercial acidic patina and aged for three years by the author to 'replicate' environmental weathering of the copper outdoors. As anticipated after cleaning, the copper surface was slightly pitted. This was useful as it would take the gold in a manner more similar to the 1906 copper sculptures.

For consistent surface treatment and similarity to the planned conservation treatment of the *Quadriga*, all surfaces were cleaned with pH neutral, non-chloride solvent gel systems (methyl 2-pyrrolidone, Back-to-Nature IV) and then de-greased with acetone.

The three primers applied to the copper panels were:

1. zinc chromate, yellow in colour, in an alkyd enamel paint system manufactured by Sherwin Williams
2. red XIM, an alkyd enamel primer, which is darker than a red lead primer
3. zinc chromate tinted to simulate the 1907 red lead colour with an intermediate light gray alkyd enamel. The colour tone was hand mixed to match the original lead-based oil primer, Munsell and CIE indexed during the analytical work.

It was the clients' decision, after viewing the sample panels in sunlight and indoors to simulate cloudy weather, that the different primer colours did not alter the colour of the gold leaf applied over it. Over time, the primer colour would become visible as the gold wears off and only then become apparent. It was decided that since the original primer would not be used, that the yellow zinc chromate be applied as it would be more visually 'forgiving' should the cyclical regilding not be able to be funded during an economic downturn.

With there being significant deterioration to the structural support system of the sculpture and a major re-roofing project scheduled for the building, the *Quadriga* was removed from its rooftop location for the conservation treatment. This enabled the work to be undertaken in the conservation facility. During the full-scale gilding, the double weight, 23.5 carat Italian Manetti gold leaf was adhered to the zinc chromate primer with LeFranc 12 hour slow size. A matte gold finish was obtained with a microcrystalline–carnauba wax mixture combined in a 4 to 1 ratio and worked to a brushable paste with gum turpentine.

The various shading samples were achieved by adding Windsor-Newton Artist Oil Colours to the wax mixture and brushing it on to the gilded

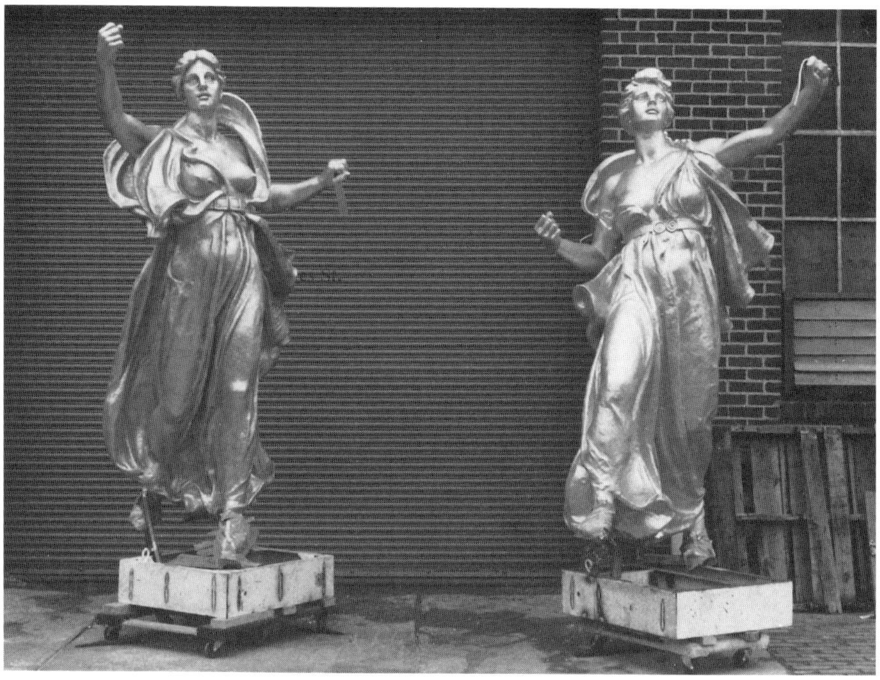

Figure 21.5 Comparison of toned, gilded lady on left and gilded lady without toning, French's *Quadriga*, during the conservation treatment. (See Plate 163.)

surface. Variations on a warm brown colour were tried, with varying amounts of raw umber, burnt umber and raw sienna. The toning did bring out the sculptural form as visible in the comparison view (see Fig. 21.5) with one female statue waxed and toned and the other gilded but before the matte wax layer was applied.

In gilding outdoor monuments, attention must be given to minimise holidays (breaks in the gold layer) and to be vigilant in maintenance of the gold leaf and/or its coating to avoid accelerated galvanic corrosion caused by gaps in this gold layer. As for all outdoor monuments, high quality workmanship and a dedicated maintenance programme is essential for preservation.

Summary of the conservation treatment

The gilding methodology for the *Quadriga* was to remove the 1949 and 1979 gilding layers to achieve a clean metal surface and to thoroughly degrease the copper. Two remote locations – the man's left ear and a recessed area by his right hand – were not cleaned as these were likely areas to have the 1907 gilding still intact. Zinc chromate primer, slow size, and 23.5 carat gold leaf were applied in the conservator's studio. After waiting several weeks to allow the new size and gold layers to dry, matting of the gold was achieved with an overall wax layer, followed by dark toning of the recesses using oil paint in wax. The materials used to achieve the dark toning followed those identified in the artist's correspondence.

The choice of zinc chromate primer in lieu of the original red lead primer was an acknowledged deviation from the original materials for two reasons: to avoid lead exposure and to avoid the problems that had been experienced when using nontraditional materials on other gilded monuments in northern US climates.

Although the selected undertone colour would differ from the original, the yellow coloured primer would be less visible should legislative funding cycles not enable regilding as soon as the gold leaf has been environmentally abraded, for example from airborne construction materials.

With the planned annual maintenance of washing off airborne-salts and replenishing weathered wax, it is anticipated that the appearance of the *Quadriga* will continue, close to the artist's intent, for several decades before regilding is needed.

Conclusion

The present condition of French's *Quadriga* supports the treatment choices that were made and the successful implementation of the annual maintenance plan. In 1999, the visual appearance of the sculptures is similar to the post-conservation treatment photodocumentation with the pigmented wax toning not having significantly weathered off. The adhesion of the gilding layers, including the zinc chromate primer to the bronze substrate, remains sound. No delamination is apparent after four years, in contrast to the two projects referenced in the text of this paper in which delamination occurred during the first year of outdoor exposure.

The effects of not regularly following the recommended maintenance of rinsing dirt and airborne salts off the surface of the sculptures early each spring has resulted in localised pitting corrosion on the undersides of the horses and drapery folds. Maintenance work has been carried out by the Minnesota Historical Society conservation staff to address this problem. The toned surface layer has not yet needed to be replenished.

Acknowledgements

A project of this scope involves numerous individuals who jointly supported such a conservation undertaking. I would like to thank:

- The client, the Minnesota State Capitol Curator, Carolyn Kompelien, and Department of Administration Architect, Richard Cottle, for supporting the research; so important in establishing historical accuracy in a conservation treatment, yet so often cut for budgetary reasons.
- Robert Herskovitz, the client's representative during the treatment, for critical reviews and constructive reconsideration of decisions as new information was revealed.
- The New York Historical Society for enabling access to the Cass Gilbert Papers during a highly understaffed time period.
- Conservation colleagues – Mark Rabinowitz, Andrzej Dajnowski and William Adair – for sharing documentation and discussions of both the successful and problematic aspects of projects which furthers all of our knowledge.
- Evergreene Painting Studios for their commitment to thoroughness in gilding and ease to work with in planning and implementing the gilding phase of the conservation treatment.
- The Conservation Technical Associates team of conservators, Chris Blum, Francis Miller and especially Joseph Sembrat, without whose enthusiastic commitment to achieve a comprehensive treatment plan and execution, this project would not have achieved its high level of success.
- And to my family, whose weekly Saturday visits to see the 'golden horses' became part of my children, Rebecca and Katrina's enjoyment for a year – as memorable to them as for the generations of Minnesota school children whose visits recommenced with the *Quadriga* reinstallation on June 21, 1995.

References

Adair, W. 1995. Telecommunication with Gold Leaf Studios, April. Washington, DC.

Bazille, J. A. 1906. Letter to Board of Capitol Commissioners, 28 December. Minnesota History Center. St. Paul, Minn.: Minnesota Historical Society.

Brooks, R. 1979. Curator's Office Files. Letter with gilding bid from Saint Paul Statuary Company, 15 March. St. Paul, Minn.: Minnesota State Capitol Building.

Carsley, G. H. 1907a. Cass Gilbert Papers. Letter to Cass Gilbert, 13 February. New York: New York Historical Society.

Carsley, G. H. 1907b. Cass Gilbert Papers. Letter to Cass Gilbert, 8 March. New York: New York Historical Society.

Carsley, G. H. 1907c. Cass Gilbert Papers. Letter to Cass Gilbert, 3 April. New York: New York Historical Society.

Dajnowski, A. and Miller, F. 1995. Telecommunication with The Chicago Parks District, April. Chicago, Ill.

French, D. C. 1903. Cass Gilbert Papers. Letter to Cass Gilbert, 5 September. New York: New York Historical Society.

French, D. C. 1906a. Minnesota History Center. Letter to Cass Gilbert, 7 April. St. Paul, Minn.: Minnesota Historical Society.

French, D. C. 1906b. D. C. French Papers Archives. Letter to Cass Gilbert, 27 September. Washington, DC: Library of Congress.

French, D. C. 1907. D. C. French Papers Archives. Letter to Cass Gilbert, 15 May. Washington, DC: Library of Congress.

French, D. C. 1908. D.C. French Papers Archives. Letter to Cass Gilbert, 27 September. Washington, DC: Library of Congress.

French, D. C. 1928. D. C. French Papers Archives. Letter to Albert W. Putnam, Esq., 20 April. Washington, DC: Library of Congress.

Gilbert, C. 1906. Minnesota History Center. Letter to D. C. French, 4 April. St. Paul, Minn.: Minnesota Historical Society.

Gilbert, C. 1907. Cass Gilbert Papers. Letter to D. C. French, 14 May. New York: New York Historical Society.

Lins, A. 1991. 'Basic physical properties of gold leaf'. In *Gilded Wood Conservation and History*, D. Bigelow (ed.), 17–23. Madison, Conn.: Sound View Press.

Marabelli, M. 1994. 'The monument of Marcus Aurelius: research and conservation'. In *Ancient and Historic Metals Conservation and Scientific Research*, D. Scott, J. Podany and B. Considine (eds), 1–19. Malibu, Calif.: Getty Conservation Institute.

Museum of American Art. no date. Vertical File in Library. Copy of an engraving of the 1893 Columbia Exposition *Quadriga*. Washington, DC.

Ogburn, F. and Passaglia et al. 1977. 'Restoration of large gilded statues using various electrochemical and metallurgical techniques'. In *Corrosion and Metal Artifacts – A Dialogue Between Conservators and Archaeologists and Corrosion Scientists*, B. F. Brown *et al.* (eds), 167–78. Washington, DC: National Bureau of Standards, Special Publication 479.

Rabinowitz, M. 1995. Telecommunication and scaffold onsite inspection, April. New York: Central Park Conservancy.

Seabury, C. 1907. Minnesota History Center. Letter to Cass Gilbert, 8 April. St. Paul, Minn.: Minnesota Historical Society.

Saint Gaudens, H. 1913. *The Reminiscences of Augustus Saint Gaudens. Two Volumes*, vol. 2. New York: The Century Company.

Styka, W. 1994. Fax transmission, 25 March. Chesterwood, Mass.: Daniel Chester French Archives.

Tator, K. B. 1983. 'Coatings'. In *Corrosion and Corrosion Protection Handbook*, P. Scweitzer (ed.), 355–88. New York: Marcel Dekker.

Thornton, J. 1991. 'The use of nontraditional gilding methods and materials in conservation'. In *Gilded Wood Conservation and History*, D. Bigelow (ed.), 17–23. Madison Conn.: Sound View Press.

Van Vlack, L. 1967. *Elements of Material Science*. Reading, Mass.: Addison-Wesley.

Vittori, O. and Mestitz, A. 1975. 'Artistic purpose of some features of corrosion on the Golden Horses of Venice'. *Burlington Magazine*, **117**: 132–9.

Welsh, F. 1994. Laboratory report for *Quadriga*, 31 May. Bryn Mawr, Penn.
Wiart, J. 1989. Conservation treatment report for the *Sherman Monument* by Saint Gaudens. Paterson, New Jersey: LMC Corporation.

22 Architectural Gilding on Exterior Metal: An Overview of Materials and Methodology

Michael W. Kramer

Abstract

Methods and materials for gilding on exterior metal have changed considerably from the historic origins of the craft to modern times. Historic techniques including flame gilding with mercury–gold amalgams proved to be highly effective in attaching gold to a copper surface, but very detrimental to the health of the gilders performing the work. Lead compounds once in common use as primers and undercoats have been legislated out of existence. The use of zinc chromate and other chromium compounds has become highly restricted in many jurisdictions. At the same time, the conditions an exterior gilded object is exposed to have become more demanding than ever before.

Environmental regulations have had a tremendous impact on all aspects of exterior gilding. From preparation to priming to finishing, the gilder needs to research and keep up with the different materials available from region to region. Beyond conventional primers and undercoats, there is a new generation of moisture cured and catalysed primers, undercoats and sealers that have been developed for exterior metal. Coupled with this is the introduction of acrylic adhesives during the past few years. The suitability and uses of these materials on actual projects will be explored.

Installations including exterior copper, lead-coated copper and bronze will be reviewed with an emphasis on how new materials and techniques incorporating acrylic resins and adhesives are expected to perform compared to more traditional methods. More changes in materials are anticipated . How the materials are changing and the impact this is having on the craft are already becoming evident in certain areas of the country.

Introduction

The gilding of exterior architectural ornament goes back hundreds of years, not only in Europe and Asia, but also in the USA. Gilded domes, spires, steeples and other ornament abound throughout France, Belgium and other western European countries. Well known examples occur across Russia in the gilded minarets and onion domes associated with the orthodox churches and government buildings in Moscow and St. Petersburg to cite some of the more well known examples. In the USA, one of the oldest gilded exterior structures is the Bullfinch Dome and lantern atop the Massachusetts State House. Dating from the late eighteenth century, this copper dome has beed re-gilded as part of the general restoration of the Massachusetts State House complex.

Gilded domes and statuary are associated with many of the state capitol buildings in this country. The Colorado State Capitol building in Denver and the Iowa State Capitol in Des Moines are two of the older gilded domes west of the Mississippi. The *Quadriga* statue on the state capitol in St. Paul, Minnesota, has recently undergone restoration and re-gilding. All in all, 13 state capitol buildings have gilded domes and others have gilded statuary associated with the capitol buildings. Some of these domes are made of copper, others are monel (a nickel–copper alloy) and still others are cast iron, lead-coated copper or combinations of different metals on different elements.

In addition to government buildings, there are many religious and commercial buildings in this country with gilded metal ornamentation that date from the eighteenth, nineteenth and early twentieth centuries. One early example is a church on Lafayette Square across from the White House.

St. John's Church in Washington, DC was built in 1816 and originally had a gilded copper dome atop the cupola over the centre of the building. In 1822 a front portico and bell tower were added. The bell tower also included a gilded copper dome. The original cupola was recently restored and re-gilded. During the course of the work it was discovered there are five different metals that make up the dome, the cross and its supporting structure. All these surfaces had been gilded. The cross is copper; the collar under the cross is tin; the ball under the collar is zinc; the foot under the ball is also tin and this entire assembly (cross, ball and foot) is anchored by a cast iron armature to the lead-coated copper dome (Figs 22.1 and 22.2). During the stripping and preparation, there was found to be zinc chromate primer underneath the size on all the elements of the dome. However, the cross also had a layer of red lead paint applied to the copper, under the zinc chromate. Not surprisingly, the gilding on the cross was in much better condition than that on the rest of the surfaces, due to the use of a lead primer on that element. Church records show the cupola had quite a bit of work done to it in 1912 and also in 1957, which is when the dome and cross were last gilded. While the original dome was copper (and the second dome on the bell tower is thought to be the original copper) it had deteriorated to the point where it was replaced with lead-coated copper in 1912.

Figure 22.1 St. John's Church, Washington DC. View of the lead-coated copper cupola showing the finial with the cross and ball before stripping and re-gilding. (Courtesy the Gilders Studio Inc.) (See Plate 164.)

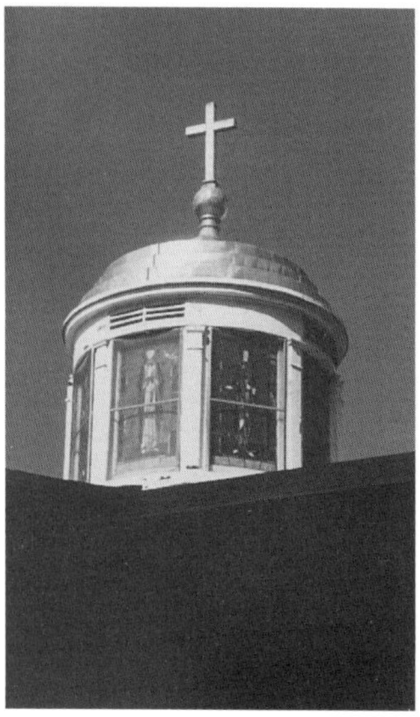

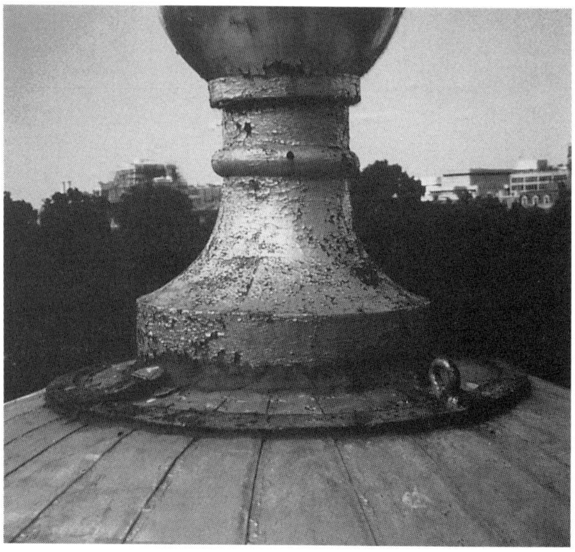

Figure 22.2 St. John's Church, Washington DC. Close up of finial showing the tin foot and base of the cast iron armature attached to the cupola dome. (Courtesy the Gilders Studio Inc.) (See Plate 165.)

Lead-coated copper came into use at the end of the nineteenth century as a roofing material. While frequently left uncoated, to weather naturally, it was also gilded on occasion, sometimes with mixed results.

10 Light St. in Baltimore, Maryland, was built in 1926. The 34 storey Aztec revival structure is topped by a 9 storey mansard roof featuring copper shingles and lead-coated copper pilasters crowned by C scrolls and wheat sheaves of the same material around the parapet. The lead-coated copper was originally gilded when the building was completed and recently restored in 1994. The original gilding was quite worn, but still well adhered to the surface (Figs. 22.3 and 22.4). In 1934 the State of West Virginia gilded its new capitol dome. Fashioned after the dome atop the church at Les Invalides in Paris, the lead-coated copper surface was over 29,000 square feet in area. Within six years the gilding failed, peeling off completely (Gauthier and Kramer 1995). More recently, the Armenian Orthodox Church built a new cathedral in New York. Completed in 1967, St. Vartan's Cathedral was capped with a lead-coated copper dome, gilded with 23 carat gold leaf. Within three years all the gilding had fallen off. One would think that more modern methods and materials would work better than the traditional technologies, especially on the exterior. Apparently, this is not always the case.

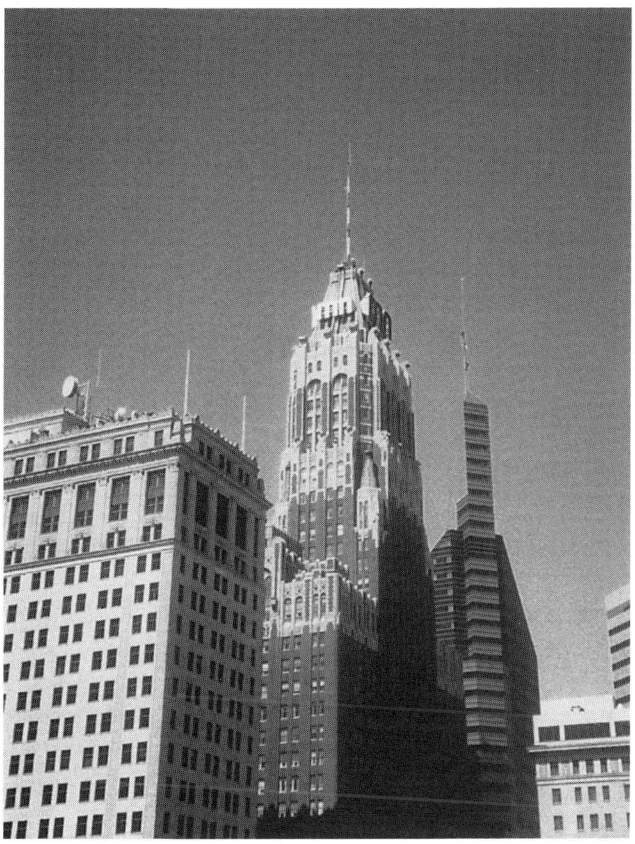

Figure 22.3 10 Light Street, Baltimore, Maryland. An overall view of the building with the 'mn' (Maryland National Bank) sign before removal and restoration of the roof. (Courtesy the Gilders Studio Inc.) (See Plate 166.)

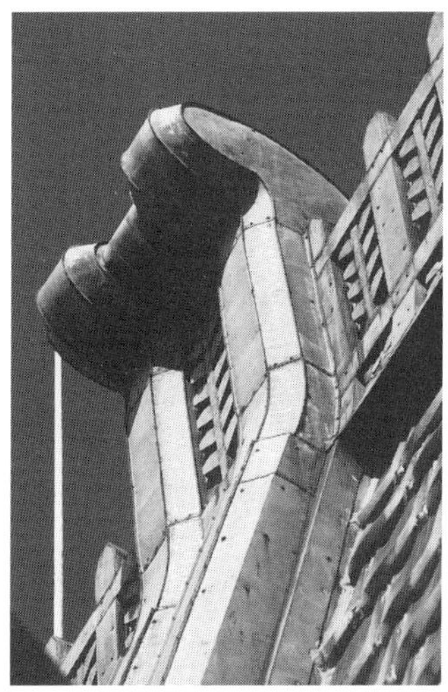

Figure 22.4 10 Light Street, Baltimore, Maryland. Close up of the 'C' scroll and wheat sheaves showing remnants of the original gilding from 1929. (Courtesy the Gilders Studio Inc.) (See Plate 167.)

Historical methods and literature

The historical literature on exterior gilding of metal since the Renaissance is very sparse to non-existent. Some of the earliest printed matter alluding to gilding on metal shows up in the *Painter's and Varnisher's Pocket Manual* published in London in 1826 (Knight and Lacey 1826). While mostly concerned with the carriage trade and varnishing, it does make reference to gilding steel by dipping the object into a solution of ether into which the gold has been dissolved (Knight and Lacey 1826: 55). This is not suitable for any large pieces in exterior environments, of course, but it is one of the earliest mentions of gilded metal. This little manual does provide a good recipe for oil size suitable for exterior gilding.

> To prepare gold size, grind some red lead with the thickest drying oil you can get, the older the better; before using it, it must be brought to a proper consistence with a little oil of turpentine. If the size is good, it will dry sufficiently in about twelve hours, to allow of the gold being applied to it. (Knight and Lacey 1826: 184)

Although almost two hundred years old, the basic formula for slow oil size has changed very little. Beyond the cursory mention of gilding with ether, there is no other mention in regards to metal in this early volume.

In the 1894 edition of the same manual, published by Baird and Co. in Philadelphia, there is somewhat more mention made of gilding on metals. While the method of dipping steel into ether is again explained (practically verbatim from the 1826 text) there is also mention of the gilding of 'inferior metals' such as copper, brass and so on (Brannt and Delf 1894: 227). This is done using an amalgam of gold and mercury applied with heat to the metal, then burnished. Whether this method would be appropriate or not for exterior use is never revealed.

Originally published in 1876 by Excelsior Publishing, *The Gilder's Manual* has a section on 'Gilding on metals' that is over 24 pages long. There is only one paragraph devoted to actually applying gold leaf to metals and no specific mention of exterior gilding. The method outlined has the metal, whether copper, brass, iron or steel, heated to a blue shade, then having the leaf applied to it while hot (Excelsior Publishing 1876: 40–67). This may be possible for large exterior objects, but is not very practical. The rest of the section is devoted to using amalgams and also electroplating. An English text published in 1915 delves a little deeper into the gilding of metals. This is the first example encountered by the author that could be practicable for exterior use. The techniques include coating new ironwork with hot boiled linseed oil or two coats of varnish. It also stresses applying two coats to copper or bronze before gilding with imitation gold leaf, silver leaf or copper to prevent tarnishing (Scott-Mitchell 1915: 124–7). It seems that the two layers of sealer are enough to forestall any galvanic reaction between the dissimilar metals. Old painted ironwork is discussed at length with removal of the paint recommended. This is followed by application of two coats of sealer. Oddly enough, there is no recommended preparation of copper for genuine gold. The practitioner is urged to gild directly on to the copper with a slow size. This would not last under exterior conditions. Generally, it seems that by 1915, the written texts are disseminating more reliable techniques in regards to metal. But even so, there is no specific mention of exterior gilding.

An early reference to gilding on exterior metal appears in a manual on oil gilding written around 1930. This manual was distributed by some of the gold leaf manufacturers in the USA. Specific procedures for chemical neutralization of zinc and galvanised iron are explained. Aluminium is mentioned, but not copper and bronze. The use of several coats of red lead and white lead primers is universally promoted for almost all types of exterior metal. Although these primers had excellent weather and corrosion resistant properties, health and legal considerations make it impossible to use them under any circumstances in the present day. Lead-based primers were not specified for use on lead surfaces. Cleaning with acid or sand blasting the lead to provide a tooth for the size was recommended, followed by using a slow oil size applied directly to the lead surface (M. Swift and Sons 1987: 3). The results of following this procedure will be explored later.

Red lead and white lead are mentioned as primers for almost any material, including wood, plaster, stone and so on. Lead primers and paints were used universally for several hundred years on exterior surfaces across the USA and overseas. Perusal of old photographs of town populations will show the painters looking very sickly, perhaps suffering from the 'painter's cholic' or more likely lead poisoning, a not uncommon ailment for painters at that time. As early as 1826, the *Painter's and Varnisher's Pocket Companion* has a large section devoted to common ailments and 'vapours' suffered by members of the trade (Knight and Lacey 1826: 200). In later editions, more space is devoted to the health hazards of the trade, with particular emphasis on exposure to lead.

In spite of the increasingly apparent risks associated with lead compounds, they were still in common use in the painting (and exterior gilding) industry until the mid-twentieth century. Examination of the Bullfinch Dome on the Massachusetts Statehouse shows evidence of white lead and red lead under the latest layer of primer, which is zinc chromate.[1] It is reasonable to assume that these traces of lead paint represent a continuation of the tradition of using lead-based primers over the copper surfaces. Coupled with the lead primers was the use of lead compounds in the size and driers associated with the varnishes frequently used on exterior gilding. By the 1950s the use of lead was proscribed in the production of paints for the housing industry and by the end of the 1960s was strictly regulated for most uses. However, for practitioners of exterior gilding, this presents a whole new set of problems. While lead hazards are not encountered daily during the application of the primers and size, lead is frequently found in the primers on old surfaces needing to be re-gilded.

Governmental regulations

Strict governmental regulations apply to the handling of lead and other compounds removed during the stripping and preparation of the surface. Lead abatement has now become a part of the gilder's work. In the specifications for the restoration of the gilding on the Bullfinch Dome on the Massachusetts State Capitol, the state specifically requires that lead abatement of the surface of the dome be included under the gilder's scope of work. This presents the gilder with complex licensing, insurance and training requirements not normally associated with the trade.

While the regulation of lead compounds safeguards the health of painters, gilders and anyone else who may come into contact with lead based paints, there were new problems presented with the outlawing of these primers. What is to be used in their stead? While use of lead primers certainly presented health risks to all involved, they worked very well on most metal surfaces to be gilded. Remember, the lead-primed cross held up much better than the other surfaces at St. John's Church in Washington, DC. Why only the cross had red lead primer and not the rest of the metal surfaces is a mystery.

When it became apparent lead would be regulated out of existence, the painting industry started exploring alternatives to lead primers for metals. One primer that came into widespread use is zinc-chromate. Developed specifically for rust inhibition, it has been widely used as a primer under gilding on a variety of surfaces including lead-coated copper, copper, cast iron, monel and others. However, the days for using zinc chromate are numbered. Already, regional environmental commissions are regulating the distribution of zinc-rich and chromium-containing products. In many areas of the country, including the tri-state area around New York and in California, the sale of zinc chromate is prohibited. Federal regulations require that removal of zinc chromate is to follow the same procedures as lead abatement. Many paint manufacturers including Benjamin Moore and Co. have stopped making zinc-rich primers altogether and are developing replacement products.

New developments

These same governmental bodies are also implementing VOC (volatile organic compound) requirements in regards to paint and primer composition. Specifically, the amount of free thinners is drastically reduced to meet these standards. This has led the paint manufacturers to introduce VOC compliant products into the affected regions only and continue distribution of the non-compliant items into the regions of the country not yet covered by the regulations. Unfortunately, the manufacturers will package the VOC compliant products and the non-compliant products with the same names and essentially the same labels. This leads to quite a bit of confusion when working in several different regions of the country, not to mention the problems it can cause when trying to use a VOC compliant primer under slow oil size. Many VOC compliant primers incorporate silicone compounds. This causes some oil sizes to reticulate. Of course the problem doesn't manifest itself until actually trying to apply the size over the already prepared and primed surface and discovering the two materials are not compatible. Then the question is to strip the surface and start over again or try to apply a barrier coat that will be compatible with both materials.

Another reaction of the paint manufacturers is to introduce a completely new genre of paints and primers known as water-borne alkyds. Unfortunately, many of these new products also contain silicone compounds and are not designed for use on metals. Furthermore, with the presence of silicone these products are generally not compatible with slow oil sizes.

However, there are other products designed for use on metals which necessarily will become useful to the trade. These are not zinc rich, do not contain chromium and are VOC compliant. Some of these are catalysed or moisture-cured urethane primers that are primarily aimed towards the roof coatings industry. Others are water-borne bonding primers for use on a wide variety of materials. There are also butyl based primers available that may fill the requirements under certain circumstances. Clearly, there are no primers available now that will act as a 'silver bullet' in the same manner as the lead and zinc chromate primers of the past. Therefore, the gilder needs to be very careful and constantly research the usefulness and compatibility of the various primers available, especially when working in different regions of the country. There is no substitute for testing and experimentation.

In conjunction with the development of new primers and paints, there are also new adhesives being developed specifically for exterior use. These are acrylic based and have undergone several transmutations since they were introduced five years ago.[2] The manufacturers feel that these are the wave of the future and will be acceptable substitutes for traditional oil size. This may well be the case, but many gilders are hesitant to try new technologies on large structures, especially when the materials have not stood the test of time. Frequently, adherence to conservation standards will preclude experimenting with new materials on an historic piece. From an economic point of view, it would be very difficult to justify applying an adhesive that has only been in production for five years and supply a ten year warranty (which is frequently demanded by the customer) on the finished product. Considering the potential liability and expense in the case of premature failure, one would be hard pressed to even consider using a product line that is virtually still in its infancy.

Examples in the field

Baltimore, Maryland

The 1994 restoration and re-gilding of the lead-coated copper elements on the roof of the Nations Bank building at 10 Light Street in Baltimore was achieved with a combination of different materials. With over 6,000 square feet of surface to be treated and re-gilded, all options had to be carefully considered. Traditionally, lead-coated copper is meant to be stripped, then washed with muriatic acid and water, which etches the surface, prior to application of the prime coat. This procedure produces waste products rich in lead sulphide and lead oxide compounds. With the strict lead abatement regulations in the City of Baltimore and the State of Maryland, the stripper by-products and the resulting acid and water run-off would have to be captured and disposed of as a hazardous waste. Financial and practical considerations precluded this from being a viable option.

The solution was to use a moisture-cured urethane primer, developed for the roof coatings industry and in use for over 20 years. Prior to application of the urethane primer, the lead-coated copper was power washed with a mild detergent to remove any loose dirt and debris. As stated earlier, the gilding from 1929 was well adhered and the coatings manufacturer felt it would act as a substantial base for the primer. Two subsequent layers of zinc-chromate were applied over the urethane primer. A slow drying size was then applied over the zinc-chromate onto which the 23 carat gold leaf was gilded with excellent results. Inspection of the gilding in 1997 showed it to be in excellent condition (Figs 22.5 and 22.6).

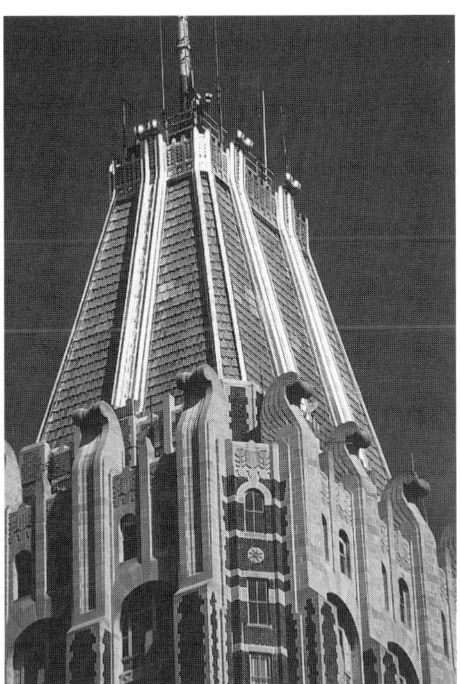

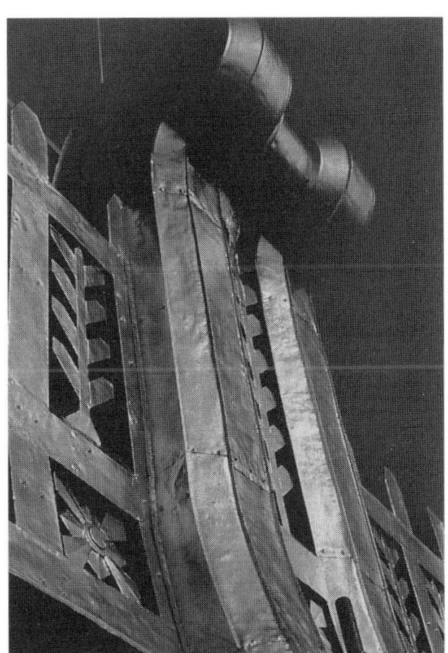

Figure 22.6 10 Light Street, Baltimore, Maryland. Detail of the 'C' scroll and wheat sheaves after gilding. (Courtesy the Gilders Studio Inc.) (See Plate 169.)

Figure 22.5 10 Light Street, Baltimore, Maryland. The roof of the building after restoration and gilding. (Courtesy the Gilders Studio Inc.) (See Plate 168.)

Dubuque, Iowa

In 1981, the copper domes atop the Dubuque County Courthouse (circa 1890) were gilded. Within five years the gilding was beginning to fail. Investigation in the summer of 1994 showed the gilding to be poorly adhered and flaking off (Fig. 22.7). Examination of the original specifications showed them to be ambiguous and left out the critical step of applying a wash primer or equivalent under the zinc-chromate (Gauthier and Kramer 1995). The domes were subsequently stripped and re-gilded. This time a urethane primer was used after the surface was etched, followed by two coats of zinc-chromate, slow size and 23 carat gold leaf (Fig. 22.8).

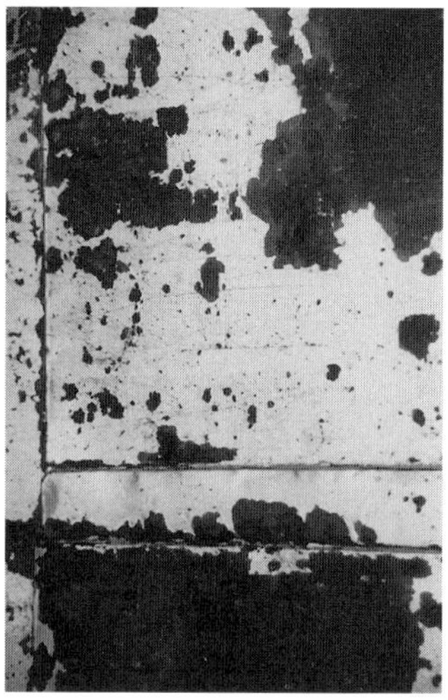

Figure 22.7 Dubuque County Courthouse. Cubuque, Iowa. Detail of lower dome showing premature deterioration of the 1981 gilding. (Courtesy the Gilders Studio Inc.) (See Plate 170.)

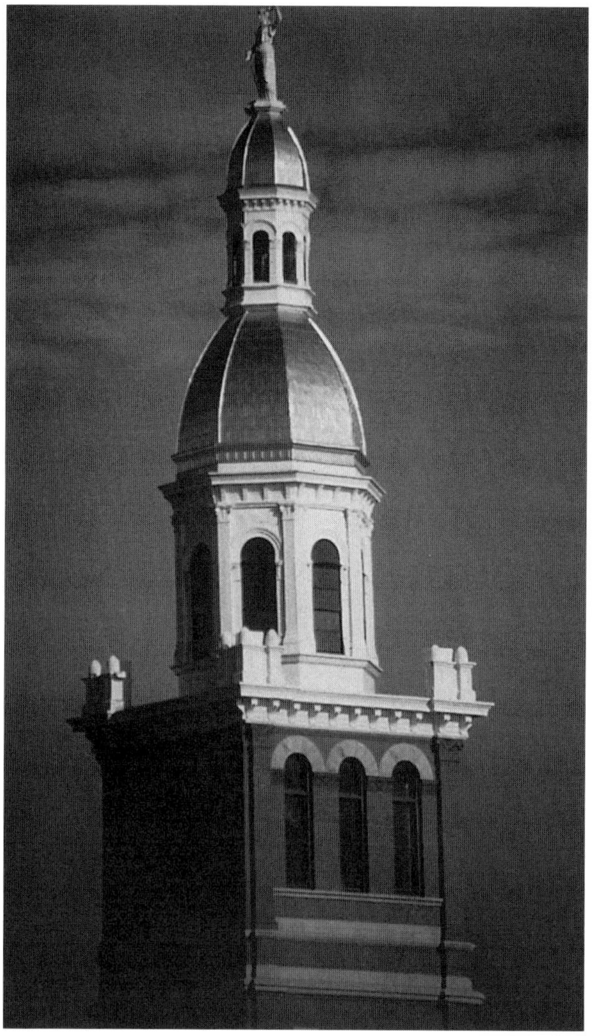

Figure 22.8 Dubuque County Courthouse, Dubuque, Iowa. Domes after completion in 1994. (Courtesy the Gilders Studio Inc.) (See Plate 171.)

Washington, DC

As mentioned earlier, St. John's Church in Washington, DC was built in 1816. After the new lead-coated copper dome was fabricated and installed atop the cupola in 1912, it was also gilded to match the original copper it was replacing, as were the cross, ball and foot above the dome. As stated before, there were a total of five different metals that needed to be treated after stripping.

After consultation with the paint manufacturers, chemists and the Copper Development Association, it was determined that several different treatments and primers needed to be used. The copper cross was etched, then coated with a urethane primer; the zinc ball was acid washed,

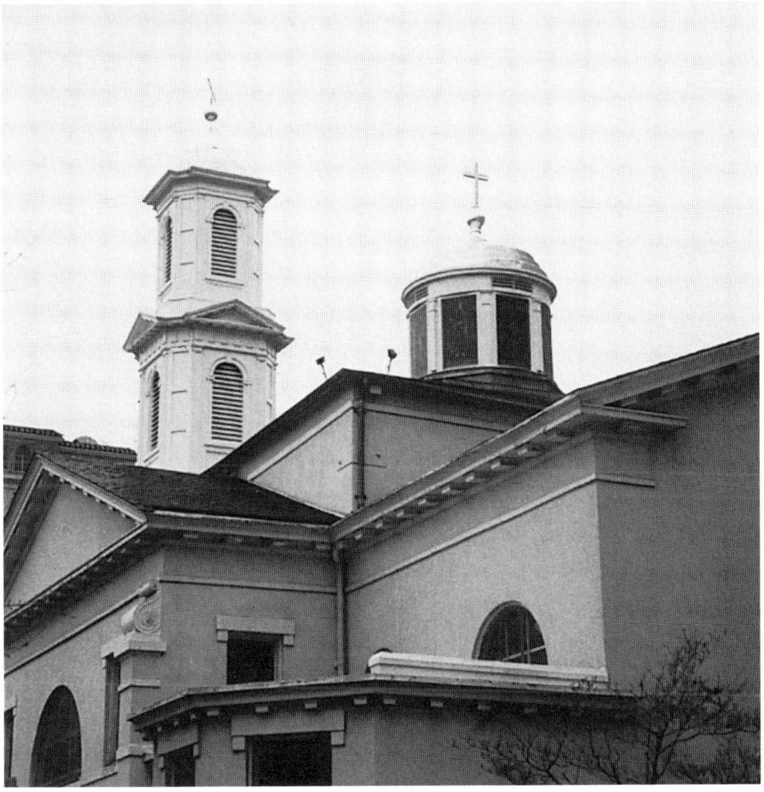

Figure 22.9 St. John's Church, Washington, DC. The two domes after completion in 1995. (Courtesy the Gilders Studio Inc.) (See Plate 172.)

then painted with a xylene-based primer; the tin collar and foot were cleaned, then primed with an alkyd rust inhibitive metal primer; the lead-coated copper dome was coated with the same urethane primer as the cross. After application of these various primers, the surfaces were painted with two coats of a yellow alkyd enamel, then slow oil size and gilded with 23 carat gold leaf. All the materials had to be researched and tested to determine the compatibility between the various components. The results are both beautiful and long lasting (Fig. 22.9).

New York

St. Vartans' Cathedral was built in 1967. The original lead-coated copper dome had a layer of 23 carat gold leaf applied to it. Within three years the leaf had all flaked off and blown away. Investigation during the spring of 1993 showed virtually no evidence of any gold leaf, primer or other substance on the surface of the dome. Physical examination yielded traces

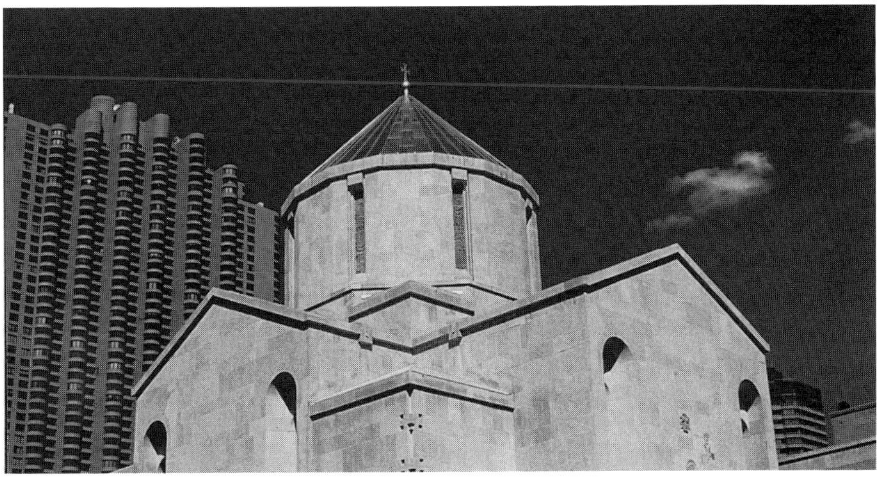

Figure 22.10 St. Vartan's Cathedral, New York. The ribbed roof after proper preparation, priming and re-gilding in 1994. (Courtesy the Gilders Studio Inc.) (See Plate 173.)

of gold leaf in crevices along the ribs. Subsequent analysis of the scrapings showed presence of gold leaf and size, but no primer. Apparently, the size was applied directly to the lead-coated copper, with no priming or other preparation. The surface was cleaned and primed with a catalysed urethane primer. Two coats of a yellow alkyd were applied to the primer, followed by a slow oil size. More than 3,500 square feet of 23 carat gold leaf was then applied to the surface(Fig. 22.10). In 1998, more than four years after completion, the gilding appears as new.

Washington, DC

In 1994, the Second Division Memorial in Presidents' Park underwent restoration. There is a 16-foot tall gilded bronze statue and hundreds of gilded letters carved into the granite monument. Erected in the 1930s the statue and monument had been last re-gilded in 1977. Examination of the statue showed the bronze to be in good condition with only minor pitting evident in areas where the water would puddle in crevasses of the sculpture.

The gilding from 1977 was in good condition with the exception of the areas that had been defaced by graffiti and damaged by individuals continually climbing up the statue. Unfortunately, the vandalism was widespread over most of the surface and caused a need for the gilding to be replaced prematurely. The statue was subsequently stripped. The primer under the gold leaf was zinc chromate, so the stripping became an abatement project. The specifications provided for the preparation were vague and actually more appropriate for cast iron, rather than bronze (Gauthier and Kramer 1995). This could have been disastrous, as sandblasting the bronze under the pressures called for would have destroyed the finish. After consultation with the architect, the specifications were modified. Once the surface was stripped and cleaned, it was glass peened to remove any residue and provide a stable surface, free of contaminants. After peening, the surface was cleaned with Scotch Brite pads, stabilised with benzotriazole and alcohol, then primed with two coats of zinc chromate. Slow oil size was used over the primer followed with 23 carat gold leaf. The

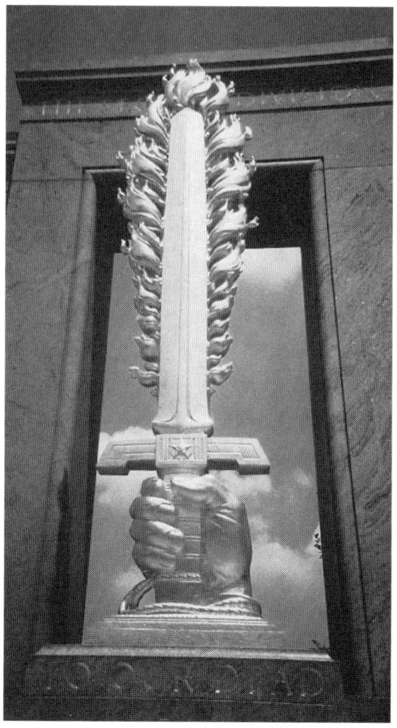

Figure 22.11 Second Division Memorial, President's Park, Washington DC. The 16-foot tall bronze statue with a new layer of 23 carat gold leaf. (See Plate 174.)

project was completed with excellent results (Fig. 22.11). Unfortunately, as soon as the barriers came down, tourists again started to climb the statue and scratch their names into the surface

Conclusions

The practice of exterior gilding has undergone radical changes in the past 50 years, and even now is in a state of flux. While the basic procedures remain the same – cleaning the surface, priming, sizing and gilding – the materials, mainly the primers, have undergone dramatic transformations. Governmental regulations have caused elimination of primers that have traditionally been used over the years. On certain projects, the gilder must adopt a new role; that of a hazardous materials abatement contractor. Sizes are also undergoing changes, with the development of new acrylic sizes that may one day become the standard. At the same time, other regulations cause some materials to be available in some regions, but not available in others.

Until environmental and VOC requirements become uniform, the potential for confusion and problems on installations will continue. The only safeguard is to continually test and experiment with the actual products available in the particular region where one is working. Careful planning, testing and research with manufacturers and other experts will ensure that projects in any region will maintain the beauty and longevity traditionally associated with exterior gilding.

Notes

1. A chemical analysis was made of the ten existing layers of material on the Massachusetts State House Dome. The layers of lead and zinc chromate were noted on page 2 in a report prepared by Conam Inspection, Inc., Natick, Massachusetts, November 1994, for Vertical Access, and made part of the bid documents by the State of Massachusetts.
2. Two varieties of exterior acrylic gold size are being distributed by Sepp Leaf Products in New York. 'Permacoll' and 'Instacoll' are manufactured in Germany by Kolner Products. 'Permacoll' is a one-part adhesive, while 'Instacoll' is two-part: a base layer and activator wash.

References

Brannt, W. T. and Delf, T. 1894. *The Painter, Gilder and Varnisher's Companion*, 27th edn. Philadelphia: H. C. Baird and Co.

Excelsior Publishing 1876. *The Gilder's Manual*. New York: Excelsior Publishing. (Reprinted by the Society of Gilders, 1990).

Gauthier, B. and Kramer, M. 1995. 'Exterior gilding: preparation is everything'. *Traditional Building Magazine*, March–April, 6, 93.

Knight and Lacey 1826. *The Painter's and Varnisher's Pocket Manual: containing rules and instructions in every thing related to the arts of painting and varnishing, with full directions for the processes of colouring foils, polishing, gilding and japanning*. London: Knight and Lacey.

Scott-Mitchell, F. 1915. *Practical Gilding, Bronzing, Lacquering and Glass Embossing, Illustrated by Actual Examples of Gilded and Lacquered Relief and Other Decorations*. London: Trade Papers Publishing Co.

M. Swift and Sons 1987. *How to Gold Leaf Effectively*. New Haven, Conn.: M. Swift and Sons.

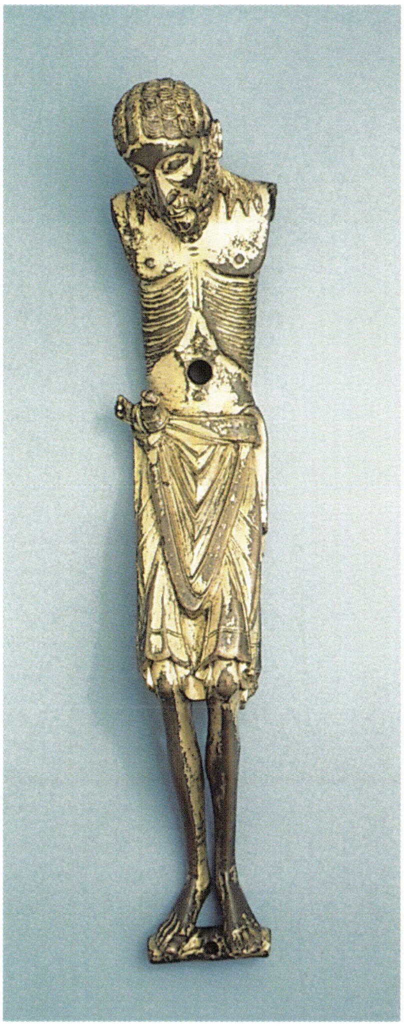

Plate 1 Romanesque crucifix (Angevin, first quarter of the twelfth century) (Copyright Trustees of the British Museum). (See Fig. 1.1, p. 6.)

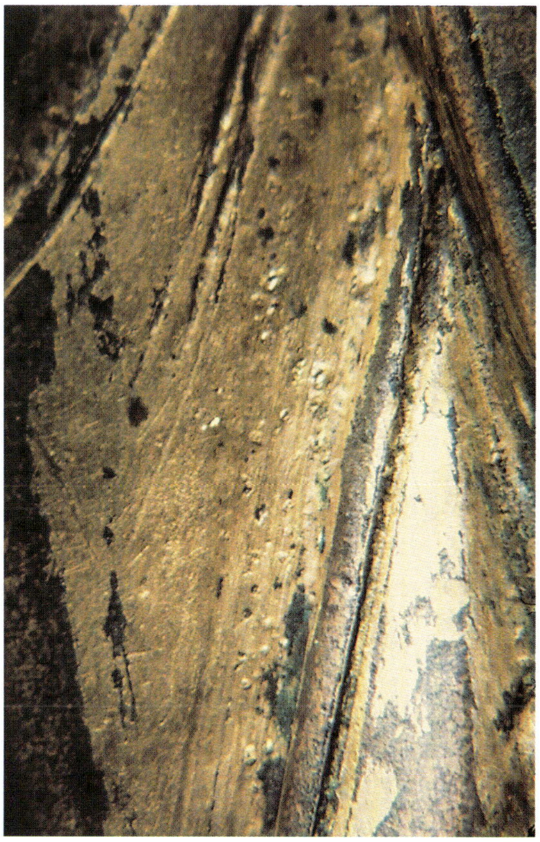

Plate 2 Detail of the surface of Figure 1.1. The crucifix is cast in 'impure' copper (containing 4.5% tin + zinc + lead) and has been fire gilded (Copyright Trustees of the British Museum). (See Fig. 1.2, p. 7.)

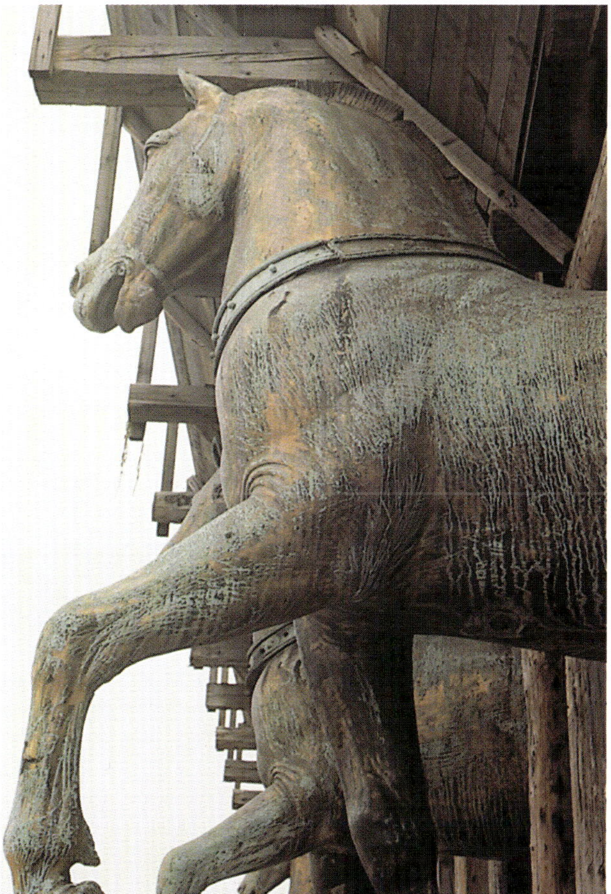

Plate 3 The gilded bronze horses on the facade of the Basilica di San Marco in Venice, photographed before their removal in the late 1970s (Copyright Trustees of the British Museum). (See Fig. 1.5, p. 12.)

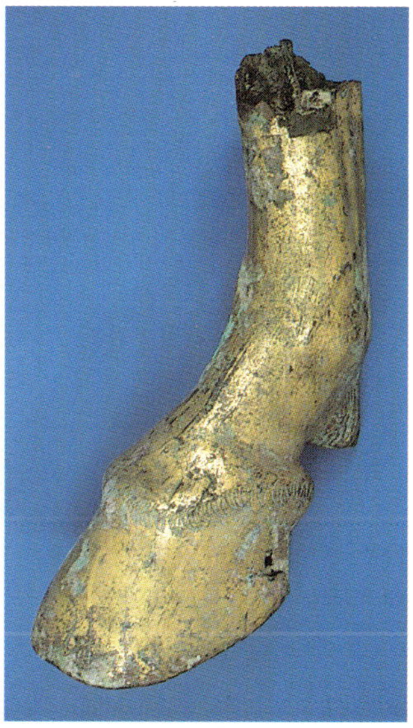

Plate 4 Life-size gilded bronze horse's hoof of the Roman period (Copyright Trustees of the British Museum). (See Fig. 1.8, p. 13.)

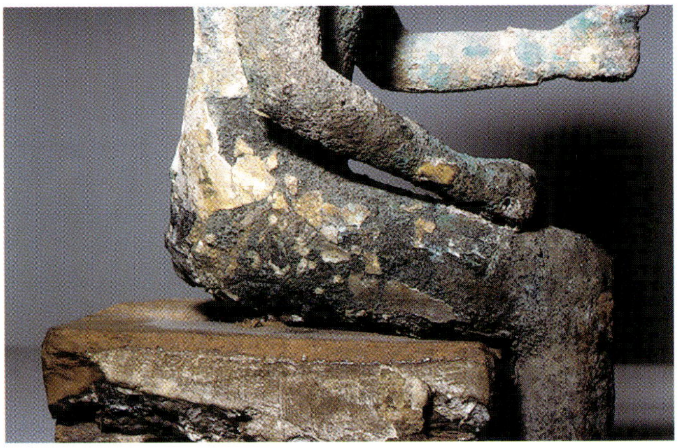

Plate 5 Detail of corrosion associated with fragmentary gilding in the area of the kilt. Note the granular green crystals of atacamite. Paratacamite and cuprite were also identified from this area. (Photo by Janet Burke, © Cleveland Museum of Art.) (See Fig. 3.4, p. 54.)

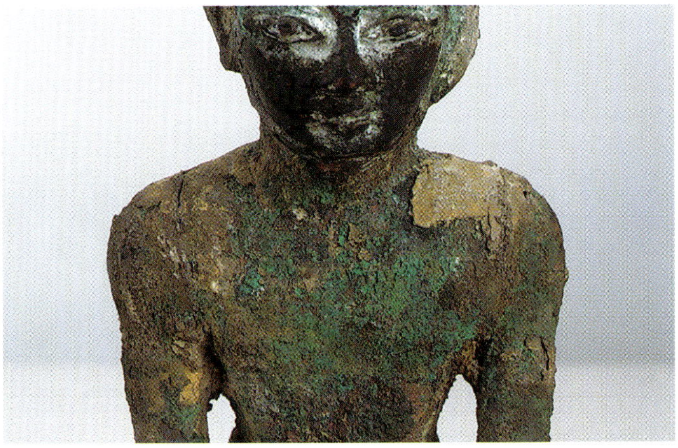

Plate 6 Detail of corrosion associated with fragmentary gilding in the area of the collar. Note the pale powdery green and blue-green corrosion. Sampleite and chalconatronite were identified in this area in addition to malachite and cuprite. Also note the dark, compact lead-rich corrosion on the surface of the face. (Photo by Janet Burke, © Cleveland Museum of Art.) (See Fig. 3.5, p. 54.)

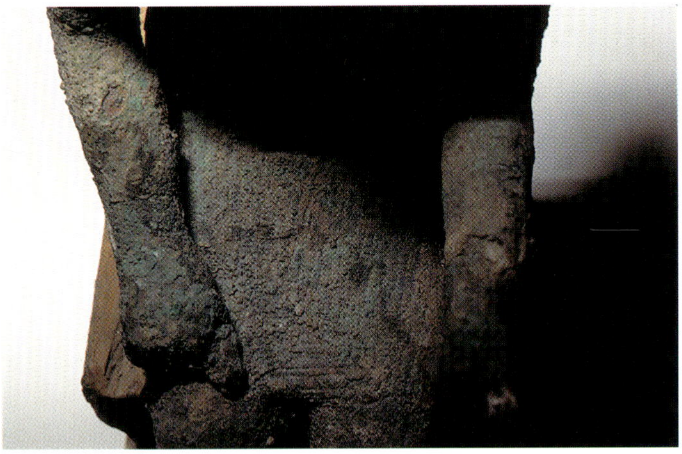

Plate 7 Detail of the striated texture preserved in the corroded bronze of the kilt. (Photo by Janet Burke, © Cleveland Museum of Art.) (See Fig. 3.7, p. 56.)

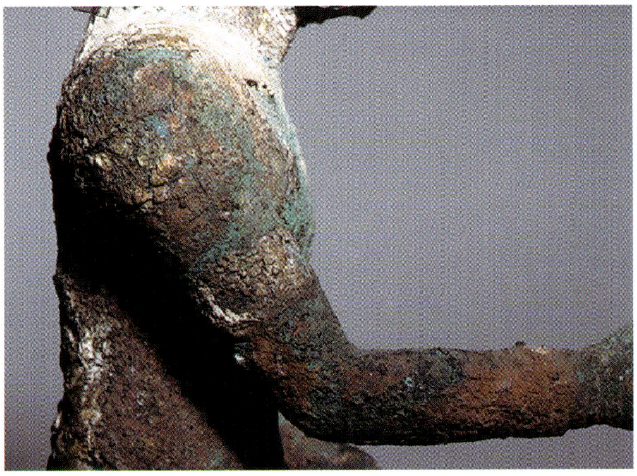

Plate 8 Detail of a plain woven fabric pseudomorphically preserved in the corrosion below the gesso and gold leaf on the proper left arm band. (Photo by Janet Burke, © Cleveland Museum of Art.) (See Fig. 3.8, p. 57.)

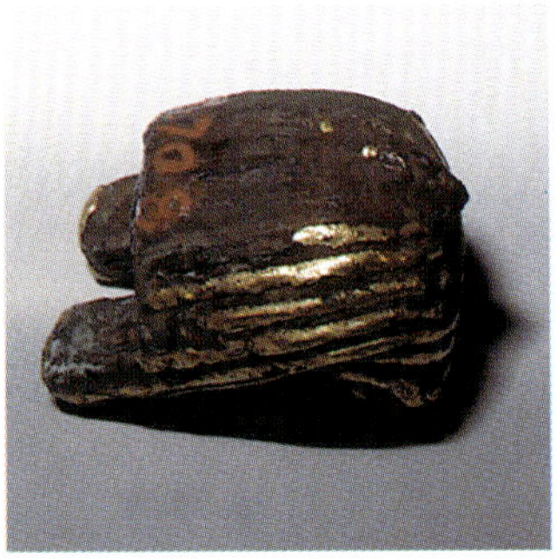

Plate 9 Detail showing the reverse of falcon head. The gilding is extant in most areas that the copper additions or inlays are preserved. Both the gilding and additions are absent on the reverse of the wig revealing the pairs of raised ribs on the surface of the wig used to hold the additions in place. Also note the powdery red corrosion on the surface of the wig and the compact red and black corrosion on the copper additions. (Photo by Janet Burke, © Cleveland Museum of Art.) (See Fig. 3.15, p. 60.)

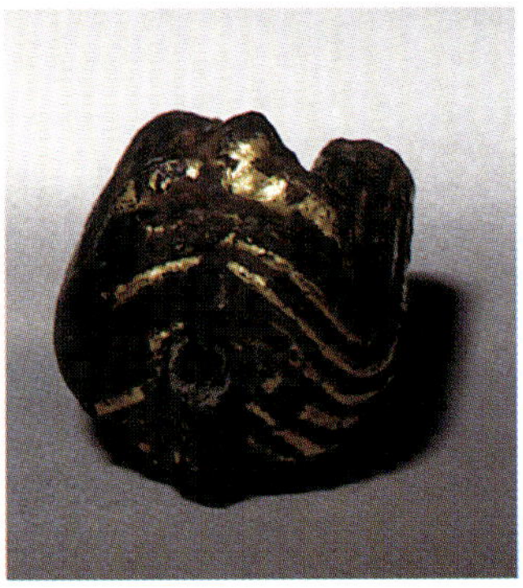

Plate 10 Detail showing the top of the falcon head. The seam between the face and wig is tight and difficult to see. The mortise which retains the tenon from the missing head-dress is readily visible. (Photo by Janet Burke, © Cleveland Museum of Art.) (See Fig. 3.16, p. 60.)

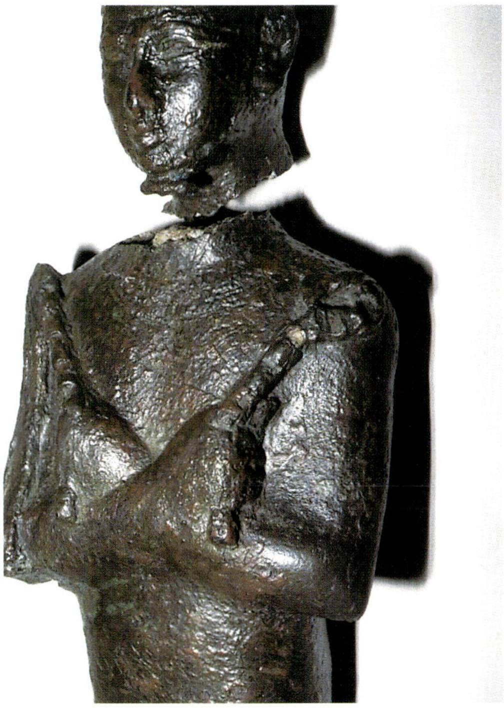

Plate 11 Detail of crook and flail on the statuette of Osiris (1914.571). White travertine inlays (est.) and the black inlays, which were probably originally silvered over a copper substrate, are visible on the proper left as a thin black surface over a compact red cuprite. Also note the uneven texture and tone of the stripped surface of the figure. A pigmented wax , coloured by carbon black according to XRD, was applied to unify the surface.
(Photo by Janet Burke, © Cleveland Museum of Art.) (See Fig. 3.17, p. 62.)

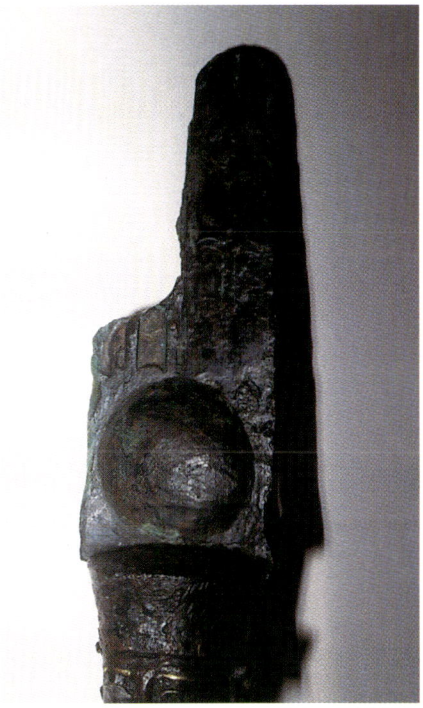

Plate 12 This detail of the double plumes surmounting the crown illustrates the use of the black copper inlays. The surface has been heavily waxed, which imparts a visible sheen to most areas.
(Photo by Janet Burke, © Cleveland Museum of Art.)
(See Fig. 3.19, p. 63.)

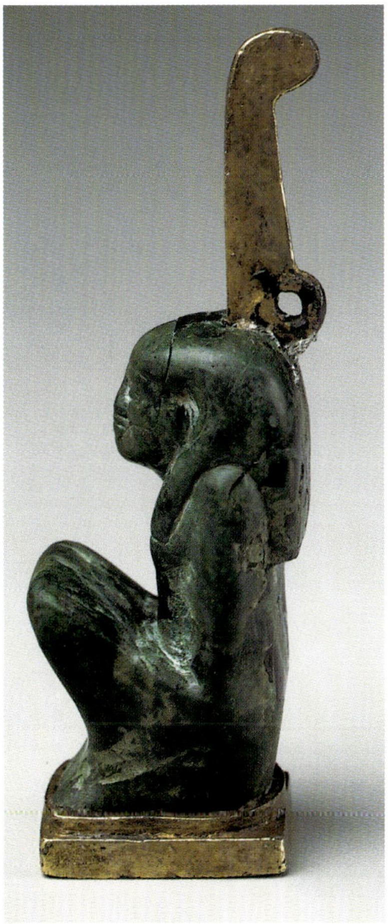

Plate 13 Maat figure amulet (*c*.716–702BC), from El Kurru, Tomb Ku. 55. Malachite, gilded silver, H: 6.0 cm, W: 2.0 cm.
(Courtesy, Museum of Fine Arts, Boston. 21.323.) (See Fig. 4.4, p. 79.)

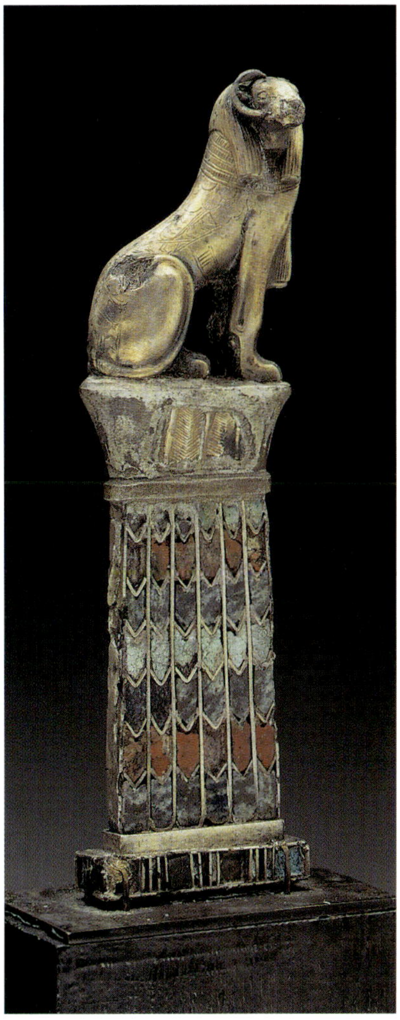

Plate 14 Ram-headed sphinx amulet on inlaid column (*c*.716–702BC), from El Kurru, Tomb Ku. 55. Gilded silver, inlays in lapis, red and turquoise glass, H: 10.3 cm, W: 3.0 cm, D: 1.5 cm.
(Courtesy, Museum of Fine Arts, Boston. 24.972.) (See Fig. 4.5, p. 80.)

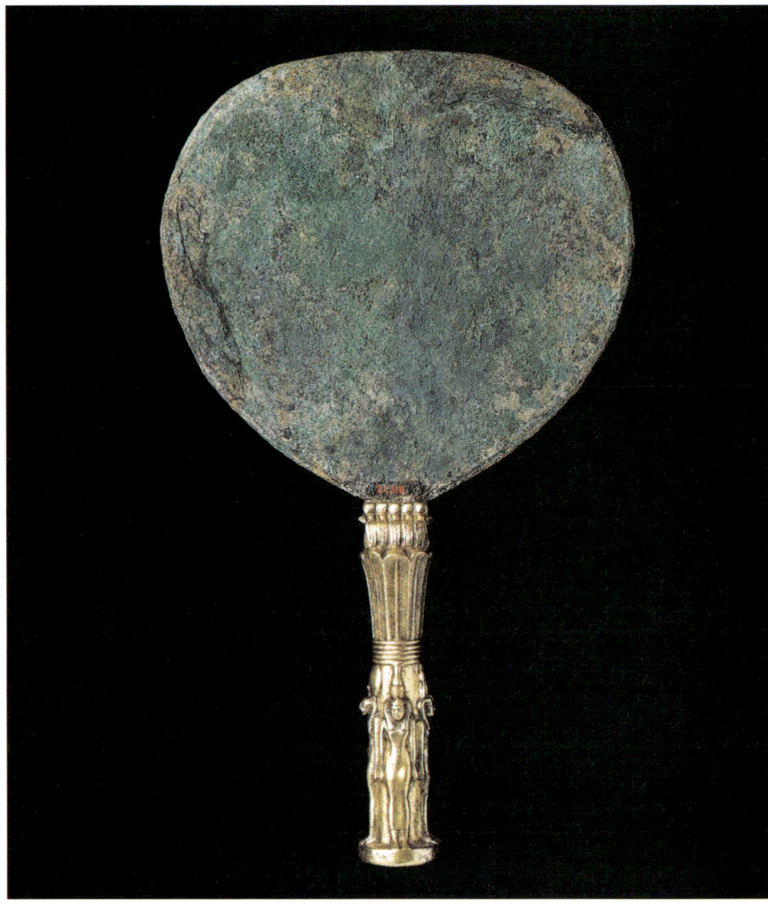

Plate 15 Mirror of King Shabako (*c.*716–702BC), from El Kurru, Tomb Ku. 15. Bronze, gilded silver, H: 33.0 cm, W: 20.0 cm, D: 5.0 cm. (Courtesy, Museum of Fine Arts, Boston. 21.318.) (See Fig. 4.6, p. 80.)

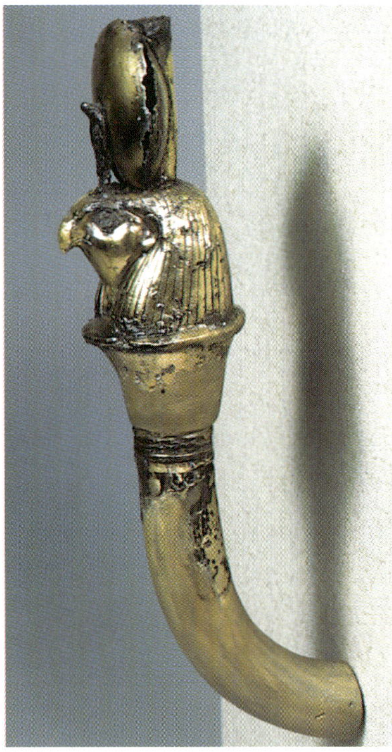

Plate 16 Plume holder from the burial of one of King Tanwetamani's horses (*c.*664–653BC), from El Kurru, Tomb Ku. 219. Gilded silver, H: 14.5 cm, W: 3.5 cm, D: 5.0 cm. Museum of Fine Arts, Boston. 23.877. (Photograph: Susanne Gänsicke.) (See Fig. 4.7, p. 81.)

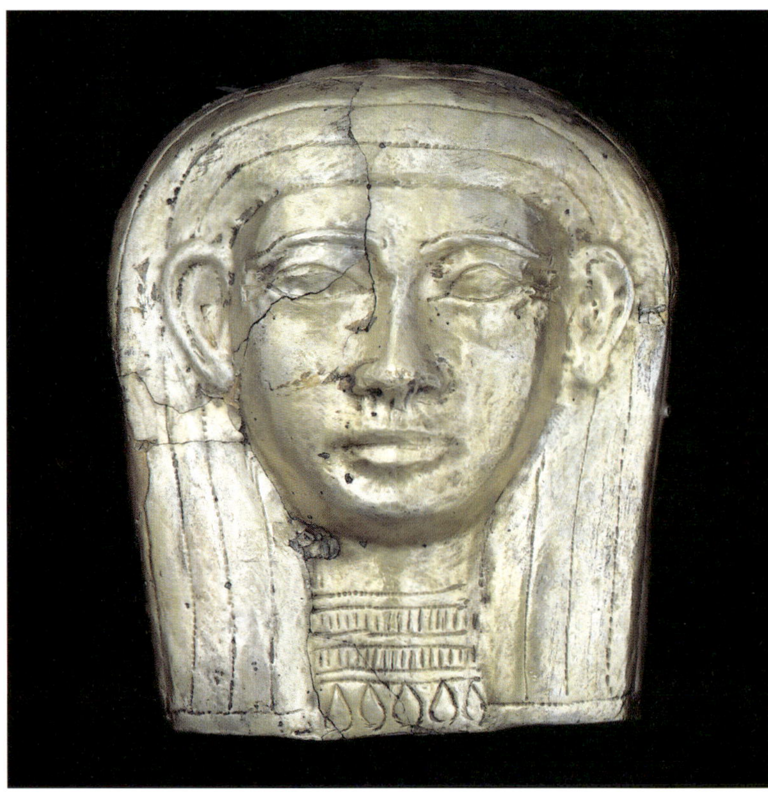

Plate 18 Mummy mask of Queen Malaqaye (*c.*664–653BC), Nuri, Pyramid Nu. 59. Gilded silver, H: 13.9 cm, W: 11.5 cm, D: 1.1 cm. (Courtesy, Museum of Fine Arts, Boston. 20.1059.) (See Fig 4.10, p. 81.)

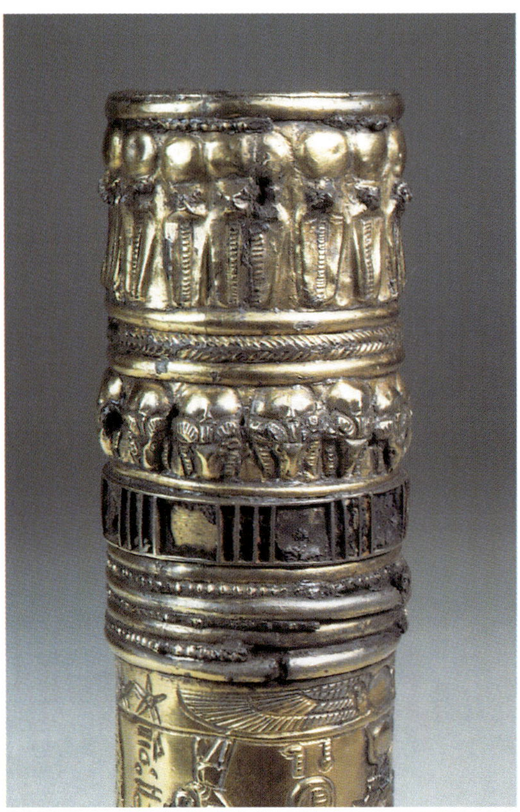

Plate 17 Detail of a cylinder from Nuri, Pyramid Nu. 10. Gilded silver, cloisons filled with paste. 20.275. (Photograph: Susanne Gänsicke.) (See Fig. 4.9, p. 81.)

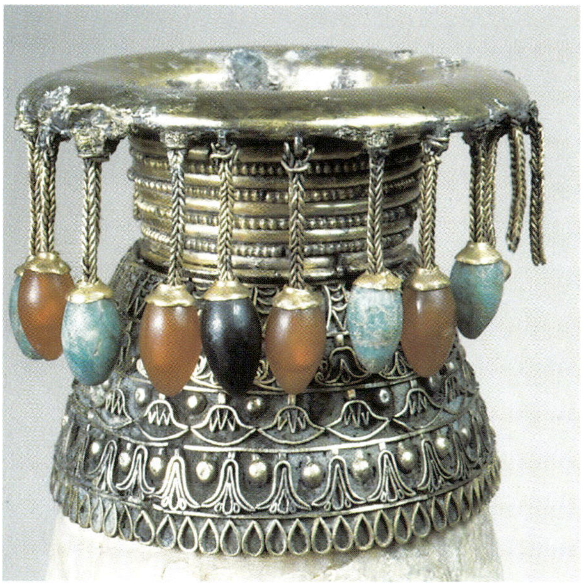

Plate 19 Alabastron of King Aspelta, detail. (Photograph: Susanne Gänsicke.) (See Fig. 4.12, p. 82.)

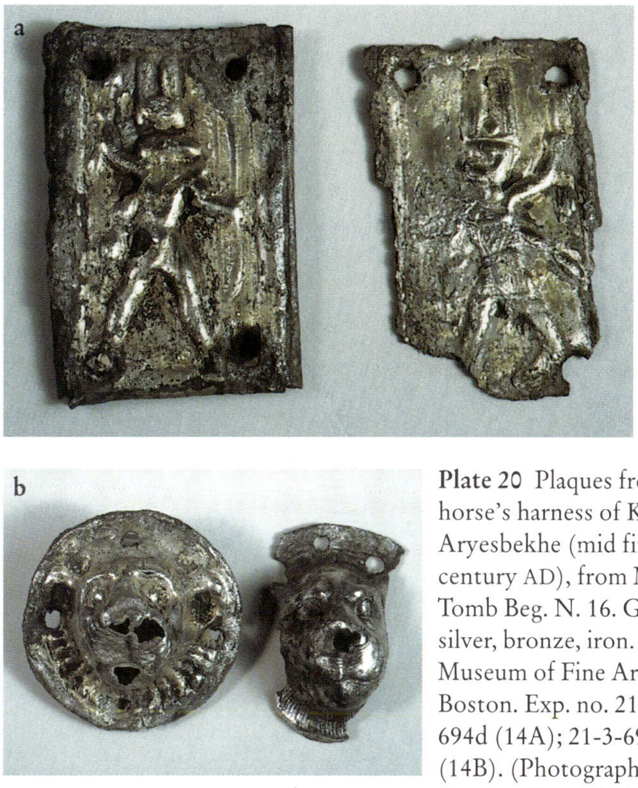

Plate 20 Plaques from a horse's harness of King Aryesbekhe (mid first century AD), from Meroë, Tomb Beg. N. 16. Gilded silver, bronze, iron. Museum of Fine Arts, Boston. Exp. no. 21-3-694d (14A); 21-3-694b,g (14B). (Photograph: Susanne Gänsicke.) (See Fig. 4.14, p. 82.)

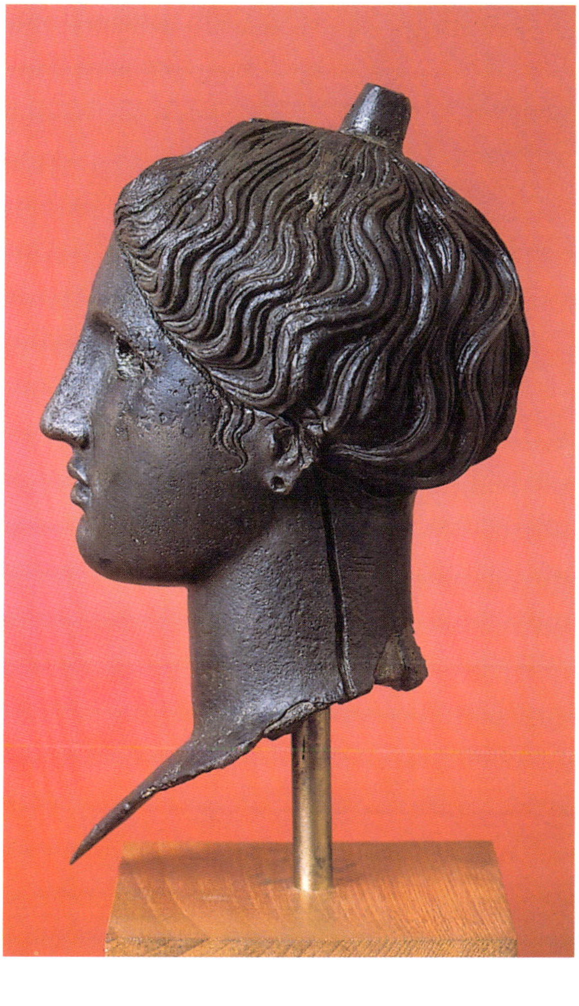

Plate 21 Nike head. (Photograph by Craig Mauzy, Agora Excavations, American School of Classical Studies, Athens.) (See Fig. 5.1, p. 97.)

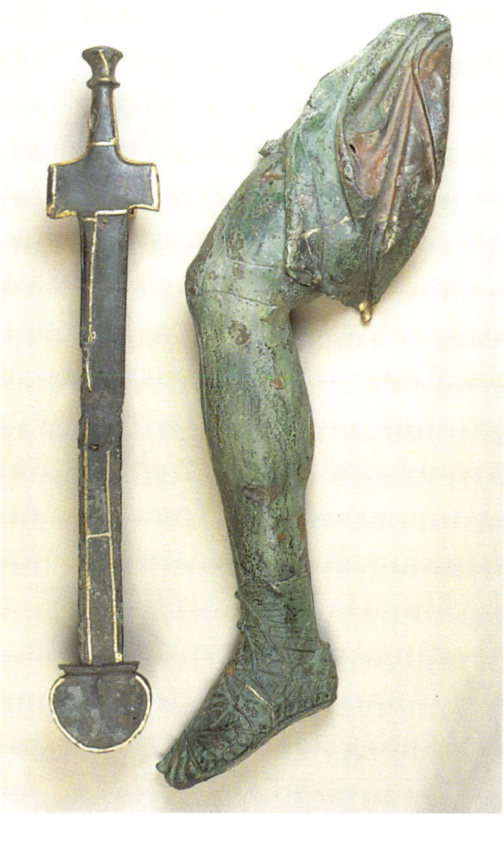

Plate 22 Equestrian leg, drapery and sword. (Photograph by Craig Mauzy, Agora Excavations, American School of Classical Studies, Athens.) (See Fig. 5.2, p. 98.)

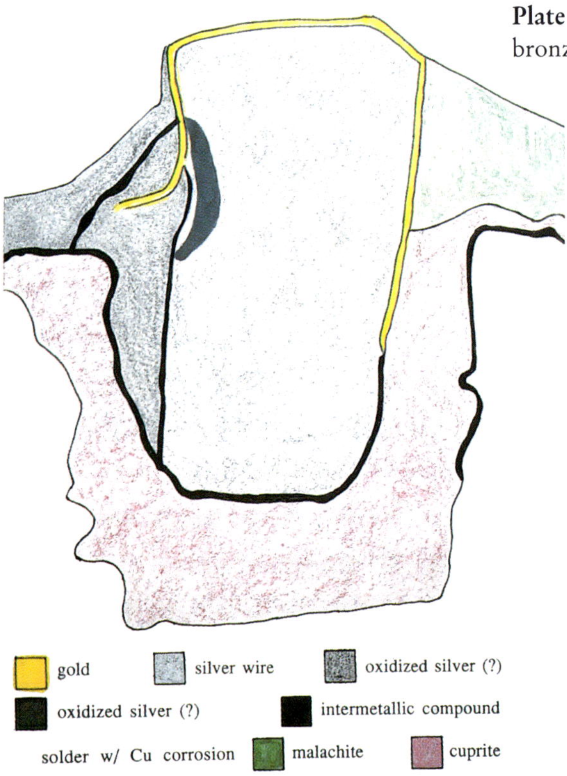

Plate 23 Drawing of a cross-section of silver wire in a bronze channel. (See Fig. 5.6, p. 101.)

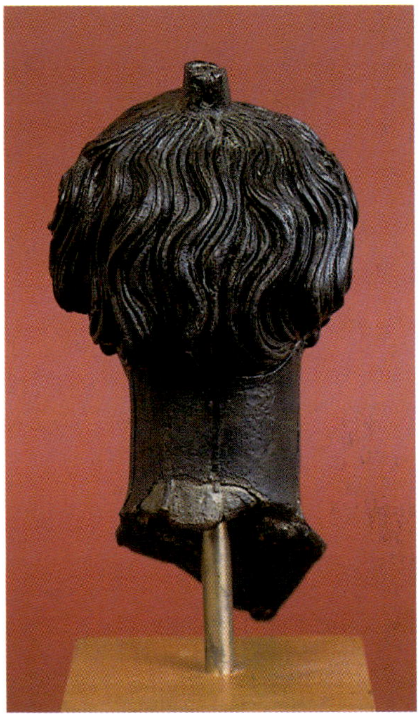

Plate 24 Rear of Nike head showing first set of grooves cut in neck. (Photograph by Craig Mauzy, Agora Excavations, American School of Classical Studies, Athens.) (See Fig. 5.8, p. 103.)

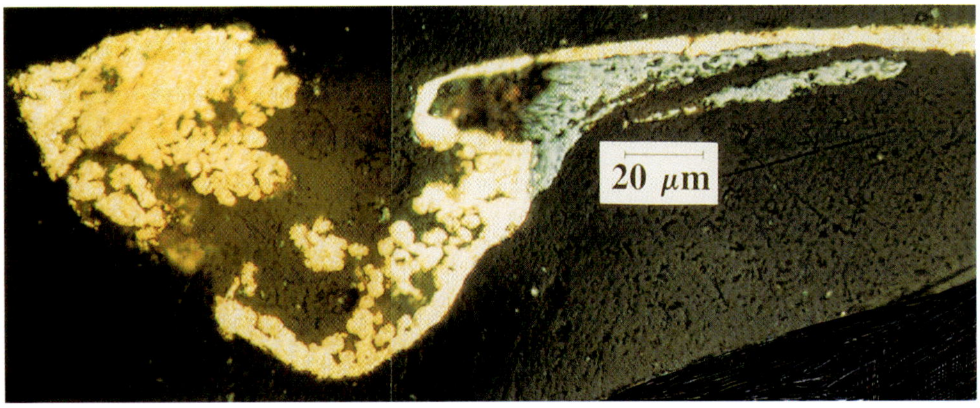

Plate 25 Fire-gilding on an Anglo-Saxon brooch (5–7th century) from Berinsfield, Oxfordshire. Ashmolean Museum, inv. no. 1989.418, cat. no. 13.12 [45]. The sample shows a cross-section of an edge of the object with porous unburnished gilding on the side (left) and compact burnished gilding on the face (top right). (See Fig. 6.1, p. 115.)

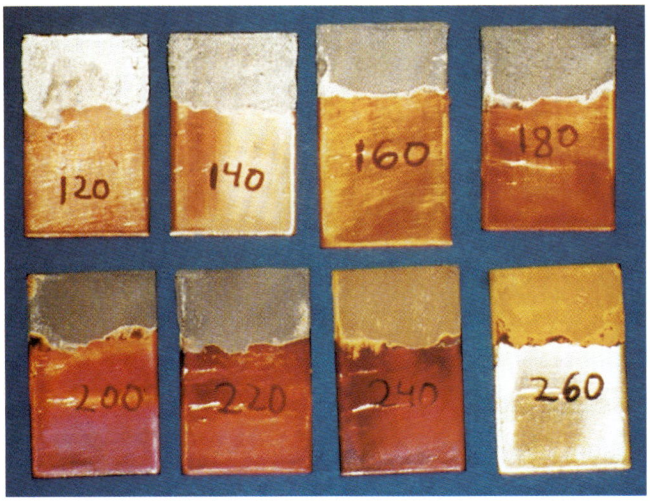

Plate 26 Replication fire gilding on copper: colour change of gold amalgam heated for 10 minutes at the temperature written on the sample (in °C). (See Fig. 6.2, p. 116.)

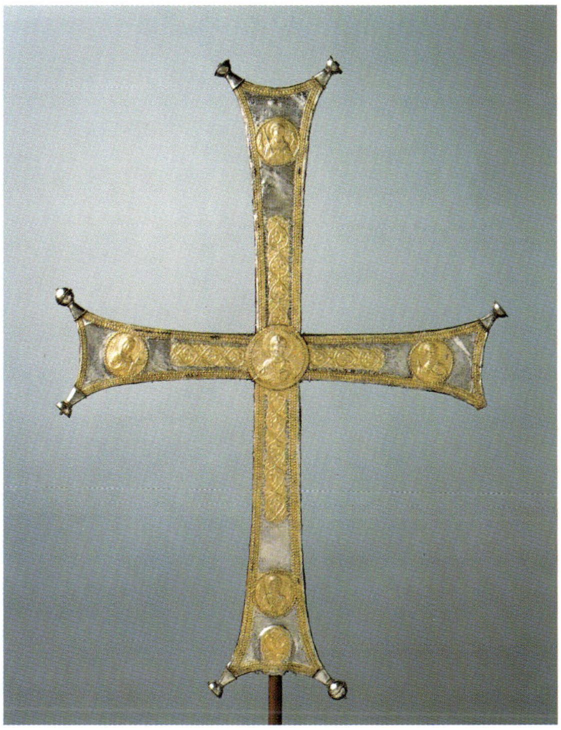

Plate 27 Processional Cross, after treatment. (The Metropolitan Museum of Art, Rogers Fund, 1993. (1993.163) (Photograph © 1996, The Metropolitan Museum of Art.) (See Fig. 7.3, p. 127.)

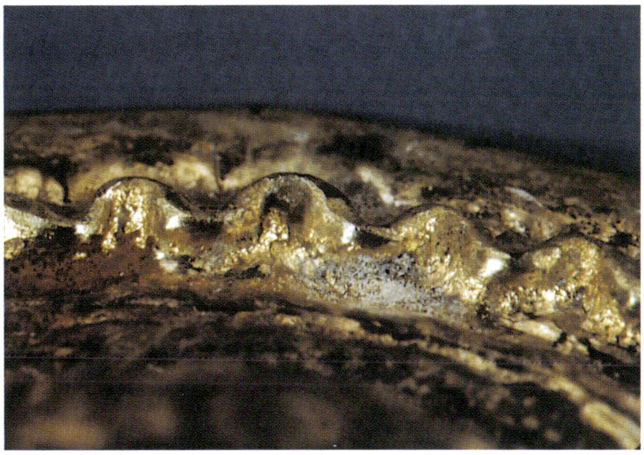

Plate 28 Detail of the border around the central medallion of Christ, before treatment. (The Metropolitan Museum of Art, Rogers Fund, 1993. (1993.163) Photograph, author, Sherman Fairchild Center for Objects Conservation, The Metropolitan Museum of Art.) (See Fig. 7.4, p. 128.)

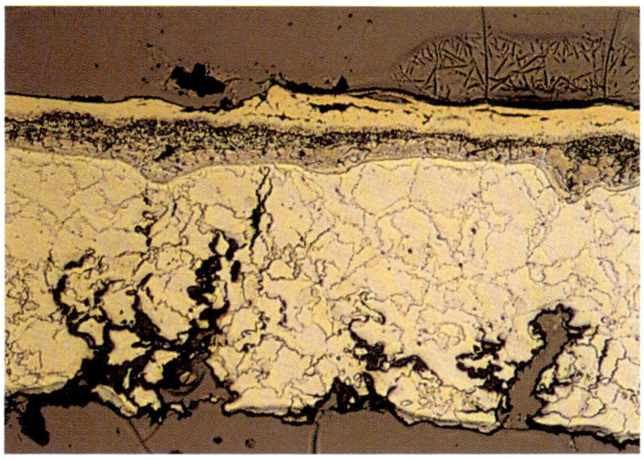

Plate 29 Detail of an etched cross-section from chalice 1986.3.6. (The Metropolitan Museum of Art, Purchase, Rogers Fund and Henry J. and Drue E. Heinz Foundation, Norbery Schimmel and Lila Acheson Wallace Gifts, 1986 (1986.3.2). Photograph, author, Sherman Fairchild Center for Objects Conservation, The Metropolitan Museum of Art.) (See Fig. 7.10, p. 132.)

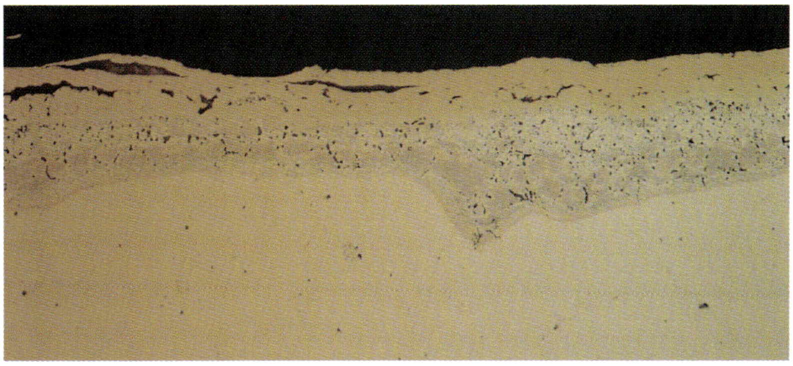

Plate 30 Detail of a polished cross-section from chalice 1986.3.6, ×1000. (The Metropolitan Museum of Art, Purchase, Rogers Fund and Henry J. and Drue E. Heinz Foundation, Norbery Schimmel and Lila Acheson Wallace Gifts, 1986 (1986.3.6). Photograph, author, Sherman Fairchild Center for Objects Conservation, The Metropolitan Museum of Art.) (See Fig. 7.11, p. 132.)

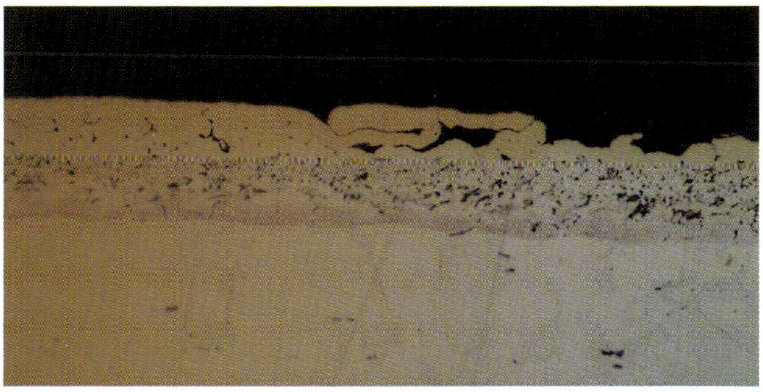

Plate 31 Detail of the polished cross-section from gilded surface of chalice 1986.3.9, ×1000. (The Metropolitan Museum of Art, Purchase, Rogers Fund and Henry J. and Drue E. Heinz Foundation, Norbery Schimmel and Lila Acheson Wallace Gifts, 1986 (1986.3.9). Photograph, A. Heywood, Sherman Fairchild Center for Objects Conservation, The Metropolitan Museum of Art.) (See Fig. 7.15, p. 134.)

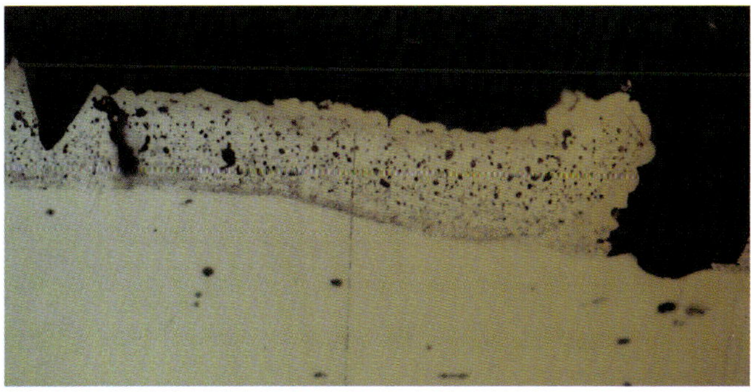

Plate 32 Detail of the polished cross-section of processional cross, 1993.163, ×1000. The diffusion of mercury well into the oxygenated zone is evinced by fingers of pale yellow colour. (The Metropolitan Museum of Art, Rogers Fund, 1993. (1993.163) Photograph, author, Sherman Fairchild Center for Objects Conservation, The Metropolitan Museum of Art.) (See Fig. 7.18, p. 136.)

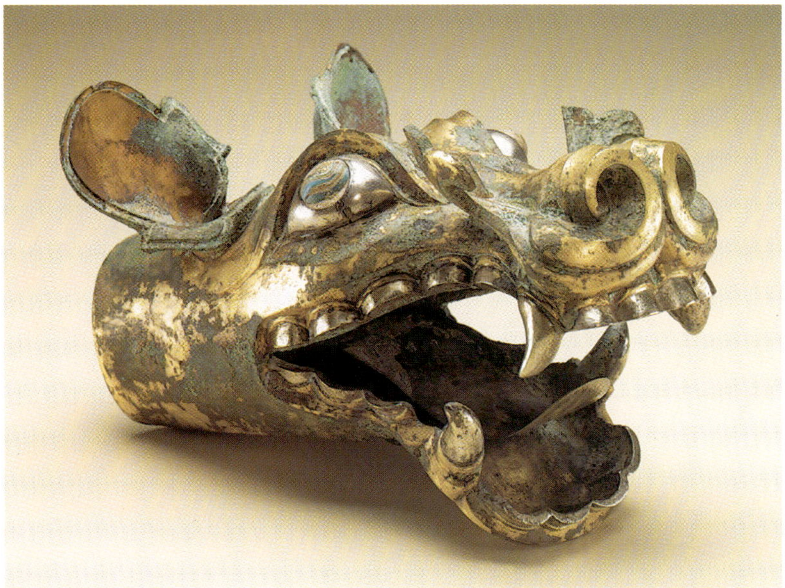

Plate 33 Terminal; bronze with gilding, silvering and inlay. (Courtesy of the Freer Gallery of Art, Smithsonian Institution, Washington, D.C., no. F1932.14.) (See Fig. 8.4, p. 149.)

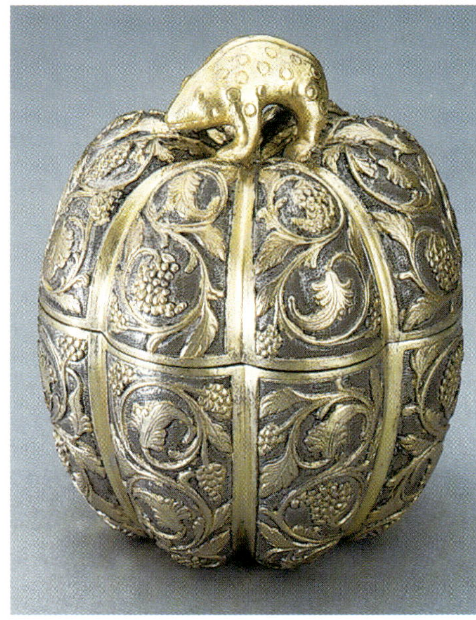

Plate 35 Container; gilt silver. (Courtesy of the Freer Gallery of Art, Smithsonian Institution, Washington, D.C., no. F1930.39.) (See Fig. 8.7, p. 152.)

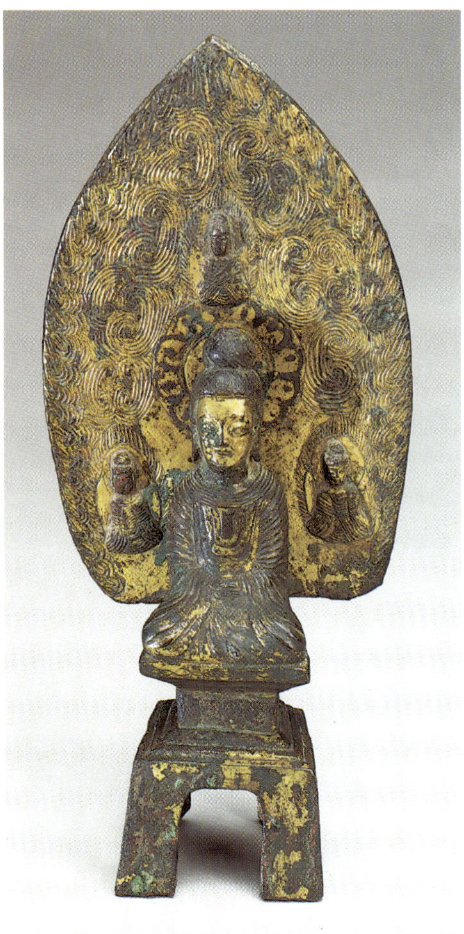

Plate 34 Buddha figure; bronze with gilding and paint. (Courtesy of the Freer Gallery of Art, Smithsonian Institution, Washington, D.C., no. F1911.121.) (See Fig. 8.5, p. 150.)

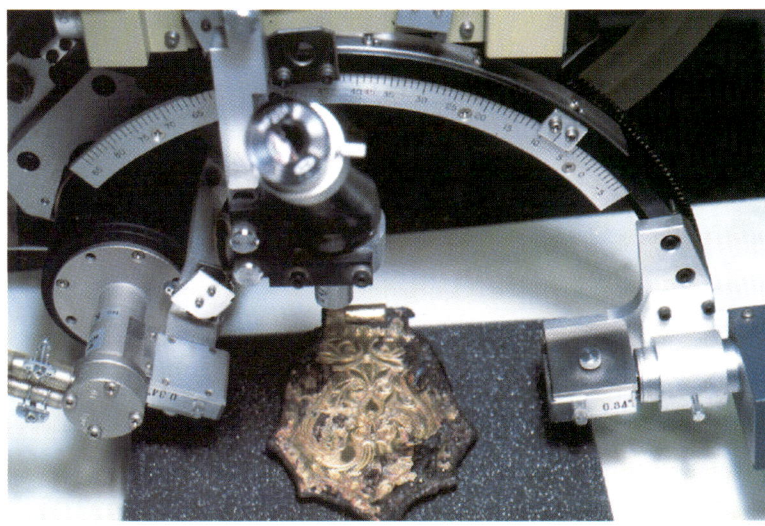

Plate 36 Non-destructive XRD apparatus. Here it is used for analysis on a horse ornament excavated in Fujinoki Tumulus. (See Fig. 9.7, p. 161.)

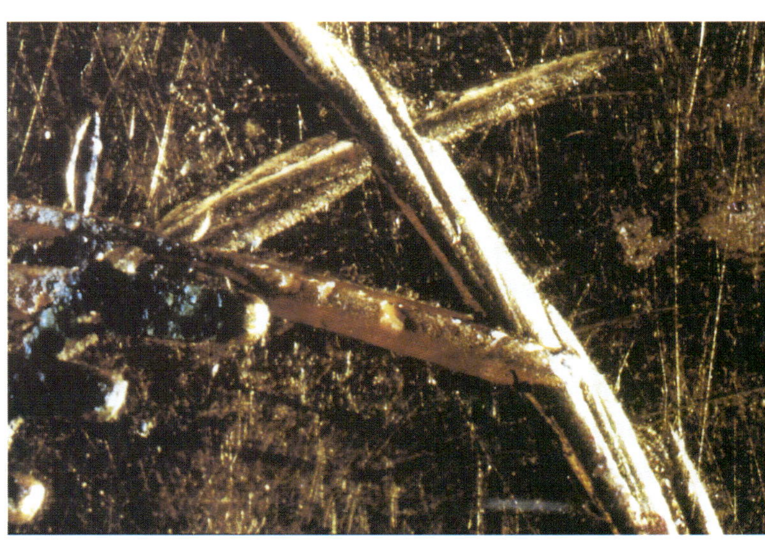

Plate 37 The detail of the gilded surface after the patina was removed. (See Fig. 9.9, p. 162.)

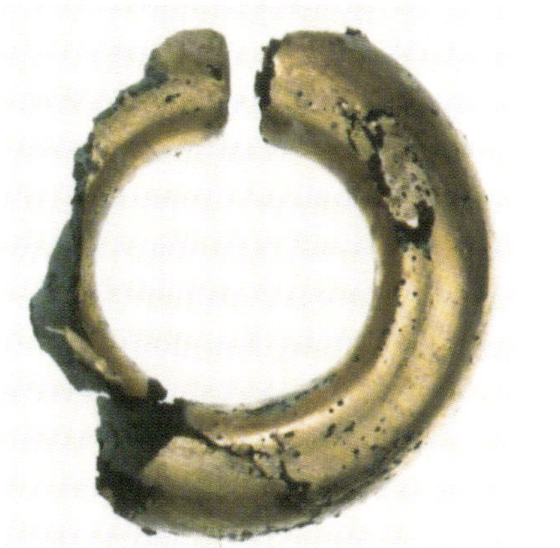

Plate 38 The earring excavated in *Takagawa* Tumulus, *Hyogo* prefecture. (See Fig. 9.10, p. 162.)

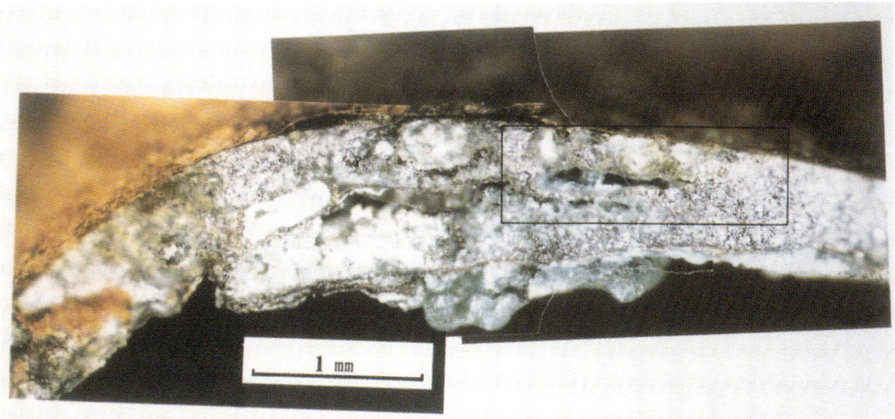

Plate 39 The area of overlapped copper on the earring is indicated by the rectangle. (See Fig. 9.11, p. 162.)

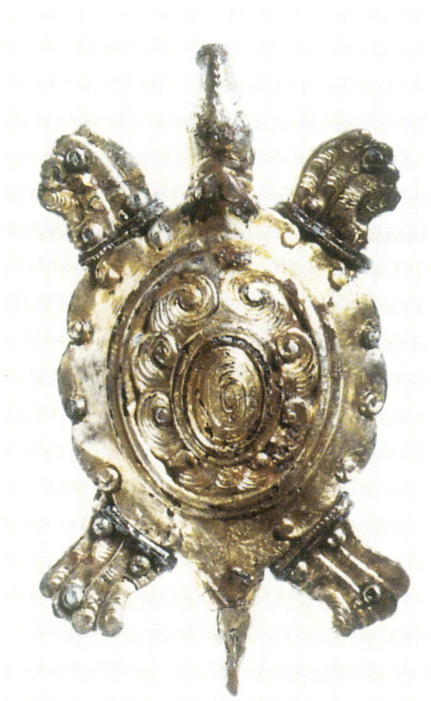

Plate 40 The turtle figure ornament excavated in *Sasatsuka* Tumulus, *Nagasaki* prefecture. (See Fig. 9.17, p. 164.)

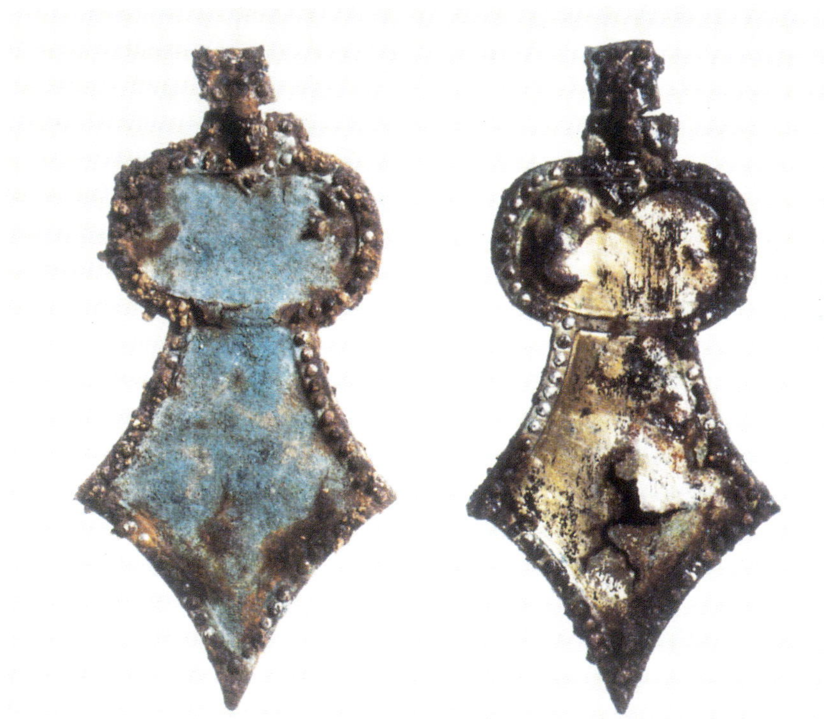

Plate 41 A pair of the horse trappings excavated in Daitsu-ji Tumulus, *Shiga* prefecture, before treatment (left) and after treatment (right). (See Fig. 9.21, p. 166.)

Plate 42 The analytical result of the cross-section of the rivet: yellow areas are iron and blue areas are silver. (See Fig. 9.22, p. 166.)

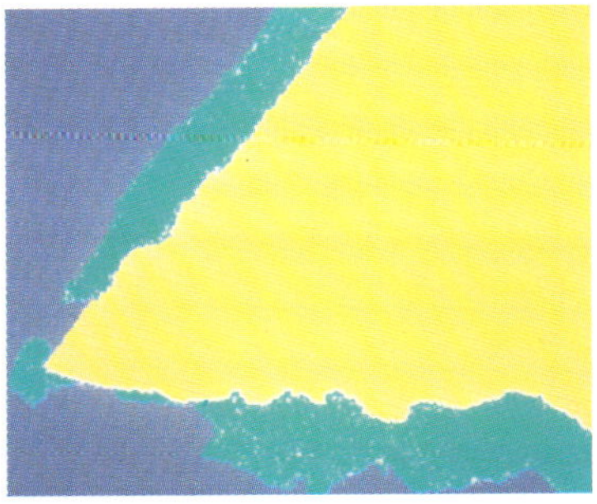

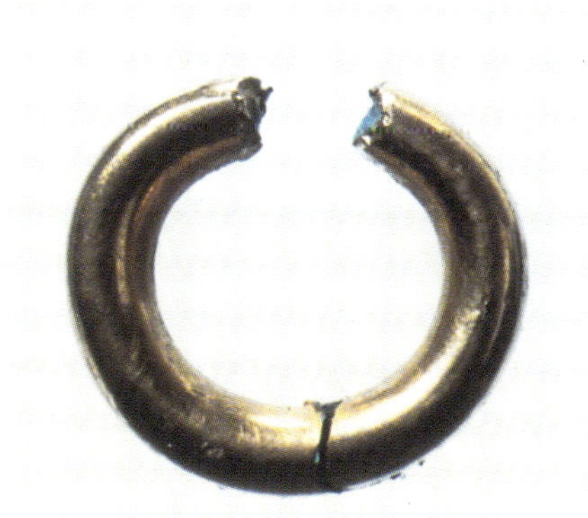

Plate 43 The earring excavated in *Oohara* Tumulus, *Shimane* prefecture. (See Fig. 9.23, p. 166.)

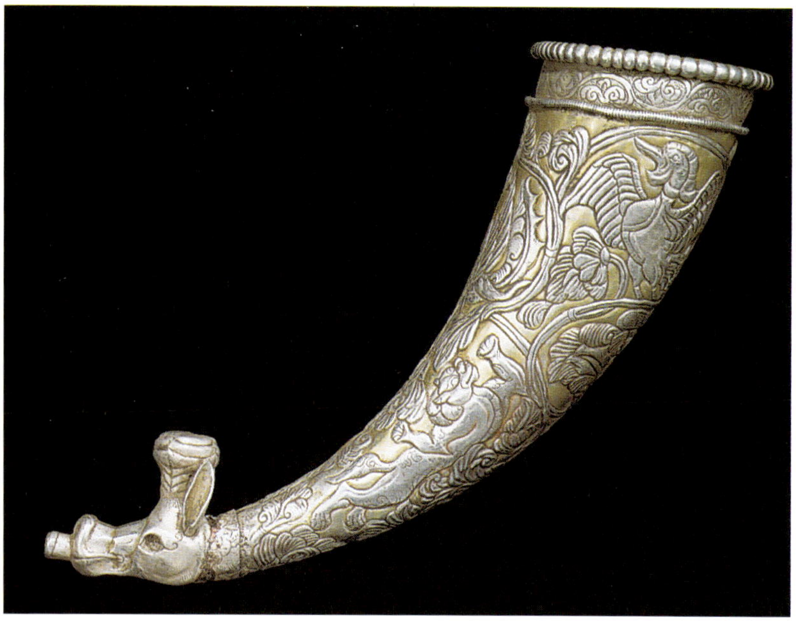

Plate 44 A gold–silver alloy foil coloured by depletion gilding was set into the recessed areas of the decoration to create the partial gilt appearance on the Tibetan rhyton (1988.67). The foil is held in place by driving the foil into the silver substrate using chasing tools around the perimeter of the gilt area. (© Cleveland Museum of Art.) (See Fig. 10.2, p. 170.)

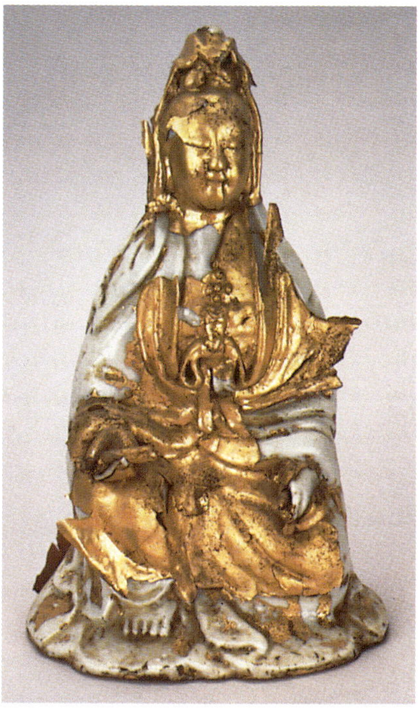

Plate 46 Vietnamese porcelain lacquer-gilded Buddha (48.2638). The smooth porcelain surface does not provide enough texture for the lacquer. It is readily lifting and curling off the entire surface. (D. Strahan, The Walters Art Gallery, Baltimore.) (See Fig. 11.2, p. 189.)

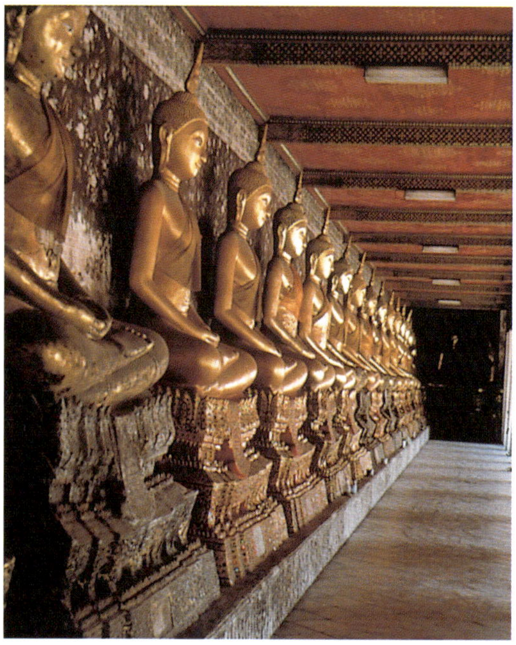

Plate 45 A gallery of copper alloy Buddhas coated with gilded lacquer at Wat Suthat in Bangkok, Thailand. (D. Strahan.) (See Fig. 11.1, p. 188.)

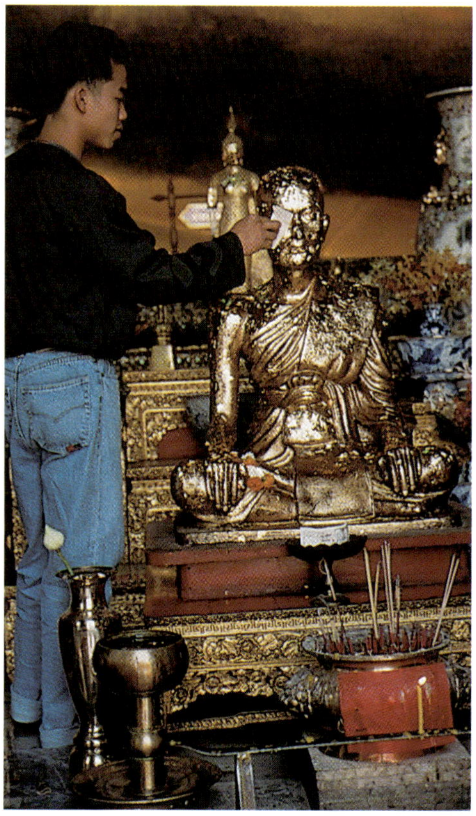

Plate 48 A devotee applying 'daily gilding' to an image in a temple in Bangkok. (D. Strahan.) (See Fig. 11.4, p. 190.)

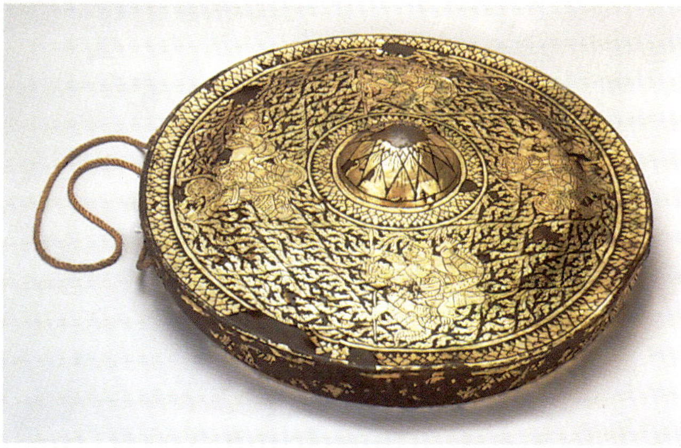

Plate 47 Nineteenth century Thai copper alloy gong (54.2836) with lifting black lacquer and gilding. (D. Strahan, The Walters Art Gallery, Baltimore.) (See Fig. 11.3, p. 189.)

Plate 49 The earliest identified lacquer-gilded Buddha in the Walters (54.2802). (The Walters Art Gallery, Baltimore.) (See Fig. 11.5, p. 190.)

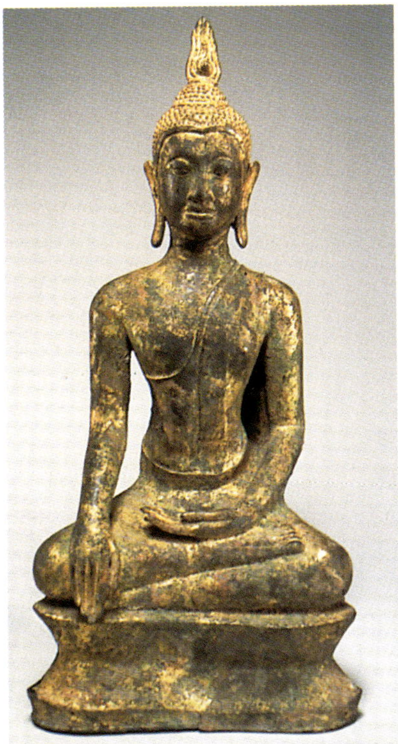

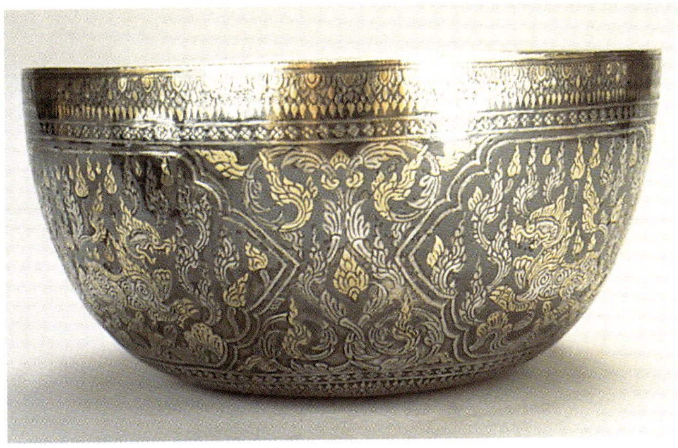

Plate 50 This silver partial gilt and niello bowl (45.21) has been gilded using mercury gilding. (D. Strahan, The Walters Art Gallery, Baltimore.) (See Fig. 11.6, p. 190.)

Plate 51 Image in Nakhon Pathom with multiple layers of gold leaf. (D. Strahan.) (See Fig. 11.7, p. 191.)

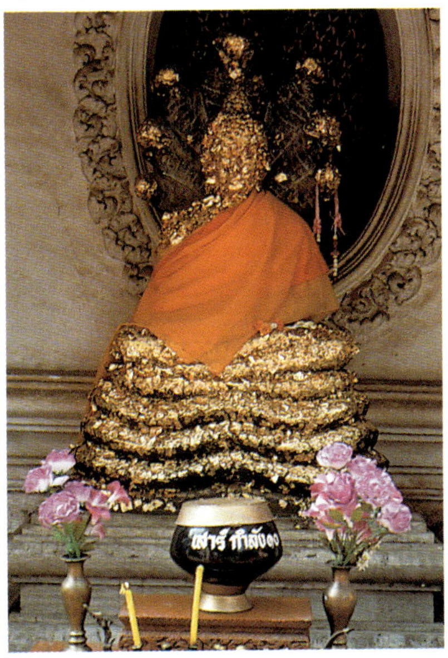

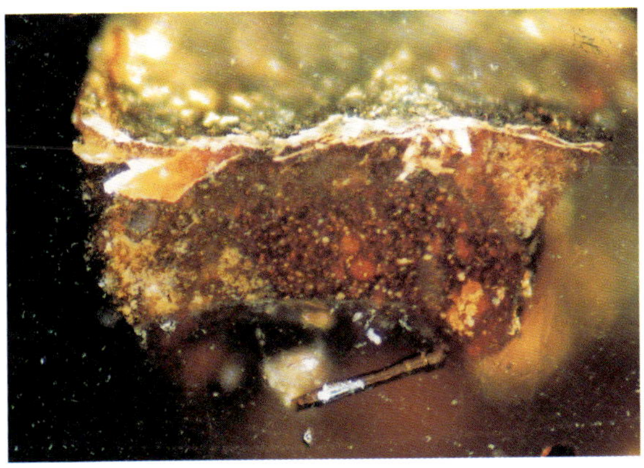

Plate 52 Optical microscopic cross-section under reflected light of Early Ayutthaya image 54.2802 (20×). Layers of bulked lacquer with gold leaf on top. (D. Strahan.) (See Fig. 11.8, p. 192.)

Plate 53→ Ultraviolet light microscopic cross-section of the same area in Figure 11.8 (54.2802). While difficult to see in visible light, the inclusions in the lacquer layer are clearly seen in ultraviolet light (20×). (D. Strahan.) (See Fig. 11.9, p. 192.)

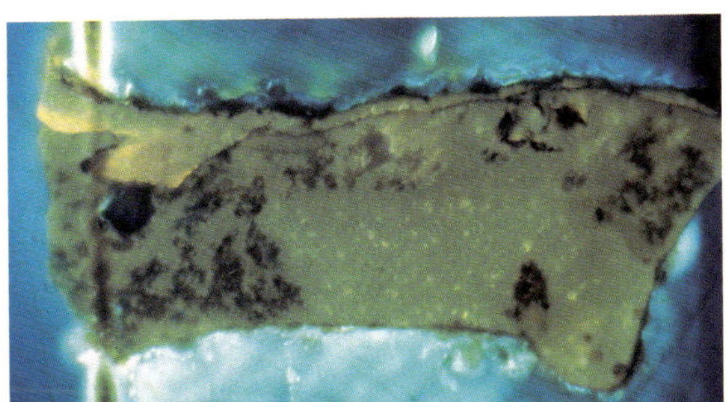

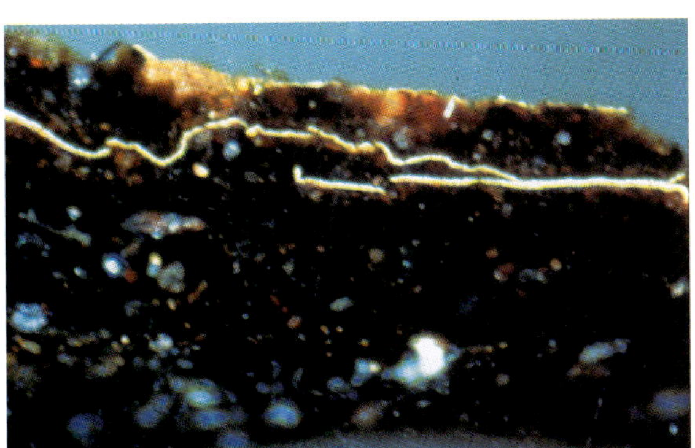

Plate 54 Optical microscopic cross-section of a Sukhothai Period image (54.2775) showing three distinct lacquer layers (20×). (D. Strahan.) (See Fig. 11.10, p. 193.)

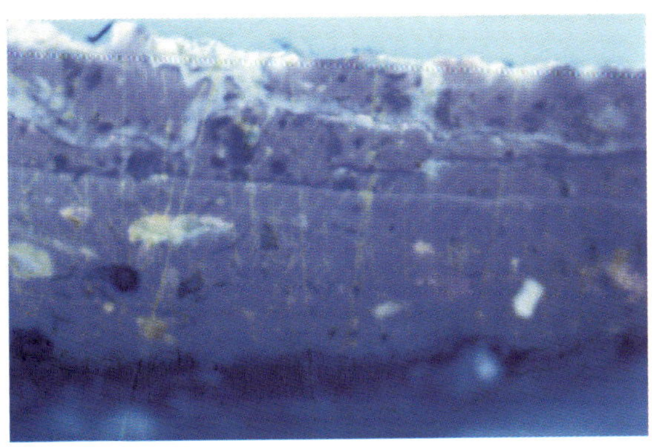

Plate 55 Ultraviolet light microscopic cross-section of the same area in Figure 11.10 (54.2775). Once again burnt plant material mixed with the lacquer becomes clearer under ultraviolet light (20×). (D. Strahan.) (See Fig. 11.11, p. 193.)

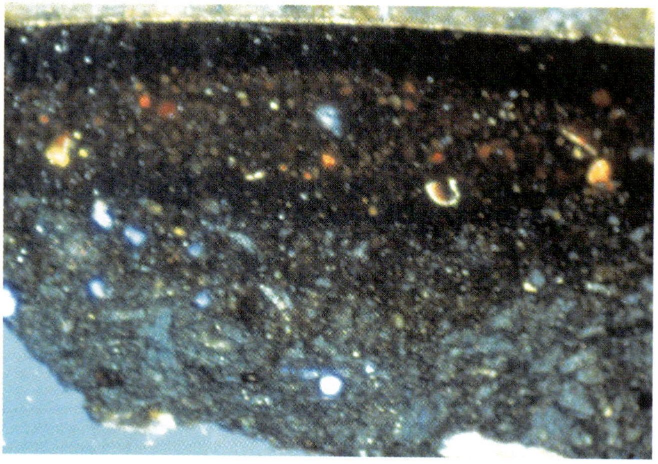

Plate 56 Optical microscopic cross-section of Late Ayutthaya image (54.2553) (20×). (D. Strahan.) (See Fig. 11.14, p. 194.)

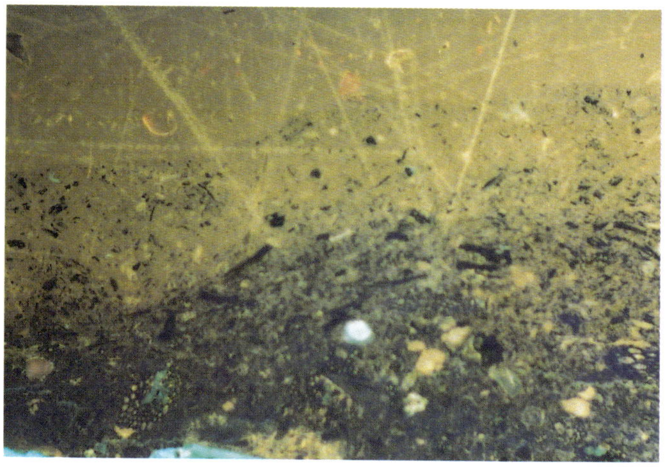

Plate 57 Ultraviolet light microscopic cross-section of the same area in Figure 11.14 (54.2553). The inclusions and gold flakes mixed in three of the four lacquer layers are clearly seen under ultraviolet light. (D. Strahan.) (See Fig. 11.15, p. 194.)

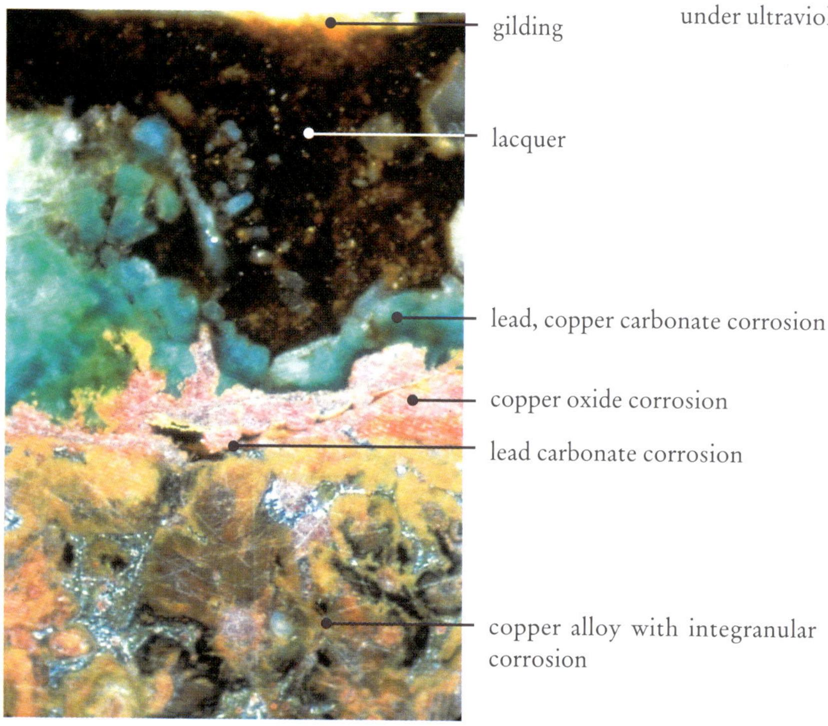

Plate 58 Dark field microscopic cross-section of 54.2684 showing metal alloy, corrosion and lacquer gilding layers (20×). (D. Strahan.) (See Fig. 11.18, p. 198.)

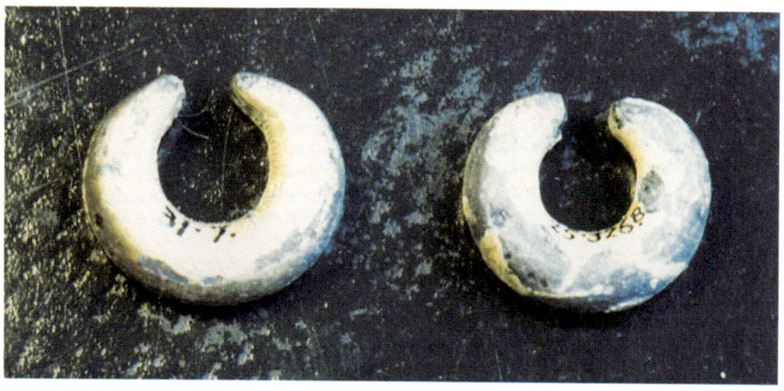

Plate 59 Two penannular nose rings from the area of Chanduy, Provincia de Guayas, Ecuador. The gilding, partially worn, has been made by wrapping gold foil over the copper. (Cambridge University Museum of Archaeology and Ethnology. 19 mm and 17.5 mm across.) (See Fig. 12.1, p. 205.)

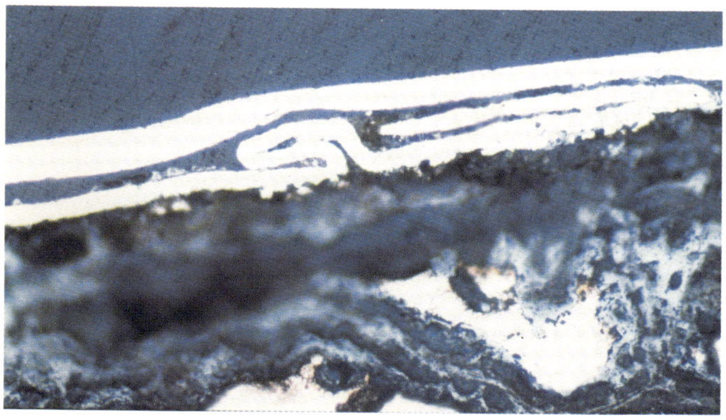

Plate 60 Foil-gilded layers over a pennanular nose ornament from La Compania, Los Rio Province, Milagro phase, Ecuador. The nose ring has been gilded by diffusion bonding of gold foil. Overlapping layers of gilding can be seen in this photomicrograph, together with a fold in the lower layer. In some areas interdiffusion is extensive and only one, apparently thicker layer of gilding can be seen. (Unetched, ×210.) (See Fig. 12.2, p. 205.)

Plate 61 Bicoloured nose ornament, Frías in style, Peru, dated to AD200–500. Eight silver crested bird-heads are arranged symmetrically around a gilt ovoid nose ornament, which has been made by foil gilding the silver substrate of which the whole ornament is made. The eye of each bird is inlaid with turquoise. (Private collection.) (See Fig. 12.3, p. 206.)

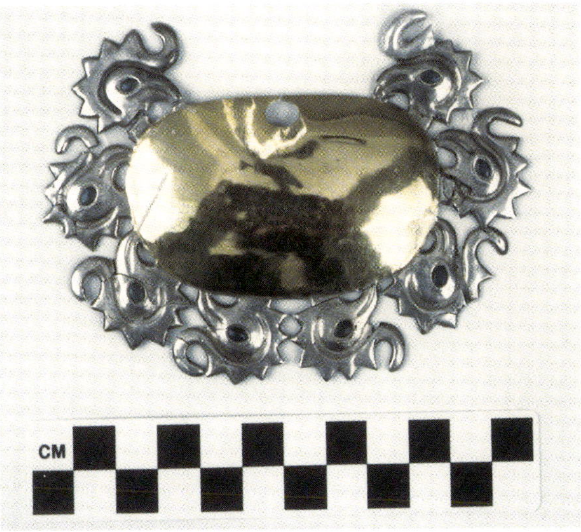

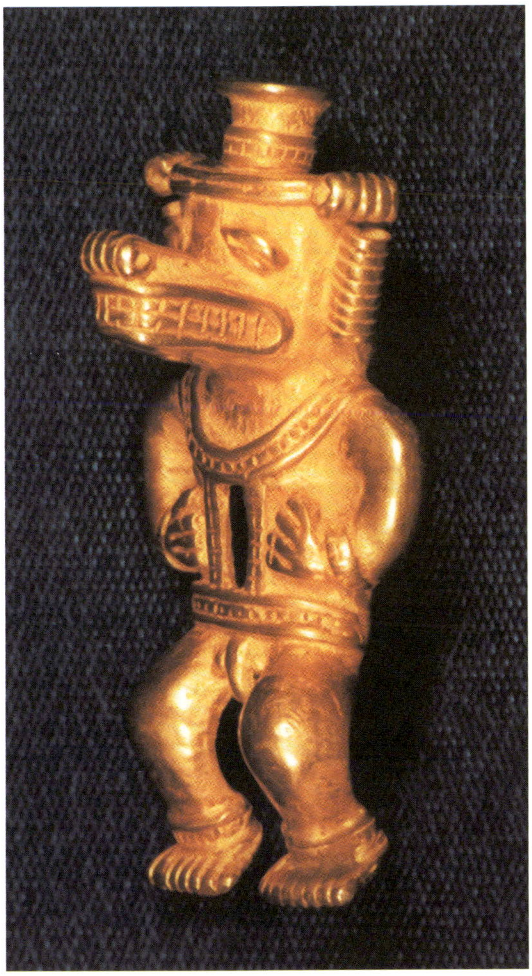

Plate 62 Tumbaga alloy pendant in the form of a human figurine with animal duality, from Costa Rica. The figurine is from the Veraguas-Gran Chiriqui area, and is dated to AD700–1500. Cast in a gold–copper alloy with 54 wt % gold, the surfaces are finished by depletion gilding. (Height 9.6 cm.) (See Fig. 12.6, p. 211.)

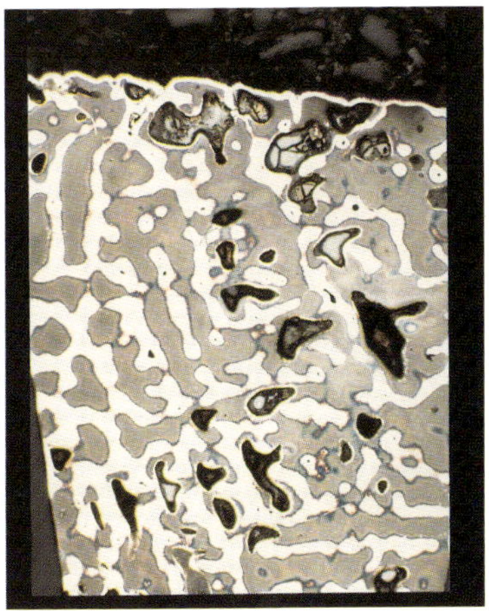

Plate 63 Cast tumbaga alloy from the Tairona area of pre-Hispanic Colombia, showing internal corrosion of copper-rich areas with lighter regions preserving gold-rich dendritic regions where the heavily cored structure preserves considerable segregation from casting. The upper surface is depletion gilded. (Unetched, ×85.) (See Fig. 12.7, p. 211.)

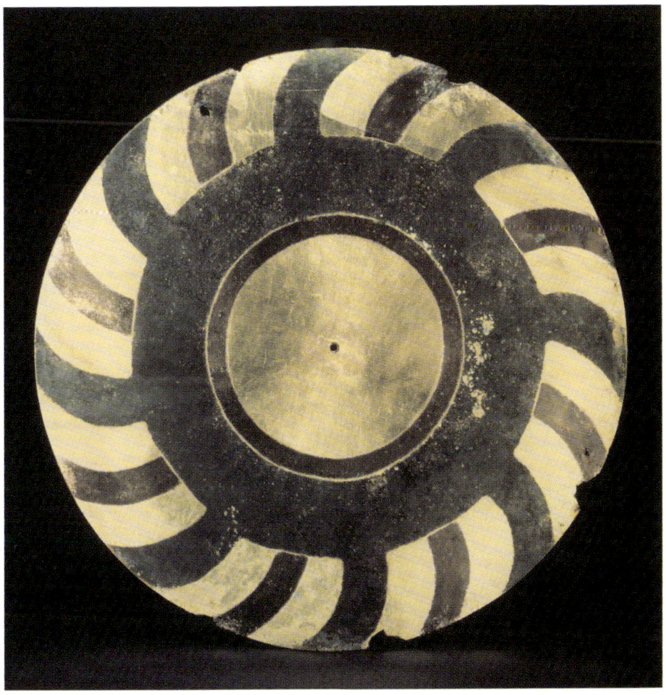

Plate 64 Bicoloured flat tumbaga disc from the Municipio of El Tambo, Department of Nariño, Colombia. The disc is 15.3 cm in diameter, and weighs 68 g. It is from the Piartal Period, dated AD800–1200. The tumbaga alloy, containing about 25 wt % gold, has been carefully smoothed and selectively finished by depletion gilding to create skilfully designed geometric colour contrast on both sides of the disc. These discs are typically 0.30–0.35 mm thick. Internal corrosion makes them very brittle. (Museo del Oro: Bogotá: 21,521.) (See Fig. 12.9, p. 212.)

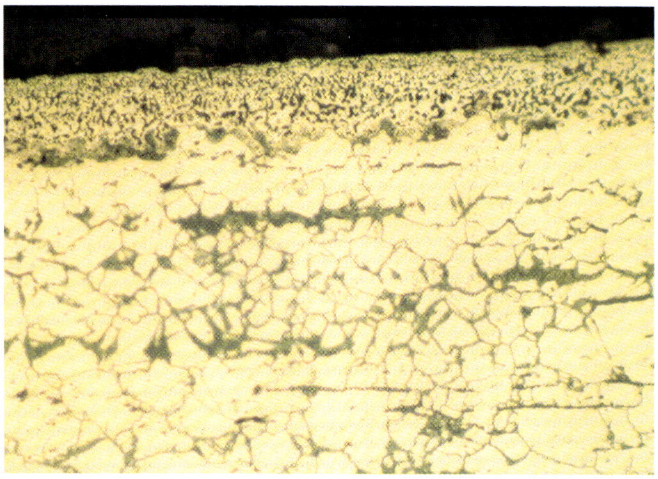

Plate 65 Photomicrograph showing depletion gilded layer on a Nariño sheet tumbaga. Note the development of island-and-channel morphology in the disturbed surface zone where copper removal has occurred during surface depletion. Some oxide stringers can be seen in the section, along with recrystallised grains of the gold–copper alloy sheet. (Magnification ×125, etched in potassium cyanide and ammonium persulphate for 15 s. Bright field illumination.) (See Fig. 12.10, p. 212.)

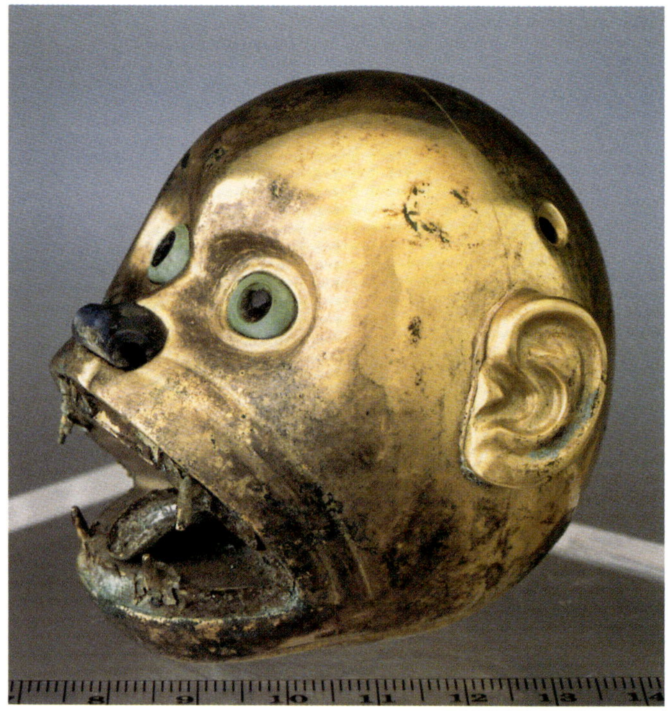

Plate 66 Monkey figurine, Moche Period, Peru. Sipan in style, probably dating to the early centuries AD. The monkey is made in a gold alloy, whose composition is given in Table 12.2, but the long teeth are made in copper that has been covered with gold by the fusion gilding technique. The nose and eyes are made in stones that have not been analysed. The monkey has been carefully made in a number of pieces that are joined together with soldering. A join can just be seen in the head, passing across the top of the skull. (Private collection.) (See Fig. 12.12, p. 214.)

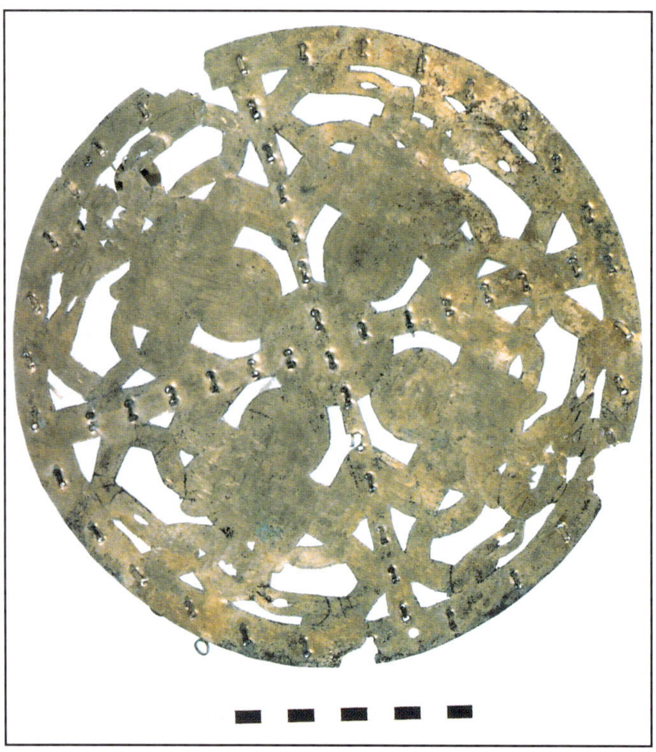

Plate 67 Large profile ornamental disc, Vicus in style, from the north coast of Peru. The disc represents four crab motifs. The flat sheet, about 1 mm thick, has been worked and annealed to shape, followed by the cutting out of the design, and gilding by the electrochemical replacement plating process. (Diameter 23.4 cm; gilded layer about 1 micron thick.) (See Fig. 12.13, p. 216.)

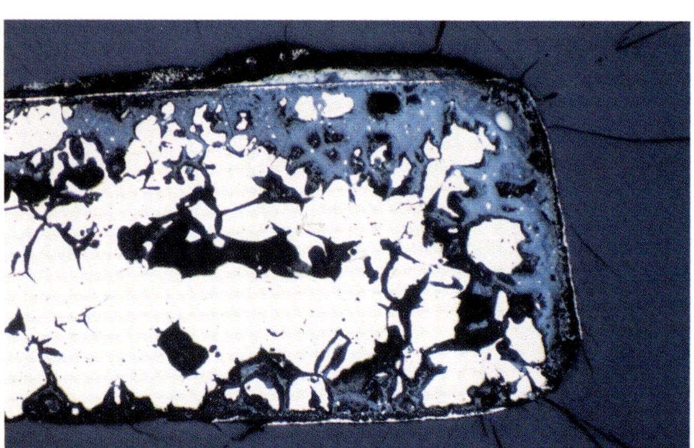

Plate 68 Photomicrograph of part of a cross-section through a Vicus dangle, showing the preservation of the electrochemical replacement plating over corroded copper interior. Note that the thin gold layer passes round the dangle and coats all surfaces of the object. (Magnification ×100. Unetched.) (See Fig. 12.14, p. 216.)

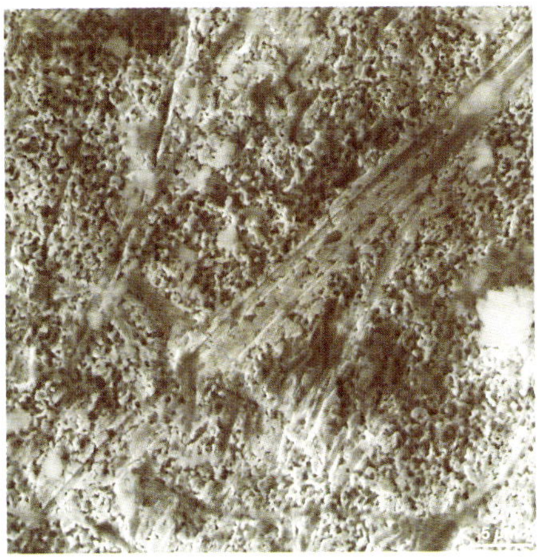

Plate 69 Surface morphology of gold wire from La Tolita, showing surface dissolution of this high gold content wire from burial. The scale bar at the bottom right is 5 microns, indicating the extensive dissolution of the surface that has occurred. (See Fig. 12.15, p. 217.)

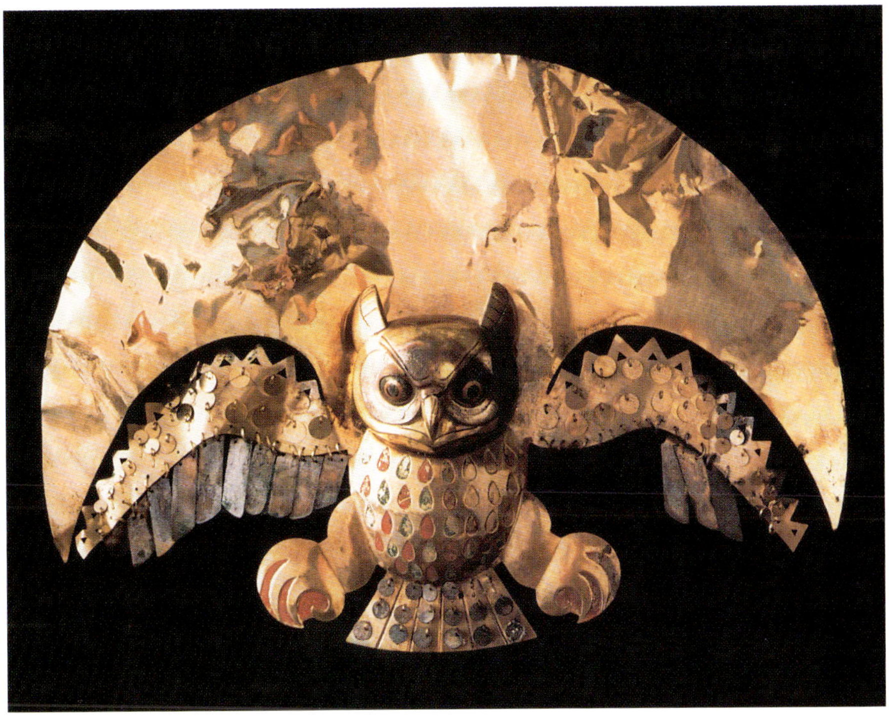

Plate 70 Spectacular Moche gold owl pectoral, site of La Mina, dated to 1st century AD. Not only does the golden owl have inlaid eyes and numerous dangles, some of which look like gilded copper, probably electrochemical replacement gilded, but the repousséed teardrop shaped feathers preserve red and green paint in alternating colours. The essentially intaglio design has helped preserve applied pigment. (Length 45 cm.) (Professor Christopher Donnan, UCLA, Fowler Museum of Cultural History) (See Fig. 12.16, p. 219.)

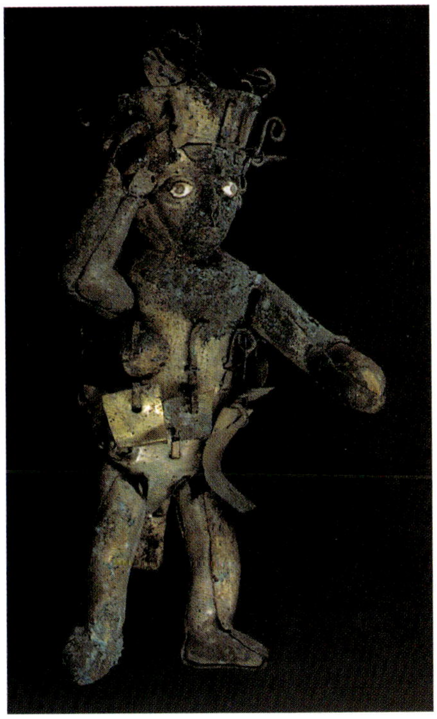

Plate 71 Gilded copper warrior figure, Moche, from Loma Negra. The Metropolitan Museum of Art, Gift of Jane Costello Goldberg, from the collection of Arnold I. Goldberg, 1981 (1981.459.32). (Photograph: E. G. Howe). (See Fig. 13.2, p. 225.)

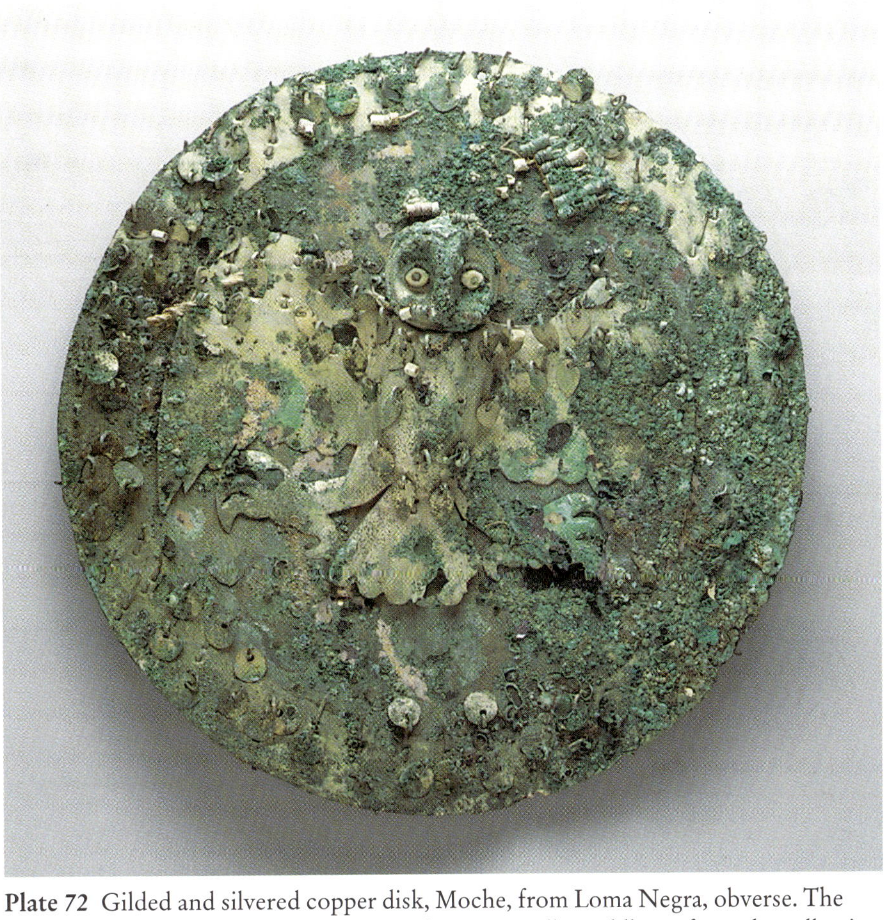

Plate 72 Gilded and silvered copper disk, Moche, from Loma Negra, obverse. The Metropolitan Museum of Art, Bequest of Jane Costello Goldberg, from the collection of Arnold I. Goldberg, 1986 (1987.394.56). (Photograph: B. J. Schwarz). (See Fig. 13.3, p. 225.)

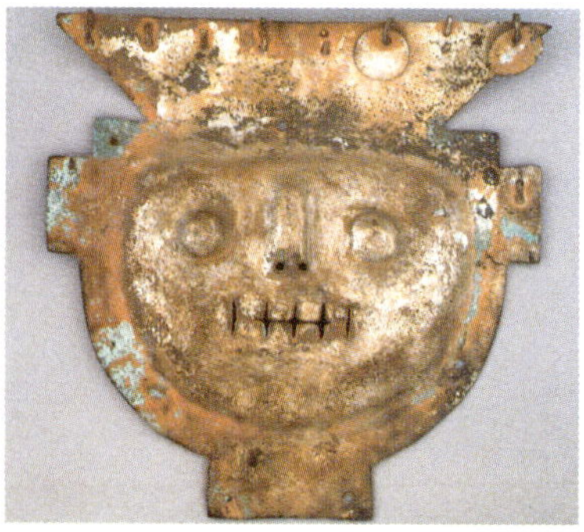

Plate 73 Gilded copper plaque, Vicús, obverse. The Metropolitan Museum of Art, Bequest of Jane Costello Goldberg, from the collection of Arnold I. Goldberg, 1986 (1987.394.169). (See Fig. 13.4, p. 226.)

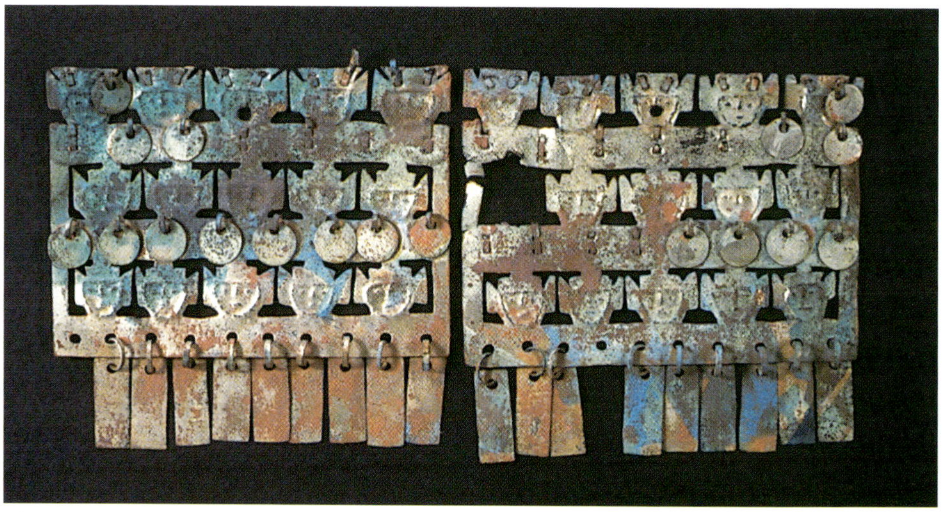

Plate 75 Gilded copper plaques, Vicús, obverse. The Metropolitan Museum of Art, Bequest of Jane Costello Goldberg, from the collection of Arnold I. Goldberg, 1986 (left: 1987.394.252, right: 1987.394.251). (See Fig. 13.6, p. 226.)

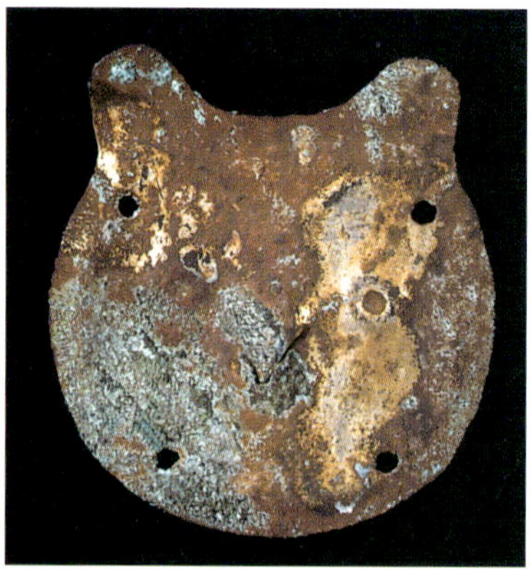

Plate 74 Gilded copper plaque, Vicús, obverse. The Metropolitan Museum of Art, Bequest of Jane Costello Goldberg, from the collection of Arnold I. Goldberg, 1986 (1987.394.599). (See Fig. 13.5, p. 226.)

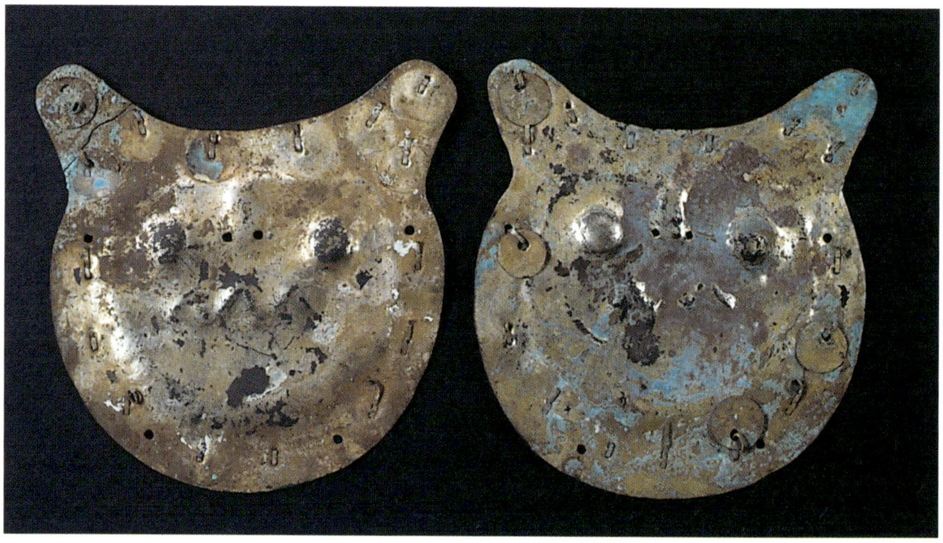

Plate 76 Gilded copper plaques, Vicús, obverse. The Metropolitan Museum of Art, Bequest of Jane Costello Goldberg, from the collection of Arnold I. Goldberg, 1986 (left: 1987.394.598, right: 1987.394.596). (See Fig. 13.7, p. 226.)

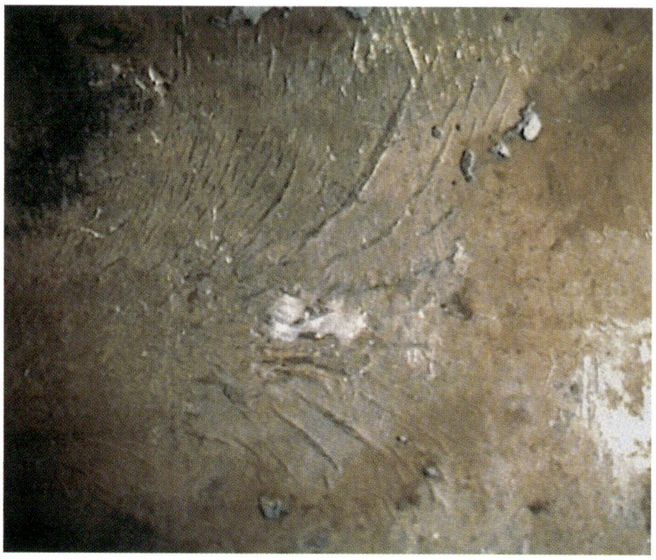

Plate 77 Gilded copper plaque, Vicús (MMA 1987.394.597), photomicrograph of the surface showing the presence of wrinkles. Original magnification: 35× (Photograph: E.G. Howe). (See Fig. 13.12, p. 232.)

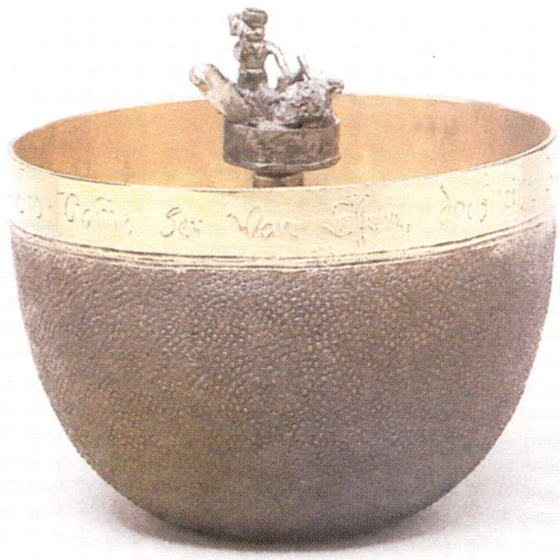

Plate 78 Herrengrun beaker (Turner 1974: plate 1) (See Fig. 14.1, p. 243.)

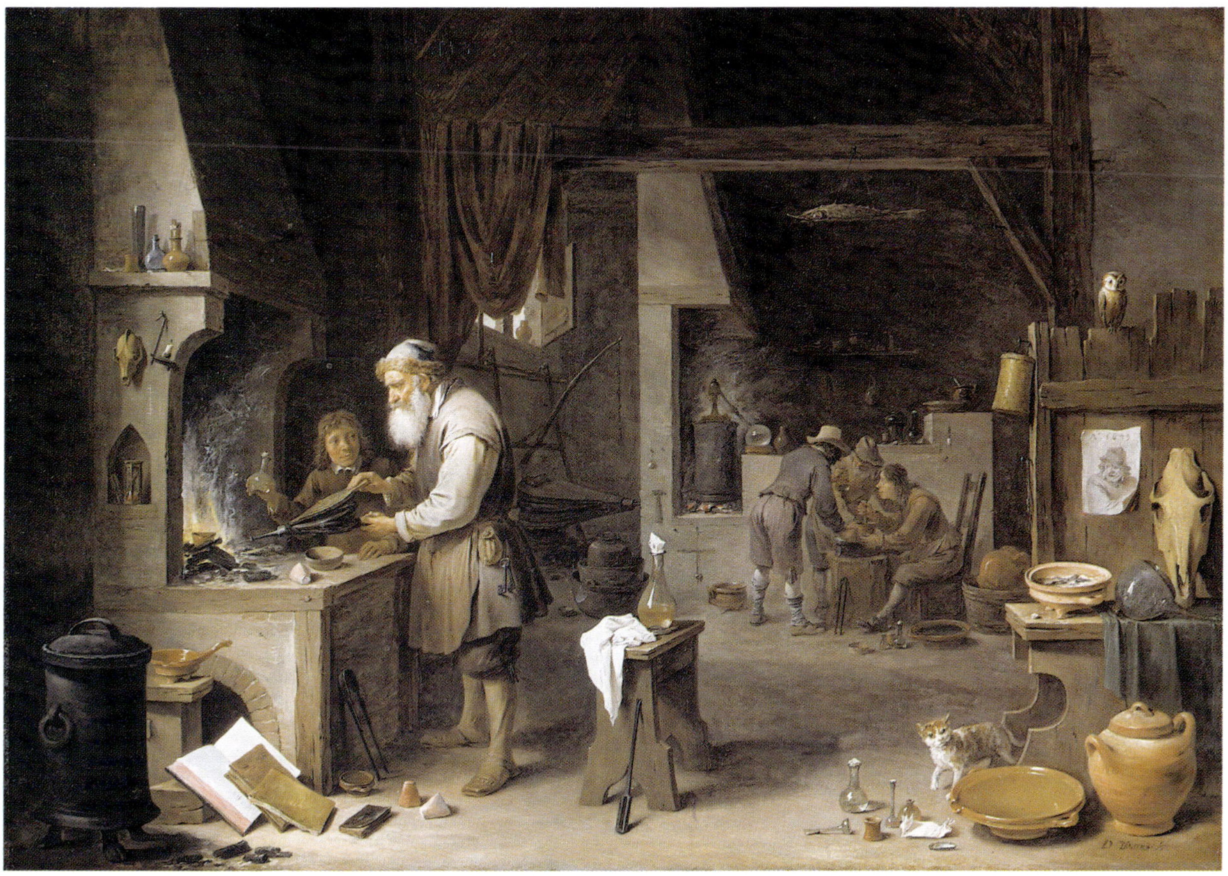

Plate 79 *The Alchemist*, David Teniers, 1649 (Philadelphia Museum of Art). (See Fig. 14.2, p. 243.)

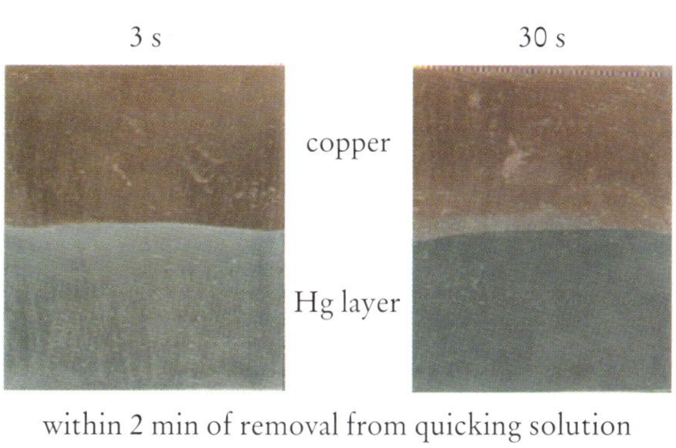

Plate 80 Copper samples A immersed in 1.0% w/w $Hg(NO_3)_2$ for 3 s and for 30 s at room temperature. Relatively little mercury (<1 μm average thickness) is deposited, although the surfaces appear even and silver-coloured to the eye.
(See Fig. 14.5, p. 251.)

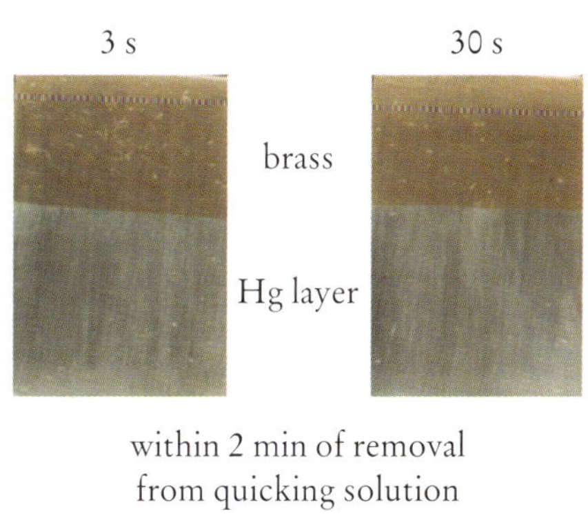

Plate 81 Brass samples B immersed in 1.0% w/w $Hg(NO_3)_2$ for 3 s and 30 s at room temperature and photographed within two minutes of depositing the mercury mirror.
(See Fig. 14.6, p. 251.)

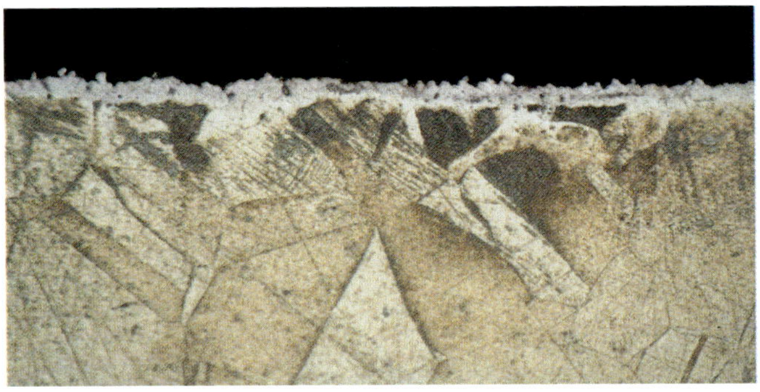

Plate 82 Brass sample C with mercury primer and leaf gilding, and heated at 500 ± 15°C for 30 min. As many as ten sheets of 23½ carat gold could be stuck/amalgamated on the brass before excess leaf sloughed off. BF 500×. The average thickness of the gold layer in this sample is 12 m; the amalgam attack along some of the grain lines penetrates 25 m. (See Fig. 14.7, p. 252.)

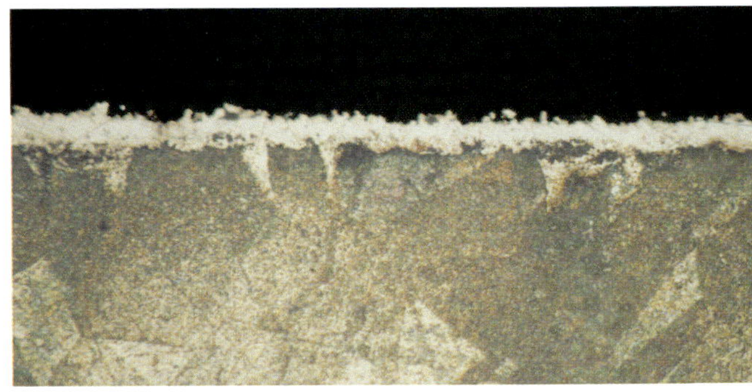

Plate 83 Brass sample D with 1% mercuric nitrate primer and leaf gilding, and heated at 500 ± 15°C for 30 min. The maximum capacity of the quicked surfaces was four leaves under these conditions. BF 500×. (See Fig. 14.8, p. 252.)

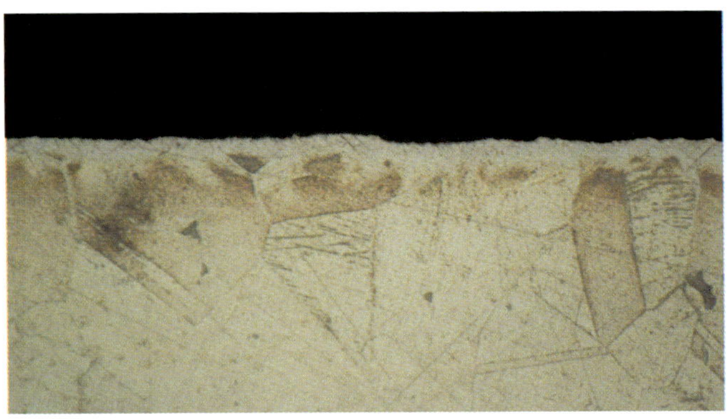

Plate 84 Brass sample E with amalgam gilding without primer. BF 500×. (See Fig. 14.9, p. 253.)

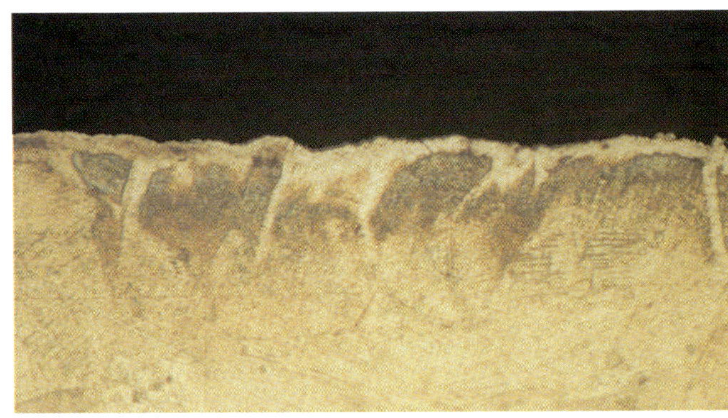

Plate 85 Brass sample F quicked in mercuric nitrate solution (as described for sample D) with amalgam gilding. BF 500×. (See Fig. 14.10, p. 253.)

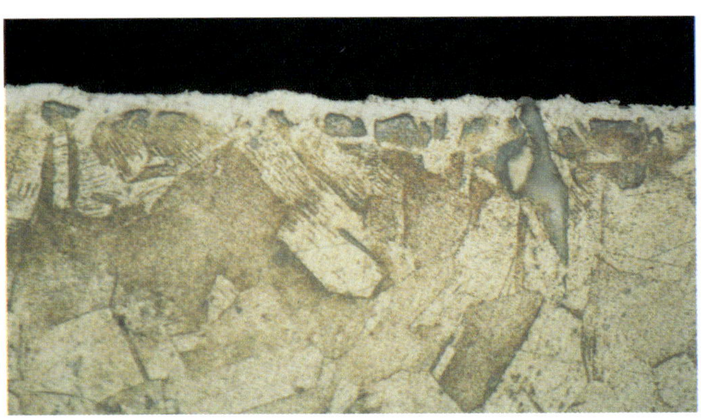

Plate 86 Brass sample G plated with mercury, plated with gold, and heated. BF 500×. (See Fig. 14.11, p. 254.)

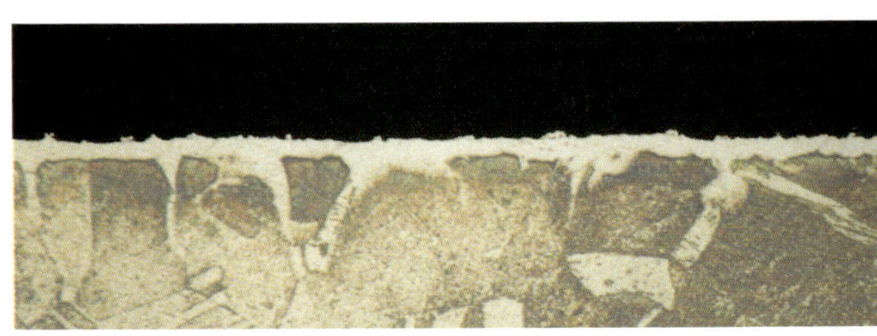

Plate 87 Brass sample H quicked in mercuric nitrate solution, gold plated, and heated. BF 500×. (See Fig. 14.12, p. 254.)

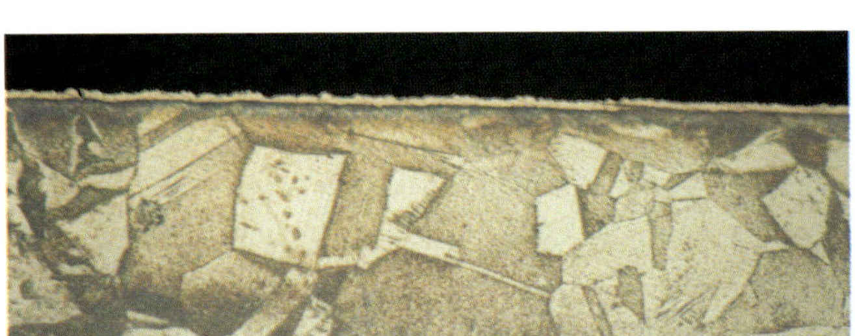

Plate 88 Brass sample I gold plated without a mercury primer or quick. The gold thickness ranges from 2 μm to 2.5 μm. BF 500×. (See Fig. 14.13, p. 254.)

quicking, 5×

immersion plating

Plate 89 Brass sample J prepared by alternately dipping in mercuric nitrate and gold cyanide. (See Fig. 14.14, p. 255.)

Plate 90 Cross-section of mercury gilding on bronze, taken from the mount illustrated in this article. (Photo: David Scott.) (See Fig. 16.2, p. 284.)

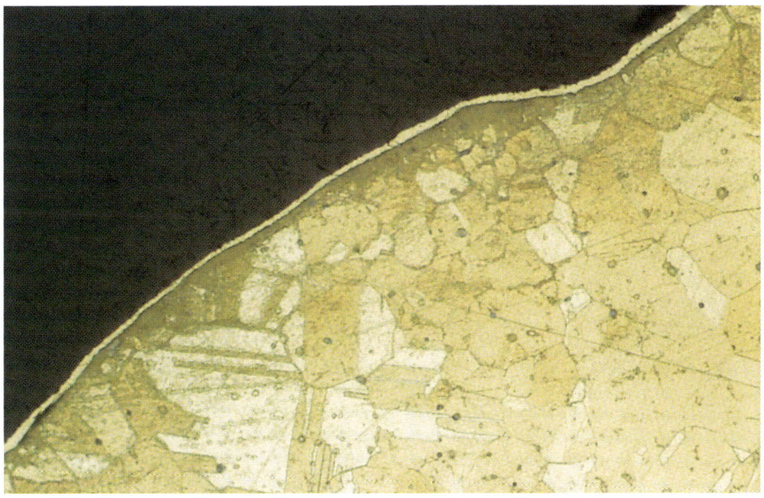

Plate 91 The mount to be reproduced, a *chute* by Dubois in the Rococo style. This foundry uses an alloy of 86.6% copper, 6.6% tin, 3.3% lead and 3.3% zinc. (Photo: Abby Hykin.) (See Fig. 16.8, p. 287.)

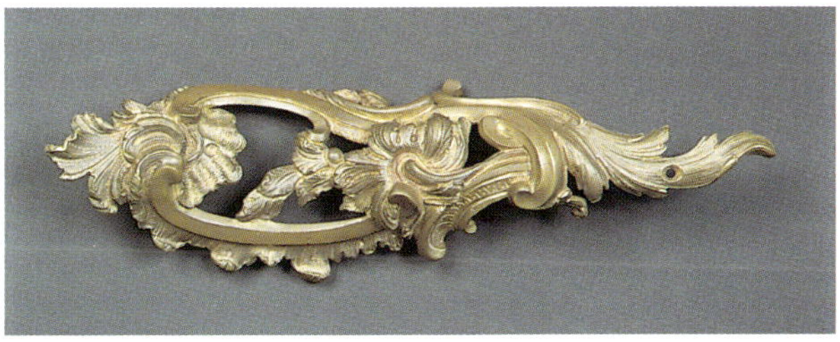

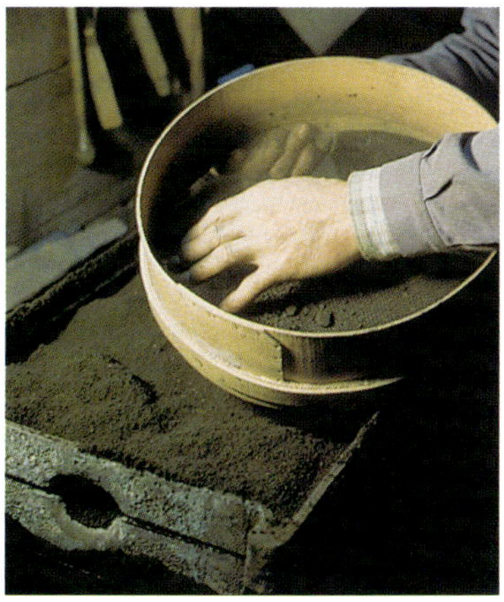

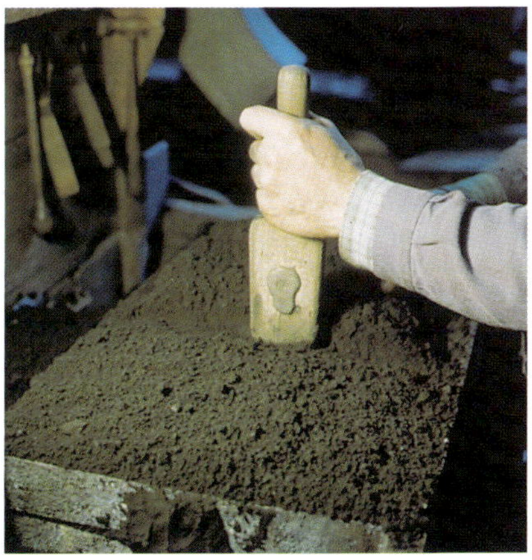

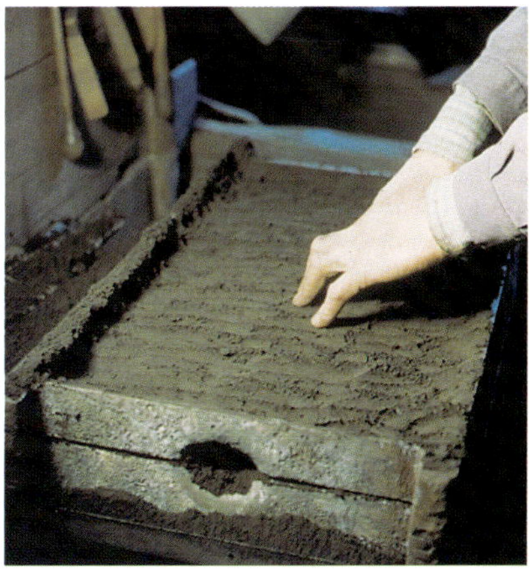

Plate 92 To cast the bronze mount, the founder fills the steel chassis with sifted sand. The pour holes for the molten bronze are visible on the near edge of the chassis. The sand is called Fountenay aux Roses. It is very fine and a yellow colour before it has been used and is composed of silica and aluminum. (All photos in Plates 92–137 are by Francis Razon, Paris, except where noted.) **Plate 93** The humid sand is compacted and will become very hard. **Plate 94** The sand is evenly spread out to have the same consistency everywhere. The bronzes to be recast will be placed on this bed of sand and then the counter caul will be placed on top. (See Figs 16.9–16.11, p. 286.)

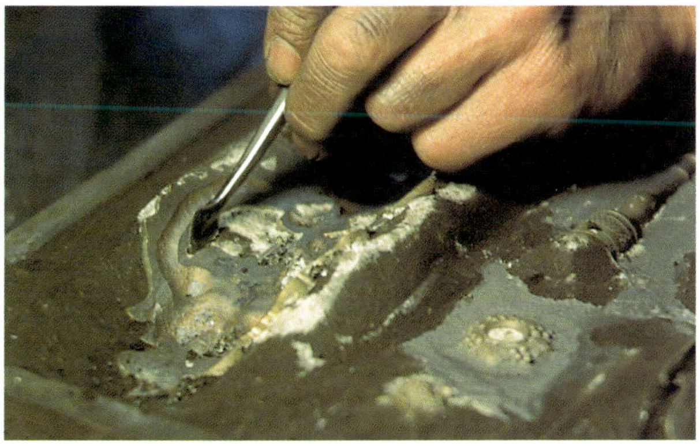

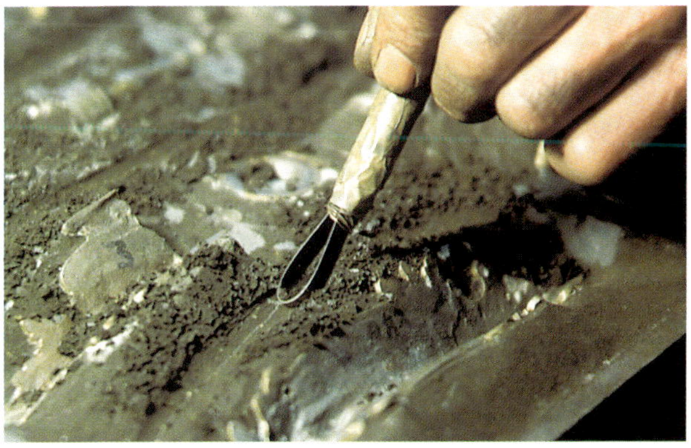

Plate 95 Having been covered with talc to facilitate their eventual removal, the models are pressed into the compacted sand face down. The founder then cleans the displaced sand from around the edges with a variety of fine scraping tools. This operation is important to avoid jagged edges on the mount at this point where the two sides of the mold meet. **Plate 96** With a scraping tool, the founder cuts the channels, called gates, through which the molten bronze will flow to the cavities left when the models are removed. A gate, or passage for the molten metal must be made for each piece. (See Figs 16.12–16.13, p. 287.)

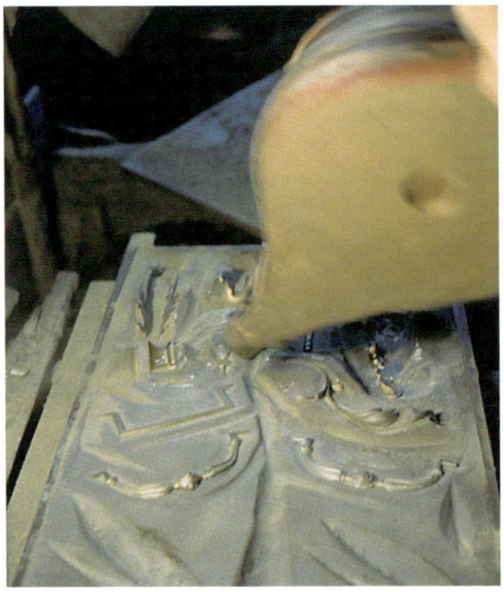

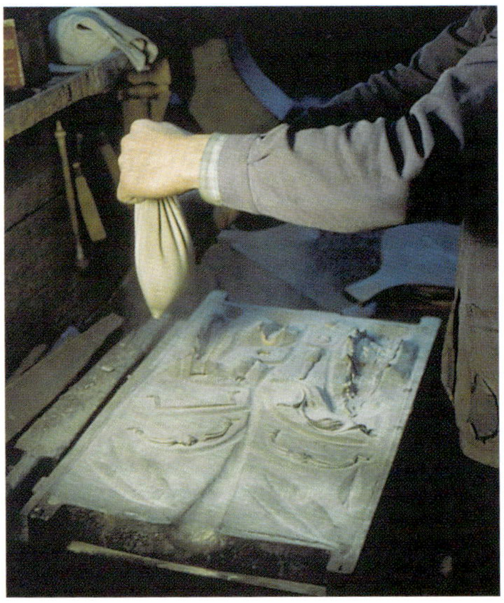

Plate 97 The final cleaning of the moulds is done with a bellows to avoid interfering with the surface of the sand. You can see all of the gates for each piece. **Plate 98** After the gates have been cleaned up, the mould is coated with wheat starch sifted through a cloth bag. This interface prevents the sand on the two sides of the mould from sticking together. (See Figs 16.14–16.15, p. 288.)

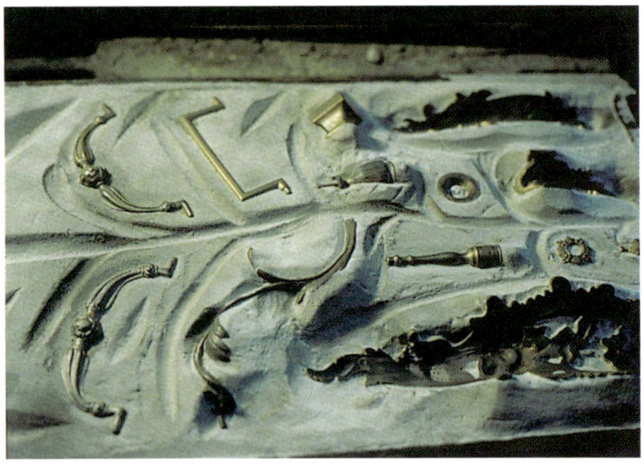

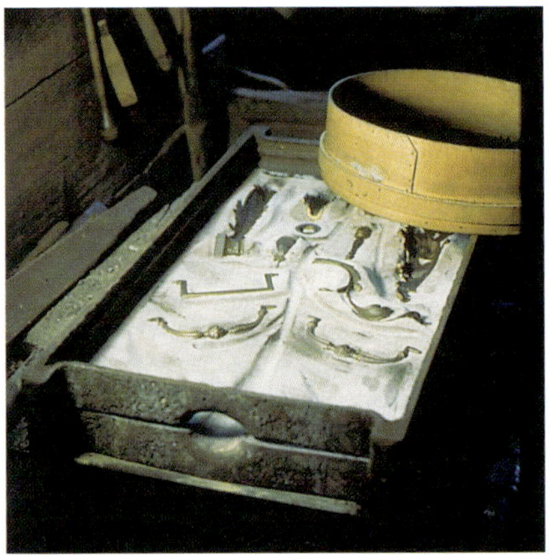

Plate 99 The first half of the mould is now ready. You can see both the gates for the flow of the molten metal and the channels cut to allow the gasses to escape. These are called the vents. The mould is now ready for the application of the counter mould, which is a faster operation since you now have only to stack another chassis on the first one and fill it with compressed sand. **Plate 100** The second steel chassis is stacked on top of the first and held in place by wooden rods. As with the first chassis, the sand is sifted into the chassis to form the counter-mould. (See Figs 16.16–16.17, p. 288.)

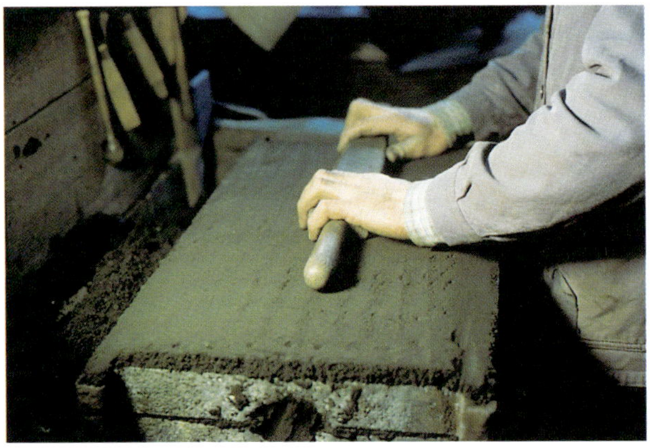

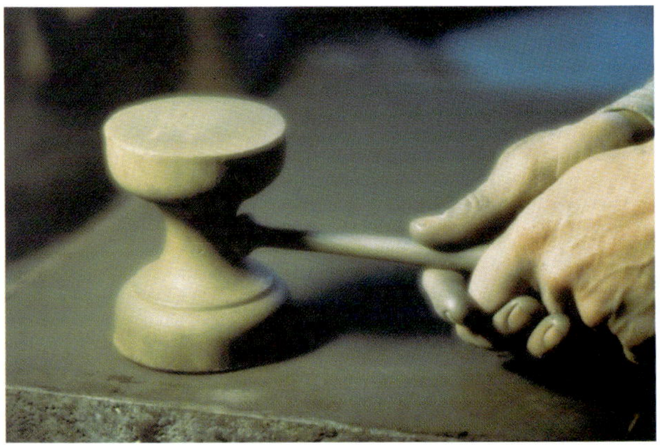

Plate 101 A wooden bat is used to pound the sand down evenly so that it marries the exact form of the back side of the bronzes underneath. Once the sand is compacted, it is scraped off level with the edges of the chassis. **Plate 102** With a boxwood mallet, the sand is tapped to release the counter mold from the mould. This is critical to separating the two halves of the mould. (See Figs 16.18–16.19, p. 288.)

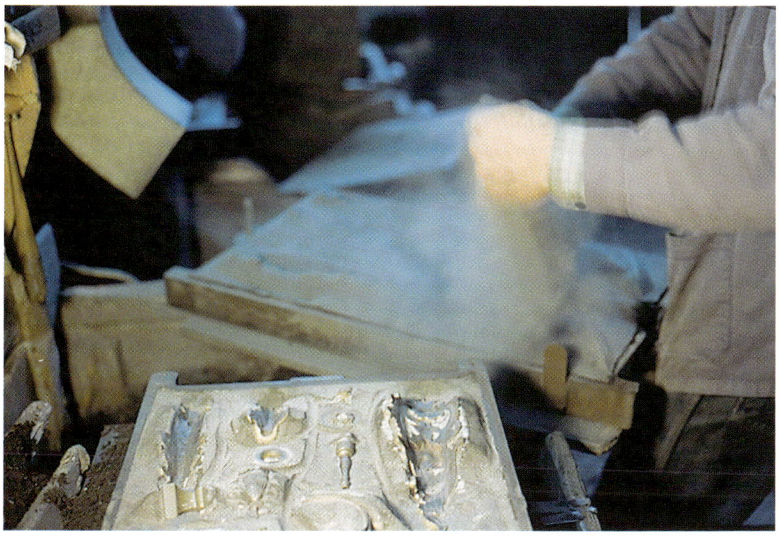

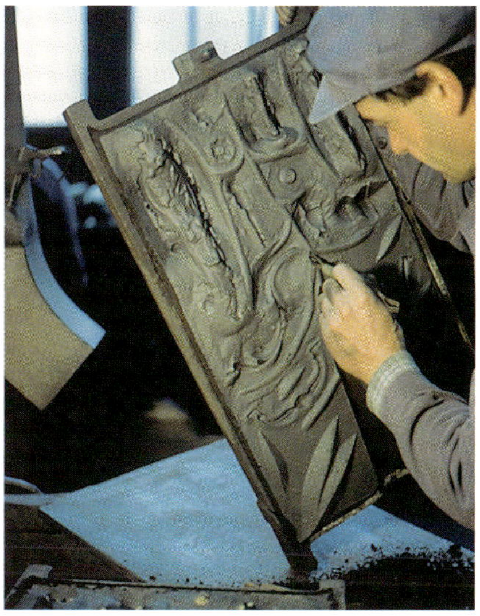

Plate 103 The two moulds are separated very delicately. **Plate 104** The gates and vents that were moulded convex in the counter mould are now scraped out in the counter molds since they need to be concave on both halves of the mould. (See Figs 16.20–16.21, p. 289.)

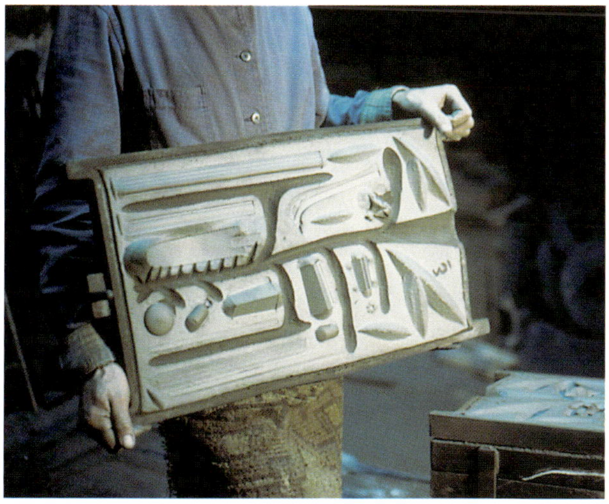

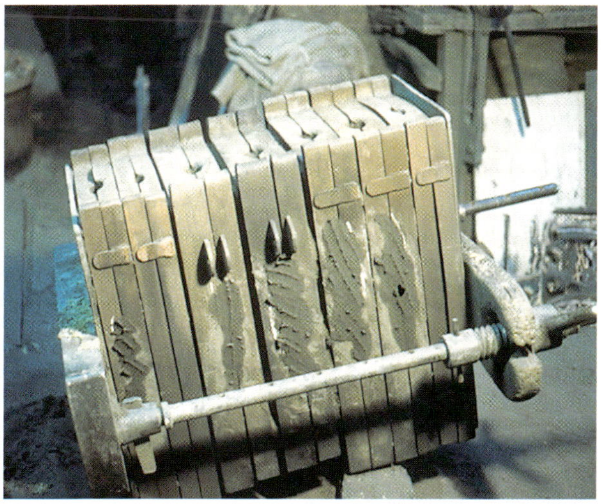

Plate 105 The moulds, showing the gates and the vents, are now ready for the pour. **Plate 106** The moulds are then stacked together and held with screw clamps to prevent deformations during the pour. The stack of moulds are put in a hot room for twelve hours to dry the sand and then stood on end to receive the poured molten metal. (See Figs 16.22–16.23, p. 289.)

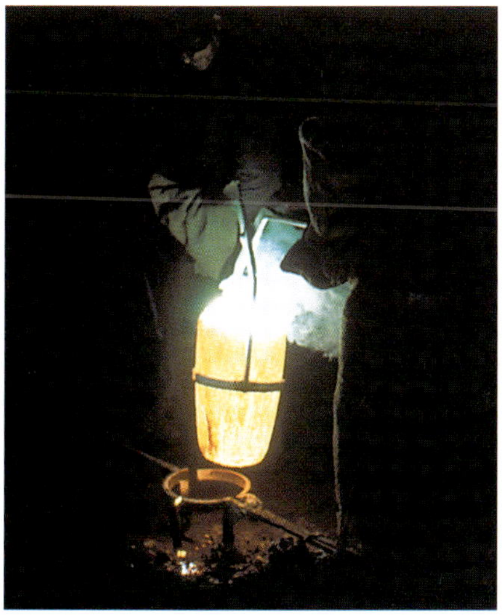

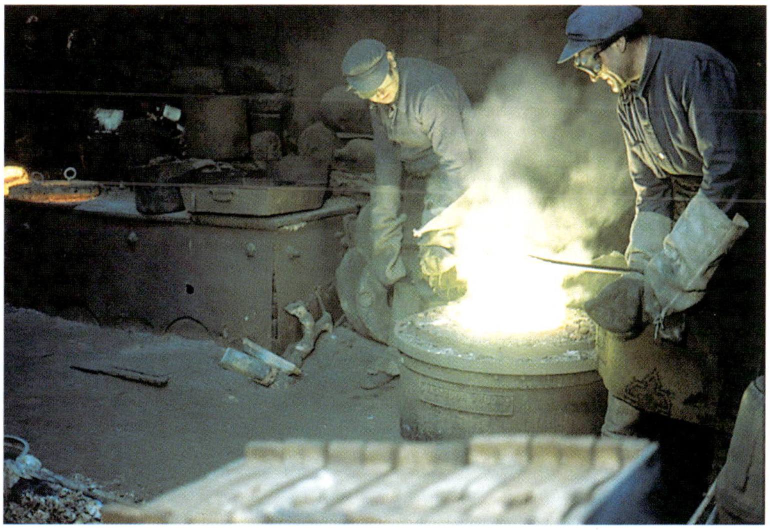

Plate 107 The alloy is melted in a crucible, which is put in a furnace. Scraps from previous pours are also added to the new alloy. **Plate 108** The two founders take the crucible out of the oven with the stack of molds in the foreground. The ambiance of the foundry has not changed since the eighteenth century. (See Figs 16.24–16.25, p. 289.)

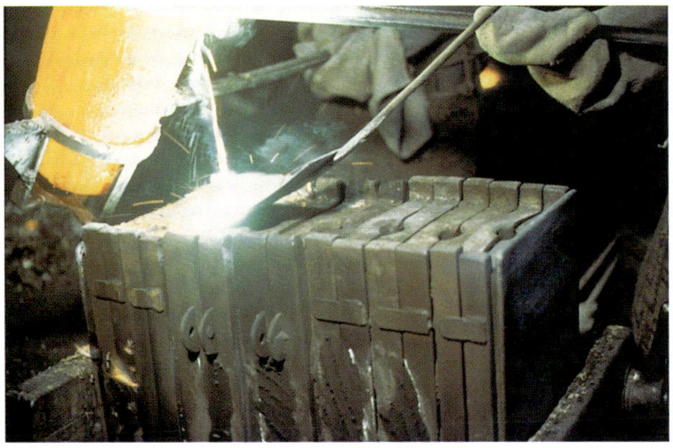
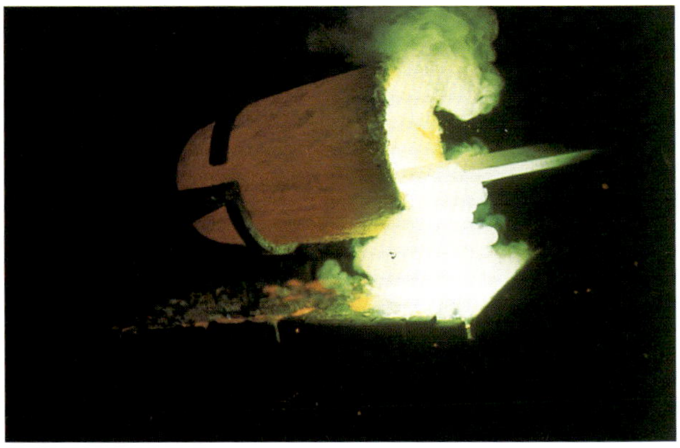

Plate 109 While two founders manipulate the lantern, a third holds a small spade to guide the molten metal, which must be prevented from splashing into the next mould to be poured. If this happens, the alloy will cool and clog the gates prematurely, thereby blocking the flow of the molten bronze. You can also see the flames of the escaping gasses burning on the sides. **Plate 110** The pour is completed. It has to progress very quickly or the metal will not flow completely into the depths of the last mould. The colour of the crucible shows that the metal has cooled considerably. (See Figs 16.26–16.27, p. 290.)

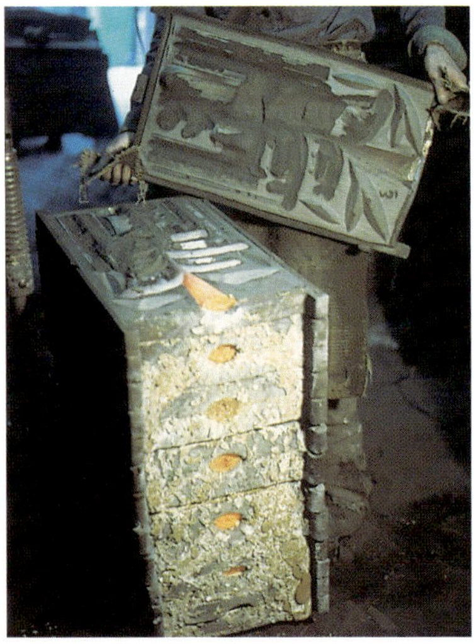
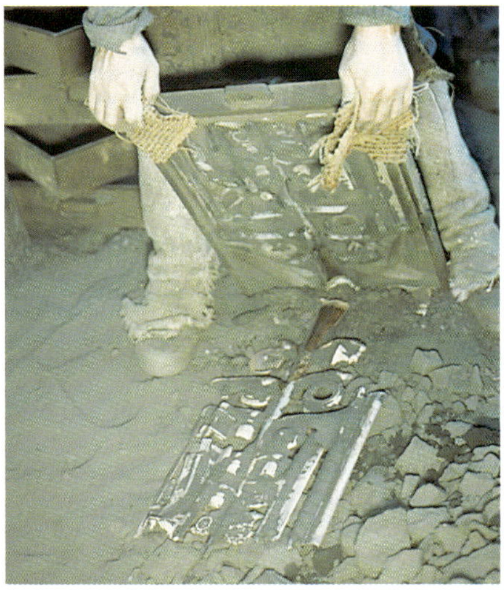

Plate 111 The moulds are broken apart immediately after the pour. You can see that the metal is still red hot. **Plate 112** The bronzes fall on the floor, but are still attached by the alloy filling the gates and vents. These will be cut to separate the bronzes. The sand moulds are only used once since they are destroyed when they are opened after the pour. The sand is then passed through a screening machine before it is reused for other moulds. (See Figs 16.28–16.29, p. 290.)

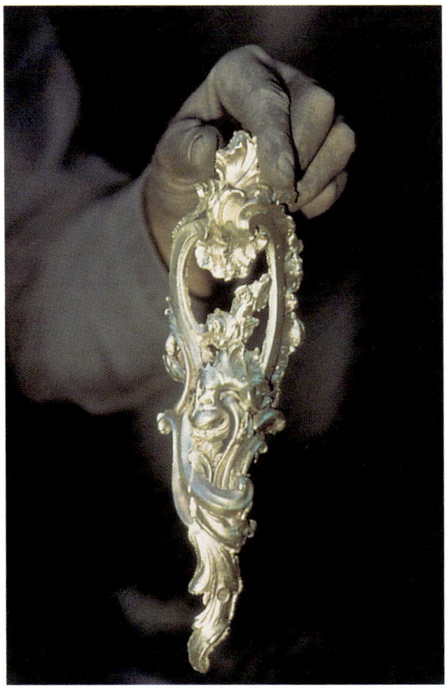
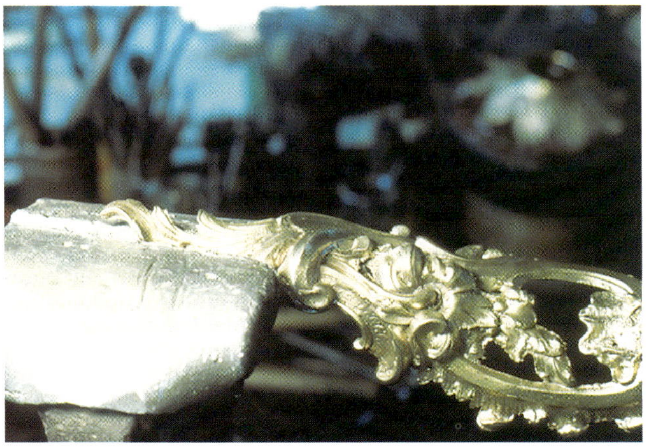

Plate 113 The bronzes are then brushed on a wheel to remove any remaining sand and the fire-skin. They are then ready to be sent to the chaser. **Plate 114** The chaser mounts each bronze in a vice on which the jaws have been lined with lead to avoid damaging the bronze. He will sharpen all of the edges and embellish the bronze with details that cannot be achieved in the casting. (See Figs 16.30–16.31, p. 290.)

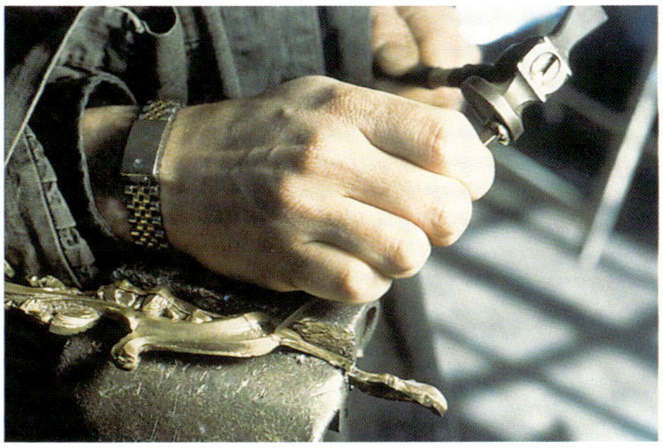

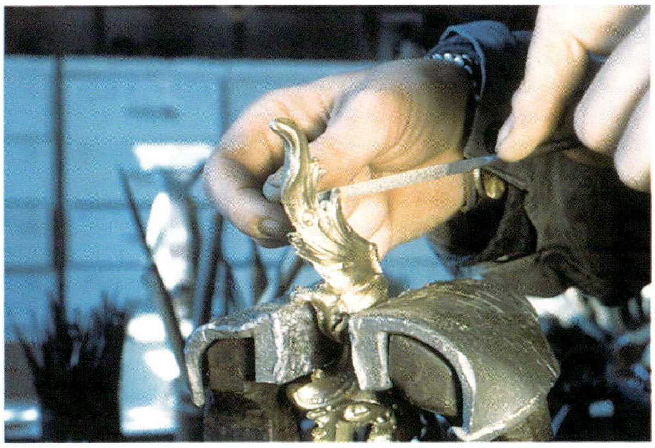

Plate 115 The burr along the edges, called the stitches, are removed with a burin and a hammer. **Plate 116** The edges are cleaned and sharpened with a file. Files with a wide variety of profiles are necessary to get into all of the tight curves of the bronzes. (See Figs 16.32–16.33, p. 291.)

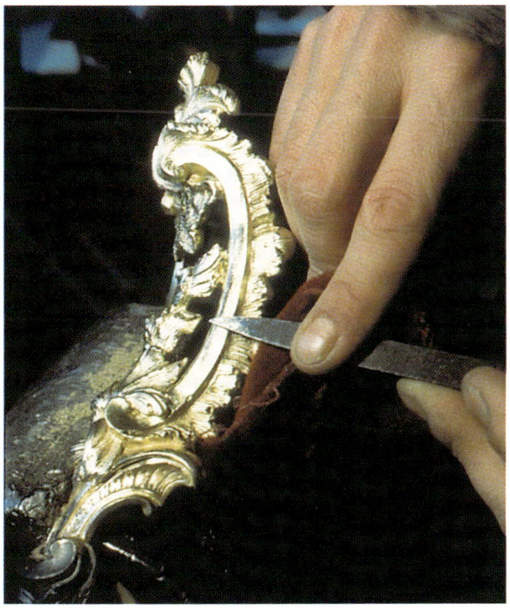

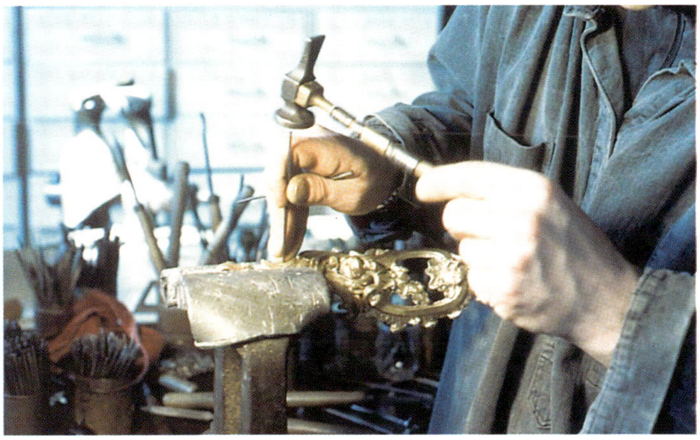

Plate 117 The chaser smoothes the flats of the mount with a riffler and then a scraper. The riffler is passed perpendicular to the direction of the strap to smooth and polish the surface that will be burnished after the gilding. **Plate 118** The veins of the bronze are sharpened with a burin. (See Figs 16.34–16.35, p. 291.)

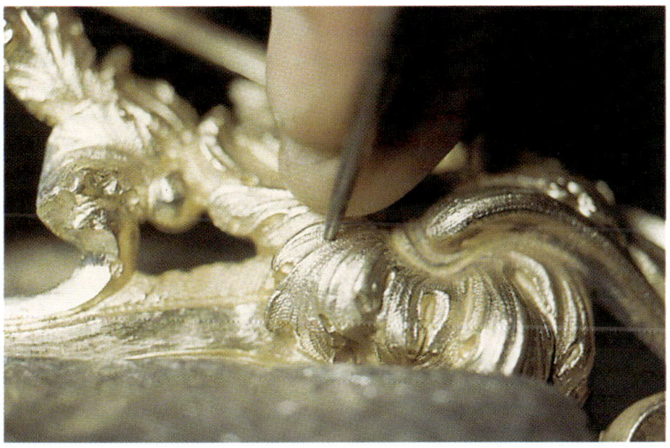

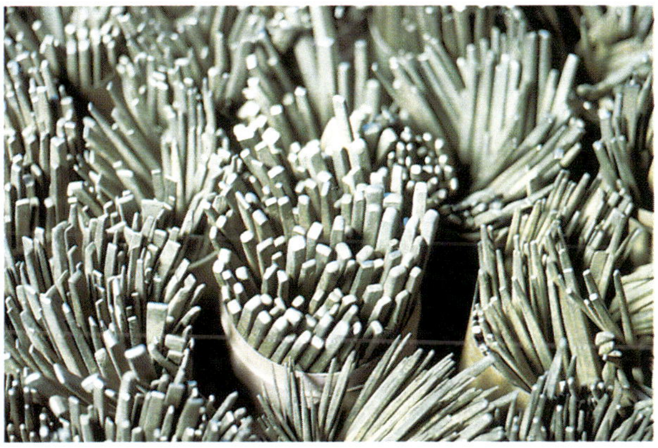

Plate 119 Punches are used to give texture. They do not remove metal, but displace it to give detailed form and volume.
Plate 120 A good chaser will have over 3,000 punches. (See Figs 16.36–16.37, p. 291.)

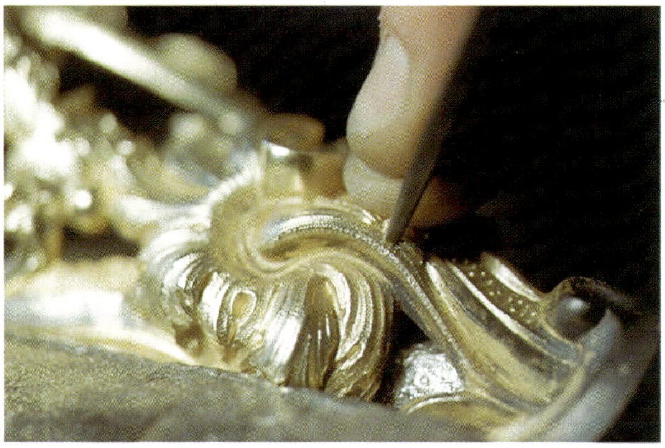

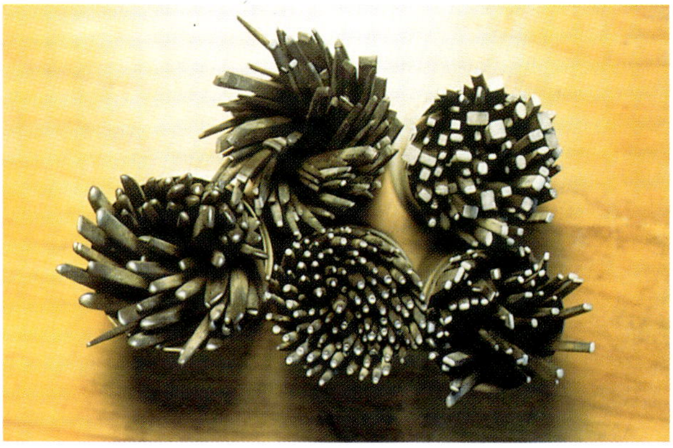

Plate 121 Every pattern or texture is produced by a separate punch, which must be the appropriate style and size. Plate 122 These are the five families of chasing tools: top left, the mats; top right, the burins; bottom left, the tracers; bottom centre, the pearlers, and *bouterolles*. (See Figs 16.38–16.39, p. 292.)

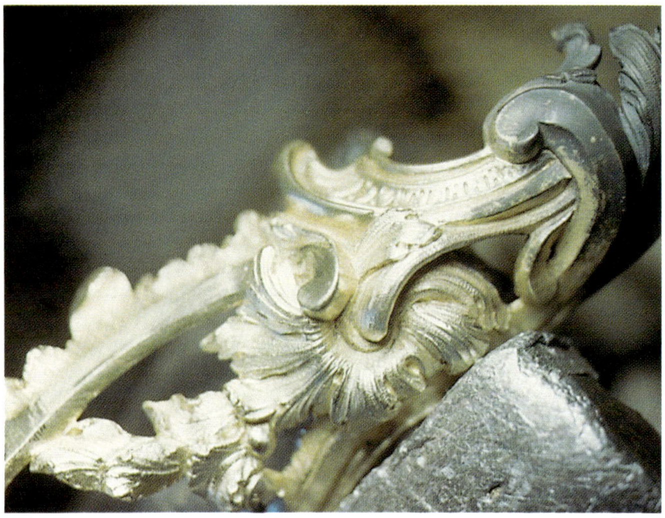

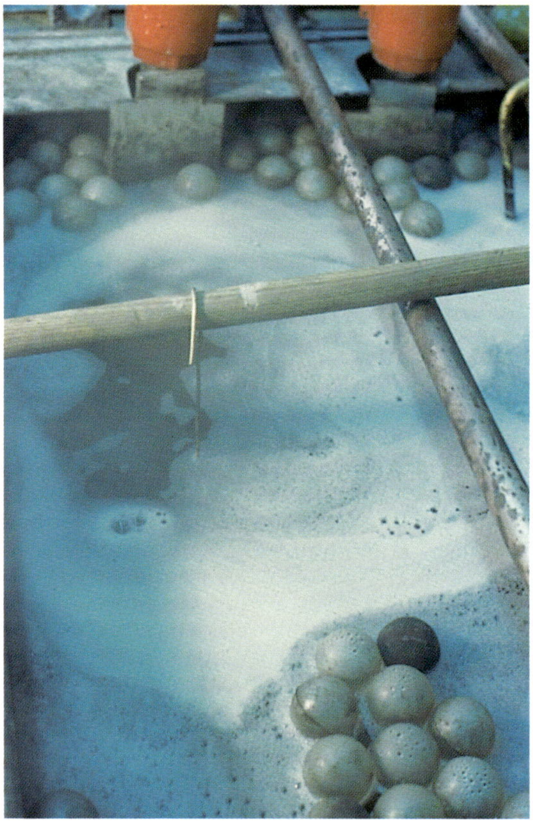

Plate 123 Every surface, both textured and smooth are produced by the chasing tools, which takes many hours. The bronze is then ready to go to the gilder. Plate 124 The first step of the gilding is to pickle or clean the bronze in nitric acid or by electrolysis. (See Figs 16.40–16.41, p. 292.)

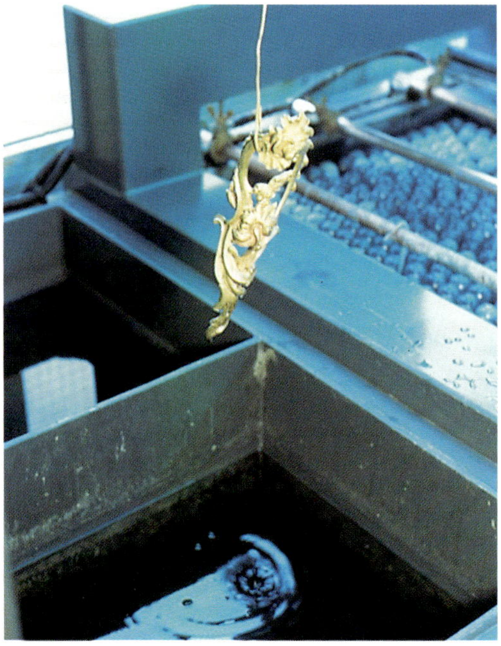

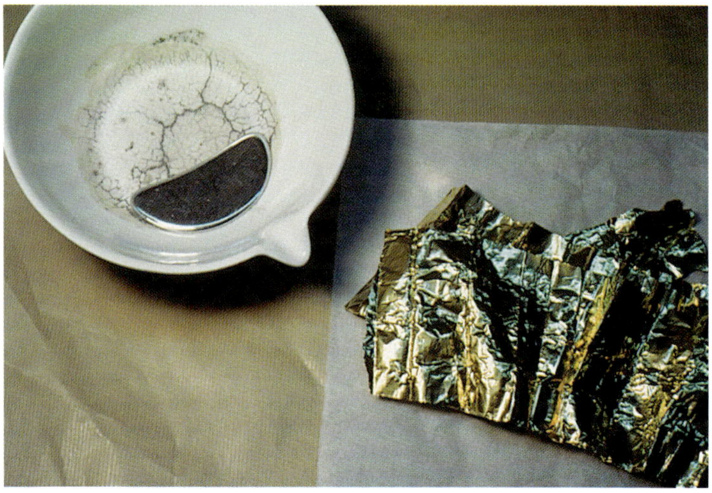

Plate 125 It is then rinsed, several times in distilled water, of any impurities that could interfere with the adhesion of the gilding. Plate 126 The amalgam is a mixture of eight parts of mercury to one part gold leaf, which is thicker than the leaf used for wood gilding. (See Figs 16.42–16.43, p. 292.)

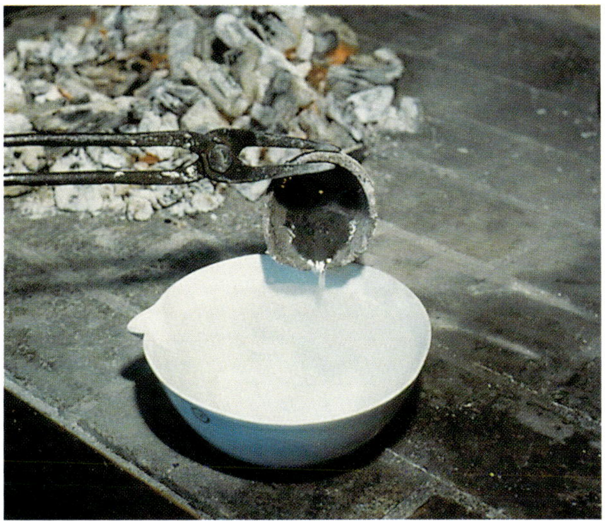

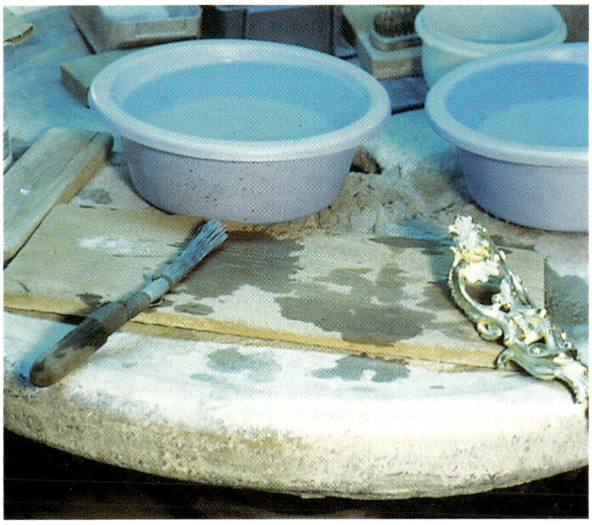

Plate 127 The mercury and gold are melted together in a crucible on a coal fire to make the amalgam, which is then rinsed several times in distilled water to remove any impurities. **Plate 128** This shows the tongs and the brush that will be used to apply the amalgam. The brush, called a bat, has brass strands. (See Figs 16.44–16.45, p. 293.)

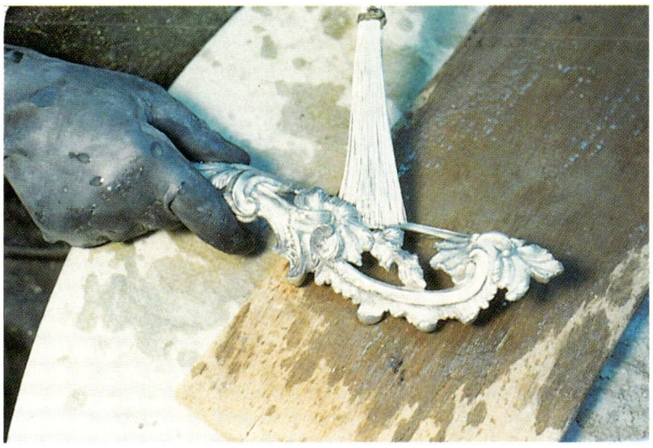

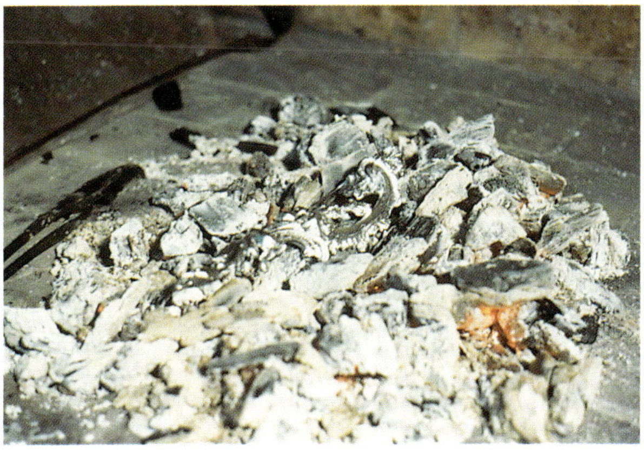

Plate 129 The amalgam is applied to the bronze with a bat. A strong tapping motion is used to drive the amalgam into the interstices of the bronze. **Plate 130** The bronze is then placed into the coal fire to evaporate the mercury. Under the influence of the heat, the metal dilates and the gold is deposited into the matrix of the metal. The bronze is recoated with the amalgam a number of times, depending on the budget and the details of form. (See Figs 16.46–16.47, p. 293.)

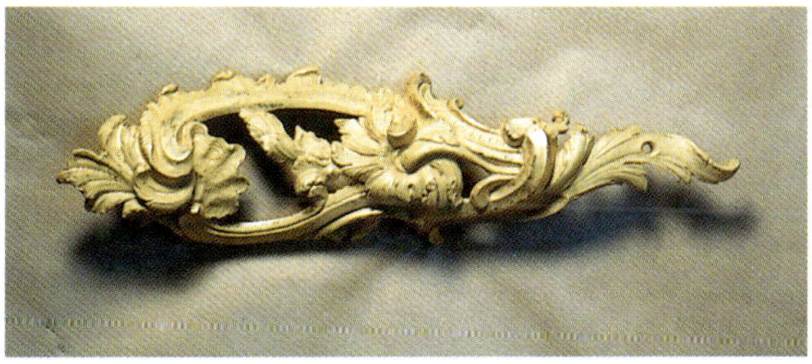

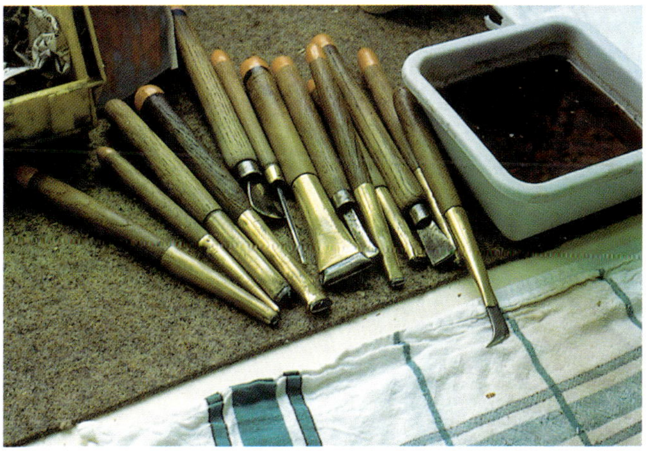

Plate 131 The gilded bronze is a dull color after the gilding and requires further finishing. It is brushed under running water to give it some life and shine and then dried in sawdust. **Plate 132** The highlights are then burnished. Here a steel burnisher is being used; they can also be made of hematite stone.
(See Figs 16.48–16.49, p. 293.)

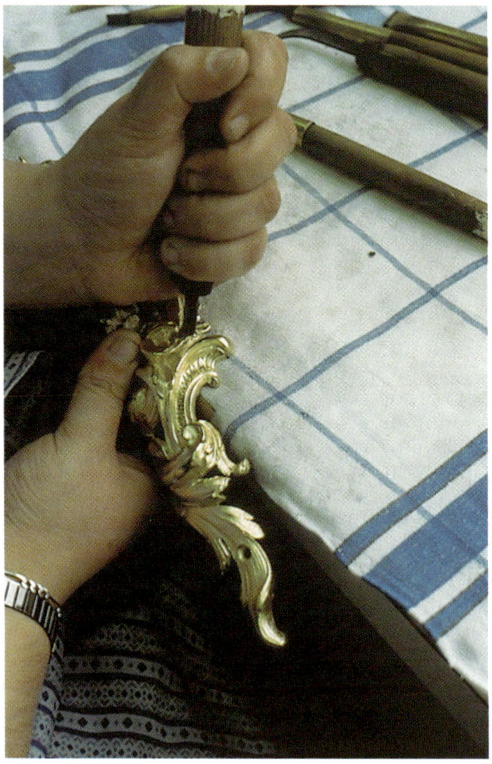

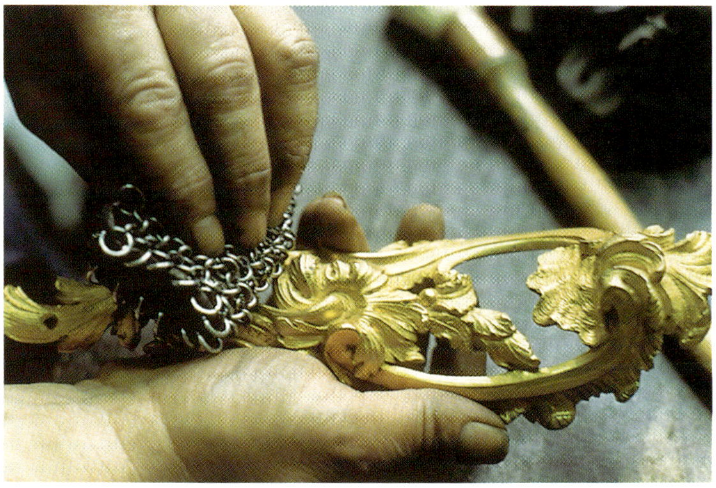

Plate 133 Polishing with the burnisher makes the flat surfaces brilliantly reflective. This is a delicate operation because you must not slip and polish a matte area, but it requires a great deal of pressure. **Plate 134** Chain mail is used to soften the edges and age the bronze. (See Figs 16.50–16.51, p. 294.)

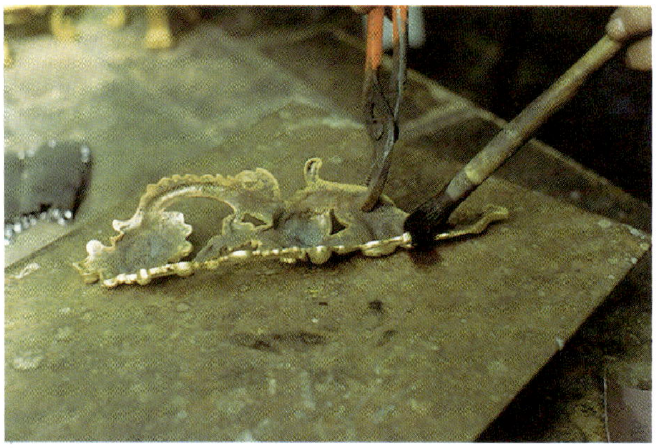

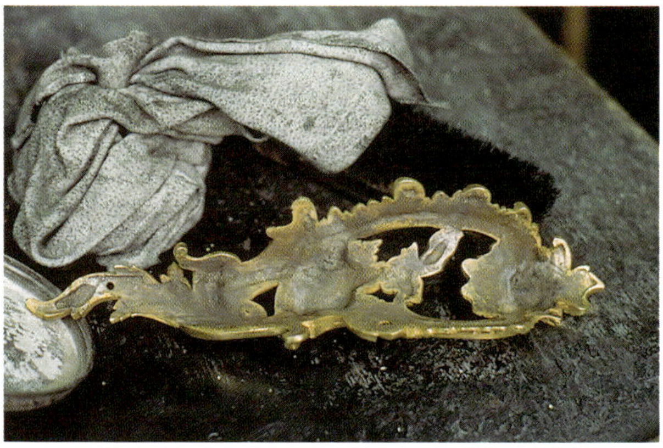

Plate 135 The backs of bronzes are patinated with acids to artificially age them. **Plate 136** The spills of mercury can be seen on the back of the bronze. (See Figs 16.52–16.53, p . 294.)

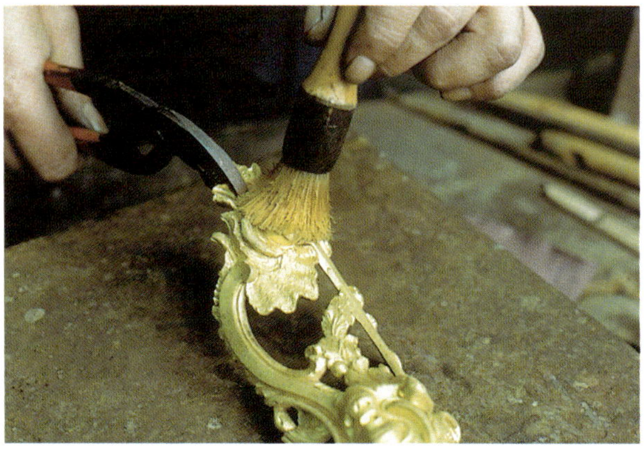

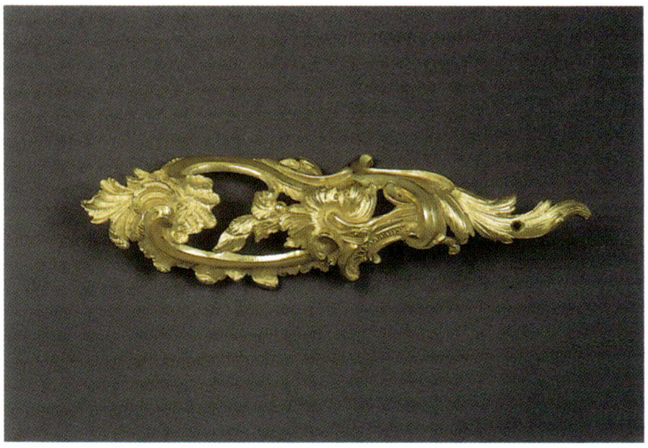

Plate 137 The bronzes are patinated or 'mis en couleur d'or' with chemicals as varied as ammonium chloride, potassium nitrate and cupric acetate to achieve the desired tone and unify the colour among a set. **Plate 138** The finished bronze. (Photo: Abby Hykin.) (See Figs 16.54–16.55, p. 294.)

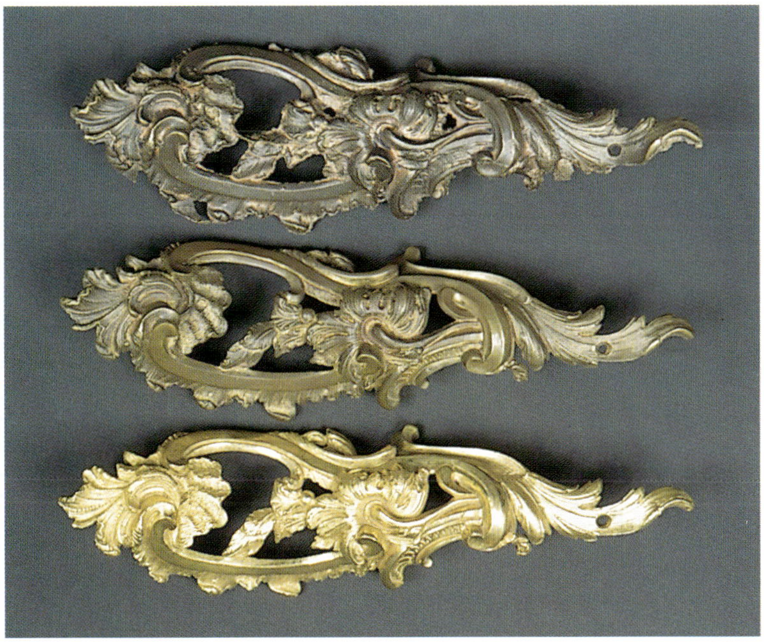

Plate 139 The three bronzes reproduced, from top to bottom, the cast before chasing, the cast and chased mount, and the cast, chased and gilded mount. (Photo: Abby Hykin.) (See Fig. 16.56, p. 295.)

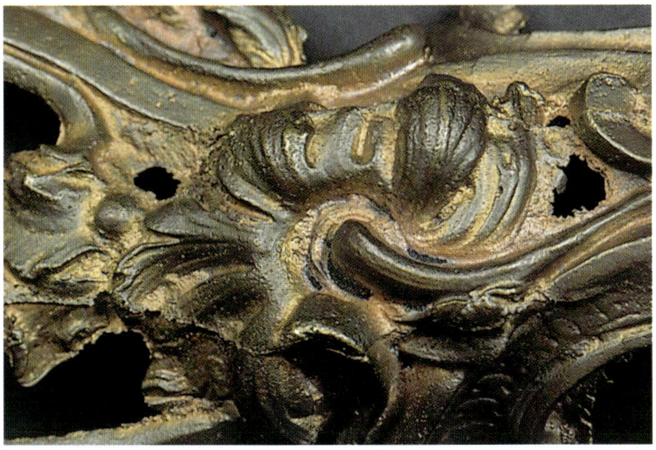

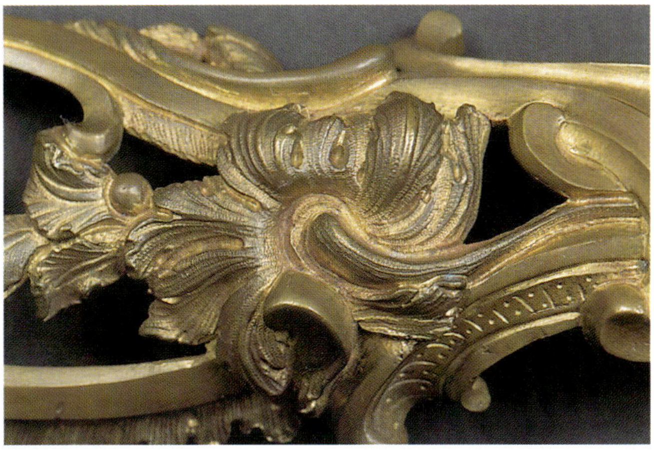

Plate 140 Detail of the cast mount without chasing or gilding. (Photo: Abby Hykin.) (See colour plates.) **Plate 141** Detail of the cast and chased mount. (Photo: Abby Hykin.) (See Figs 16.57–16.58, p. 295.)

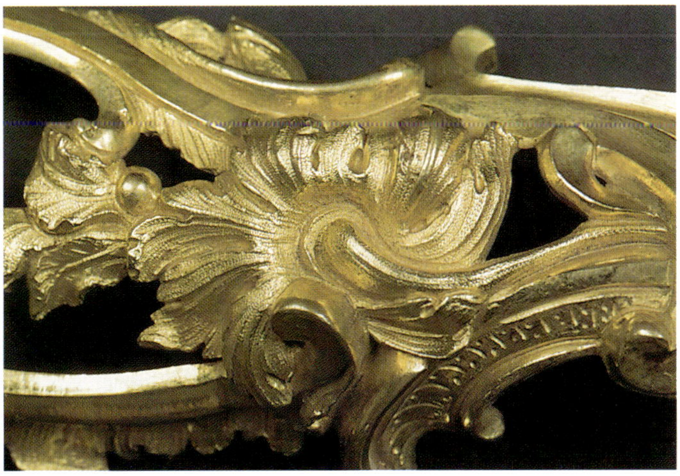

Plate 142 Detail of the cast, chased and gilded mount. (Photo: Abby Hykin.) (See Fig. 16.59, p. 295.)

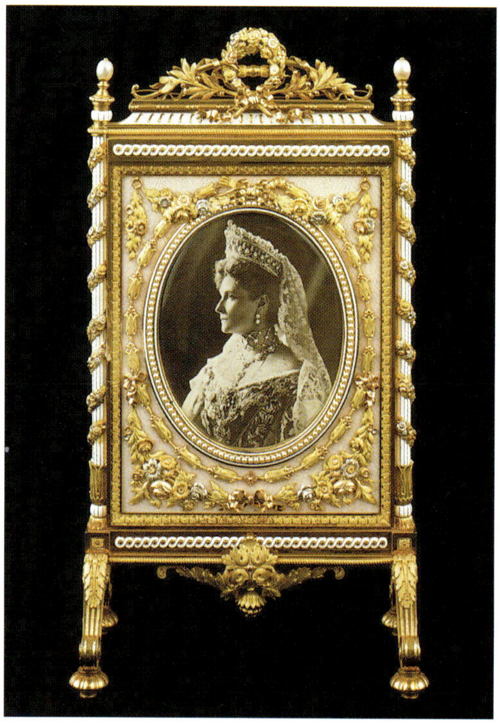

Plate 143 Imperial Fire-Screen Frame. Double-sided frame with photographs of Czar Nicholas II and his wife Czarina Alexandra Feodorovna. 7 2/8" tall. (Photograph used by permission of The Forbes Magazine Collection, New York. All rights reserved. Photo by H. Peter Curran.) (See Fig. 17.1, p. 299.)

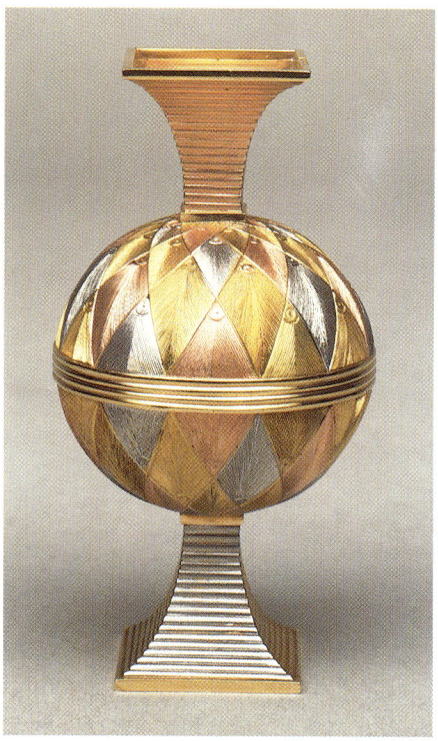

Plate 144 Double Marriage Cup. 3 5/8" tall. (Photograph used by permission of A La Vieille Russie, Workmaster Michael Perchi, St. Petersburg, before 1896.) (See Fig. 17.2, p. 299.)

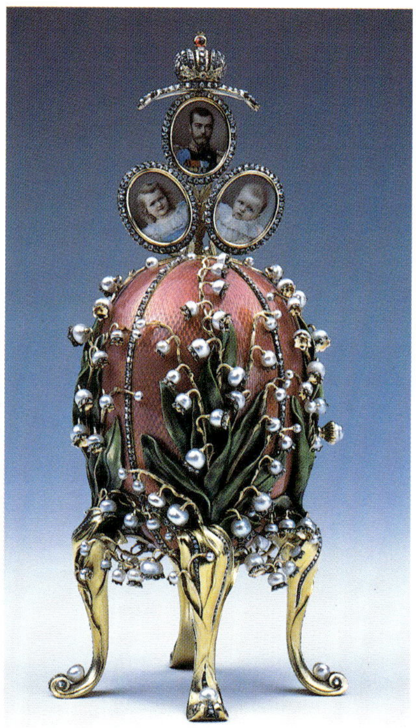

Plate 145 Imperial Lilies of the Valley Easter Egg. 7 7/8" tall. (Photograph used by permission of The Forbes Magazine Collection, New York. All rights reserved. Photo by Joseph Coscia, Jr.) (See Fig. 17.3, p. 300.)

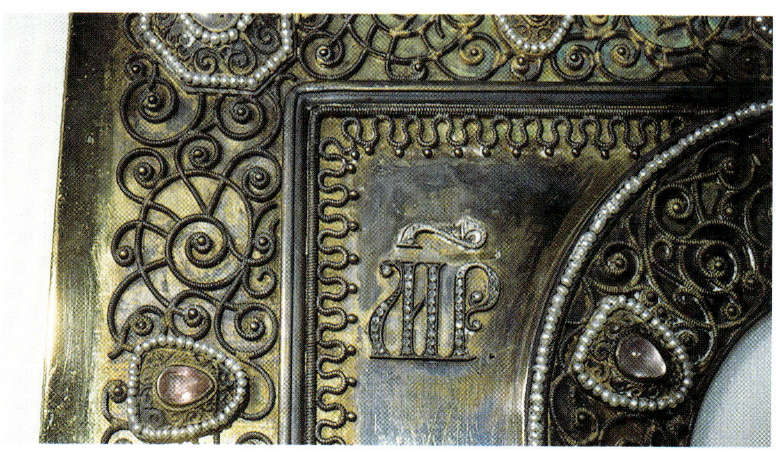

Plate 146 Detail of damaged gilding, before cleaning, on oklad of Fabergé Icon of Our Lady of Iberia. 11 7/8 × 10 1/8". (Virginia Museum of Fine Arts, Richmond, VA. Bequest of Lillian Pratt Thomas.) (See Fig. 17.4, p. 302.)

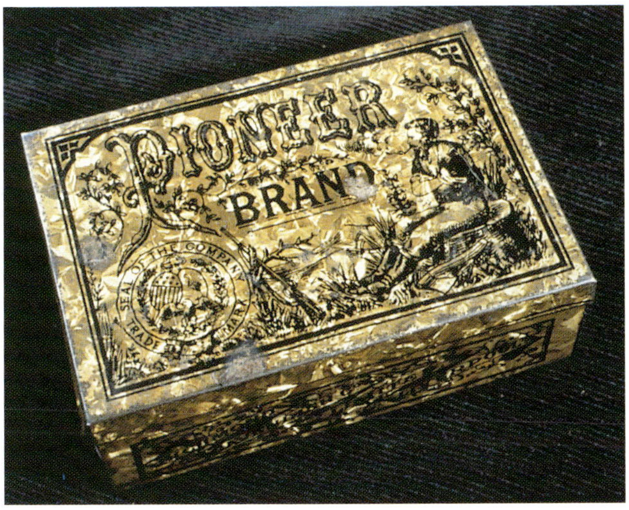

Plate 147 Tobacco tin made of sheet iron coated with crystallised and etched tin (moiré metallique), laquered with a dyed organic coating. (See Fig. 18.1, p. 308.)

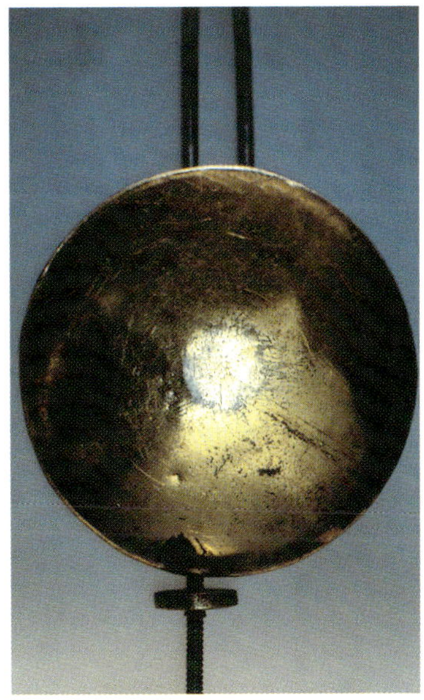

Plate 148 Clock pendulum bob of brass, covered with a gold-coloured lacquer. (See Fig. 18.2, p. 310.)

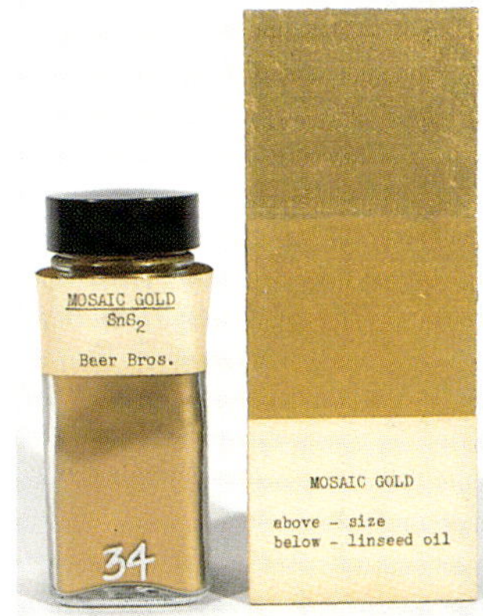

Plate 149 Mosaic gold and sample tile from the Edward Forbes Pigment Collection of historic pigments and media. (Straus Center for Conservation, Harvard University Art Museums, Photograph Eugene Farrell.) (See Fig. 18.3, p. 313.)

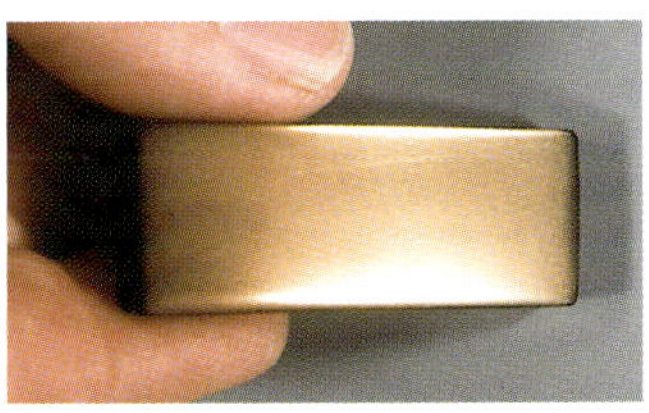

Plate 150 Steel coupon with 'straw' colour produced by a thin film of magnetite. (See Fig. 18.4, p. 314.)

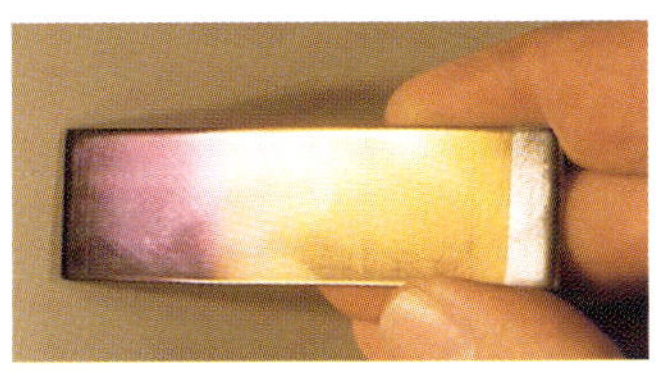

Plate 151 Anodised titanium metal showing gold colour. (See Fig. 18.5, p. 315.)

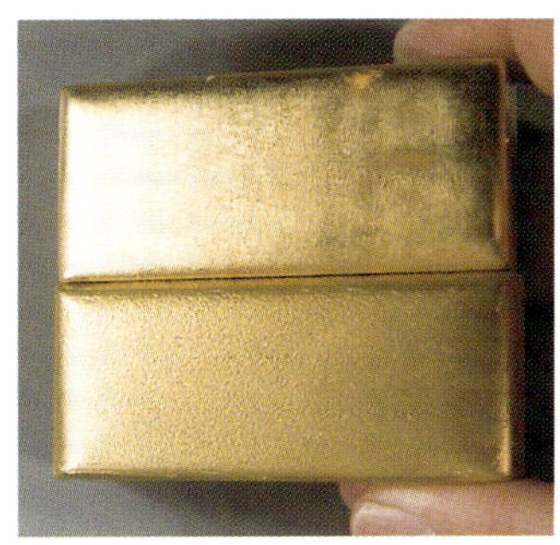

Plate 152 Two 'gilded' sample tiles. The top one is gold leaf on oil size. The bottom one is mica-based lustre pigments over acylic emulsion underpainting. (See Fig. 18.6, p. 315.)

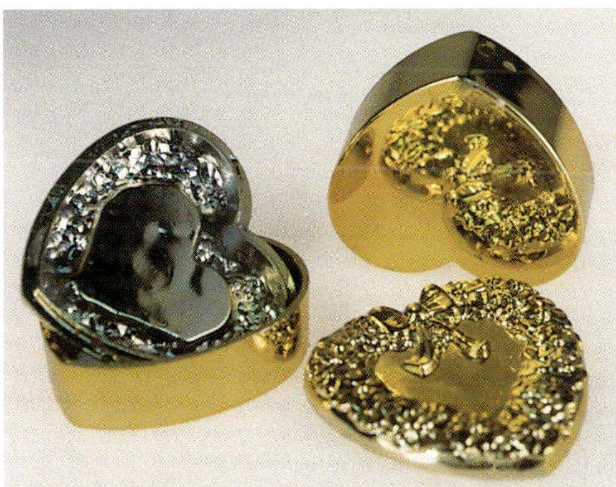

Plate 153 ABS plastic boxes with vapour-deposited aluminium, 'gilded' with a dyed organic coating. (See Fig. 18.7, p. 315.)

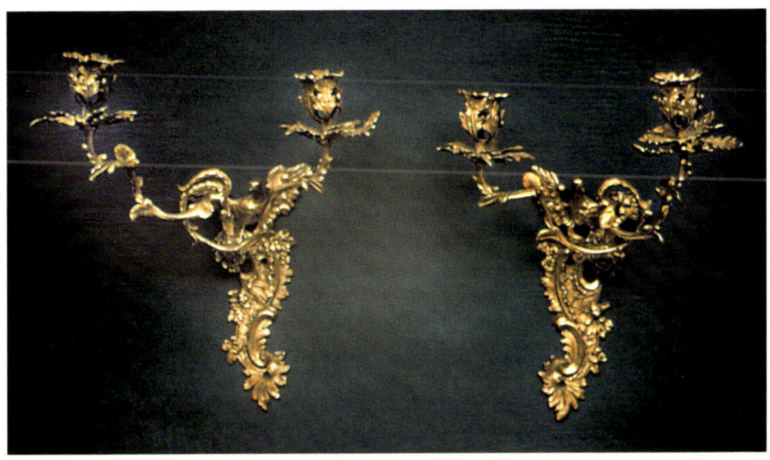

Plate 154 Pair of lacquered brass wall sconces. (Courtesy, Winterthur Museum 57.126.26 & 27.) (See Fig. 19.1, p. 320.)

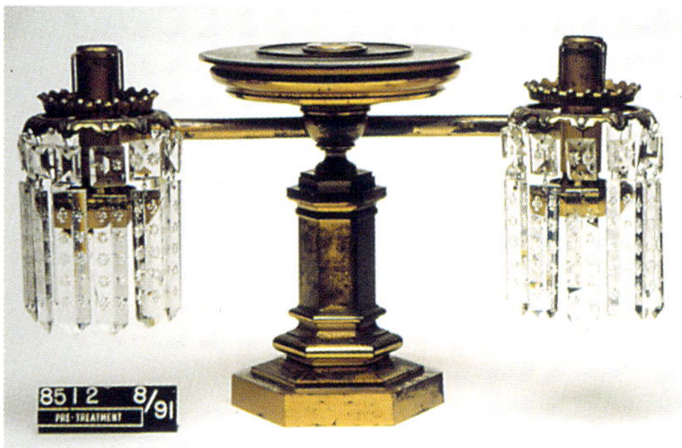

Plate 155 One of a pair of argand lamps by Henry Hooper, before treatment. (Courtesy, Winterthur Museum 85.1.2.) (See Fig. 19.2, p. 322.)

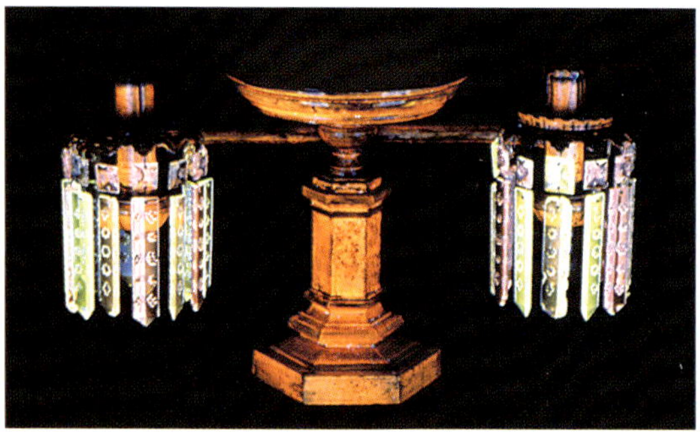

Plate 156 Argand lamp illuminated with ultra-violet light. (Courtesy, Winterthur Museum 85.1.2.) (See Fig. 19.3, p. 323.)

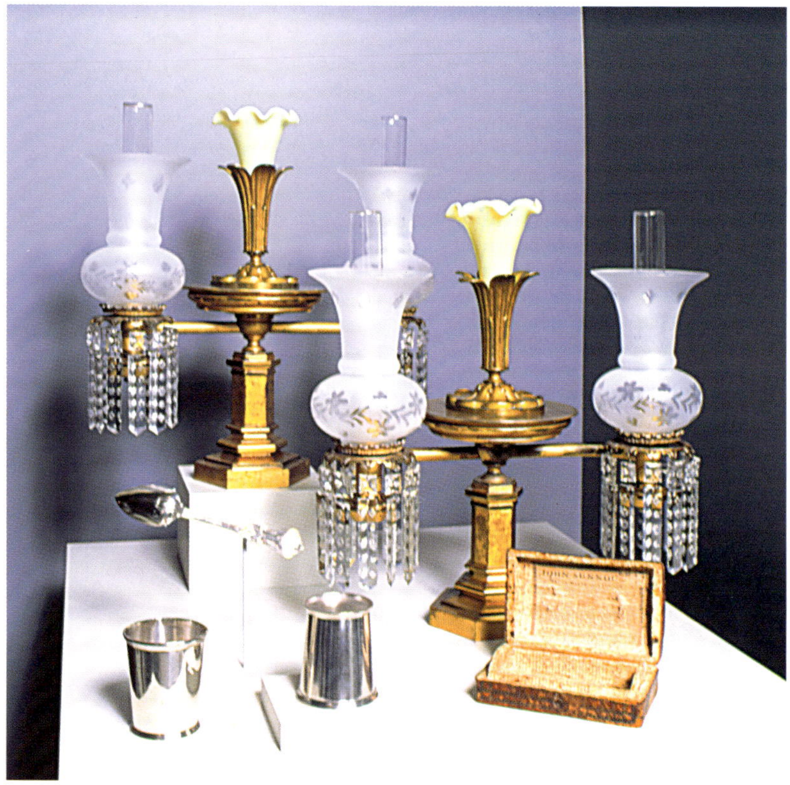

Plate 157 Pair of argand lamps after treatment. (Courtesy, Winterthur Museum 85.1.1 & 2.) (See Fig. 19.5, p. 326.)

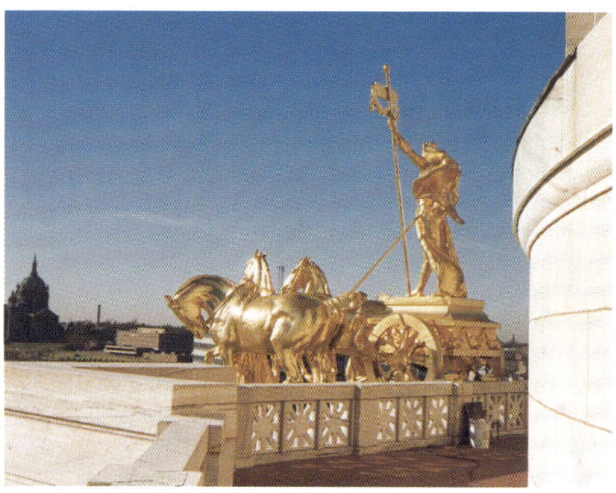

Plate 159 View of the *Quadriga* on a sunny day in 1993. (See Fig. 21.1, p. 338.)

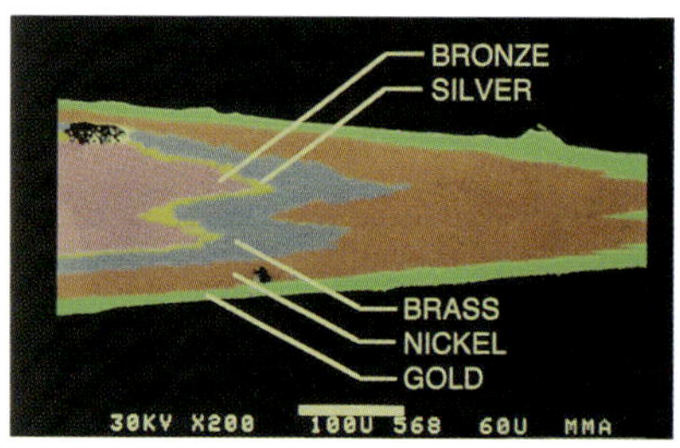

Plate 158 SEM photograph with added colour: taper section showing layers of plating. (Photo: Mark Wypyski.) (See Fig. 20.2, p. 331.)

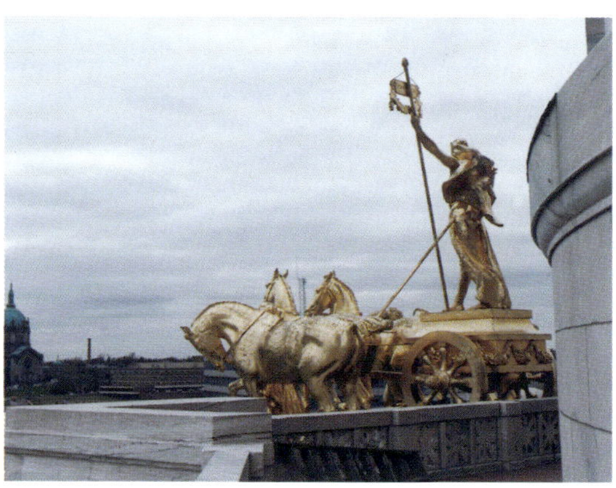

Plate 160 View of the *Quadriga* on an overcast day in the same week in 1993. (See Fig. 21.2, p. 338.)

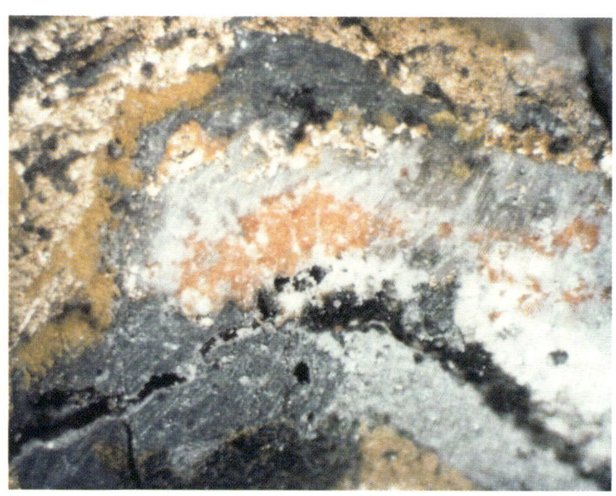

Plate 161 Cross-section from the *Quadriga*, removed from the proper right lady's drapery, bottom rear edge. (Source, Frank S. Welsh.) (See Fig. 21.3, p. 341.)

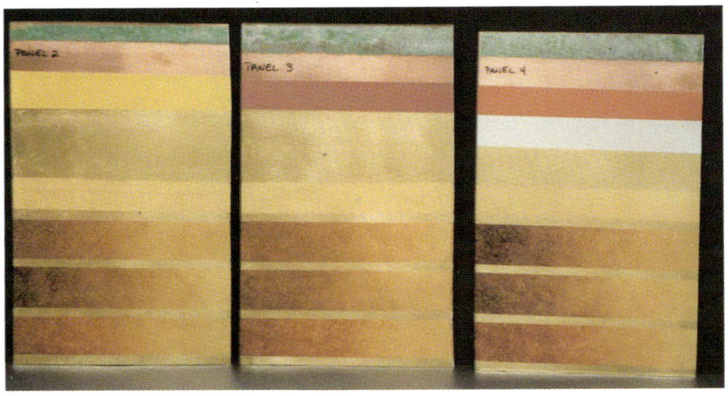

Plate 162 Gilding sample test panels prepared for the client: 1. corroded surface; 2. cleaned surface with residual pitting; 3. primer; 4. gold leaf; 5. wax as matting agent; 6. three colours as toning choices graded from dark to light. (See Fig. 21.4, p. 345.)

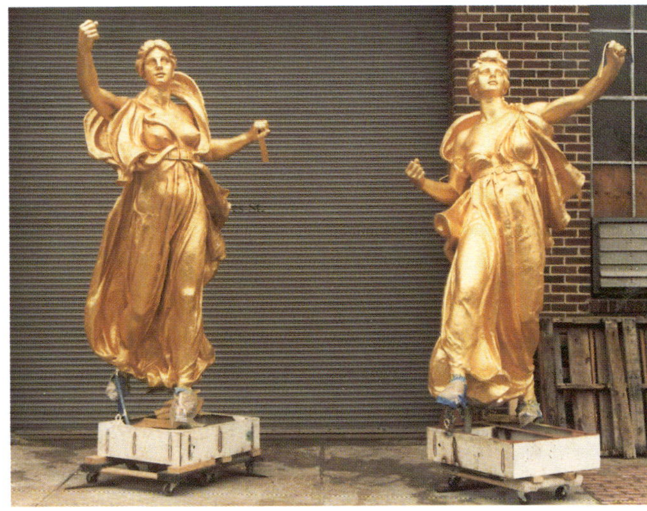

Plate 163 Comparison of toned, gilded lady on left and gilded lady without toning, French's *Quadriga*, during the conservation treatment. (See Fig. 21.5, p. 346.)

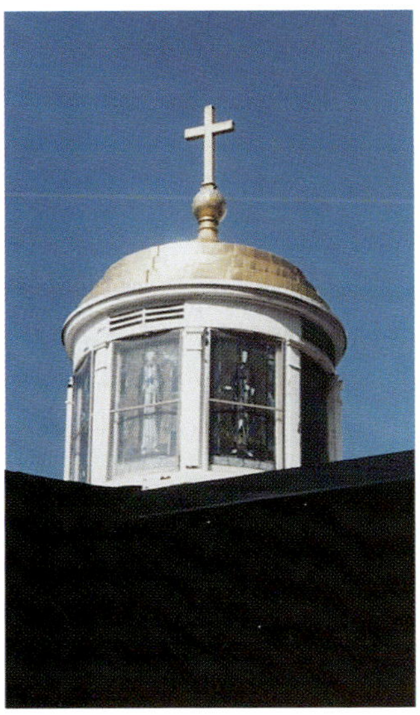

Plate 164 St. John's Church, Washington DC. View of the lead-coated copper cupola showing the finial with the cross and ball before stripping and re-gilding. (Courtesy the Gilders Studio Inc.) (See Figs 22.1, p. 352.)

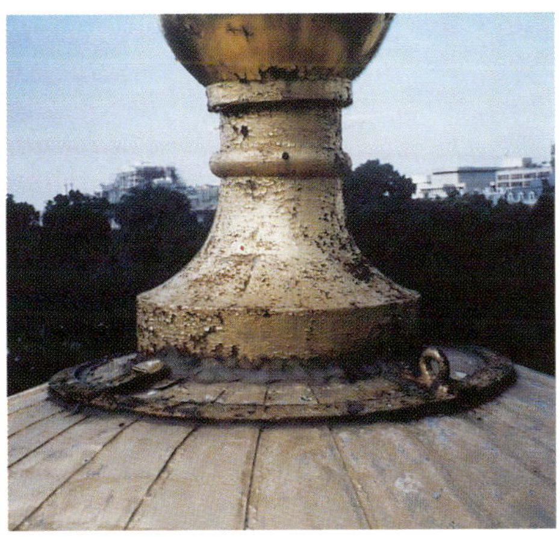

Plate 165 St. John's Church, Washington DC. Close up of finial showing the tin foot and base of the cast iron armature attached to the cupola dome. (Courtesy the Gilders Studio Inc.) (See Fig. 22.2, p. 352.)

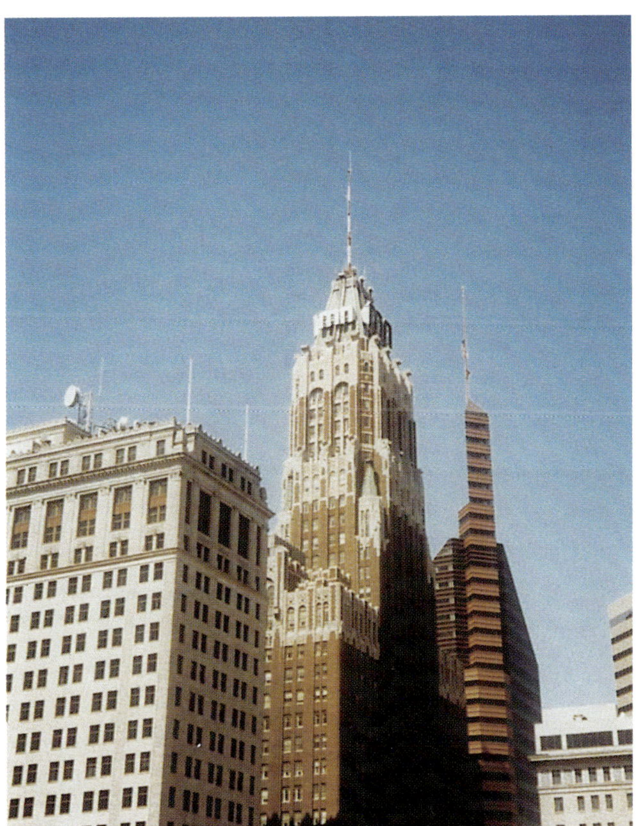

Plate 166 10 Light Street, Baltimore, Maryland. An overall view of the building with the 'mn' (Maryland National Bank) sign before removal and restoration of the roof. (Courtesy the Gilders Studio Inc.) (See Fig. 22.3, p. 353.)

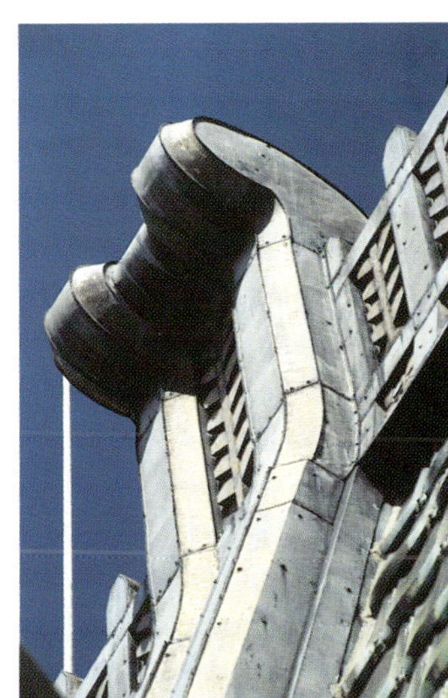

Plate 167 10 Light Street, Baltimore, Maryland. Close up of the 'C' scroll and wheat sheaves showing remnants of the original gilding from 1929. (Courtesy the Gilders Studio Inc.) (See Fig. 22.4, p. 353.)

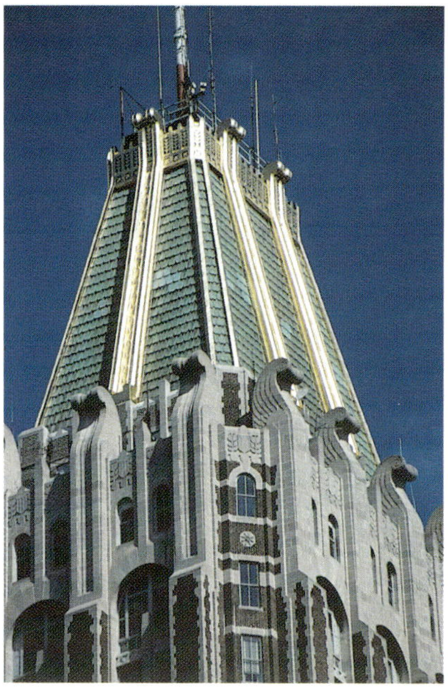

Plate 168 10 Light Street, Baltimore, Maryland. The roof of the building after restoration and gilding. (Courtesy the Gilders Studio Inc.) (See Fig. 22.5, p. 357.)

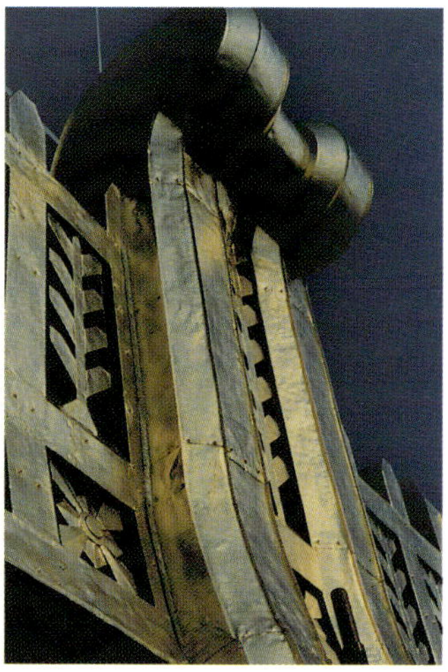

Plate 169 10 Light Street, Baltimore, Maryland. Detail of the 'C' scroll and wheat sheaves after gilding. (Courtesy the Gilders Studio Inc.) (See Fig. 22.6, p. 357.)

Plate 170 Dubuque County Courthouse. Cubuque, Iowa. Detail of lower dome showing premature deterioration of the 1981 gilding. (Courtesy the Gilders Studio Inc.) (See Fig. 22.7, p. 358.)

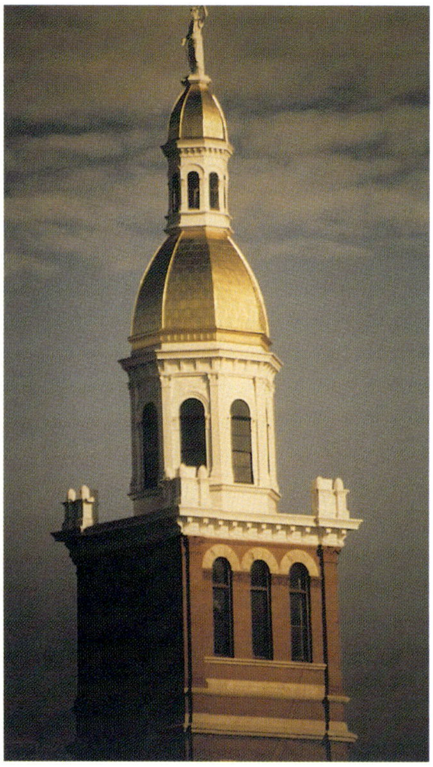

Plate 171 Dubuque County Courthouse, Dubuque, Iowa. Domes after completion in 1994. (Courtesy the Gilders Studio Inc.) (See Fig. 22.8, p. 358.)

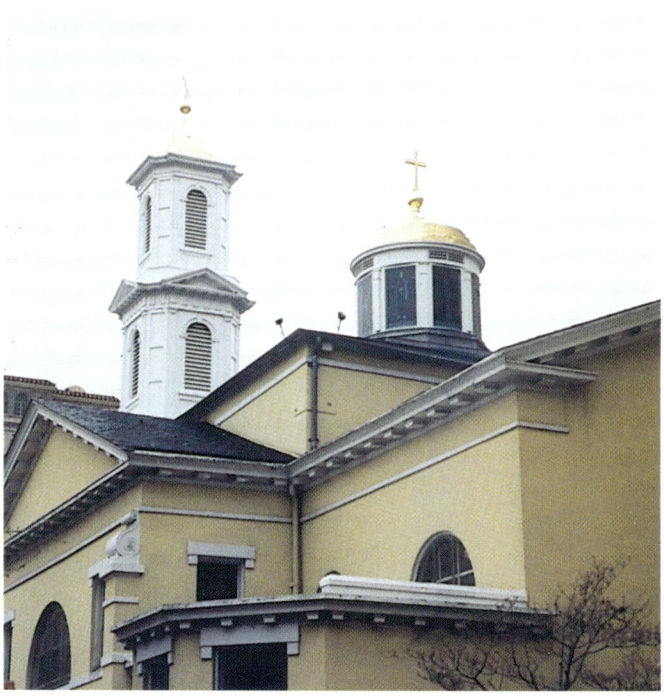

Plate 172 St. John's Church, Washington, DC. The two domes after completion in 1995. (Courtesy the Gilders Studio Inc.) (See Fig. 22.9, p. 359.)

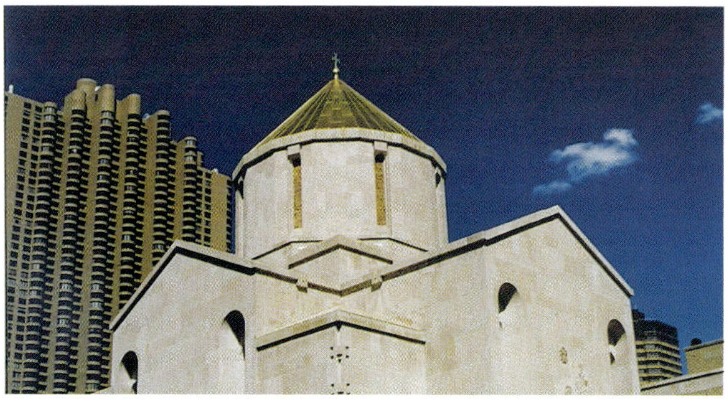

Plate 173 St. Vartan's Cathedral, New York. The ribbed roof after proper preparation, priming and re-gilding in 1994. (Courtesy the Gilders Studio Inc.) (See Fig. 22.10, p. 359.)

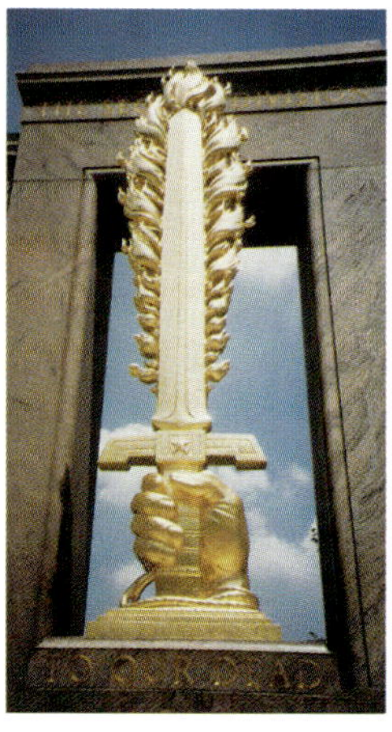

Plate 174 Second Division Memorial, President's Park, Washington DC. The 16-foot tall bronze statue with a new layer of 23 carat gold leaf. (See Fig. 22.11, p. 360.)